T0258945

TAILINGS AND MINE WASTE '00

PROCEEDINGS OF THE SEVENTH INTERNATIONAL CONFERENCE ON TAILINGS
AND MINE WASTE '00/FORT COLLINS/COLORADO/USA/23-26 JANUARY 2000

Tailings and Mine Waste '00

A.A.BALKEMA/ROTTERDAM/BROOKFIELD/2000

Organized by Geotechnical Engineering Program, Department of Civil Engineering, Colorado State University

ORGANIZING COMMITTEE

John D. Nelson	Conference Chair, Colorado State University, Fort Collins, Colorado
Debora Cave	HSI Geotrans, Inc., Westminster, Colorado
William A. Cincilla	Golder Associates, Lakewood, Colorado
Cary L. Foulk	M.F.G., Boulder, Colorado
Linda L. Hinshaw	Colorado State University, Fort Collins, Colorado
Robert L. Medlock	Engineering Consultant, Fort Collins, Colorado
Debora J. Miller	ESA Consultants, Fort Collins, Colorado
Sean C. Muller	Brown and Caldwell, Denver, Colorado
Charles D. Shackelford	Colorado State University, Fort Collins, Colorado
William P. Van Liew	US National Park Service, Fort Collins, Colorado
Dirk J.A. Van Zyl	Mackay School of Mines, Reno, Nevada

The texts of the various papers in this volume were set individually by typists under the supervision of each of the authors concerned.

Published by
A.A. Balkema, P.O. Box 1675, 3000 BR Rotterdam, Netherlands
Fax: +31.10.413.5947; E-mail: balkema@balkema.nl; Internet site: www.balkema.nl
A.A. Balkema Publishers, Old Post Road, Brookfield, VT 05036-9704, USA
Fax: 802.276.3837; E-mail: info@ashgate.com

ISBN 90 5809 126 0
© 2000 A.A. Balkema, Rotterdam
Printed in the Netherlands

Table of contents

Geotechnical considerations

Liners, covers and barriers

Groundwater and geochemistry

Surface water quality

Remediation and reclamation

Tailings and Mine Waste'00 © 2000 Balkema, Rotterdam, ISBN 90 5809 126 0

Preface

This is the seventh annual Tailings and Mine Waste Conference held at Colorado State University in Fort Collins, Colorado. The purpose of these conferences is to provide a forum for discussion and establishment of dialogue among all people in the mining industry and environmental community regarding tailings and mine waste. Previous conferences have been successful in providing opportunities for formal and informal discussion, exhibits by equipment and instrumentation companies, technical exhibits, and general social interaction.

This year's conference has over 60 papers. These papers address the important issues faced by the mining industry today. These proceedings will provide a record of the discussions at the conference that will remain of value for many years.

Design, operation and disposal

Design, operation and disposal

Tailings and Mine Waste'00 © 2000 Balkema, Rotterdam, ISBN 90 5809 126 0

Upstream constructed tailings dams – A review of the basics

M. P. Davies & T. E. Martin
AGRA Earth and Environmental Limited, Burnaby, B.C., Canada

ABSTRACT: Upstream tailings dams are a technically feasible option for many mining operations. Upstream dams are those, regardless of other names provided, that have the crest of the dam move in a pond-ward direction relative to the initial (starter) dam. Upstream tailings dams are typically less expensive than centreline or downstream dams due to the lesser "dam" fill volumes required. When designed with sound engineering principles and constructed and monitored with observation of those principles, this economical advantage can be realized without compromising the stewardship requirements of modern tailings management.

Upstream tailings dams have unnecessarily become the "poorer" cousins to other tailings dam geometries with many designers and regulatory jurisdictions. The reluctance to consider upstream dams appears to be at least partially due to upstream tailings dams providing more than one half of the more dramatic mine tailings impoundment failures. Notwithstanding the historically higher percentage of tailings dams that are upstream constructed, there are a significant number of upstream dam failures/incidents that can be directly attributed to poor design, construction and/or stewardship of the facility.

This paper presents the "basics" of appropriate upstream tailings dam design, construction and stewardship. The paper includes a case history that clearly demonstrates the potential ramifications of ignoring these basics; the 1991 static liquefaction failure at the Sullivan Mine. It also discusses two designs that satisfy the basics.

1 UPSTREAM TAILINGS DAMS - GENERAL CONSIDERATIONS

Upstream constructed tailings dams are common in the mining industry and remain a viable storage alternative provided a number of design issues can be met in appreciation of prevailing site conditions. At the same time, upstream constructed dams have contributed to a majority percentage of all tailings dam "failures" (notwithstanding that they are also the most abundant tailings dam type) and essentially all of the dramatic static/transient load induced liquefaction flowslide events. Figure 1 demonstrates, for those cases "publicly" documented, the ratio between upstream dam failures and failures from other tailings dam geometries. Figure 2 demonstrates the wide range of failure types which have been identified as the prime cause of each of the documented upstream dam failures. This seemingly checkered history has many regulatory jurisdictions extremely hesitant to consider upstream methods (e.g. Chile, California) whilst others tend to show reluctance in accepting upstream dams though their reasons cited are not always consistent with the actual site conditions and design presented.

Prior to detailing the basics of upstream tailings dam design and construction, the seemingly trivial question of "what defines an upstream dam" requires clarification. In spite of their historical and present popularity, it is surprising that there remains difficulty in naming these structures. A number of inventive, curious and sometimes purely misleading names have been provided to upstream constructed dams. There is no judgment of the reasons used by those making this error but the damage it can do by creating confusion amongst owners, regulators and even designers is real (e.g. the latter start to believe if you call it something else, it will behave like something else! - case histories defend this assertion). Without any need for special cases, the following definitions should be adopted industry wide:

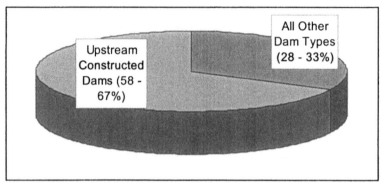

Figure 1. Failure of Tailings by Construction Geometry (data from USCOLD, 1994, UNEP, 1996 and WISE, 1999).

Downstream Construction - the dam crest during construction stages moves downstream from the crest of the initial raise (starter dam).

Centreline Construction - the dam crest, upstream and downstream edges, remain fixed relative to upstream and downstream directions as the dam is sequentially raised.

Upstream Construction - the dam crest, including the upstream edge of the crest, moves over beached tailings at any point in its sequential raising.

The above are simple and universal. The best test for an upstream constructed facility is whether a vertical line extended from the upstream edge of the crest ever touches beached tailings; if the answer is yes, regardless of how much it touches, the facility is an upstream constructed facility.

Rule 1 - If there are any beached tailings beneath the upstream crest of the tailings dam, then the structure has been developed using upstream construction.

This paper presents a series of basics but, as above, several are considered so fundamental as to deserve recognition as "rules." All rules are meant to be broken but this should only be the case with upstream tailings dams when considerable, and appropriate, thought has been applied, and multiple lines of defense are provided, in terms of design, construction, operation, and monitoring.

There are cases where combinations of raise geometries are used during the life of a facility. There is no question that there are degrees to how much of a facility may be developed in an upstream fashion. However, even a modest amount of upstream construction invokes the requirement for the highest level of design skill, appropriate construction and operational stewardship that can be made available.

As described by Martin and McRoberts (1998), there are "good" and "poor" manners with which to develop an upstream tailings dam. For the most part, ensuring good drainage and a wide beach above water (BAW) zone is "good" practice whereby having tailings near the dam crest deposited directly into water (beach below water - BBW) is "poor" practice. Maintenance of a wide, drained beach, as described in the next section, provides an upstream facility with the most resistance from static and transient loads as well as usually providing (dependent upon geometry) more freeboard.

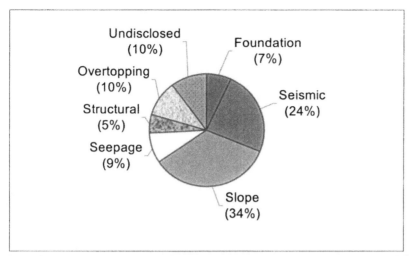

Figure 2. Suggested Prime Failure Mode of Documented Upstream Tailings Dam Failures (data from USCOLD, 1994; UNEP 1996 and WISE, 1999).

> Rule 2 - Maximizing the BAW zone is good practice when developing an upstream tailings dam.

Based upon available literature, there are at least 3500 tailings dams worldwide. Although not a precise estimate, the authors suggest that more than 50% of these are upstream constructed. As shown in Figure 1, there have been fewer than 100 documented "significant" upstream failures suggesting a failure rate of no more than 5 or 6%. Many upstream facilities, particularly those with wide BAW zones, have performed well in significant seismic events, when subject to intense statistically infrequent rainfall events and in-spite of questionable stewardship. In other words, if a set of conditions can be met, it appears as though upstream tailings dams can still be considered an option for modern tailings facilities.

> Rule 3 - Upstream constructed tailings dams are not necessarily bad engineering practice.

The following section presents the basic set of conditions desirable prior to contemplation of designing and operating an upstream tailings dam.

2 DESIGN AND CONSTRUCTION CONSIDERATIONS

From a "basics" perspective, the following list presents a handy checklist of issues to consider when contemplating an upstream tailings dam.

- ☑ foundation conditions;
- ☑ climatic conditions;
- ☑ tectonic setting;
- ☑ tailings properties;
- ☑ geotechnical considerations;
- ☑ geochemical considerations;
- ☑ operating criteria;
- ☑ monitoring and ongoing evaluation;
- ☑ closure plans; and
- ☑ regulatory requirements

Foundation Conditions

For foundation conditions, pervious foundations with a proven (by site investigation and design) resistance to liquefaction and/or extensive strain-softening under design static and seismic loads are preferred. Where foundation conditions can lead to rapid (e.g. centimeters per day or faster) movements and or brittle strain softening, upstream dam development is contraindicated. In these poorer foundation conditions, a more strain-independent dam is advisable.

> Rule 4 - Upstream dams require a good foundation - preferably well-drained and of adequate constitutive behaviour.

Climatic Conditions

Upstream dams are far more likely to succeed in arid conditions (where long beaches can be truly subaerial) than in temperate conditions though there are many successful examples in the latter. At the same time climatic conditions contributory to tailings dam "successes" from, for example South Africa or Arizona, do not necessarily translate as sound ideas for coastal North America.

Tectonic Setting

Understanding the seismicity of the tailings dam site is extremely important when contemplating upstream construction. As shown in Figure 2, seismic events have been shown to be responsible for roughly one quarter of upstream dam failures (conversely, there have been essentially no dramatic failures of other types of tailings dams due to seismic loading alone).

As noted by McLeod et al. (1991), there are really no aseismic regions in the world. However, it is the regions that can provide mechanisms which create several significant cycles of peak ground accelerations within the tailings of greater than 0.1 to 0.2 g which are of greatest concern. Analyses appropriately assessing the attenuation or amplification of background seismic waves by the tailings dam are required to determine seismic design parameters.

Tailings Properties

All tailings are not identical in mechanical behaviour for a number of reasons. It is essential that the gradation, fabric (including grain angularity) and bulk density be determined. At the design stage, the likely pore pressure conditions need to be appropriately modeled and during operations they need to be appropriately measured (e.g. Stauffer and Obermeyer, 1988, and Martin, 1998), plotted, and interpreted. Where pore pressure increases due to load additions exceed ratios greater than about 0.8, shear-induced pore pressures due to contractant response are a likely explanation and present a warning that "the end may be nigh". A proper combination of pore pressures and bulk density is required to correctly estimate in-situ stresses and, hence, in-situ state. The non-conservatism of using incorrect values is touched upon by Ladd (1991), Carrier (1991) and Martin and McRoberts (1998), and should be obvious to all designers. This item is listed herein as, from the authors' experience, too many facilities are being designed with blanket $S_G = 2.67$ to determine bulk density and hydrostatic conditions to describe pore water distribution. This complacency has been directly attributable to at least one "failure" where the designer's calculations showed there was "no way" the dam should have failed.

> Rule 5 - Correct tailings bulk densities and in-situ stress conditions need to be used when estimating in-situ state.

Geotechnical Considerations

For upstream tailings dams, transient loads leading to brittle response are the most important geotechnical consideration. Often, the most critical transient load to consider is seismic loading.

Tailings facilities in several of the major jurisdictions in the world are subject to specific guidelines with respect to addressing the potential for seismic liquefaction. For seismic liquefaction to occur, the number of significant cycles that are imparted by the earthquake must generate sufficient pore pressure rise and resulting brittle behaviour. The importance of defining the tectonic setting was discussed above. The resulting behaviour of the tailings in response to the seismic loading depends upon the initial state of the tailings and the size and nature of the seismic event.

Non-seismic transient loads can also lead to liquefaction. Static liquefaction is often a more difficult phenomenon to describe and/or anticipate than its seismic counterpart. However, it is a prime failure mode for upstream constructed tailings dams. There is limited mention of static liquefaction as a phenomenon in governance literature. As with seismic liquefaction, the most common manner to address static liquefaction design issues is to use empirical relationships. For example, Plewes et al. (1988) note approximate rates and construction lift thickness required to initiate a static liquefaction event in mine tailings deposited in BBW conditions. These approximate rates for saturated tailings are in the order of 5 m/day or higher.

To demonstrate whether a static liquefaction trigger exists for a given tailings facility, a generic slope configuration(s) and probable stress loading paths for the tailings deposit should first be appreciated. Figure 3 presents such a generic slope. The value of "S" is often termed the slope of the tailings facility (or as the horizontal component in H:V ratios). Both compression and extension stress path triaxial data and simple shear data on the tailings are required to do a specific evaluation. From this data, at least conceptually, the "collapse surface" can be approximately located within the lines of phase transformation (steady state) as shown on Figure 4. In practice, it is often difficult to establish a collapse surface in compressive loading. The lines of phase transformation are identical in either compressive or extensive stress space. However, this isotropy is not evident with the collapse surface that is steeper in compressive loading than in extension. This anisotropy is largely due to fabric/grain imbrication (almost always preferential to the horizontal plane) and increases with grain angularity and elongation; two characteristics common to ground mill tailings. The imbrication, due mainly to the hydraulic deposition processes, results in elongated grains being aligned preferentially in horizontal to sub-horizontal layers. Typically, this horizontal plane is normal to the maximum principle stress resulting in additional cross-plane anisotropy in triaxial loading conditions.

Also shown on Figure 4 are equivalent tailings embankment slope values "S". Stresses within a slope are very complex and general rules of thumb are guidelines at best. With this in mind, as a guideline, values of "S" greater than the collapse surface value are much more susceptible to "spontaneous" liquefaction; that is to experience a brittle liquefaction event with essentially negligible trigger application. For many tailings, a compressive loading slope of about 2H:1V to 3H:1V and an extensive loading slope of about 3.2H:1V to 3.9H:1V results; the implication being slopes flatter than these values have overall lower shear bias and hence lower risk of large-scale spontaneous/static liquefaction given equivalent in-situ states. It is interesting to note from the international database (e.g., USCOLD, 1994) that no tailings impoundment of either overall or large intermediate slope flatter than 4 to 5H:1V has failure attributable to a

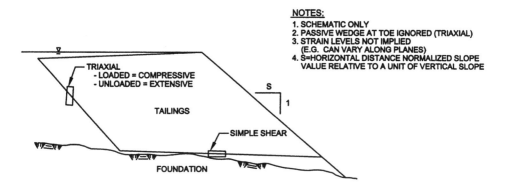

Figure 3. Generic Slope and stress Loading Paths for Upstream Tailings Dams

7

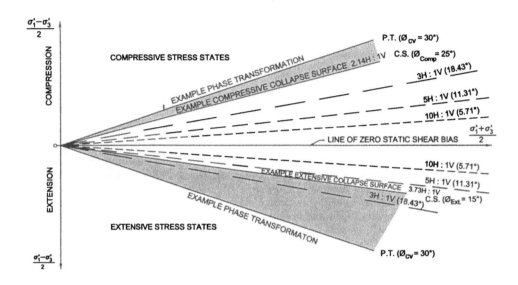

Figure 4. Tailings Slopes in Comparison to Typical Phase Transformation Angles (adapted from Davies, 1999)

"spontaneous" static liquefaction event (i.e. negligible trigger application). On the other hand, there are several upstream constructed tailings impoundments of overall or intermediate 2H:1V to 3H:1V slopes which are considered to have failed in this manner. Kramer and Seed (1988) demonstrated in the laboratory that there is a marked increase in static liquefaction susceptibility with increase in principal effective stress ratio. This type of soil behaviour has been observed by many other researchers and described in literature at least as far back as Bjerrum et al. (1961).

> Rule 6 - If the impoundment has an upstream constructed dam, and the dam slopes are steeper than about 4H:1V, the facility has a greater susceptibility to static liquefaction with minimal trigger application.

Another geotechnical consideration, given full discussion in Carrier (1991) and Martin and McRoberts (1998), is the appropriate analytical framework utilized to assess slope stability. As shown on Figure 2, slope failures are the leading cause of documented tailings dam failures. Effective stress analysis (ESA) is only appropriate with upstream dams if the pore pressures within the dam and foundations are known *accurately* at all times, including during shearing. For potential failure conditions, this involves estimation techniques as the actual pore pressures at failure can only be measured once it is too late for the given dam. Though regulators do not want to hear this, it is nearly impossible to monitor pore pressures accurately in many instances with one of the most difficult cases being within an upstream dam undergoing raise-induced pore pressures and shear strains (e.g. due to foundation or internal dam deformations) which are also inducing pore pressures. Further, there is little benefit in obtaining data on shear-induced pore pressures if the dam has to approach a state of failure to provide this data. The use of undrained strength analyses (USA) should always be included in the assessment of upstream tailings dams. Although the parameters need to be developed on a site-specific basis, "rule of thumb" suggested values for USA tailings analyses are presented in Table 1.

The values in Table 1 are typically conservative. The steady state strengths should only be used for "confined" analyses; that is when the tailings do not flow (Davies and Campanella, 1994). For flowslide conditions, e.g. where runouts are to be predicted, a stress-normalized approach is not appropriate (Davies, 1999) and estimated values such as these presented by Vick (1991) are suggested.

8

Table 1 - Basic USA Parameters for Upstream Tailings Dams

CONDITION	USA
Peak Strength Coarse Tailings ($\geq$ 50% coarser 74 µm)	0.28
Peak Strength Fine Tailings (< 50% coarser 74 µ m)	0.22
Steady State for Tailings With < 30% or > 60% fines (74 µ m)	0.10
Steady State for Tailings With $\geq$30% and $\leq$ 60% fines (74 µ m)	0.06

> Rule 7 - USA should be used for upstream dams where the tailings are uncompacted and largely to wholly saturated.

Another key geotechnical consideration is the potential for internal erosion. Incompatible filter designs with actual construction materials must be avoided. The use of pipeworks in dams should also be discouraged as these features rarely offer any economical or technical advantages yet open the dam to a range of extra failure modes. However, these can be used (eg. through dam decants) if multiple lines of defense to internal erosion, especially filters, are present, and the decant is not subject to potential rupture.

Geochemical Considerations

Geochemical issues are of paramount importance in the selection of tailings dam type and operating criteria. The past 20 years has brought significant enlightenment to geochemical issues and manners with which to minimize/mitigate potential downstream impacts from stored mine tailings. Seepage rich in heavy metals can be just as severe a failure as a physical dam breach from an environmental impact persective. Probably the single most important issue related to upstream tailings dams is, particularly for sulphide ores, that facilities are often required to be maintained in a flooded condition upon closure. When this requirement is present, it can invalidate the candicy of the upstream geometry in many cases. The alternative is typically "perpetual" collection and treatment.

Operating Criteria

No amount of good design will overcome poor tailings impoundment stewardship. This poor stewardship typically comes from one of two sources:

1. operating criteria which are dichotomous with the geotechnical design; and
2. facility operations which ignore the basic fundamentals of the design.

Obviously when both sources are present, the dam is provided with a much better opportunity to fail. In the first case, the rule is simple: if the design requirement is a necessary component of structural integrity, do not create an operating scenario that makes achieving the design requirement impossible. As an example, one of the authors was recently involved in reviewing an upstream dam in South America that failed and created a flowslide due to excessive encroachment of the pond onto the beach. The designer's requirement of a wide BAW zone was combined with these same designers allowing for a pond size completely inadequate to allow settling of fines prior to decanting.. Another anecdotal example involves a facility where the design of a BAW zone was carried out using laboratory derived beach slopes which, during regular tailings line flushing, were reduced to 0.5% thereby causing pond encroachment and subsequently instability. The authors suggest this is common and should not "surprise" operators or experienced designers. The companion paper (Martin and Davies, 1999) presents stewardship trends and suggestions. Herein, rules 8 and 9 are provided.

> Rule 8 - Design geotechnical requirements and actual operating criteria must be compatible.

> Rule 9 - Laboratory-scale and literature predicted beach slopes tend to be incorrect. Do not rely upon steep beach slopes to meet operating criteria.

Monitoring and Ongoing Evaluation

An upstream tailings dam, by its very nature, is a "work in progress". More than any other type of tailings dam, detailed monitoring and review of the facility is required throughout its life to check that its performance and condition are satisfactory. A designer of such a facility is placed at the mercy of the operators, and must make it clear that the design is but the first step in an ongoing process of monitoring, evaluation, and design adjustment as required. Where the designer will not maintain an ongoing involvement in the monitoring and evaluation of the upstream tailings dam, this design should not be chosen.

> Rule 10 – Designers should avoid designing upstream tailings dams where the designer will not (or cannot) be able to maintain a high level of involvement in the monitoring and evaluation of the facility.

Closure Plans

Modern day facilities should all be designed and operated with closure in mind. If a wide BAW zone is required for perpetual structural integrity under static/seismic loading, consider this prior to selecting an upstream geometry for, as an example, sulphide rich tailings that will require a closure water cover due to oxidation concerns (see Geochemical Considerations). It is less palatable to have to rehabilitate a dam at mine closure once the revenue stream is gone than to choose the appropriate dam geometry for closure conditions at the outset.

Regulatory Requirements

Regulators worldwide are becoming increasingly educated about tailings dam design and stewardship requirements. However, upstream dams remain very unpopular with many regulators even when all of the technical checklist points to success with the method. Improved communication (e.g.. not renaming upstream dams) and disclosure of all issues on a technical checklist can go a long way to demonstrating necessary partnership with a regulator. Of course, there are jurisdictions where upstream construction is more difficult to gain approval and the cost of such a process, including potential non-permitting, need to be weighed against another storage geometry at a probable higher unit storage cost.

Final Design Consideration Thought

If one becomes a student of tailings dam case histories, an interesting conclusion arises. These failures, each and every one, were entirely predictable in hindsight. There are no unknown loading causes, no mysterious soil mechanics, no "substantially different material behaviour" and definitely no acceptable failures. There is lack of design ability, poor stewardship (construction, operating or closure) or a combination of the two, in each and every case history. If one or more of the "rules" are broken and the basic design and construction considerations ignored, candidacy for a case history is immediate.

> Upstream Tailings Dam Axiom - Tailings dam failures are a result of design and/or construction/management flaws - not "acts of god".

3 WHEN THE RULES ARE BROKEN - SULLIVAN MINE CASE HISTORY

The Cominco Metals/Sullivan Mine located near Kimberley, British Columbia is an underground lead/zinc mine which is currently processing an average of about 8000 tons per day of feed. The mine was established in 1905 and is scheduled for closure in 2001. Active closure activities have been taking place at the mine throughout the 1990s.

From the start of operations, all of the mine tailings have been hydraulically transported to an area southeast of the concentrator. A series of separate areas have been developed over the years by constructing a system of containment dykes as necessary. The only presently active tailings pond is the Active Iron Pond (AIP) which is surrounded by several inactive tailings ponds.

The AIP is formed by about 1500 m of earthfill dykes which presently have a maximum height of approximately 21 m. The starter dykes for this pond were constructed in 1975 and were incrementally raised using the upstream method of construction. The exterior shell of mechanically placed and compacted tailings was progressively "stepped" upstream directly onto the previously spigotted tailings beach (suspect foundation conditions - see checklist). Portions of the dykes were founded on the native ground, while other portions were constructed over previously deposited tailings. Design and regular inspection of the AIP facility to 1991 was carried out by a consulting engineering firm with the last inspection prior to the event happening within weeks of the event.

On August 23, 1991, about 2:00 p.m., the southeastern portion of the dyke between approximately stations 29+00 and 39+00 suddenly moved (slumped) during construction of a 2.4 m incremental raise of the dyke. Figures 5 and 6 show aerial and ground views, respectively, of the movement two days after the event. The event occurred relatively quickly, just as the final lift of the dyke raise was being placed. Construction at the time was such that loaded scrapers from the borrow pit located north of the site would travel along the dyke crest to spread their loads. Borrow for dyke raises up to and including 1991 was from coarse siliceous tailings. Empty scrapers would return to the borrow pit by running down a construction road on the dyke slope and onto a lower access road at the toe of the iron dyke. It was reported by eye witnesses that an empty scraper had punched through the lower return road, immediately preceding the event.

The area of the slump event was about 300 m long and 12 m high, and involved up to an estimated 75,000 m^3 of materials. The toe area was observed to have been displaced about 15 m to 45 m downstream of its original position although there was clear evidence of disturbance for a distance of up to about 100 m downstream of the original toe. The final slopes of the slumped mass averaged about 10H:1V to 15H:1V. Visual observations indicated that the compacted dyke sections and roadways basically remained intact, but they did experience some brittle rupturing. Numerous sand boils were observed issuing from the ground surface just after the failure and for several hours later standpipe piezometers became flowing artesian wells. In addition, the surface of the toe area which was mainly dry immediately after the event, became wetter over the next 24 hours as water seeped from the movement zone. Fortunately, there were no casualties resulting from the event. The pooled water in the AIP was located sufficiently far from the failed section to remain unaffected and, as the slide materials were adequately contained within the No. 2 Siliceous Pond, there were no spills to the environment. Monitoring of survey points that were immediately installed within the area of the event showed little or no additional movements following the initial event.

Based upon a review of key visual observations, data records and the nature of the movement, it was concluded that the tailings within the foundation of the AIP dyke had liquefied. It is probable that triggering of liquefaction was contributed to by the additional loadings on the crest due to fill placement during the dyke raising, and/or by the dynamic loadings imposed near the toe due to construction equipment traffic. However, there is stronger evidence indicating that the lower portions of the dyke slope may already have been in a state of incipient failure at the time. It is likely that the liquefaction event might have eventually occurred even if construction of the latest dyke increment had not taken place. More details can be found in Davies et al (1998) who fully describe the case history.

A very interesting aspect of this failure is that, prior to the occurrence of the failure, a number of standpipe piezometers (hardly appropriate for measuring construction pore pressure response in fine tailings) in the area were flowing. This was considered a nuisance and the offending piezometers were capped off. The review engineers obviously did not understand the significance of, or the potential for, the high pore pressures.

Figure 5 - Aerial View of Sullivan Static Liquefaction Event

Figure 6 - Ground View of Sullivan Static Liquefaction Event

From a geotechnical perspective, the key conclusions from this case history include:

- the pre-failure dyke slope was in the range of 2.5H:1V to 3H:1V, which was likely in excess of the contractant and/or extensive collapse surface for the fine-grained cohesionless tailings (Rule 6 broken);

12

- excess pore pressure rise immediately prior to the event clearly occurred, though only ESA with hydrostatic conditions were applied (Rule 7 broken);

- back-calculated strengths for the liquefied tailings yielded a S_u / σ'_{vo} value of about 0.08 (consistent with Table 1);

- in-situ shear wave velocity measurements *were not* consistent with the brittle behaviour when published shear-wave velocity-based liquefaction susceptibility criterion were applied; and

- forensic limit-equilibrium analyses, using pre-event available data, showed the tenuous nature of the dyke when correct ESA and USA were used in the dyke assessment.

The failure did not result in any off-site releases due to the fortunate fact that no pooled water was near the failure area; correct stewardship by the owner/operator of a flawed facility. All of the broken rules here appear to be with an insufficient regard to a basic design checklist and to a lack of adherence to the basic rules of upstream dam design.

4 WHEN THE RULES ARE FOLLOWED

4.1 *Inco R4 Tailings Dams*

Martin and Tissington (1996) describe the design evolution of upstream tailings dams at Inco's Copper Cliff tailings facilities, in use since the 1930's, in Ontario. A typical design section for the perimeter dams for the latest expansion of the facility, denoted the R4 area, is illustrated in Figure 7. The dams will have a maximum height of about 150 ft. Key design features as they relate to the 10 "rules" described above are as follows:

Rule 1 – these dams were named for what they were; upstream dams
Rule 2 – the BAW zone (500 ft wide) is maximized, with the outer 300-ft wide shell being dozer-compacted to achieve a suitably dense, and dilatant, in-situ state to preclude any potential for liquefaction irrespective of trigger mechanisms.
Rule 4 – the dam is founded on competent foundations (till or bedrock) – where lacustrine clays and silts exist below the outer 300-ft wide shell, these are excavated and replaced by coarse rockfill causeways
Rule 5 – tailings properties are determined from field and laboratory testing, and appropriately selected piezometers are used to determine pore pressure profiles
Rule 6 – the outer dam slope is 4H:1V which, together with the compaction of the outer shell, addresses the potential for static liquefaction
Rule 7 – USA analyses (ESA for the outer shell, USA for the beached tailings) were used as the basis for design, with the uncompacted tailings assigned an appropriate high strain steady state strength
Rule 8 – design and operating criteria are compatible with operators very much involved in the design process and in the ongoing monitoring and evaluation of the facility
Rule 9 – beach slopes assumed for design were based on extensive beach surveys, with no reliance on laboratory or theoretical beach profiles
Rule 10 – the R4 tailings dams are closely monitored using piezometers, gradation and field density tests are taken, and the designers are involved in ongoing evaluation of the performance of the dams.

4.2 *New Tailings Dam Proposed For a Silver Mine in Mexico*

Figure 8 illustrates, in section, the design for a proposed upstream dam (Rule 1) for a silver mine in a relatively arid setting in Mexico. This impoundment will be lined with geomembrane (due to cyanide in the tailings effluent), and includes an extensive underdrainage system (perforated

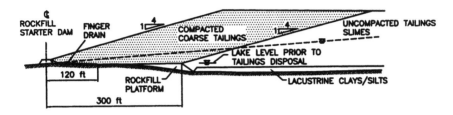

Figure 7. Typical R4 tailings dam design section.

pipe encapsulated in drain rock and filters) which discharges via a reinforced concrete-encased decant (surrounded by filters) through the dam to an underdrain pond downstream of the dam. The underdrain system is intended to maintain low heads over the 60-mil geomembrane liner (to minimize leakage through liner defects) and to maintain low pore pressures (below hydrostatic) in the tailings deposit (Rule 4 – well-drained foundation), particularly in the uncompacted tailings beach that will underlie the compacted shell. The starter dam will be constructed of compacted tailings borrowed from other impoundments that have been inactive for several decades, due to no other suitable materials being readily available. The outer shell will be raised of select coarse tailings from these other impoundments, compacted to achieve a dilatant state. Field density testing and grain size tests will be carried out to confirm that the dam is constructed according to design specifications. An extensive network of vibrating wire piezometers will be installed to confirm pore pressure profiles (horizontal and vertical) in the dam, with standpipes used as observation wells to confirm saturation levels. Design is based on ESA parameters for the compacted outer shell, and steady state USA parameters for the spigotted tailings beach (Rule 7). Design is also conservatively based on hydrostatic pore pressure conditions, whereas pore pressures considerably less than hydrostatic are expected (Rule 5 – conservative approach to estimating in-situ stress state).

A significant portion of the design for this facility includes, and is consistent with, operating criteria (Rule 8), and requirements for field and laboratory testing, monitoring, and ongoing evaluation and review of the performance of the facility as it is constructed (Rule 10). Field density tests will be carried out on the spigotted tailings beach as well as on the compacted outer shell. These measures will allow any design changes that may be necessitated by conditions less favorable than assumed in design, to be identified and implemented on a timely basis.

5 SUMMARY

Upstream tailings dams can be a technically valid option for many projects provided the basics and rules listed in the paper are followed from project concept through mine closure. When the

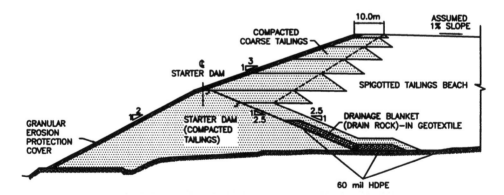

Figure 8. Proposed tailings dam design section - Mexican silver mine.

basics and rules are ignored, the collective case history database is provided an excellent opportunity to expand.

REFERENCES

Bjerrum, L., S. Kringstad and O. Kummeneje (1961). "The shear strength of fine sand." *Proceedings of the Fifth International Conference on Soil Mechanics and Foundation Engineering,* Paris, France, pp. 29-37.

Carrier, W.D. (1991). Stability of tailings dams. *XV Ciclo di Conferenze di Geotecnica di Torino,* Italy, November.

Davies, M.P. (1999). *Piezocone technology for the geoenvironmental characterization of mine tailings.* Ph.D. thesis, University of British Columbia.

Davies, M.P., B.G. Chin and B.B. Dawson (1998). "Static liquefaction slump of mine tailings - a case history. *Proceedings of 51st Canadian Geotechnical Conference,* Edmonton.

Davies, M.P. and R.G. Campanella (1994). "Selecting design values of undrained strength for cohesionless soils". *Proceedings of the 47th Canadian Geotechnical Conference,* Halifax, pp. 176-186.

Kramer, S.L. and H.B. Seed (1988). *"Initiation of soil liquefaction and static loading conditions."* ASCE Journal of Geotechnical Engineering, Volume 114, Number 4, April, pp. 412-430.

Ladd, C.C. (1991). Stability evaluation during staged construction. The 22nd Karl Terzaghi Lecture, Boston, 1986, *Journ. Of Geotech. Eng.,* ASCE, Vol. 117, pp. 537-615.

Martin, T.E. (1998). "Characterization of pore pressure conditions in upstream tailings dams. *Tailings and Mine Waste '99,* Fort Collins, Colorado, pp. 303-314.

Martin, T.E. and M.P. Davies (1999). Trends in the stewardship of tailings dams. *"Tailings and Mine Waste '2000,* Fort Collins, Colorado.

Martin, T.E. and I. Tissington. (1996) Design evolution of tailings dams at Inco Sudbury. *"Tailings and Mine Waste"* '96, Fort Collins, Colorado, pp. 49-58.

Martin, T.E. and E.C. McRoberts (1998) "Some considerations in the stability analysis of upstream tailings dams. *Tailings and Mine Waste '99,* Fort Collins, Colorado, pp. 287-302.

McLeod, H.N., R.W. Chambers and M.P. Davies (1991). "Seismic design of hydraulic fill tailings structures." *Proceedings IX Panamerican Conference on Soil Mechanics and Foundation Engineering,* Viña del mar, Chile, PP. 1063-1081.

Plewes, H.D., E.C. McRoberts, and W.K. Chan (1988). Downhole nuclear density logging in sand tailings. *Hydraulic Fill Structures,* ASCE Spec. Publication No. 21, pp. 290-309.

Stauffer, P.A., and J.R. Obermeyer (1988). Pore water pressure conditions in tailings dams. *Hydraulic Fill Structures,* ASCE Spec. Publication No. 21, pp. 924-939.

Troncoso, J. (1990). Failure risks of abandoned tailings dams. *Proc. Int. Symposium on Safety and Rehabilitation of Tailings Dams,* ICOLD.

UNEP (1996) "United Nations Environment Programme. - Tailings Dam Incidents 1980-1996," a report by Mining Journal Research Services, UK, May.

USCOLD (1994). *Tailings dam incidents.* U.S. Committee on Large Dams, 82 p.

Vick S.G. (1991). "Inundation risk from Ttilings dam flow failures", *Proc. 9th Panamerican Conf. Soil Mechanics and Foundation Engineering,* Viña del Mar.

Vick, S.G. (1990). *Planning, design, and analysis of tailings dams.* Bitech Publishers, ISBN 0-921095-12-0.

WISE (1999). "Chronology of Major Tailings Dam Failures," WISE Uranium Project, WWW access.

Tailings and Mine Waste'00 © 2000 Balkema, Rotterdam, ISBN 90 5809 126 0

Slurry distributor for tailings dam construction

John E. Lemieux & Claude Y. Bédard
Journeaux, Bédard and Associates Inc., Dorval, Qué., Canada

ABSTRACT: In the process of extracting and concentrating iron ore from the Mont-Wright mine in Northern Québec Canada the Québec Cartier Mining (QCM) company has successfully stored over 18-20 million cubic meters of tailings and 36 million cubic meters of process water annually since 1976.

The tailings impoundment area extends over 8 square kilometres with the south and southeast limits retained by a 6.5 kilometres long permeable tailings dam reaching 100 meters high.

The annual upstream construction of this tailings dam has been achieved using a conventional spigotting technique parallel to the dam. In 1996, in collaboration with QCM, a program to test three (3) new distributor concepts and designs for tailings dam construction was organised. Two (2) of the 3 designs were modifications to the conventional spigot technique. The test results on this third and completely original design confirmed significant advantages over the existing system, including: over 50% smaller starter dykes, lower fabrication costs, quicker installation, less manpower to operate and a 15-20% reduction in required pumping capacity.

This article will present a brief review of the designs tested during the trial program and the advantages and disadvantages of the selected new distributor, which is in use at QCM since 1998.

1 INTRODUCTION

QCM required a more efficient and cost effective method of placing the material needed for the bi-annual raising of the tailings retention dams. Several new ideas were proposed by Journeaux, Bédard & Assoc. Inc. (JBA) and in collaboration with QCM these methods of depositing and distributing tailings along the entire length of a 6.5 km tailings retention dam were tested. One method proved outstanding, and over a 3 year period from 1996 to 1998 was completely developed and put into production in 1994, hence replacing all previous methods used.

2 EXISTING METHOD

At QCM, the winter months do not lend themselves well to the construction of the retaining dykes. Hence, the tailings are simply open pipe dumped into the tailings impoundment from a single discharge point. However, during summer months, the retaining dams must be raised to allow storage of an entire year's volume of tailings and the accumulation of winter ice inside the solids storage area.

The material used to construct the dams consists of the coarse tailings separated from the total mill discharge by a standard thickening procedure. The fine material is deposited elsewhere in the impoundment area.

The standard method used conventional spigots leading off a header to distribute the coarse material parallel to the length of the dam. The header consists of a 225m long pipe reducing in size from 500 to 150mm. Each of the spigots is equipped with a pinch valve allowing work crews to stop flow in certain spigots and lengthen them perpendicular to the dam in 10m sections. In this way, horizontal beaches were deposited parallel to the dam with widths of up to 100m. The beaches are used for both erosion protection and as a source of borrow material pushed by tract mount tractors to the elevation required for the following year's starter dykes.

The spigot header is installed the following year on this mechanically constructed starter dyke. Dyke height is based on the location of required volume inside the tailings impoundment area to contain a minimum of one year of tailings material, which is calculated based on trends observed from historical fill data. On average, the solids storage area will fill to a height of 1.8m annually over its entire surface, being 5 m high at the discharge point.

3 DESIGN REQUIREMENTS

A substantial portion of the cost of raising the dam is due to the starter dyke that must be constructed mechanically. The principal need that had to be addressed by the new distributor was to minimise the size of the starter dyke required. Typically, the cross section of this starter dyke had a 10 meter crest with upstream and downstream slopes of 3H:1V and 2H:1V respectively.

In addition, it was mandatory that the new methods have less labour intensive operational requirements.

4 METHODS TESTED

Three (3) different approaches to the main problem of minimising the starter dyke size were suggested. Two (2) were based on the existing distributor concept using reducers and spigots, see figure 1.

Figure 1: One of the methods tested showing a small starter dyke with the distributor being manipulated from a lower level.

In one proposed method, the existing distributor was installed and maintained from a lower lift and the spigots themselves were constructed so as to bring the material up the newly constructed small starter dyke and then continue horizontally towards the interior of the tailings impoundment thus producing the required beaches.

The second method investigated was to simply conceive a machine that could install the distributor on the small starter dyke without actually having to operate on the dyke.

Both of these methods were tested and proved technically successful. Both addressed the need for smaller mechanically constructed starter dykes but effectively needed an increased labour requirement to ensure their safe and efficient operation.

The third method surpassed the original design considerations and expectations. The following section provides a description of the new slurry distributor.

5 NEW SLURRY DISTRIBUTOR

The third distribution method is a complete departure from the traditional methods, see figure 2. The square section distributor (0.5m x 0.5m x 70m), with paired openings at intervals along its length, was installed parallel to the dam structure and located a given distance from the starter dyke. However, a certain amount of sedimentation occurs in the slurry mix as it is transported by pipeline to the deposition area. This coupled with the fact that the natural deposition angle of the slurry being pumped increases with the grain size of the material were critical characteristics which required field adjustments.

By designing the distributor with a square cross section and creating adjustable openings along the full height of the side walls, it was possible to overcome these difficulties. In this way it is possible to ensure that the total slurry, from the bottom of the square distributor to the top, is used to construct the beaches. The conical deposition mounds thus formed at each opening all have constant deposition angles.

The material exiting the side openings accumulates in a conical fashion under the opening, see figure 3. Eventually, the accumulation of this material will restrict flow through the opening and material will be forced to the following pair of openings. The entire process is repeated until all the openings are obstructed and practically all the flow is directed to the end of the distributor. When the end of the distributor is completely filled and the entire assembly is buried in the tailings it is necessary to advance the entire apparatus forward to the next deposition position.

It is possible to adapt the distributor, by adjusting the opening locations, opening dimensions, pipeline cross-sectional area and other key parameters to other slurry flows with differing characteristics.

Figure 2: Taken from the crest of the starter dyke looking at the adopted distributor at a distance parallel to the starter dyke.

Figure 3: Taken from the top of the distributor looking in the direction of flow, note the height and offset of the distributor with respect to the starter dyke at extreme left.

6 IMPLICATIONS AND ADVANTAGES

The advantages of this deposition method address both requirements of QCM including a drastic reduction in both the size of the mechanically constructed starter dyke and the operational labour requirement.

With this method, the distributor must be placed at an elevation above and at a distance parallel to the starter dyke. This immediately suggests that the deposited beach will have a high point that is higher than the starter dyke. In this way, the height of the starter dyke required is reduced and the total mechanical construction volume is further reduced. For example, if it is determined that a 4 meter lift is required it may be possible to obtain the 4 meter height with only a 2.5 meter high starter dyke, as shown in figure 4.

This distributor functions without restricting the flow through the slurry pipeline. This

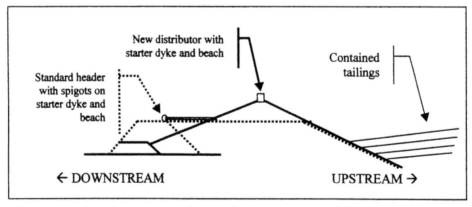

Figure 4: Cross sectional sketch showing the differences in starter dyke size and raised height.

translates directly into additional savings at many levels including: lower pumping costs (approx. 15 to 20% at QCM), lower maintenance costs on pumps and piping due to reduced wear, lower operational costs due to reduced number of leaks in the pipeline.

Another major advantage of this equipment is that it can be dismantled and moved to the next position using a single standard loader, and this on a regular daily frequency, if required.

7 CONCLUSION

As part of the tailings impoundment operations at QCM the tailings retaining dam must be raised every year. A need was expressed for a more efficient and cost effective method of placing the material required.

Both JBA and QCM proposed several new ideas and in collaboration with QCM these methods of depositing and distributing tailings along the entire length of a 6.5km tailings retention dam were tested. One of which proved outstanding and over a 3 year period from 1996 to 1998 was completely developed and put into production replacing the existing method in 1999.

The new tailing distributor meets all the objectives set by QCM. It has been demonstrated that starter dykes are 50% smaller and that set-ups are easier and quicker requiring, fewer man-hours. The new distributor results in a significant decrease in required pumping capacity (ranging from 15 to 20% for QCM) and a significant reduction in maintenance costs of pumps and pipeline breaks. In addition, this distributor can be adapted to many different slurry characteristics and can be built at relatively low costs.

Finally, the tailings dam construction program using the new distributor has been accomplished without modifying any operating parameters within the concentrator mill.

Tailings and Mine Waste'00 © 2000 Balkema, Rotterdam, ISBN 90 5809 126 0

Sand transport and placement

S. Barrera & C. Riveros

Geotécnica Consultores S.A., Santiago, Chile

ABSTRACT: The present paper analyzes how design and construction of a sand dam impose conditions on the design of the sand transport and discharge systems of the dam wall, and also the inter-relationship with the operational aspects involved, which are: maximum admissible slurry flow when discharging to avoid erosion of the dam wall, adequate thickness of sand lifts for compaction, operation period for discharge, water management for pipe flushing, and selecting the most suitable pump type for sand slurries.

1 INTRODUCTION

The construction of dam walls for tailings impoundments often considers the use of the coarse fraction of the tailings (the sand). This sand is obtained from the tailings by means of a classification process and is then transported and placed on the dam wall. This entire process is often done hydraulically due to the nature of the tailings. For this reason, determining the geometry of the dam wall, the dam raising method, the placed sand density and the fines content of sand, should bear in mind aspects such as: the content of solids in the sand discharge, the maximum amount of sand available from the tailings, the drainage rate of the sand recently placed and the permeability of the sand once it has been compacted. All the above factors not only influence the design of classification, transportation and discharge systems for the sand, but also influence on operational aspects of the dam wall construction. It therefore follows that both design and construction must be absolutely linked in order to reconcile both aspects, and to obtain a safe structure.

The use of tailing sand for the construction of dam walls in tailing impoundments, in general, requires the following steps:

- Tailings classification, consisting in separating the tailings in two fractions: a coarse fraction (sand) with which the dam is constructed, and a fine fraction, which is deposited in the impoundment. This process is done through individual cyclones throughout the length of the dam crest or grouped together in stations, according to their production level.

- Sand transport towards the dam wall. This process only takes place if the cyclone stations are located away from the wall.

- Sand placement in the dam wall to obtain the design geometry of the wall.

In the construction of large impoundments, these processes are associated to large quantities of solids and, in many cases, to important transport distances. In these conditions, the use of water as a transport and placement system for the sand presents evident advantages. This latter because it presents a continuos transport medium of low operating costs (when water is not an extremely scarce resource).

The transport of solids goes back to the last century and has been the object of various studies, with the development of theories, empirical models, experimental verifications, design standards, etc. (Wasp, Faddick, Newitt, Durand among others), which has resulted in a know-how in present times that allows for safe and reliable designs.

Just as the tailings management by hydraulic means presents the previous mentioned advantages, the effect of the water in the sand transport must be considered in the design, operation, and construction of the dam wall. This effect is obviously more significant when the volumes of water required in the transport system increase.

It is clear from the previous statement that the construction aspects of the dam wall must be considered in the design of the sand transport and placement system.

The present paper does not have the objective of dealing with aspects related to the hydraulic design of sand slurries nor with geotechnical aspects associated to dam wall stability but rather to emphasize the need of including the construction and design aspects of the dam wall in the design of the piping or transport systems.

2 SAM DAM CONSTRUCTION

The construction of dams with tailing sands is generally done by hydraulic placement of the sand in inclined layers on the downstream slope of the dam, discharging the sand from the dam crest. The sand placed in such manner flows over the slope, draining the transport water towards the drainage system located at the toe of the dam, leaving a remnant of water in the placed sand (8-10%). These layers are mechanically compacted once their placement is complete and once a minimum bearing capacity is achieved for the safe operation of the compaction equipment.

The dam design requires the sand to have a permeability greater than a 10^{-4} cm/s to allow for a quick drainage of the transport water in the sand; this is achieved by limiting the amount of fines. In the case of copper tailing, this permeability limit means that the fines content cannot be greater than 15 to 25% depending on the compaction degree and the geotechnical characteristics of the tailings (specific gravity, clay content among others).

In the light of only this geotechnical criteria, it would seem convenient to use sands with less fines content since they would be more permeable this way. But a lesser fine content in the sand implies:

   i.   Less sand production and greater classification facilities (classification system in two stages).

   ii.   Coarser sands, which would produce greater wear in the pipes and more friction, with the corresponding increase in head loss.

   iii. Faster water drainage on the dam, which requires a greater compaction effort since the optimum moisture content is not present or, as an option, placement of thinner sand layers, which implies an additional cost in compaction and sometimes operational difficulties.

These aspects together with the fact that a limited percentage of fines does not represent a significant reduction in the sand resisting parameters (cohesion and friction angle), make it highly convenient to search for a balanced sand that satisfies the permeability limits and also facilitates the operations of transport and placement without jeopardizing the dam wall stability.

With a permeable sand, with a granulometry that facilitates the compaction and a drainage system that is able to evacuate the sand transport water, the downstream slope of the dam is determined guaranteeing an appropriate stability for these civil works under seismic conditions. In general for dams of tailing sands, the stability calculations indicate that a downstream slope in the order of 2.5 to 3.5:1 (H:V) is sufficient to achieve a stable dam. In practice, the hydraulic placement of the sand is done with concentrations in the order of 65 to 70% with which a slope between 2.5 to 4.5:1 (H:V) is naturally achieved. This slope is flatter than the one required for slope stability reasons, which implies that sand dams constructed this way are very stable provided the sand meets the minimum permeabilities and the dam has an adequate drainage system.

The construction of a sand dam constitutes a continuous task, which depending on the need for sand, may extend for 24 hours a day, every day of the year, being interrupted only for plant maintenance needs. This means that the sand production and placement is also a continuous task, to meet the constant need for dam raising. Considering that the sand placement pipes are located on the dam crest and downstream slope, it is evident that the tasks of relocating pipelines elevation-wise, manage discharge locations, extend pipelines, spread and compact the sand on the dam, and flushing pipelines, turn out to be the most important aspects to be considered in the design of the sand transport and placement system to construct sand dam walls.

3 DESIGN

The transport and placement of the sand not only must comply with the hydraulic criteria proper to slurry transport such as criteria for limiting velocities, head loss analysis, erosion of the pipelines, pump systems, etc., but must also include those characteristics that facilitate the operation and construction of the sand dam walls. It is imperative to know the operational process involved in the construction of sand dam walls.

The sand transport towards the wall (in the case where there is a centralized cyclone station) is done generally by pipelines under pressure that are located throughout the crest of the dam. These pipelines must be extended constantly as the length of dam increases.

When the sand is transported by pipelines to the dam, the sand placement on the dam may be done primarily by the next two methods:

i.   By a group of small diameter spigots that extend from the pipeline along the crest, through which the sand is discharged to the wall simultaneously. This method is used preferably when the sand is pumped to the wall by positive displacement pumps since the pressure produced by these pumps does not allow for silting up of the pipelines since the flow diminishes gradually after each discharge.

ii.   By perforated pipes that originate from the pipeline located on the crest of the dam which extend throughout the downstream slope of the dam. The sand is discharged on the dam through the perforations directly on the dam slope. These pipes may change position on the slope spanning a greater area of influence facilitating the spreading of the sand. This system is used generally when the sand transport towards the dam is done by gravitational flow or by centrifugal pumps.

The facilities described previously require an important operational management whose success depends mainly on the operator's experience and on the proper design making those facilities easy to operate. With regards the proper design, the designer must take into consideration the aspects that influence the operation. Among those aspects, the following are worth mentioning:

a) Concentration of the sand in its transport towards the dam. Generally the sand leaves the classification units with concentrations between 70 to 72%, which corresponds to a high value which may cause silting in the pipelines, specially in the sections located on the crest, where the pipeline is horizontal or with very small longitudinal slopes. To avoid deposition and plugging problems, water must be added to dilute the concentration to an order of 65%. Much lower concentrations mean a greater water content affecting the inclination of the slopes of the dam.

b) Energy dissipation systems on the pipelines. When there is an excess of hydraulic head to transport the sand to the dam, it is necessary to dissipate such energy to reduce the discharge velocities of the sand in the dam. Energy dissipation systems must be available to dissipate the excess energy. These systems must consider that the energy to be dissipated at the start of the dam construction is greater (greater elevation difference) and that as the dam increases in height, the energy to be dissipated decreases. For this reason, a solution that works well at the start of operations may not be suited for latter stages of operation, such as the drop boxes system for example. What must also be considered is that the sand is transported in high concentration and being an abrasive slurry, systems that produce head loss through the strangulation of flow sections are not recommended. Such is the case of the energy dissipating ceramic rings, which may become blocked and would not last long due to the excessive wear produced by this type of slurries. Energy dissipation by reducing the pipes' diameter would produce a greater wear in the pipes but has the advantage of being able to change the pipes for ones of greater diameter as the energy dissipation need decreases.

c) Backup transport systems. When the sand is produced by one fixed cyclone station, it is sometimes convenient to have more than one pipeline, instead of only one of larger diameter to transport the sand on the dam wall that makes the task of relocating the pipelines as the dam increases unmanageable. In this way it is possible to continue sand distribution, meanwhile one of the pipelines is being relocated, specially when the dam to be built, requires classification 100% of the time. This may also be complemented with a pipeline that discharges sand on the dam's abutment for later spreading without interrupting the classification so it may operate when discharging sand from the dam's crest is not possible. With more than one pipeline operating simultaneously, the flow discharging from each location is also reduced with the following advantages: the discharge velocity is reduced avoiding erosion on the dam, the segregation of the material on the slope is reduced avoiding the migration of fines to the drainage system and thus clogging it and a faster drainage is obtained by discharging less water per sector. Appropriate sand lift thicknesses for compaction are also obtained.

d) Telescoping pipeline systems. The telescoping pipeline system, with progressive diameter reductions, is used when there is need for discharges in more than one location at a time and there is only one pipeline in operation. The reduction in diameters avoids deposition in sectors of small flows or excessive wear in sectors of greater flows. This system incorporates an additional operational complexity to the overall system in addition to fixing the discharge locations to be operated.

e) Intermediate discharges on the slope. When the wall is too high, the sand discharged from the crest may not reach the lower part of the slope making it necessary to locate another pipeline at an intermediate elevation on the dam slope. This could represent an additional advantage, since it will be possible to have two independent workfronts that can operate simultaneously or alternating.

f) Flushing water system. When the transport of sand is interrupted, it is necessary to flush the pipes to avoid deposition, for which a system that adds water immediately after the operation has stopped must be available. This system must operate automatically even when there has been a power outage (i.e. when sand is pumped to the dam). The pipe flushing water must be discharged into the impoundment and not on the dam.

g) Pump seals. Pumps that do not require water seals (for centrifugal pumps) must be used in order not to add water to the sand that may diminishe its concentration.

h) Instrumentation. Generally the system that incorporates water for dilution and flushing the pipelines must consider using automated systems that control the density of the sand (densimeters) and flows (flowmeters). From theses instruments' processed information, either valves that incorporate water into the system or valves that flush the pipelines are opened as required. In this way, the amount of water being added to the sand is controlled and thus plugging problems of pipelines and head boxes is avoided.

i) Pipe unions. Considering that the pipelines located on the crest of the dam must be constantly relocated, it is necessary that their unions allow for an easy mantling and dismantling. It is for this purpose that vicatulic unions are recommended. Moreover, it is important that the pipeline's length not be excessive so that it may be easily removed.

4 FINAL COMMENTS

It is clear from what has been presented in this article that the designer of the sand transport and placement systems must have a deep knowledge and understanding of the operational and constructive aspects of a sand dam wall. These concepts can be summarized as follows: Due consideration of constructability and operational aspects associated to the sand dam construction is a key factor in the design process, in order to produce a safe structure through a simple and cost effective construction.

REFERENCES

Barrera S, Lara J. *Characterization of Cycloned Sands for the Seismic Design Deposit*, 1998.
ICOLD. Tailings Dam. *Transport Placement and Decantation*, 1995.
Rayo J, *Aplicación del Transporte Hidráulico de Sólidos por Tuberías en la Industria Minera*, 1974.

Tailings and Mine Waste'00 © 2000 Balkema, Rotterdam, ISBN 90 5809 126 0

Subaerial tailing deposition - Design, construction and operation for facility closure and reclamation

B. Ulrich & D. R. East
Knight Piésold and Company, Denver, Colo., USA

J. Gorman
Jeritt Canyon Joint Venture, Elko, Nev., USA

ABSTRACT: The Big Springs tailing impoundment located in northeastern Nevada was commissioned in 1988. Initial construction of the four-sided facility included a low- permeability soil liner with a full drainage blanket and zoned, earthfill embankments. Subsequent facility expansion was accomplished using a combination of upstream and downstream construction. Tailing was spigotted from approximately one-half of the perimeter dike using managed, subaerial deposition techniques.

During the initial construction, vibrating wire piezometers were installed at various points within the facility. Based on the data collected from the piezometers throughout the operational life of the impoundment, the pore pressures within the facility remained low, indicating that proper drainage of the subaerial mass was occurring. Periodic inspection by trench excavation and sampling of the subaerial have also indicated that the facility was functioning as designed.

The subaerial subaerial deposition system proved to be a cost-effective and operationally-flexible design. The subaerial mass also is currently well-drained and capable of supporting earth-moving equipment necessary for the final closure and reclamation work. Information presented herein is provided to illustrate the importance of the various aspects initially designed into the facility to create a facility that is readily amenable to final closure and reclamation.

1 INTRODUCTION

The Big Springs Project was a gold mining and milling complex located in northern Elko County, Nevada, approximately 90 km (55 miles) north of Elko. The mill and subaerial impoundment were constructed on alluvial sediments along the eastern flank of the Independence Mountains, typical to the Basin and Range Province. Tailing was produced at the mine from 1988 through 1994 at an average rate of approximately 900 tonnes per day (1,000 tpd). The footprint of the final tailing impoundment is approximately 20 hectares (50 acres).

Foremost in the design process was the desire to develop an environmentally-efficient tailing storage facility which would minimize capital expenditures and operating costs. Based on the specific site conditions, the subaerial concept of tailing management, together with a combination of upstream and downstream construction, was employed into the design of the Big Springs tailing impoundment.

2 TAILING DISPOSAL METHODOLOGIES

As discussed in East (1990), the general engineering challenge is to develop solutions to tailing management problems that are cost-effective and take advantage of the positive engineering properties of the material itself.

Presently, several tailing disposal methods are receiving attention which, in addition to subaerial tailing disposal, include thickened tailing, dewatered tailing, and paste tailing. While each of the alternative methods may offer some benefits in highly specialized applications, few meet the cost-effective nature and operational flexibility that the subaerial method accomplishes.

3 SUBAERIAL TAILING DEPOSITION

3.1 *General Concepts*

Central to the success of the Big Springs tailing impoundment was the use of the subaerial method of tailing deposition. Subaerial tailing deposition was described by Knight and Haile (1983) as a technique by which tailing is deposited in thin layers and allowed to settle, drain, and partially air dry prior to burial by subsequent layers. Through the incorporation of a blanket drain, increased densities can be realized with air drying serving to further increase the density. The process of subaerial tailing deposition is implemented as follows:

• Tailing is discharged onto a portion of a gently sloping beach by means of multiple spigotting points, thereby forming a uniformly thick layer.

• Upon completion of this layer, the discharge is moved to another section of beach, and the newly deposited layer is allowed to settle and drain. Liquid released by settling flows over the sloping beach surface and collects in the supernatant pool for recovery.

• Liquids draining from the new layer are absorbed by the air voids in the unsaturated layer underlying the newly deposited layer.

• As the layer is left exposed, additional moisture is removed from the newly deposited layer through evaporation and, if allowed to dry for a sufficient period of time, the newly formed layer becomes unsaturated.

• Free-draining fluids are intercepted by a previously constructed drainage blanket within the facility and conveyed from the facility.

Natural factors such as gravity, settling velocities, evaporation, and drainage are therefore utilized to effectively manage the tailing deposit.

3.2 *Additional Considerations*

The subaerial deposition method provides additional advantages to tailing management. By reducing the discharge velocity from turbulent flow to laminar flow at the multiple spigotting points, the coarser particles will settle near the embankment. The greater drainage potential of the coarser particles together with a slightly steeper tailing beach slope aids in the removal of fluids from tailing deposited in the proximity of the embankment. The resulting drained tailing mass provides significant structural integrity that can be incorporated into the structure of the containment facility (East, 1990).

Due to the laminated structure of the tailing, downward migration of fluids is restricted, thereby reducing the potential to overwhelm the capacity of the blanket drain. The resulting drained tailing mass of a properly designed and operated subaerial tailing impoundment has a low propensity toward liquefaction and is amenable to upstream construction. As indicated by Knight and Haile (1983), the subaerial method of tailing deposition is designed to result in a tailing mass exhibiting low pore pressures. At decommissioning, the tailing should be fully consolidated and drained.

Since tailing is typically a relatively fine-grained material, high negative pore pressures can develop within a properly designed and operated subaerial tailing deposit (East, 1990); in fact, the subaerial method appears to be the only manner in which a significant portion of the deposit can develop high negative pore pressures since the desiccation of relatively thin layers is required to achieve this state. This is not only beneficial in densifying the tailing but also in increasing the peak shearing resistance of the material, reducing the propensity toward liquefaction, decreasing the permeability, and allowing for construction of embankment raises using the upstream method.

Another benefit of the underdrainage system is to reduce the hydraulic head on the primary liner, which can lead to significant cost savings and the possibility of allowing the use of a natural clay liner as was selected for the Big Springs project.

4 INITIAL TESTS ON TAILING SLURRY

At the commencement of the design process, a series of bench-scale tests was carried out to assess the settling behavior of the Big Springs tailing under a variety of operational conditions. These tests were conducted using laboratory-manufactured tailing material deemed to be representative of the intended production tailing. The tests included the following:

• Sedimentation under undrained conditions (essentially the environment encountered with subaqueous tailing deposition)

• Sedimentation with underdrainage (representative of a freshly deposited layer of subaerial tailing prior to the onset of evaporation)

• Settled and air-dried (representative of a layer of subaerial tailing after desiccation by evaporation).

Based on several years of experience, these tests have been found to be helpful in establishing the settling behavior of tailing. A summary of the test results is shown in Figure 1, which indicates the variation of dry density versus solids content for the three depositional environments.

The test results indicate that a settled and air-dried slurry would be on the order of 40 percent denser than a slurry settled under undrained conditions. Accordingly, the facility could be sized considering the 40 percent decrease in required storage volume, and thus a substantial savings in construction costs could be realized over the life of the facility. For preliminary sizing of the facility, an average tailing dry density of 1.4 kg/m³ (90 pcf) (representing some additional density increase due to consolidation as the tailing becomes buried) was adopted.

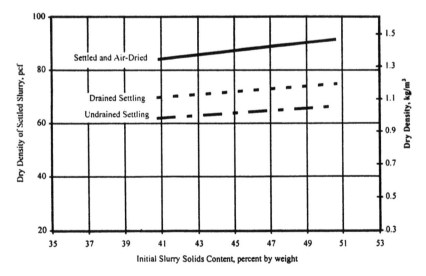

Figure 1. Results of Laboratory Sedimentation Tests on Tailing

31

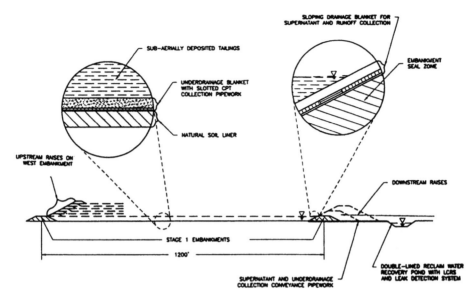

Figure 2. General Profile Through Facility

5 FACILITY DESIGN

The four-sided embankment was built on terrain that gently slopes toward the east and was designed as a zoned earthfill structure consisting of a low-permeability seal zone and a compacted structural zone. Longitudinal drains constructed along the embankment axis provided a means to intercept the phreatic surface. An idealized profile through the facility is shown in Figure 2.

A low-permeability soil liner was constructed within the basin of the impoundment, and a suitably-graded blanket drain was constructed upon the basin seal. Drainage of the tailing mass was also enhanced by a network of slotted pipes installed in a herringbone pattern within the blanket drain. Fluid collected by the underdrainage system was conveyed by HDPE pipe beneath the embankment to the externally located HDPE-lined reclaim water pond.

A supernatant water collection system was constructed along the inner slopes of the topographically lower east embankment. This system consists of a sloping drain and pipework to collect and convey fluids by gravity to the external HDPE-lined reclaim water pond.

Tailing from the mill was pumped as slurry to a distribution system located along the inner slopes of the westerly embankments. Tailing was deposited in thin layers in a controlled, rotational sequence to develop a large beach gently sloping toward the east. Expansion of the facility was planned to occur biennially with the easterly (pool side) embankments being raised using downstream construction and the westerly (deposition-side) embankments being raised using the upstream technique with the drained and consolidated tailing as foundation for the upstream raises.

6 STAGED CONSTRUCTION

Although originally designed to be constructed in four stages, the Big Springs tailing impoundment was constructed in five stages to accommodate additional storage requirements. Each expansion was sized based on mill production forecasts at the time of the design; thus, the design was quite flexible.

During each expansion, the westerly embankments were raised using the upstream method of construction. The upstream toe of the raises spanned a maximum of approximately 15 m (50 feet) onto the tailing beach. This construction was accomplished with minimal problems; the tailing surface was found to serve as a suitable foundation for the construction. Ultimately, the impoundment was constructed to a maximum height of approximately 20 m (70 feet).

This manner of staged construction not only delayed capital expenditure, but the total construction costs were also decreased due to the notable reduction in necessary fill volumes required for raising approximately one-half of the embankment using the upstream construction method.

7 FACILITY PERFORMANCE

The performance of the facility in terms of its design as a drained and consolidated subaerial tailing deposit has been monitored via the observation of vibrating wire piezometers that were installed in the tailing basin. Performance has also been monitored using inspection trenches excavated into the tailing beach.

7.1 Vibrating Wire Piezometers

The following description of the piezometer readings during the facility start-up was provided by East (1990). Figure 3 shows the layout of vibrating wire piezometers. Piezometer P1 was installed in the supernatant recovery system to monitor the head imposed by the supernatant pool.
Piezometer P2 was installed within the underdrainage blanket in the vicinity of the pool and is isolated from P1 by a compacted clay liner and very fine tailing. The data recorded during the first 16 months of operation are shown in Figure 4. Piezometer P1 indicates an early increase in head due to the storage of mill water prior to tailing production. Piezometer P2 indicates a similar response while the underdrainage valves were closed to contain the mill water within the tailing impoundment.

Approximately three months after start-up, the underdrainage valves were opened and, as indicated by piezometer P2, the head within the underdrainage system exhibited an immediate decrease while the pond level remained relatively constant. During this period, piezometer P3, located immediately above the underdrainage blanket, indicated saturated conditions to a depth of about 2 feet at P3. Similar to P2, when the underdrainage valves were opened, piezometer P3 exhibited an immediate decrease in head. Piezometers P4 and P5 were buried within the tailing mass but did not record significant heads, which suggests the tailing mass was draining as designed. The significant negative pore pressures as discussed in Section 3.2 can be observed in the data shown on Figure 4.

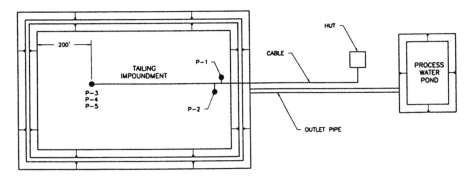

Figure 3. General Instrumentation Schematic

33

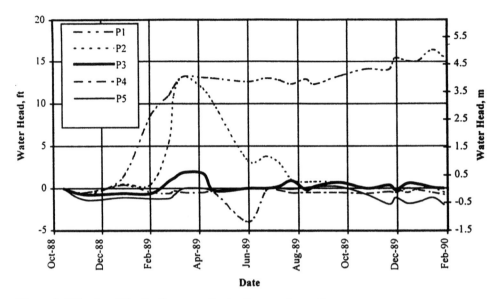

Figure 4. Vibrating Wire Performance During Impoundment Start-up

These piezometers were monitored periodically throughout the operational life of the impoundment. Long-term assessment of these piezometers indicated that the phreatic surface within the tailing was relatively low (typically on the order of a few feet) and that downward flow of the process waters was taking place within the tailing. The piezometer installed in the drainage blanket typically indicated that a relatively free-draining base existed.

7.2 Inspection Trenches

In 1993, after five years of operation, eight test pits located at various points throughout the tailing beach were excavated using a track-mounted backhoe to allow examination of the tailing deposit. This assessment included the recovery of six Shelby samples at depths of up to 5 m (16 feet) below the tailing surface. At the time of the investigation, the average tailing thickness was approximately 15 m (50 feet).

In addition to the Shelby samples, a total of 11 moisture-density tests were conducted on the surface and near surface tailing using either a nuclear densiometer or sandcone. Analysis of these tests and samples are summarized as follows:

Parameter	Range	Average
Dry Density, kg/m^3 (pcf)	1.1 to 1.8 (70 to 114)	1.4 (85)
Moisture Content, pct.	11 to 43	26
Saturation, pct.	42 to 84	67

The average dry density of the tailing was essentially equivalent to the original design assumption and, as indicated above, the tailing exhibited relatively low saturation levels, indicating that an acceptable degree of drainage and drying of the deposit was taking place. Further confirmation of the drainage condition was provided by piezometer data which indicated that the phreatic surface was close to the base of the tailing.

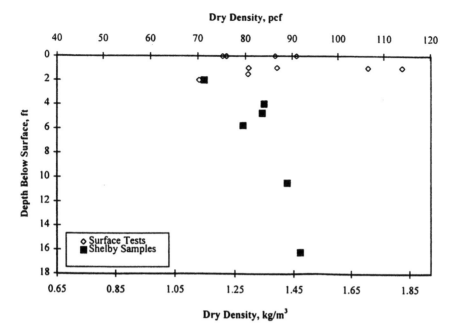

Figure 5. Results of Field Sampling

These results are also summarized in Figures 5 and 6. Figure 5 represents the relationship between tailing dry density and depth. It may be observed from the information given in Figure 5 that the density is relatively constant through the depth tested. Figure 6 represents the relationship between tailing dry density and moisture content. The information given in Figure 6 indicates that the degree of saturation is well below 100 percent.

8 CLOSURE CONDITIONS AND CONSTRUCTION CONSIDERATIONS

8.1 *Conditions*

A key feature of the design, incremental construction, and operation of a subaerial tailing impoundment is the resulting well-drained, stable, low water inventory, and "full" facility at the time of closure. Well-drained tailing has greater density and is more geotechnically stable. Fluid volumes in the facility are minimized through optimal water recovery and recycling through the mill during operation. Low water inventories at the time of closure also provide greater flexibility in scheduling the closure program. The extensive beach development on the west side of the Big Springs facility in the post-closure configuration has provided opportunity for forced evaporation of the remaining water in inventory and the water accumulated over the winter months, thus minimizing the need for water treatment and/or other handling measures. Continued circulation of the drain-down fluids collected in the reclaim pond through the carbon columns in the mill has yielded modest gold recoveries that offset a portion of mill operating costs. By carefully staging the dam lift construction to meet the forecasted mill throughputs, a "full" tailing impoundment can be realized at the time when milling ceases. A "full" facility can result in lower final recontouring and earthmoving costs.

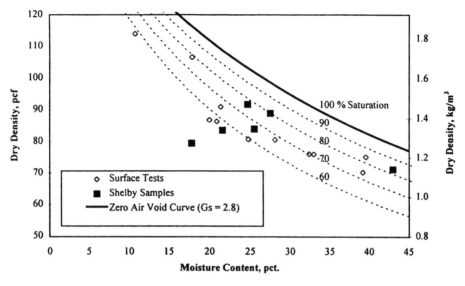

Figure 6. Moisture-Density Relationship of Field Samples

8.2 *Design*

A critical closure design criterion for the tailing impoundment is to minimize the infiltration of precipitation through the facility. To achieve this, the tailing will be recontoured and then covered with a low-permeability barrier layer and a growth medium layer. The facility will be recontoured to achieve a southwest aspect where possible to enhance sublimation of snow in the winter months. The facility also will be recontoured to promote runoff of incident precipitation and spring snowmelt.

The engineered cover will be installed over the recontoured surface. It will be composed of a low- permeability barrier layer overlain by a growth medium layer. The low-permeability barrier layer is designed to slow the vertical infiltration of water into the tailing material while the growth medium layer provides a substrate for development of the vegetative cover. Evapotranspiration losses of stored water in the growth medium by grasses and forbs established in the cover are a significant water loss process. Evapotranspiration losses and runoff from the engineered cover can reduce the net infiltration of water through the tailing facility to less than 5 percent of the incident precipitation.

Water balance modeling of the cover design to determine the runoff, evapotranspiration, and infiltration values are underway utilizing the Hydrologic Evaluation of Landfill Performance (HELP) model (Schroeder, et al., 1994). The water balance model runs will utilize site-specific data collected during geotechnical investigations during the previous construction phases, data collected from recent investigations undertaken as part of the closure investigation, as well as typical values for the climate and vegetative conditions at the Big Springs site.

8.3 *Implementation*

The implementation of the final closure design is again benefited by the well-drained nature of the facility. The well-drained tailing can accommodate heavier equipment which in turn can provide greater flexibility for the contractor to implement the recontouring design. This can result in lower overall reclamation costs. In order to assess the working surface and identify potential construction issues, in the autumn of 1998 four trenches were excavated to a depth of approximately 15 feet in the western one-half of the tailing facility. The excavations allowed visual inspection of the near-surface deposits and a better evaluation of the equipment limitations for the regrading. No water was observed in the pits, and the material was relatively dry. Based on the performance of the excavator, the tailing surface was judged to be a good working surface for low ground pressure equipment.

9 CONCLUSIONS

As the reclamation and closure of the Big Springs tailing impoundment approaches, it is clear that the subaerial tailing method performed well in every regard and as originally intended. The following original goals were accomplished successfully due to proper design, construction, and operation practices:
• The tailing management needs were fulfilled in a cost-effective manner, taking advantage of the positive engineering properties of the tailing material.
• The stored density of tailing was increased, thereby reducing the overall impoundment size and construction costs.
• Potential long-term seepage from the facility has been minimized through effective drainage during operation, thereby protecting both short-term and long-term groundwater and surface water flows.
• A greater efficiency of water usage was provided.
• A drained and consolidated tailing mass is more amenable to regrading and reclamation.
In addition to these goals, the facility also performed as a means to recover additional gold which would have gone unrecovered if any other method had been employed for tailing disposal. Based on all the above factors, it is clear that the subaerial method proved to be a very effective manner in which to impound the tailing at Big Springs and that this method also resulted in a deposit that is readily amenable to reclamation.

REFERENCES

East, D.R. 1990, Performance achievements of drained managed tailings storage facilities. *Proc. Western Regional Symposium on Mining and Mineral Processing Wastes, Berkeley, May*: 229-233.
Knight, R.B. and J.P. Haile 1983. Sub-aerial tailings deposition, *Proc. 7th Pan-American Conference on Soil Mechanics and Foundation Engineering. Vancouver, B.C., Canada, June*: 627-639.
Schroeder, P.R., T.S. Dozier, P.A. Zappi, B.M. McEnroe, J.W. Sjstrom, and R.L. Payton 1994. *The hydrologic evaluation of landfill performance (HELP) model, engineering documentation for version 3*. Cincinnati, Ohio: U.S. Environmental Protection Agency, EPA\600\R-94\168b, September, 116 p.

Tailings and Mine Waste'00 © 2000 Balkema, Rotterdam, ISBN 90 5809 126 0

Sustainable development in disposal of tailings

E. I. Robinsky
E. I. Robinsky Associates Limited, Toronto, Ont., Canada (Formerly: Department of Civil Engineering, University of Toronto)

ABSTRACT: Where the mining industry has too often failed the sustainable development goal is in tailings dam failures. The weakness of conventional tailings dams is that they retain very loose unconsolidated tailings and a large quantity of liquid. If the dam fails, the contents liquefy and flow through the breach. Environmental concerns led the author, in the 1960's to the concept of Thickened Tailings Disposal - the TTD system. The system requires that tailings be thickened to a heavy, but pumpable, slurry. Released from an elevated position, the thickened tailings form a self-supporting ridge or cone, designed to attain 2 to 6 percent side slopes. The principal aims of the TTD system are to eliminate the conventional settlement pond, reduce or eliminate perimeter dams, and allow the deposit to consolidate to permit progressive reclamation even as mining continues. These aims can be satisfied by: 1. providing adequate thickening, 2. not limiting the choice of disposal area to the nearest valley, and 3. planning the discharge layout for maximum natural drying and for progressive reclamation. This paper is based on 70 world-wide TTD studies by the author.

1 INTRODUCTION

The concept of Thickened Tailings Disposal (TTD) was introduced to the mining industry by the author in the late 1960's, Robinsky (1975, 1978, 1982, 1986), Salvas (1989), Paradis (1992), Williams (1992). However, it is only in the last 10 years that the system and its advantages is being given serious consideration. Originally referred to as 'Thickened Discharge', or, 'Central Discharge' the author adopted the more descriptive term, the 'Thickened Tailings Disposal (TTD) system'. Although other descriptive terms such as 'dry stacking', 'subaerial deposition', and 'paste deposition' have more recently been introduced by others, they all follow the basic principles laid down by the TTD system - thicken the tailings to a heavy slurry consistency and discharge from an elevated position to form a slope and obtain maximum surface·evaporation. This is a major step towards sustainable development - the elimination of high vulnerable dams and super-imposed sedimentation ponds of the conventional disposal approach.

This paper has been written to encourage anyone faced with the problem of tailings disposal, either for a new mine or for extending the life of an existing disposal area, to consider the TTD system as a viable alternative. The paper reviews some of the positive features of the TTD system, drawing examples from a number of TTD studies.

2 SYSTEM DESCRIPTION

The basic concept of a TTD system , compared to the conventional approach, is illustrated in Fig. 1. Briefly explained, in the TTD system the tailings are thickened at the plant, conveyed to the disposal area by pipeline and discharged from an elevated natural or man-made position. In the latter case, the discharge pipeline is advanced progressively over the tailings deposit to form a ridge. The pipeline may be founded directly on the previously discharged tailings, if necessary supported by a roadway consisting of gravel underlain by geofabric. Reclamation of the tailings surface can commence, even as mining continues. A small pond beyond the toe of the tailings deposit is used to recycle precipitation and the liquid extruded from the tailings. The necessity of a tailings retention dam is determined by the topography of the site. Figure 1 illustrates topographic conditions that do not call for a tailings dam.

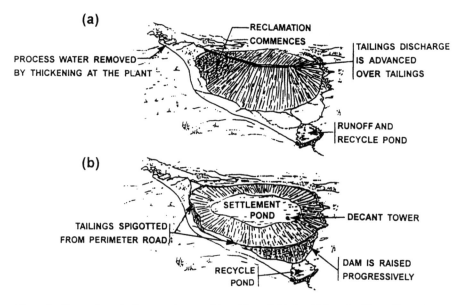

Figure 1. Comparison, in the same topographic setting of, (a) the thickened tailings disposal system and, (b) conventional tailings disposal.

3 AREAL DIMENSIONS

It may appear, that the TTD system requires a larger disposal site than conventional discharge. Generally this is not true, as shown schematically in Fig. 2. Within fixed boundaries, the TTD system will always accommodate more tailings than a conventional system. Because thickened

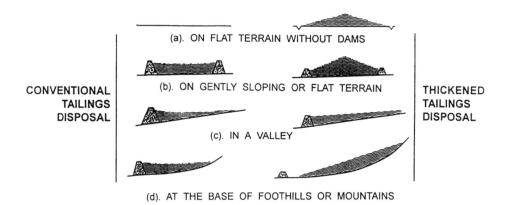

Figure 2. Comparison of area requirements for both the thickened and conventional disposal schemes. Note that dams are smaller for the TTD system.

tailings have shear strength or viscosity they will stack at a positive slope, creating a convex surface. Note conditions (a) and (b). In all cases note that the confining, or perimeter dams are smaller, using the thickened tailings disposal system. Perhaps the most important fact brought out in Fig. 2(c) and (d), is that a flatter slope is often a better choice for the TTD system than a steep slope. However, the system is versatile and can be accommodated in any topographic setting. This is contrary to conventional disposal where a valley is always the preferred choice.

Fig. 3 shows a disposal site where the confining dam, constructed for the TTD system was one-half as high, one-third as long, and contained one-ninth as much fill as the projected conventional dam. Fig. 4 illustrates a proposed disposal solution in the Andes Mountains.

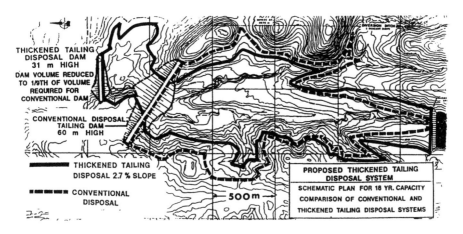

Figure 3. Plan view of a TTD tailings disposal system in the Andes, compared to originally proposed conventional plan.

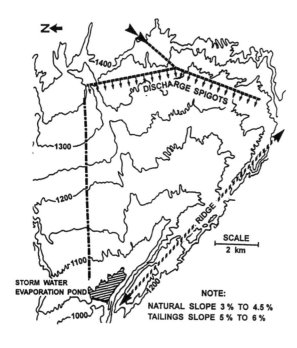

Figure 4. Proposed TTD disposal plan for a 60,000 TPD mine in the Andes. Design eliminates tailings dam and need for water recycling.

41

The climate is very dry, rainfall is rare, and the natural evaporation rate is extremely high throughout the year. The TTD system was proposed to eliminate the tailings retention dam by discharging tailings through many spigots. The small precipitation collection pond was positioned sufficiently far from the discharge spigots to ensure thickened tailings flow would not reach this pond. By this means, this design eliminated the need for recycling pumps, most of the process liquid having been taken off by the thickeners. In this case, the TTD solution required a much larger surface area than a conventional disposal system. However with the elimination of a large dam and the elimination of a water recycling system, it is considered that the TTD approach would be far superior both environmentally and economically.

4 ROLE OF THICKENING

The key to a successful TTD system is the transformation of the plant mixture of tailings solids and liquid to a heavy slurry consistency. The progressive change that occurs in thickeners is illustrated in Fig. 5. Discharged with only minor thickening, the tailings immediately segregate. The coarse fraction settles out close to the discharge line, and the liquid with the fines fraction flows to the sedimentation pond - this is conventional disposal. As thickening is increased, there is less slurry to carry the coarse fraction, and it begins to stack closer to the discharge point (B-B in Fig. 5). The fines fraction still runs out. Finally, as thickening is increased further, in this case to 67.5 percent solids the mixture reverts to a non-segregating slurry (point C). The voids in the coarse fraction are now filled with a matrix of fines resulting in a homogeneous mixture. From this point onward the behaviour of the slurry is suitable for disposal by the TTD system. It should be noted that the TTD example of tailings chosen to illustrate the behaviour shown in Fig. 5 was exceptionally coarse, 75 percent being retained on the No. 200 mesh sieve. For more typical tailings the reaction would not be as pronounced. Every mine produces unique tailings, and these will exhibit different relations between the percent solids and the deposition slopes. The degree of thickening required to attain the desired slope may be obtained directly from curves such as illustrated in Fig. 5.

It is not recommended to commence a TTD system without adequate thickening. Partial thickening achieves very little. The fines settlement pond cannot be avoided, and, a tailings confining dam for the inadequately thickened tailings would have to be provided. The principal TTD advantages are lost. If the TTD system is to be used, it is necessary to thicken the tailings

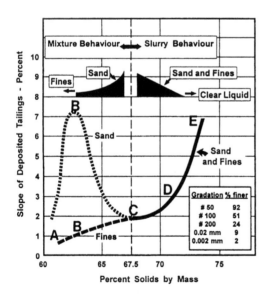

Figure 5. Example of progressive change in tailings behaviour with increased thickening.

beyond the absolute minimum requirement. A factor of safety is essential. Plants cannot be guaranteed to produce a uniform product, and variations in the composition of the ore may occur.

The basin thickener has been the primary piece of equipment for thickening tailings. Originally it's use was limited to clarifying the tailings process water to permit recycling of the overflow water back into the process. With the introduction of the thickened tailings concept, and the scarcity of water in many parts of the world, thickener designs have been improving over the years. The use of synthetic flocculants, deeper basins or tanks, and special raking mechanisms all help in producing a denser underflow at a faster rate. It is also possible to filter part of the thickener underflow and recombine it with the rest of the underflow, to attain a higher percent solids before discharge. Thickening of large quantities of tailings still remains a costly operation. In estimating the economics of adopting the TTD concept, the elimination of large dams, elimination of the sedimentation pond, and the substantially lower environmental impact should be taken into consideration. In addition, because the volume of tailings slurry discharged, and the volume of recycled water are considerably reduced, thereby saving costs in pipeline installation and pumping energy, the search for a more suitable disposed area should be extended geographically without penalty. This is a most important advantage of the TTD system. The nearest valley may not be the optimum location when taking into account long term costs, future reclamation expenditures, and the preservation of the environment.

In order to increase the tailings deposition slope, it is possible, in some cases, to increase the intergranular attraction of the tailings particles, thereby increasing the viscosity of the discharge. Fig. 6 illustrates the effect of adding a small quantity of lime to a tailings containing a considerable amount of clay minerals. In the example shown the residual fines fraction of the tar sands washing process takes on a steeper slope with the addition of 11.4 kg of lime per 10 tonnes (dry) of tailings.

A similar effect has been obtained at the Falconbridge copper/zinc Kidd Creek mine in Ontario. A lime station has been installed near the discharge end of the system. Fig. 7 reveals laboratory tests results that show that maximum improvement of deposition slope occurs with the most concentrated slurries. Also, to be noted, is that there is a practical limit to the addition of lime. Beyond 0.1 percent addition of lime the effect on the deposition slope becomes negligible.

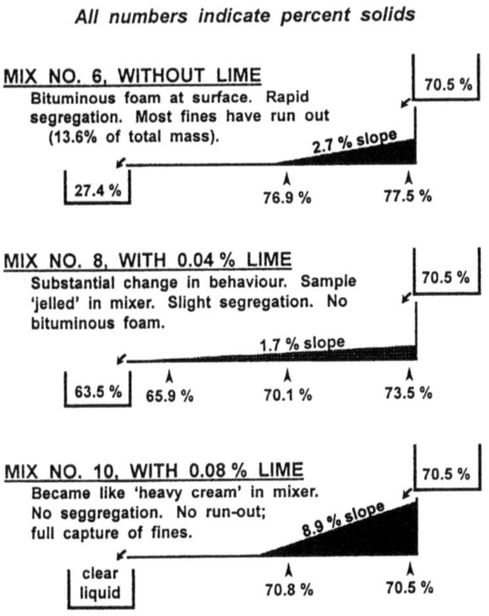

All numbers indicate percent solids

MIX NO. 6, WITHOUT LIME
Bituminous foam at surface. Rapid segregation. Most fines have run out (13.6% of total mass).
2.7 % slope
70.5 %
27.4 % 76.9 % 77.5 %

MIX NO. 8, WITH 0.04 % LIME
Substantial change in behaviour. Sample 'jelled' in mixer. Slight segregation. No bituminous foam.
1.7 % slope
70.5 %
63.5 % 65.9 % 70.1 % 73.5 %

MIX NO. 10, WITH 0.08 % LIME
Became like 'heavy cream' in mixer. No seggregation. No run-out; full capture of fines.
8.9 % slope
70.5 %
clear liquid 70.8 % 70.5 %

Figure 6. Example of effect of adding a trace of lime to tailings containing clay minerals.

43

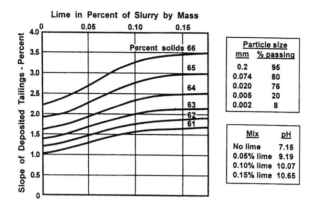

Figure 7. Example of the effect of a trace amount of lime to a typical copper/zinc operation tailings.

5 STABILITY OF DEPOSIT

In most TTD metal processing operations, excepting bauxite and other very fine-grained residues, the tailings, after discharge, will consolidate and reach a stable condition within a short time. Consolidation under self-weight commences immediately after discharge. In addition, because of the homogeneous nature of the discharged tailings, the fines fraction within the mass develops high capillary rise. This rise may be well above the water table within the inclined deposit. The high suction causes compression between the tailings particles. This results in further consolidation. Fig. 8 reveals, schematically, the additional equivalent loading that will occur if the water table falls to 3 m below the tailings surface. With these metal processing operations, hydraulic conductivity of the tailings would generally be sufficient to allow the water table to drop, thereby developing this additional consolidation mechanism.

Maximum capillary or matric suction develops as a result of evaporation of the tailings liquid from the inclined TTD surface. The drying of the surface causes the menisci to retreat into the fines of the tailings mass. The result is the development of maximum matric suction, Barbour (1992, 1993). These tension stresses draw the particles together, causing shrinkage cracks in the deposit. At the same time, where the rate of desiccation is greater than the rate of re-supply of liquid from within the deposited mass, the tailings will reach the shrinkage limit. This is a very desirable state that nature provides. The shrinkage limit is a geotechnical term that indicates the state of the material at its minimum volume, as caused by desiccation. In most metal processing

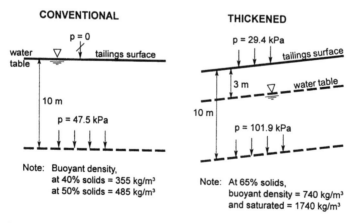

Figure 8. Schematic view showing the additional consolidation caused by capillary suction in a homogeneous TTD sloping deposit.

44

operations, it is expected that, as the deposit builds layer by layer, the shrinkage limit will be reached, resulting in a dense deposit throughout the depth of the deposit. In particular, with fine-grained tailings that contain a considerable amount of liquid, such as bauxite residue in alumina operations, every attempt should be made to allow the shrinkage limit to develop. This is best achieved by discharging the residue through several spigots at the same time. Tests indicate that it is just as effective to discharge the tailings from several spigots simultaneously, as discharging a larger quantity, alternating between several discharge spigots. Fig. 9 illustrates the advantage of reaching the shrinkage limit in the deposit, compared to simple consolidation by self-weight. The test results shown are for the upper 2.4 m of two continuous sampling operations carried out to a depth of 12 m at the Falconbridge Kidd Creek Mine (temperate climatic conditions). The average percent solids over the entire depth was 80.5 with a water content of 24.2 percent. The shrinkage limit by laboratory determinations averaged 80.4 percent. Discharge was originally through many spigots, operating simultaneously. Later, the system was altered to a single discharge point.

6 EATHQUAKE RESISTANCE

Concerns regarding stability in an earthquake environment are much reduced if the following basic elements of a TTD system have been observed.
 a) the settling pond on top of the tailings has been eliminated, thus the deposit is much denser,
 b) the tailings surface is sloping and cannot collect water, and,
 c) the highly concentrated homogenous tailings slurry is discharged to a topographic plan that will promote good drying conditions.
Of utmost importance in the design is to provide a large drying surface to encourage the attainment of the shrinkage limit, layer by layer as the tailings accumulate. Having reached the shrinkage limit, the tailings under shear stress would revert to a dilatent state, creating a high negative pore pressure that will oppose the strain and liquefaction. In the initial stages of a new project, it may not be possible to attain the shrinkage limit because of the rapid accumulation of tailings near the discharge point. However, with time, the tailings surface will expand and desiccation conditions will improve. In any event, the implementation of the TTD system will prevent an ecological disaster. Such a conclusion can be drawn on the basis of a paper by Dobry and Alvarez (1967), wherein they reported their findings of the conditions of 22 dams (at 10 mines) after an earthquake in Chile 1965. These were all conventional tailings dams. The

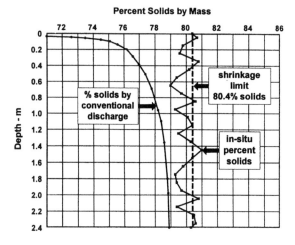

Figure 9. Illustration of the effectiveness of reacting the shrinkage limit. Natural desiccation provides substantial densification of tailings.

quake registered a magnitude of 7 to 7.25 on the Richter scale. Most of the dams failed. Ten were working dams, containing settling ponds, the remaining were either abandoned or emergency dams, with little or no liquid. The worst disaster was the collapse of the El Cobre dam where 1.9 million tonnes of tailings flowed to a maximum distance of 12 km. It was noted in the study that a few of the abandoned or disused (dry) tailings dams underwent local slide failures or exhibited fractures, but the damage in all cases was only a local dam collapse with little or no ecological damage. It may therefore be concluded that any system that eliminates water containment and promotes drying and consolidation of the tailings deposit would not produce an ecological disaster, even if the dam itself failed. A study by Poulos (1985) revealed that the Alcan TTD deposit in Quebec at 2.9% slope would be safe against liquefaction under an earthquake that induces a 0.1 g peak ground acceleration. In the thickened tailings disposal system, the liquid settling pond is eliminated and the sloping surface of the tailings deposit allows rapid runoff and surface drying. Furthermore, the system requires substantially lower confining dams, if any, which also helps to reduce the possibility and magnitude of dam failure.

7 REDUCTION IN HANDLING LIQUID

Other than the physical advantages of the TTD system, discussed above, Fig. 10 indicates the reduction in handling of the water that is associated with the discharge of tailings. The table is based on numerous studies of tailings, grouped approximately in accordance with particle sizing. Generally, it was found that the liquid discharged for a TTD system was between one-fifth and one-third of that normally discharged in a conventional disposal approach. This reduction will influence the cost of purchasing and installing the discharge and recycle pipelines. In some cases (Fig. 4) the reduction is enough to consider the elimination of water recycling pumps altogether.

8 PROGRESSIVE RECLAMATION

The ideal topographic layout for the progressive reclamation of a tailings disposal area has one long dimension relative to its width, as shown schematically in Fig. 1. Fig. 11 illustrates, as an example, the layout for a TTD system that will extend the life of an existing copper mine disposal area for another 20 years at a tailings slope of 6 percent. Regulatory authorities will not permit further raising of the 600 m long, 30 m high dam along the south side of the area. Thickened discharge is to be carried out sequentially from the points designated by numbers 1 to 14.

Suggested particle size description for tailings	Particle size (mm) at median and / at finest 10%	Concentration expressed in percent solids, and / fluid quantity discharged in tonnes per tonne (dry) of tailings		Quantity of fluid discharged by using TTD system as percentage of quantity discharged by conv. disposal
		Conventional	Thickened	
Very Fine	0.002 / <0.0005	15 / 5.7	46 / 1.18	21
Very Fine	0.0035 / <0.0005	12 / 7.3	50 / 1.00	14
Fine	0.02 / 0.0009	21 / 3.76	61 / 0.64	17
Medium	0.02 / 0.002	40 / 1.5	65 / 0.54	36
Medium	0.027 / 0.0015	40 / 1.5	67 / 0.49	33
Coarse	0.055 / 0.0045	36 / 1.78	68 / 0.47	26
Coarse	0.07 / 0.005	35 / 1.86	70 / 0.43	23
Very Coarse	0.08 / 0.010	20 / 4.0	73 / 0.36	21
Very Coarse	0.06 / 0.013	35 / 1.86	72 / 0.39	9

Figure 10. Comparison of the quantity of liquid discharged by TTD and conventional approach.

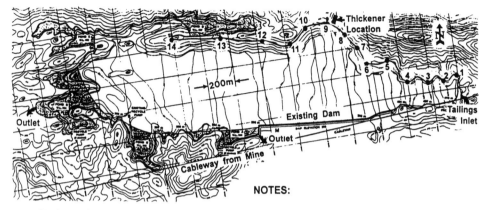

NOTES:

1. Discharge, followed by progressive reclamation, is to be carried out sequentially from points denoted by large-sized numbers 1 through 14.
2. The above points indicate the approximate desired upper limits of the tailings deposit.
3. Discharge ramp (road) is to be located by management, but must be no less than 4 m higher than the upper limit of tailings (denoted by points 1 through 14).
4. The elevations denoted thus, '394 m', indicate the maximum allowable elevation of tailings along the southern limit of the deposit. When these elevations are attained by the tailings, the discharge point must be moved westward.

5. If it is desired to eliminate the use of positive displacement pumps at the concentrator plant, an alternative position for the thickening plant is shown at the top of this plan. The latter would allow discharge of the heavy underflow by gravity, aided possibly with a centrifugal pump.
6. Because of the substantial reduction in water accumulation in the disposal area, it may not be warranted to install recycle pumps. During the wet season it is expected that some rain water will accumulate in the indicated ponds. However, the plan provides two permanent overflow outlets at invert elevations of 394 m, whether recycle pumps are installed or avoided altogether.

Figure 11. Approved layout for a TTD system. The tailings are to be discharged from the north wall of the valley.

The thickeners are being positioned at a high elevation along the north side. This will avoid installing positive displacement pumps for the highly concentrated tailings slurry. Reclamation is to follow close behind the discharge as it is advanced westward.

9 SUMMARY

This paper suggests that there are ways to improve the disposal of mine tailings wastes to better conform with the aims of 'sustainable development'. By the elimination of the settlement pond situated on the top of the tailings deposit, and by reducing, and sometimes even eliminating, the perimeter dams, the tailings deposit is no longer subject to collapse and cause an environmental catastrophe. This is made possible by discharging a thickened tailings slurry from an elevated position to form a self-supporting slope, well-drained and subject to natural consolidation and desiccation. Thorough testing of the slurry and careful planning of the disposal site are essential.

10 REFERENCES

Barbour, S.L., Yang, N., and Wilson, G.W. 1992. A laboratory characterization of the properties of the Kidd Creek thickened tailings required for evaporative flux modelling. Geo-Environmental Engineering Group, Department of Civil Engineering, University of Saskatchewan, Saskatoon, Saskatchewan.
Barbour, S.L., Wilson, G.W., St-Arnaud, L.C. 1993. Evaluation of the saturated-unsaturated groundwater conditions of a thickened tailings deposit. *Canadian Geotechnical Journal.* Vol. 30:6:935-945.
Dobry, R., and Alvarez. 1967. Seismic failures of Chilean tailings dams. *JSMFD*, ASCE, 93:SM6:237-260.

Paradis, R.D. 1992. Disposal of red mud using wet stacking technology. *Proc. International Bauxite Tailings Workshop*. November, Perth, Western Australia.

Poulos, S.J., Robinsky, E.I., and Keller, T.O. 1985. Liquefaction resistance of thickened tailings. *Journal of Geotechncial Engineering*. ASCE, Vol. 111:12:1380-1394.

Robinsky, E.I. 1975. Thickened discharge - A new approach to tailings disposal. *Canadian Mining and Metallurgical Bulletin*. 68:47-53.

Robinsky, E.I. 1978. Tailing disposal by the thickened discharge method for improved economy and environmental control. *Tailing Disposal Today*. 2:75-92, 1979 by Miller Freeman Publications Inc., San Francisco, California, *Proceedings of the 2nd International Tailings Symposium*, May, Denver, Colorado.

Robinsky, E.I. 1982. Thickened tailings disposal in any topography. *Light Metals 1982, Proceedings of the 111th AIME Annual Meeting*, February, Dallas, Texas, pp. 239-248, Published by the Metallurgical Society of AIME.

Robinsky, E.I. 1986. Current status of the sloped thickened tailings disposal system. *Bauxite Tailings, Proceedings of an International Conference*, Kingston, Jamaica, October 1986, pp. 91-99, The Jamaica Bauxite Institute and the University of the West Indies, Kingston, Jamaica.

Salvas, R.J. 1989. Benefiical characteristics of sloped tailings deposits. *Proceedings, International Symposium on Tailings and Effluent Management*. Metallurgical Society of the Canadian Institute of Mines and Metallurgy. 14:215-226, Pergamon Press Inc..

Williams, M.P.A. 1992. Australian experience with the central discharge method for tailings disposal. *Proc. Second Int. Conf. On Environmental Issues and Waste Management in Energy and Mineral Production*. Calgary, Alberta, 1:567-577.

Tailings and Mine Waste'00 © 2000 Balkema, Rotterdam, ISBN 90 5809 126 0

System aspects of pumping and co-disposal of waste rock and tailings

A. Sellgren
Luleå University of Technology, Sweden

G. Addie
GIW Industries Incorporated, Grovetown, Ga., USA

ABSTRACT: Effective use of co-disposal of waste rock and tailings depends among other things of the use of a high slurry solids concentration so as to minimize segregation of the finest particles. Generalised large-scale loop-results from the GIW Hydraulic Testing Laboratory, U.S.A., have here been used to demonstrate the economical potential of pumping at high solids concentrations with centrifugal pumps thus reducing the costs for pipelines, energy and water recycling.

1 INTRODUCTION

Today coarse waste rock is normally handled on a dry basis in separate transportation systems, such as conveyor belts etc. and disposed of in a separate area. An alternative to the conventional disposal of waste rock and the fine-grained tailings is disposing of a mixture of the products by the so called co-disposal method, Williams et al. (1992), Van Rooyen (1992). For example, the washing of coal results often in a coarse waste product (<50 mm) and silt-sized tailings finer than about 100 µm. Economic and environmental benefits described by Morris et al. (1997a, b) from field investigations showed less storage volume, large slope angles and increased strength and stiffness.

The co-disposal concept uses the fact that a tailings slurry fills the voids in the coarse waste rock, giving the favourable properties mentioned above.

Coarse mineral-particle slurries with particle sizes of 100-200 mm are now pumped effectively in dredging and mining operations. The potential for combined slurry pumping of waste rock and tailings in the metal ore industry with co-disposal in one disposal area was pointed out by Sellgren et al. (1994, 1995), based on the favourable pumping conditions for products with a broad particle size distribution. The integrated handling of combined coarse particle and tailings and co-disposal possesses great economical potential, Sellgren (1997). Sellgren et al. (1998) compared waste rock belt conveying (420 t/h) and conventional fine tailings pumping (420 t/h). It was found here that the investment for an integrated pumping system was about half of the total investment for the separate systems and that the operating costs were approximately the same.

The coarse particles in the disposed mixture form a geotechnically stable deposit. It drains quickly and can support heavy traffic almost immediately. Stability is achieved, not only during operation, but also after decommissioning and closure of the mine.

Conventionally deposited waste rock have a slope of about 35 to 40°. In the rehabilitation and decommissioning stage, the deposit often must be reshaped to a flatter final slope and capped with, 0.3 to 1.3 m covering etc., dependent on the environmental sensitivity of the material.

Conventional tailings disposal areas with slopes of about 1:100 create large areas which may be geotechnically unstable and require special capping. Multi-layer capping may be very expensive, over 10 USD per m² (ICOLD, 1996). Integrated handling with co-disposal mainly forms slopes at 1:10 and only one disposal area is required. The reduced rehabilitation cost associated

49

with the reduced surface area may be substantial. Co-disposal gives gently sloping contours which facilate revegetation.

Key problems with integrated handling and co-disposal are segregation (wash-out) of fine particles and the pump and pipeline wear according to Williams (1997). In this case parts of the fine particles segregated out of the mixture are carried in suspension to the end of the deposition area where a delta is formed similar to a conventional tailings delta.

Reported operating experience in the coal industry with combined pumping with max. particle sizes up to 50 mm and co-disposal seems mainly to have been carried out at a solids concentration by mass of about 35%. The relatively low solids content of the pumped slurry was reported as being accompanied by severe pump and pipeline wear. Pipeline blockages were also reported at higher concentrations Williams (1997).

A low solids concentration encourages erosion and fine particle segregation in the disposal area. For efficient co-disposal the solids concentration should therefore be as high as possible. Furthermore, low solids concentrations result in large volumes of water which in turn necessitates efficient handling and recovery of water.

The water content or the solids concentration is a key parameter in evaluating the effectiveness of co-disposal and the associated coarse particle slurry pumping.

The objective is here to demonstrate how the water content influences the energy consumption and pipeline diameters for the pumping of coarse particle slurries to the co-disposal site and to show the costs of recycling of the water used.

2 CHARACTERISATION

Typical particle size distributions of a tailings product and a waste rock product from metal ore processing are shown in Figure 1 together with the distribution of a mixture of the two products in mass ratios 1:1.

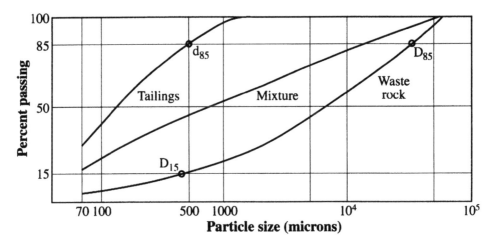

Figure 1. Particle size distributions of typical tailings and waste rock products. Shown is also the distribution of a mixture of the products in mass ratio 1:1. The value of d_{15} is about 40 μm

A mixture of two industrially handled products are in principle bimodal, which results in a gap-grading, as experienced in the field investigations with coal tailings by Morris et al. (1997a). The less pronounced the gap is, the more efficient is the trapping of fines which means that segregation is reduced. In practise, there is of course a limit as regards to reducing segregation by modifying the particle size distributions by extra crushing, milling and/or detailed recipe specifications.

50

The mixture curve in Figure 1 fell within curves which specify material suitable for the construction of the core in large water power earth dams, giving maximum density and minimum porosity. For dam safety, it is essential that no internal erosion or segregation take place and that the material drains effectively.

The following relationships developed for large water power earth dams from Sherard et al. (1963) have also been used by Morris et al. (1997b) as design criteria for co-disposal of waste rock and tailing

Segregation:
$$\frac{D_{15}}{d_{85}} < 4-5 \tag{1}$$

Drainage:
$$\frac{D_{15}}{d_{15}} > 4-5 \tag{2}$$

where D_{15}, d_{85} and d_{15} were defined in Figure 1.

It can be seen that eqs. (1) and (2) are well fulfilled by the products in Figure 1.

Some segregation or wash-out of fines inevitably will take place, the proportion increases with decreasing tailings particle size. Fines are also generated by solids degradation in the pumps and in the pipeline. Well-planned shifting movement of the discharge point and preferably upslope discharge are some factors that can minimize the impact of segregation Morris et al. (1997a). The slope according to Morris et al. (1999) seems to be independent of the water content in the pumped slurry.

2.1 Slurry pumping

Coarse particle slurry pumping design data has been generated in recent years at the Hydraulic Testing Laboratory, GIW Industries Inc., U.S.A. Here pipeline friction losses and pump efficiencies can be investigated in loops with pipe diameters of up to 0.5 m and installed pump power of up to 1500 kW. Experimental experience has shown that slurries with broad and even size distributions can be pumped energy-effectively at solids concentrations by mass ("weight") of over 60%, see for example Sundqvist et al. (1996). Operating conditions and pump efficiencies etc. used in the examples below are based on generalised experimental results from the GIW Hydraulic Testing Laboratory.

The solids concentration by volume may be the most representative parameter for expressing the slurry pumping character of the slurry. The water requirement for transporting solids particles in slurry form however is best expressed by the ratio of mass flow rate of water MO (kg/s) and the mass flow rate of solids, MS (kg/s). If the solids content is expressed by the concentration by mass, CW, then MO, MS and CW are interrelated in the following way:

$$CW = \frac{MS}{MO+MS} \tag{3}$$

Expressing the water requirement parameter MO/MS by M* then eq. (3) gives

$$M^* = \frac{1-CW}{CW} \tag{4}$$

Note that M* is non-linearly related to CW. For example an increase in CW from 34% with a factor 1.6 to 55% decreases M* from 1.94 to 0.82, m^3/tonne, i.e. the water required for the higher concentration is only 42% of the water needed at the lower concentration.

The flow rate of mixture, Q, in a pipeline is given by the following expressions:

51

$$Q = \frac{MS}{SS} + QO$$

where QO is the flow rate of water in the slurry $\qquad$ (5)

$$QO = \frac{MO}{SO}$$

where SS and SO are the densities of solids and water, respectively, in kg/m^3. The density of the mixtures, SM, can be calculated with the following relationship.

$$SM = \frac{SS \cdot SO}{SS - (SS - SO)CW}$$ $\qquad$ (6)

Considering transportation of MS kg per second of dry solids to a disposal area located L m away from the concentrator with a static lifting of $i \cdot L$ where i is the average slurry pipeline slope. The power, P(W), required to pump the slurry to the disposal area is $(i \cdot L \ll L)$:

$$P = \frac{SM \cdot g \cdot L(I + i)Q}{\eta}$$ $\qquad$ (7)

where η is the total efficiency of the pumping (including motors etc.) and where I is the pipe-line friction loss gradient (hydraulic gradient) in metres of slurry per metre of pipe. Q and SM were obtained from eqs. (5) and (6), respectively.

The power required, PO, to recirculate the corresponding water downhill is $(i_o \cdot L \ll L)$:

$$PO = \frac{SO \cdot g \cdot L(IO - i)QO}{\eta_0}$$ $\qquad$ (8)

where η_0 is the total efficiency for pumping water and i_o is the average water pipeline slope. IO is the hydraulic gradient (m of water/m) for pumping water. The total power requirement is then:

$$PTOT = P + PO$$ $\qquad$ (9)

The total specific energy consumption, e, for transporting the tailings is:

$$e = \frac{PTOT}{MS}$$ $\qquad$ (10)

An example will now be given for the pumping of 3.5 Mtonnes of waste rock per year and the same amount of tailings with particle size distributions according to Figure 1. Assuming about 7900 h of operation per year then the tonnage MS is 885 tonnes/h. Two CW-values will be studied and it is assumed that SS = 2900 kg/m^3, SO = 1000 kg/m^3, L = 3200 m and slopes (i, i$_o$) of 0,+0.02 and -0.02, corresponding to uphill and downhill distances of 64 m, respectively.

Calculated pipeline diameters and matching velocities, friction losses and pumping efficiencies are given in Table 1 for a horizontal pipeline.

Table 1 Calculated slurry pumping data for two solids concentrations at 885 tonnes/h of solids. Horizontal pipeline and an assumed water velocity of 1.5 m/s in the return pipeline.

Conc. by mass CW (%)	34	55
Water requirement, QO (m³/h)	1726	726
Mixture flow rate, Q, m³/h	2034	1017
Slurry pumping		
Pipeline diameter, (m)	0.4	0.3
Velocity, (m/s)	4.5	4.0
Hydr. Gradient, I (m/slurry/m)	0.055	0.073
Pumping efficiency, η (%)	55	48
Water pumping		
Pipeline diameter, (m)	0.55	0.36
Pumping efficiency (%)	65	65
Hydr. Gradient, IO (m water/m)	0.005	0.008

With the values from Table 1 in eqs. (5 to 10) together with the length, L = 3200 m, then the following power consumptions were calculated for horizontal pipelines, Table 2.

Table 2 Calculated energy consumptions for the example and data given in Table 1

Conc. by mass, CW (%)	34	55
Water requirement, QO (m³/h)	1726	726
Slurry pumping power req., P (kW)	2270	2134
Water pumping power req., Po(kW)	127	83
Total pumping req. PTOT (kW)	2397	2217
Spec. energy consumption, e (kWh/tonne)	2.71	2.50

The values presented in Table 2 will now be compared to powers calculated in slurry pumping systems with slopes of -0.02 (downhill) and +0.02 (uphill), see Table 3.

Table 3 Energy consumptions for the data in Table 1 with inclined pipeline systems slopes, - 0.02, downhill slurry pumping and +0.02, uphill slurry pumping.

CW%	34		55	
Slurry pipeline slope	-0.02	+0.02	-0.02	+0.02
P (kW)	1444	3095	1549	2719
PO (kW)	589	-	277	-
PTOT (kW)	2033	3095	1826	2719
e (kWh/tonne)	2.30	3.50	2.06	3.07

3 DISCUSSION AND CONCLUSIONS

It can be seen from Table (2) that the specific energy consumption decreases slightly from 2.71 to 2.50 kWh/tonne when the concentration is increased from 34 to 55%. The power to recirculate water is practically negligible in the case of horizontal pipeline systems.

With uphill slurry pumping water is recycled by gravity without any energy consumption independent of the solids concentration, see Table (3). With downhill slurry pumping and static lifting for the recycled water about 30% of the total energy consumption comes from the water pumping at CW = 34%. With the high solids concentration, 15% is spent on recycling water.

The leading cost reduction factor while pumping at a high solids concentration is the smaller pipeline diameters required, 0.3 m instead of 0.4 m for the slurry pumping, and 0.36 m instead of

0.55 for the water recycling. The substantial decrease in pipeline diameters gives an indication of the reduction potential in capital cost for pipeline material.

High solids concentrations strongly improve the effectiveness of co-disposal. Large-scale loop investigations have shown favourable operating conditions with coarse particle pumping at higher solids concentrations. In practice the once-through systems must be effectively controlled, however.

Experience reported by Williams (1997) indicated that the limitation of CW to about 35% was partly related to pump and pipeline wear. Wear in pumps and pipelines are related to the solids concentration in a complex way. For example, with high solids concentrations pipeline wear may be more evenly distributed.

In summary, the effectiveness of co-disposal relies strongly on high solids concentrations which also lower pipeline investment costs and pumping power requirements per tonne. Reliable operation with high solids concentrations, however, require effective control systems and application and development of suitable wear resistant materials and special pump designs.

REFERENCES

ICOLD 1996. Tailings dams and environment. Review and recommendations. Bulletin 103.

Morris, P. and D.J. Williams 1997a: Results of field trials of co-disposal of coarse and fine wastes. *Trans. Instn. Min. Metall (Sect. A: Min. industry)*, 106:A38-41.

Morris, P. and D.J. Williams 1997b. Co-disposal of washery wastes at Jeebropilly colliery, Queensland, Australia. *Trans. Instn. Min. Metall (Sec. A. Min. industry)* 106:A25-34.

Morris, P.H. and D.J. Williams 1999. Some comments on the prediction of mine waste beach slopes. *Int. J.of Surface Mining*, Reclamation and Environment 13:31-36.

Sellgren, A. 1994. System Aspects on the Pumping of Waste Rock. *Proceedings, Nordic Conference in Mineral processing. February 14-16*, Luleå, Sweden (in Swedish).

Sellgren, A. and G. Addie 1997. Cost-effective pumping of coarse mineral products using fine sands. *Powder Technology* 94:191-194.

Sellgren, A. and G. Addie 1998. Effective integrated mine waste handling with slurry pumping. *Proceedings Fifth Int. Conf. On Tailings and Mine waste 98' Ft. Collins, Col., January 26-28*, U.S.A.

Sellgren, A. and Å. Sundqvist 1994. System approaches to effective mine waste handling based on slurry pumping. Poster manuscript, *Third International Conference on the Abatement of Acidic Drainage, Pittsburgh April* 244-29, U.S.A.

Sellgren, A. and Å. Sundqvist 1995. Integrated approaches to mine waste handling with slurry pumping. Hynes, T.P. and Blanchette, M.C. (Eds). *Proc. of Sudbury '95 - Mining and the environment. May 28 - June 1*, Sudbury, Ontario, Canada. Volume 3:1089-1094. CANMET, Ottawa.

Sherard, J.L., R.J. Wooward, S. Gizienski, and W. Clevenger 1963. *Earth-Rock Dams. Eng. Problems of Design and Construction*. John Wiley and Sons Inc.

Sundqvist, Å., A. Sellgren and G. Addie 1996. Slurry pipeline friction losses for coarse and high-density industrial products. *Powder Technology*, 89.

Van Rooyen, K.C. 1992. An integrated method of coal discharge and slurry disposal to reduce environmental impact from coal residue. *Proc. 2nd Int. Conf. Environmental issues and management of mine waste in energy and mineral production*, Calgary (Rotterdam: Balkema, 1992), vol 1:419-25.

Williams, D.J. and V. Kuganathan 1992. Co-disposal of fine and coarse grained coal mine washery wastes. *Int. J. Environmental issues in minerals and energy industry*, 53-8.

Williams, D.J. 1997. Effectiveness of co-disposing coal washery wastes, *Proceedings: Tailings and Mine Waste 97*, Ft. Collins U.S.A. (Rotterdam Balkema 1992), 335-341.

Tailings and Mine Waste'00 © 2000 Balkema, Rotterdam, ISBN 90 5809 126 0

Principles of tailings dewatering by solar evaporation

G. E. Blight & L. Lufu
Witwatersrand University, Johannesburg, South Africa

ABSTRACT: Solar energy is the only completely cost-free source of energy available on earth. However, because the characteristics of solar energy are not everywhere fully understood, and because it is often regarded as unpredictable and unreliable, it is virtually ignored as an energy source. This paper will explore the potential for the conscious application of solar energy to the drying of tailings. It will show that the effects of solar drying can rationally and with advantage be taken into account when considering the design and performance of tailings deposits.

1 INTRODUCTION

In arid and semi-arid regions, solar radiation has a considerable capacity to evaporate water from the earth's surface, whether from free water surfaces, vegetated or bare soil or the surfaces of waste deposits such as tailings dams. The evaporative capacity in humid regions is also not inconsiderable, but is usually more than balanced by precipitation. Solar radiation nevertheless operates effectively in all climates and its contribution to overall water losses can be recognized and exploited with benefit.

A prime objective of tailings or other waste management should be to maximize the mass of solids stored within a given volume of waste impoundment. This in turn will maximize the utilization of airspace, and in the case of a tailings dam, also maximise shear stability of the outer wall of the tailings impoundment as well as allowing for easier and quicker eventual rehabilitation. In order to achieve the above ideal, hydraulic fill tailings beaches should, to the greatest extent possible, be allowed to drain and dry out, and, weather and climate permitting, to become sundried between successive depositions of tailings.

If a strategy of keeping the beaches drained and sundried and minimizing the pool area is followed, problems at closure and with rehabilitation of the impoundment's top surface will be minimized. The beaches should become trafficable to low ground pressure mechanical equipment within a relatively short time from closure. The only intractable area then is the pool. If the pool has been kept small throughout the operating life of the impoundment, this problem will also have been minimized.

The objective of the present study was to demonstrate the capability and gain an understanding of the process of sun-drying of a tailings surface, so as to be able to predict drying rates at design and during operation.

2 CONVERSION OF SOLAR ENERGY INTO LATENT HEAT OF EVAPORATION

At the surface of the soil or a waste deposit, the net incident solar radiation R_n is converted in accordance with the surface energy balance.

$$R_n = G + H + L_e \qquad (1)$$

Where G is the soil or waste heat flux, the energy consumed in heating the near surface soil or waste,

H is the sensible heat flux, the energy consumed in heating the air above the surface, and

L_e is the latent heat flux for evaporation, consumed in evaporating water from the surface.

R_n can be measured directly and G can be estimated from changes in the temperature and temperature gradient below the surface together with the specific heat capacity for the soil or waste. H, the sensible heat may be calculated from temperature and relative humidity gradients above the surface.

The process has been fully set out and explained by Blight (1997).

3 INVESTIGATION OF SOLAR DRYING OF TAILINGS

To investigate the behaviour of mineral tailings subjected to solar drying, a series of drying tests was carried out on specimens of three tailings materials:

1. a high clay content heavy minerals tailings, produced by washing beach sand deposits to separate heavy minerals (mainly titanium);
2. non-plastic gold tailings produced by milling quartzite, and
3. a power station fly ash.

Figure 1 shows the particle size analyses of the three materials for which the particle specific gravities and Atterberg limits are shown in Table 1:

Table 1. Properties of mineral tailings

Material	Particle S G	Liquid Limit	Plastic Limit
heavy minerals	2.81	76	30
gold	2.69	non plastic	
fly ash	2.19	non plastic	

These three materials were selected as it was expected that on drying, the heavy minerals tailings would shrink and crack extensively, the gold tailings shrink and crack moderately, and that the fly ash would not shrink or crack. Furthermore the heavy minerals tailings are an example of a clayey plastic material that will not form an hydraulic fill beach on which particle size segregation occurs; the gold tailings are fine, but non-plastic and do form a segregated hydraulic beach; the fly ash is coarser than the gold tailings and poses no unusual problems when deposited hydraulically (Blight, 1994). The drying experiment was set up in a greenhouse so that natural solar radiation could be used, but extraneous

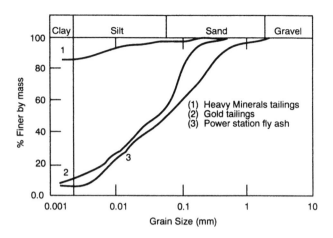

Figure 1: Particle size distributions of three mineral tailings.

effects such as wind and rain would be excluded.

The tailings were deposited in layers in square 0.5m x 0.5m trays 0.3m deep supported above the floor on 150 mm high wooden pallets. The layers of heavy minerals and gold tailings were 100 mm in initial thickness while the fly ash was placed in a single 300 mm thick layer. The experiment was carried out twice, using two different initial water contents. Initial percentages of solids and water contents for the two stages of the experiment were as shown in Table 2:

Table 2. Initial conditions in drying experiment

Material	Percent solids*		Water content*	
	at placing			
heavy minerals	20%	29%	390%	240%
gold	61%	35%	65%	189%
fly ash**	56%		78%	

*   by dry mass
**   placed with light compaction

Measurements were made daily of the mass of each tray, and weekly of the radiation balance . Measuring the radiation balance entailed making measurements, during daylight hours of the incoming and outgoing solar radiation using a hand-held radiometer to ascertain R_n (equation (1)), of temperatures in the tailings in each tray, by means of buried thermistors, to ascertain G, and of relative humidities and temperatures at two heights above each tray (100 mm and 1m) to determine humidity and temperature gradients in the air above each tailings sample. A hand-held wet and dry junction thermocouple psychrometer was used for this.

After deposition, each layer was allowed to dry until the rate of water loss declined to a low level, whereafter the next layer was added. In the case of the fly ash, water was added to the single layer of ash once the rate of water loss had declined.

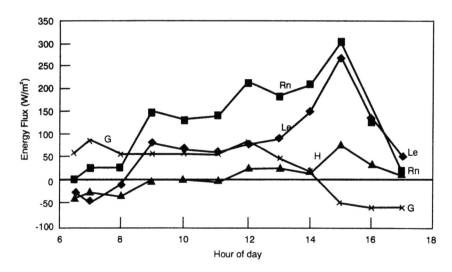

Figure 2: Typical radiation balance measurements taken in greenhouse during solar drying experiment

Figure 2 shows the results of a typical day's set of radiation balance measurements. During the night R_n becomes negative and there might be slight condensation on the tailings surface. Generally, though exchanges of energy during the night are small. In Figure 2, the fluctuations in the R_n curve are caused by clouds passing in front of the sun. Note that in the early morning H is negative, showing that the air is cooling and heating the tailings surface (G is positive). L_e is also negative, showing that condensation is occurring in the air and on the tailings surface. At the end of the day, as R_n declines, the near-surface tailings gives up its heat to the air, G declines and H increases. The area under the L_e curve in MJ/m² of tailings surface represents the energy for that day expended on evaporating water from the tailings. If this is divided by the latent heat of evaporation of water (2.47 MJ/kg) the result is the mass of water evaporated per m² of tailings surface, in kg/m² which is equivalent to litres/m² or mm of evaporation.

4 RESULTS OF SOLAR DRYING EXPERIMENT

Figures 3, 4 and 5 present the results of the solar drying experiment for the second (higher initial water content) stage (there was only one stage for the fly ash). The results for the first stage are very similar.

Figure 3 shows the results in terms of overall water content of the tailings in the trays. The steps in the water content-time record represent the addition of new layers of tailings slurry. Although each successive layer was of the same thickness, the effect on the overall water content decreased because of the accumulation of solid material. It will be noted that each successive layer dries rapidly at first, but that the rate of loss of water content decreases with time and exponentially approaches a very low value.

The corresponding suction values based on the average water content have been superimposed on the drying diagrams for the three tailings. It will be seen that in terms of overall water contents, the rate of loss of water falls to a low value, even though the suction remains relatively low. However, because of the moisture content gradient with depth that

58

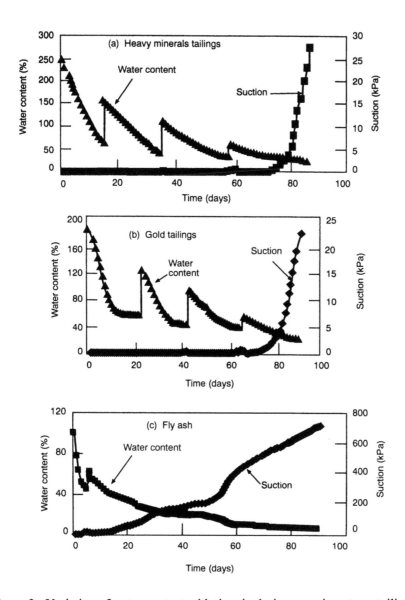

Figure 3: Variation of water content with time in drying experiments on tailings.

existed in the tailings, the suction at the very surface of the tailings must have been much higher than the values shown in Figure 3. This is shown by the steep climb of the suction-time curves at the end of the experiment when the tailings were allowed to dry to a lower water content than previously.

Figure 4 shows the results of Figure 3 expressed in terms of evaporation rate in mm of water per day. The measured values have some scatter because, apart from inevitable errors in weighing the trays, the weather varied from day to day and hence so did the energy causing the drying. The data show that the evaporation rate immediately after placing a layer started at about 4 mm/day which was close to the A pan (American standard A evaporation pan) rate of evaporation, and then rapidly declined and approached zero

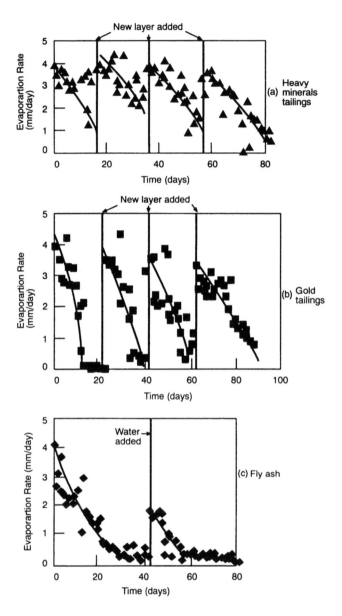

Figure 4: Variation of evaporation rate with time in drying experiments on tailings.

after about 15 days. This applied to both the heavy minerals and gold tailings. The fly ash took somewhat longer to dry out.

Cracking of the surfaces occurred after about 10 to 14 days, and the cracks continued to extend and widen thereafter, but there is no indication in either Figure 3 or 4 that the increase in surface area as a result of cracking had any effect on the rate of drying.
As mentioned earlier, the radiation balance for the trays of tailings was evaluated weekly and the theoretical evaporation determined. It is thus possible to compare the "actual evaporation" from the tailings surfaces (ie that established by direct weighing) with that calculated from the surface energy balance. The results of these comparisons are shown

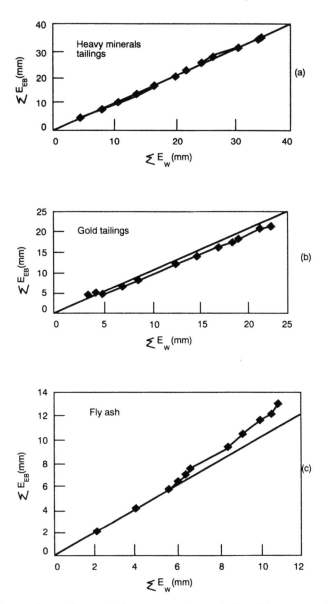

Figure 5: Comparison of cumulative evaporation calculated by energy balance (ΣE_{EB}) with evaporation established by weighing (ΣE_w)

in Figure 5. Figure 5 shows that the cumulative evaporation calculated from the surface energy balance (E_{EB}) for the heavy minerals tailings was very close to that measured by weighing. The calculations slightly underestimated the evaporation from the gold tailings. While the surface of the fly ash was moist the energy balance predicted evaporation from its surface very closely. Once the ash had dried out at the surface, the energy balance overestimated evaporation. This is probably because the humidity gradient above the ash surface was controlled by average conditions in the greenhouse which did not accurately represent conditions above the ash.

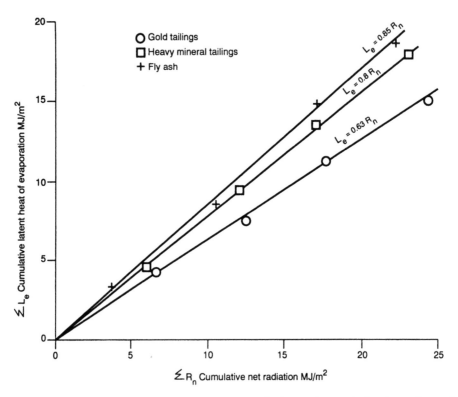

Figure 6: Relationship between cumulative net radiation and cumulative latent heat of evaporation in greenhouse drying experiment.

Hence it was confirmed that the energy balance measurements realistically estimate rates and quantities of evaporation from a tailings surface, at least under sheltered conditions.

It is apparent from Figure 2 that L_e, the latent heat of evaporation, represents a substantial proportion of R_n, the net incident radiation. Figure 6 shows cumulative measurements of L_e (ΣL_e) plotted against corresponding cumulative measurements of R_n (ΣR_n). The figure shows that for each type of tailings, L makes up a fairly constant fraction of R_n. Once curves like those shown in Figure 6 have been established, it is only necessary to measure R_n (a very simple procedure) to establish the value of L_e.

5 CONCLUSIONS

The process of solar drying of tailings has been investigated for three types of tailings(Figure 1). Solar energy striking the surface of the tailings is partly converted into latent heat of evaporation. The proportion so converted is approximately constant for each type of tailings, and in this series of tests varied from 63% to 85% of the net incoming radiation (Figure 6). Initially, the rate of drying is approximately the same as the rate of evaporation from an A pan, but the rate progressively falls to a low value (Figures 2 and 3). If the net radiation is measured, the evaporation from the tailings can be predicted with good accuracy (Figure 5)

6 ACKNOWLEDGEMENT

The research reported in this paper was supported by the South African National Research Foundation and Knight Piesold, consulting engineers, Johannesburg. The authors thank both of these organizations for their assistance.

REFERENCES

Blight, G.E. 1997. Interactions between the atmosphere and the earth. *Geotechnique*, 42, 715-766.

Blight, G.E. 1994. The master profile for hydraulic fill tailings beaches. *Proc. Instn Civ. Engrs (London) Geotech. Engng.*, 107, Jan, 27-40.

Tailings and Mine Waste'00 © *2000 Balkema, Rotterdam, ISBN 90 5809 126 0*

Slurry impoundment internal drain design and construction

R.E.Snow
D'Appolonia Engineering, Monroe, Pa., USA

J.L.Olson
PDC Technical Services Incorporated (Formerly: Turris Coal Company)

W.Schultz
Turris Coal Company

ABSTRACT: As part of expansion plans for a coal refuse disposal impoundment at the Elkhart Mine in central Illinois, an internal drainage system was designed and constructed along 11,000 feet of perimeter dikes. The objective of the drainage system design and construction included control of phreatic levels and seepage considering an anisotropy of the coarse coal refuse/combustion waste dike materials of approximately 50. The challenge was to develop, design and construct methods which could achieve this objective without resorting to expensive chimney or blanket drains. The lower drain system was constructed in 1996 and 1997, requiring placement of 24,000 tons of aggregate on the existing 40 foot-high, 2.5:1 (horizontal to vertical) slopes using a mobile conveyor system and wheel loader.

1 SITE BACKGROUND

The Elkhart Mine slurry impoundment is located in Logan County, Illinois, and commenced operation in 1981. The initial stage of the impoundment was constructed of native soils (clays and silts), and was expanded arealy and raised in height using initially coarse coal refuse, and subsequently coarse coal refuse and coal combustion waste (CCW primarily consisting of fluidized bed combustion ash from nearby power plants). Fine coal refuse slurry is pumped into the impoundment area to settle, and clarified water is returned to the plant for coal processing. Coarse coal refuse and CCW are used to construct the perimeter dikes downstream of the initial dikes, resulting in a "downstream constructed" embankment. The coal refuse disposal facility currently occupies about 200 acres adjacent to the coal preparation plant, which processes about 2.4 million tons per year of clean coal. Approximately 11,000 feet of perimeter dike contains the slurry impoundment. Figure 1 shows a typical cross section of the dike at the time of the design.

The addition of CCW to the coarse refuse is achieved in the loadout bin, and the material is placed in lifts of approximately 1 foot, spread and compacted with haul equipment and a self-propelled sheeps foot compactor. The CCW material reacts with the moisture in the coarse coal refuse resulting in some cementation with significant increase in strength. The coarse refuse as well as the coarse refuse/CCW mixture is subject to development of compaction planes when compacted, and despite scarifying activities, produces anisotropic conditions. The amount of CCW is limited by regulatory permits to less than 25 percent of the coarse refuse by weight, and the continuation of some coal contracts which currently require haul back of CCW from the power plant. Accordingly, the design has not relied on the strength improvement from the CCW, although the affects of the ash were considered in the seepage analysis.

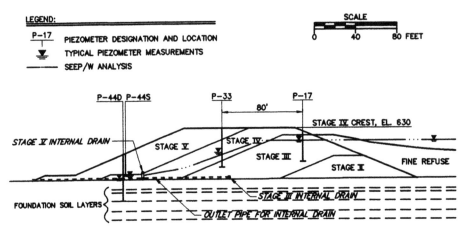

Figure 1 – Cross Section of Dike Prior to Expansion

Expansion plans for the facility were developed in 1996 to increase the ultimate embankment height from 50 to over 100 feet and thus provide several additional years of production capacity. Raising of the dikes will be accomplished with "upstream construction," where the coarse refuse/CCW fill is placed over settled deposits (deltas) of fine refuse slurry along the inside edge of the perimeter dikes. This results in decreasing the impoundment area as the perimeter dikes are raised. To design the expansion and upstream construction dikes, the stability of the embankment was analyzed, including consideration of seismic loading conditions, and the requirements for control of the seepage and the piezometric level within the dam established. Based on these requirements, analyses were conducted to design the internal drain system, which included lower drain features for the final downstream constructed stages, and middle and upper drain features within the upstream constructed stages. The lower drain consists of intermittent fingers parallel to the slope connecting to a continuous longitudinal (or collection) drain near the toe to intercept seepage planes within the embankment without being a continuous blanket. The middle and upper drains are smaller longitudinal drains positioned to control the piezometric level.

2 INTERNAL DRAIN DESIGN

The internal drain design was based on performance and design criteria, considering the existing conditions and parameters of the slurry impoundment, using two dimensional analyses of seepage to develop the location, size, and configuration of the drains.

2.1 Performance criteria

The objective of the internal drainage system was to control phreatic levels and seepage to maintain stability and facilitate construction and operation of the slurry impoundment. Stability analyses of the downstream slopes considering static and seismic loading cases were performed based on preliminary estimates of the piezometric level within the embankment. These results indicated the portions of the embankment cross section for which the stability is most sensitive to the phreatic surface, and the piezometric levels which should be maintained to achieve acceptable factors of safety. The drain location and geometry could then be determined which would control the phreatic surface.

Design criteria for the drains also were based on standard filtration and clogging criteria. The geotextile filter fabric was designed to meet two fundamental criteria: (1) the fabric must prevent piping of coarse refuse into the drain, and (2) the fabric must allow free drainage into the drain over time without clogging. Based on the embankment material grain size, as presented in Figure 2, the design criteria considering the use of geotextile (Koerner 1986) were applied and a fabric with as large an apparent opening size as possible was selected. A significant fraction of the fines in Figure 2 is composed of the CCW mixture and not completely hydrated. Borehole samples indicated more cementation then the surface samples shown in Figure 2, such that the design criteria were applied using the mean and mean-standard deviation curves.

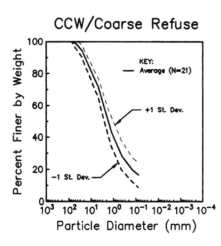

To assist in evaluating potential for clogging with the use of geotextile, gradient ratio tests following ASTM-D5101-90, were also performed on the geotextile using fresh samples of coarse coal refuse and coarse coal refuse/CCW mixture. The tests indicated that some period of stabilization occurred before equilibrium conditions developed under the test's large gradients. Following redistribution, the tests indicate that the equilibrium flux rate is in excess of the required design rate.

Figure 2 - Grain Size Analysis

2.2 Seepage Parameters and Existing Conditions

Seepage parameters required for the analysis of the proposed design included permeability (horizontal and vertical) of the embankment materials. At the time of the design, the existing impoundment had been constructed to a height of about 40 feet, with monitoring piezometers located along several cross sections. Available permeability data from field and laboratory testing provided initial estimates of vertical and horizontal values, which were then used in analyses to simulate the existing piezometric level at each cross section. This permitted development of consistent permeability values, and provided the most reliable estimate of anisotropy.

The field permeability testing program on embankment materials included: constant and falling head tests performed in auger holes completed in the early stages of the coarse refuse embankment, and rising head tests performed in the piezometer casings completed in both zones of coarse refuse and coarse refuse/CCW mixture. Standard laboratory tests were also performed on the coarse refuse materials, which yielded more of a vertical permeability. Both laboratory and field testing were performed on the foundation soils, consisting of silt and clay and generally low in permeability. For the proposed upstream construction stages, the fine coal refuse permeability plays a significant role in the development of the embankment phreatic surface, and the fine refuse permeability was estimated based on grain size distribution and dissipation data available from seismic piezocone soundings in the existing fines deposits.

Water levels in the embankment piezometers and the reservoir level have been recorded on a monthly basis, and measurements of the original drain discharges are periodically

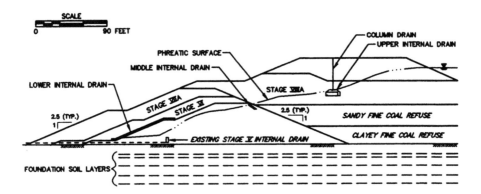

Figure 3 - Cross Section of Proposed Dike Expansion

performed. Review of the records of dike construction, piezometric level and reservoir level indicate that piezometric responses occur relatively quickly (within a month) of reservoir level increase, with embankment construction at a much slower rate. Accordingly, steady state analyses were used to simulate recorded piezometer levels based on the reservoir level and existing embankment geometry. Using the measured drain discharge rates and the estimated permeability for the foundation materials and fine coal refuse, trial ranges of horizontal permeability and anisotropy were selected for the embankment materials to simulate phreatic surfaces at the monitored cross section exhibiting the most critical conditions (i.e., higher piezometer levels and seepage rates). For each combination of horizontal conductivity and anisotropy ratio assessed for the embankment materials, a steady state phreatic surface for the existing facility was estimated. Each phreatic surface was rated qualitatively based on its approximation of the known water levels in piezometers within the embankment and foundation soils. Based on this qualitative rating, a range of horizontal permeabilities and anisotropy ratios was selected as generally representative of those in the embankment. The analysis indicate a representative horizontal permeability of approximately 2.7×10^{-4} cm/sec and anisotropy ratio of 50 (i.e., vertical permeability of approximately 5.5×10^{-6} cm/sec) for the embankment materials. Figure 1 presents the cross section, monitored piezometer levels, and predicted phreatic surface.

2.3 Analysis

The computer program, SEEP/W by GEO-SLOPE International Ltd., was used to perform two dimensional, steady state analyses of the phreatic surface considering dike geometry, impoundment levels, and material properties. SEEP/W is a finite element program which solves the coupled equations of mass continuity and flow using a hydraulic conductivity function for unsaturated conditions dependent on the saturated permeability (Green and Corey, 1971). In addition to analysis of the existing conditions to validate the horizontal permeability and anisotropy of the coarse refuse, SEEP/W was used to analyze the performance of various internal drain configurations in controlling the phreatic level to meet stability requirements.

Two separate drain structures were evaluated: one within the downstream final stages to control the phreatic surface near the toe of the embankment, and one in the expansion upstream stages to intercept seepage for future site development. Figure 3 illustrates the general embankment and drain configuration. The two dimensional analyses represented the cross section of the long perimeter dikes well for the upper portion of the drain system; however, to evaluate the use of a finger drain system extending up the lower embankment

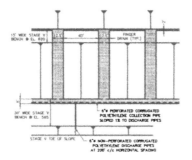

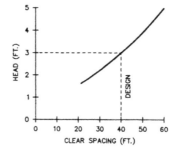

Figure 4 – Finger Drain Plan

Figure 5 – Piezometric Head at Midpoint between Drains

slope, in place of a continuous blanket drain, a supplemental two dimension analysis was also conducted. This analysis simulated the inclined phreatic surface, extending from the continuous upper drain to the toe collection drain. The permeability values were transformed for the approximate 13 degree inclined seepage plane and a weighting procedure used to account for the differing volumes of coarse refuse and drain material represented in the plane of simulation. By specifying constant head boundary conditions at the upper continuous drain and the discharge locations of the toe drains, corresponding to their elevations, seepage along the phreatic surface could be approximated and the influence of varying the clear spacing between finger drains evaluated. Figure 4 shows the plan of the lower finger drain system, which was used in the analyses, and Figure 5 shows the affect of clear spacing between finger drains, on the computed mid-point head above the finger drain piezometric level. These analyses assisted in selecting a 40-foot clear spacing between fingers.

2.4 Drain Features

The lower internal drain consists of a continuous toe collection system, fingers which extend up the embankment slope, and a continuous upper drain system. A coarse aggregate was specified for the drain material, based on its availability and grain size characteristics, with 6-inch perforated collection pipe in the continuous toe section which tied into solid discharge

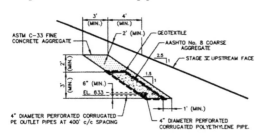

Figure 6 - Middle Drain Design

pipe running to the downstream embankment toe on 200 foot spacings. A geotextile with an apparent opening size of 0.212 mm was selected for the filter media of the lower drain to minimize clogging and maximize the permeability of the fabric. A fine aggregate (sand) was selected to supplement the geotextile at the upper drain components. Figures 6 and 7 illustrate the upper drain system

design. A series of piezometers were proposed within the embankment to monitor the phreatic surface and performance of the drain system.

3 DRAIN CONSTRUCTION

Construction of the lower internal drain began in April 1996 and was completed in August 1997. For the most part, the drain was constructed by Elkhart Mine personnel. Outside

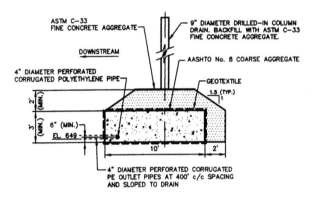

Figure 7 - Upper Drain Design

contractors were used to place drainage media during early stages of construction. The drain was constructed on the downstream face of Stage V in the following general sequence:

1. Lateral Installation
2. Finger Drain Trench Excavation
3. Geotextile Placement
4. Perimeter Pipe Installation
5. Coarse Aggregate Placement
6. Drain Coverage

Lateral Installation. The laterals were 6-inch diameter solid corrugated polyethylene (CPE) pipe. Prior to installation, the Stage V buttress was extended to the permitted impoundment boundary to provide additional working area for the drain installation. Laterals were installed by trenching through the buttress with a rubber-tired backhoe equipped with a 1-foot wide bucket. Following placement of the lateral, the trench was backfilled with the excavated coarse refuse. The lateral outlets discharged to the impoundment perimeter ditches.

Finger Drain Trench Excavation. A Caterpillar D-6 dozer was used to excavate the drain fingers. The dozer was equipped with a 13-foot wide blade that cut a 2-foot deep trench from the Stage V bench down to the Stage V buttress. The excavation spoils were spread out across the buttress. Once the fingers were excavated, the buttress area was graded in preparation for the installation of the perimeter pipe and gravel connector for the continuous toe collection system.

Geotextile Placement. The excavation and grading work was followed by placement of a nonwoven geotextile that served as filter protection for the gravel drainage media. The geotextile was used to line each finger trench, and the top and bottom gravel connectors. Sufficient geotextile was laid out to allow covering of the finger trenches and connectors following gravel placement. The geotextile was placed using manual labor, which provided significant challenges in windy conditions. Typically, the laborer would cut a piece of fabric measuring 15'W x 35'L and carry it up the slope for placement. The fabric was secured with 12-inch nails.

Perimeter Pipe Installation. Following placement of the geotextile, the perimeter pipe was installed and connected to the laterals. The perimeter pipe was a 6-inch diameter perforated CPE pipe. Six-inch tees were used at the lateral connections, which were reinforced with duct tape.

70

Coarse Aggregate Placement. Once the perimeter pipe was installed, placement of the coarse aggregate drain material began. Approximately 24,000 tons of 3/8-inch washed gravel were used in the lower internal drain construction. Equipment used to place gravel included a Caterpillar 966 front-end loader, a Rotec Superswinger mobile conveyor system, and a semi-dump trailer.

Gravel in the lower portion of the drain, including the entire lower gravel connector, was placed by mine personnel with the 966 front-end loader shown in Figure 8. To maximize the efficiency of the 966, gravel stockpiles were strategically placed near the work areas to minimize haul distance. In addition, the 966 was equipped with an oversized coal loading bucket that could transport 8 to 10 tons of gravel per trip. Initially, the 966 was used to place gravel in the finger trenches about mid-way up the slope.

For the upper portions of the internal drain (the upper finger trenches and connector), gravel was conveyed from the buttress level to the upper slope face and bench using the Rotec Superswinger. An outside contractor provided the truck-mounted conveyor, conveyor operator, and semi-dump trailer. The Superswinger was equipped with a hopper and telescoping arm that had a reach of approximately 100 feet. In addition, the conveyor arm was mounted on a swivel to allow 360-degree access. Loading material onto the conveyor consisted of using the 966 to load gravel from the stockpile into the semi-dump trailer. Gravel was then dumped into the hopper set up at the rear of the conveyor truck. Figures 9 and 10 show the conveyor in operation. The conveyor arm was remote controlled, allowing the operator to extend and swing the arm as necessary to place gravel in the correct locations.

Approximately 300 to 400 tons of gravel could be placed with the conveyor operating for a full day. However, conveyor availability became one of the biggest challenges of the project since its availability was subject to the contractor's schedule. Full days of operation were rare. As the project proceeded, conveyor availability became more erratic and unreliable. To maintain drain construction on a continuous basis, alternate methods of placing aggregate in the upper portion of the drain had to be developed.

About the time that conveyor use began to diminish, the drain was being constructed on shorter slopes. This presented an opportunity to test the 966 and determine how far up the 2.5H to 1V slope it could climb and place gravel. In dry conditions, the 966 was able to reach the top of the shorter fingers without rutting or damaging the geotextile. As drain construction continued into

Figure 8 - Cat 966 Loader placing aggregate in finger drain trench.

Figure 9 - Semi-dump trailer loading gravel onto conveyor.

Figure 10 - Conveyor placing coarse aggregate in finger drain trench.

areas consisting of longer slope distances, the 966 was further tested in its ability to climb the slopes. Not only was the 966 used to place gravel in each finger, it was used to complete the upper connector. In all, the 966 was used to complete about 80% of the coarse aggregate placement, resulting in efficient construction and minimizing outside contractor costs.

Drain Coverage. Once the gravel was in place in a given area, it was covered with geotextile. Seams were overlapped and bound with hog rings. Once the gravel was covered, scraper operators were instructed to place coarse refuse at the base of the drain. The refuse was then dozed up the slope in order to cover the geotextile and underlying gravel. Figure 11 shows several finger trenches in various phases of construction, including coverage by a dozer.

72

Figure 11 - Lower internal drain under construction.

4 REFERENCES

Green, R. E. & J. C. Corey 1971. Calculation of Hydraulic Conductivity: A Further Evaluation of Some Predictive Methods. Soil Science Society of America Proc. 35:3-8.

Koerner, R. M. 1986, Designing with Geosynthetics. New York: Prentice Hall.

Tailings and Mine Waste'00 © 2000 Balkema, Rotterdam, ISBN 90 5809 126 0

Paste dewatering techniques and paste plant circuit design

R.A.Tenbergen
Golder Paste Technology Limited, Sudbury, Ont., Canada

ABSTRACT: The number of paste backfill operations in the mining industry has been increasing rapidly since the 1990's. Paste fill operations, when designed and operated properly, supply an engineered, low strength structural fill material for underground backfill purposes. There is a great interest to apply paste technology for surface disposal of wastes in the base metal, gold mining and industrial minerals sectors. All pastefill operations need to de-water the mineral wastes to produce a paste. This paper presents some of the product quality considerations, practical aspects, benefits and drawbacks that go into the selection and design of a particular dewatering circuit and paste plant.

1.0 WHAT IS PASTE

The definition of paste is generally used very loosely in the mining industry. Paste is a densified uniform material of such mineralogical and size make up, that it will bleed only minor quantities of water when at rest, experience minimum segregation and can be moved in a pipeline at line velocities well below that of critical velocities for similar sized materials at lower pulp densities. Pastes can remain sitting in a pipeline for extended periods of time when no cementitious material is present, and its slump can be measured. The slump is normally measured using an ASTM 30.5 cm (12 inch) slump cone, a standard tool used in the concrete industry. Paste can generally be produced from materials with a wide range of size distributions, however they usually contain a minimum of 15% by weight of minus 20μ material (Figure 1). Mineralogical make up is very important as not all materials within the outlined size distribution may make paste.. Hence rheological testing is required.

2.0 GENERAL PASTE BENEFITS

Most mines need some back filling of mined areas to maintain structural integrity and to allow maximum extraction of the ore. Historically this might have been done through filling with waste rock or sand fill and since the 1950's through hydraulic fill.

Hydraulic fill, whether derived from the classified coarse portion of mill tailings or from alluvial sand requires relatively large quantities of water for pipeline transport underground. A coarse product has to be used to meet preset "percolation rate" requirements to prevent hydraulic heads from building up behind fill barriers and prevent liquefaction. The binder added to hydraulic fill tends to segregate during the settling / de-watering in the fill area, while the high water content does not promote strength development. The "water" drainage from hydraulic fill operations usually carries cement and other fines requiring special settling

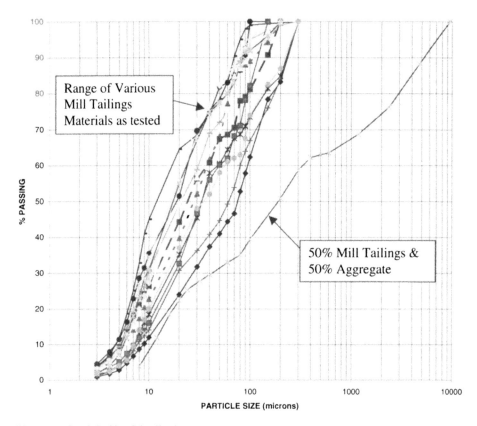

Figure 1 – Particle Size Distribution

and pumping facilities. Especially with deeper mining operations rock fill and hydraulic backfill become more costly to operate.

Development of paste backfill for routine underground fill applications became a reality in the late 1980's. The major benefit is derived from being able to send unclassified mill tailings or tailings of generally finer size consistency underground through a borehole and pipeline distribution system at a much higher density than is normally the case with hydraulic fill. The product can be engineered to provide the required structural strength underground. Early strength development allows for a faster mining cycle.

The table below shows a typical comparison for the water contents of hydraulic, high density and paste fill applications with fill derived from relatively coarse base metal tailings. With paste fill the water going underground is essentially tied up with the solids and there is virtually no drainage. With hydraulic fill twice as much water goes underground and some 55% of this will have to be returned to surface.

| Fill type | % Solids | | Tons of Water / |
	by Weight	by Volume	Ton of Solids
Hydraulic Fill	68	44	.47
High Density Fill	75	52	.33
Paste Fill	82	63	.22

One of the major economic benefits is that paste fill generally produces the required fill strength with about half the binder addition of that required for hydraulic fill, even though paste will consist of full plant tailings and hydraulic fill is derived from the classified coarse portion of the same mill tailings.

Paste for surface disposal offers many opportunities for a simplified and cost effective disposal method as compared to slurry disposal. Paste disposal may be quite effective in postponing the expansion of an existing slurry disposal system.

3.0 PASTE CHARACTERISATION

Each paste application is unique due to the mineralogical and size distribution characteristics of the tailings, process history and final application of the paste. Each application should be subject to a rigorous review to properly assess local conditions so that a practical and cost effective paste application may be designed.

3.1 Backfill strength considerations

Paste used for backfill purposes normally requires that certain strengths develop after placement underground. Early and long-term strength requirements will depend upon the particular mining method. Strength is gained through the addition of cementitious materials, such as Portland cement and/or fly ash. Obviously it is desirable to dispose of as much tailings underground as possible versus the use of imported material, such as alluvial sand. However, it may not be economic to only use very fine grained materials underground, such as might be the case with some gold tailings, as it may become very expensive to use large quantities of binder to obtain the required strength. Under these circumstances it may be advantageous to supplement the fine tailings with aggregate, such as alluvial sand or crushed waste rock fines or a combination thereof. Quite often minus 1 cm material can be added to tailings paste up to levels of 50% by weight with very beneficial results with respect to strength and binder requirements and flowability of the paste.

Figure 2 shows typical % solids by weight versus slump relationships for relatively coarse gold tailings, base metal tailings and blended gold tailings with aggregate. As the coarseness increases the % solids by weight will increase accordingly.

Figure 3 shows 28-day uniaxial compressive strengths of test cylinders made from gold tailings paste of various slumps. The significant impact of aggregate addition on strength is clear. A review of this alternative is usually in order once the binder addition exceeds 5 % by weight, especially in some remote areas where cement costs are over $100 /ton. Aggregate can usually be delivered for $ 2 / ton.

3.2 Paste friction loss / placement considerations

In the design of a paste plant and delivery system, be it for underground backfill or surface disposal, it is necessary to assess pipeline friction losses for different slump pastes. Friction losses will decrease as the slump of the paste increases. Friction losses however, can vary widely between pastes from different mines even though the slumps may be similar. A review of the physical layout of the orebody is required, to assess whether fill can be distributed to the extremities of the orebody, from a central borehole, using gravity flow alone or whether pumping is required or possibly through the use of more than one borehole from surface. This is an iterative process, the outcome of which also influences the design requirements of the surface paste preparation plant.

Figure 4 shows friction losses for a particular gold tailings paste, at various paste slumps, with and without aggregate addition. Friction losses varied by a factor of five in pump loop tests between pastes varying in slump between 17.8 and 25.4 cm. Surface disposal of paste would be possible with a slump of 25.4 cm..

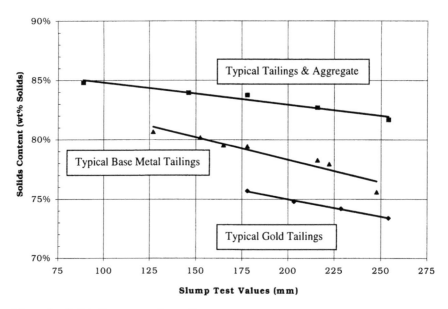

Figure 2 – Solids Content vs. Paste Slump

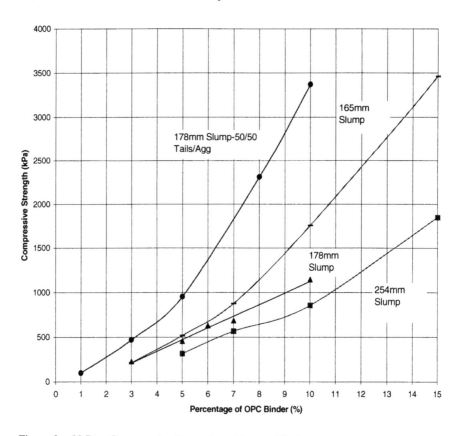

Figure 3 – 28 Day Compressive Strength vs. Percent Binder

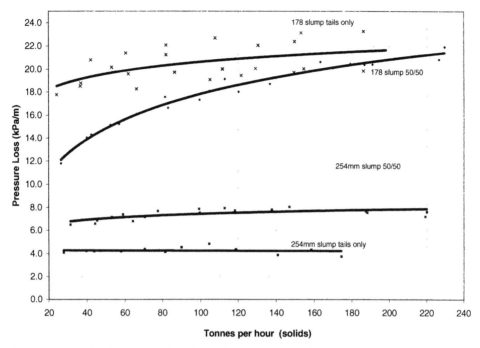

Figure 4 – Friction Losses vs. Flow Rate Through a 150mm Pipe

The high friction loss experienced with the low slump paste limits the horizontal distance to which paste will flow under gravitational forces, even though a lower slump is desirable from the paste strength point of view. The addition of aggregate to the gold tailings lowers the friction losses significantly and allows a wider distribution. Friction loss data from pump loop tests are used to superimpose a gravitational distribution cone over the orebody that indicates the maximum extent to which various paste mixtures can flow by gravity under various slump and paste recipe conditions (Figure 5).

Results from several iterations will determine the most economic means of providing backfill for the mine and whether pumping with positive displacement pumps is required. Pumps are a significant additional capital and operating cost to the project and one tends to use them only as a last resort.

4.0 DEWATERING

All tailings paste preparation plants use some form of dewatering. Tailings slurry dewatering becomes more aggressive as one uses a thickener, deep tank thickener, vacuum filter, pressure filter or centrifuge. In the selection of the appropriate dewatering step, one has to assess the particular conditions and objectives and go through the iterative process mentioned previously. It is obvious that the first two dewatering techniques produce a product that is flowable. The latter dewatering techniques require the addition of some water to their products to make them flowable again. Being able to mix some water back with a filter product gives one complete control over the slump, which could be quite advantageous in an underground gravity flow distribution system. This allows one to match the static head of the distribution system with the friction losses and thus run with a full pipeline system for the different pour points. These

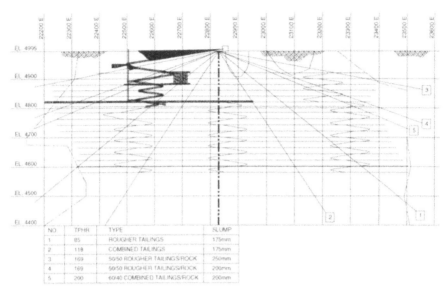

NO	TPHR	TYPE	SLUMP
1	85	ROUGHER TAILINGS	175mm
2	118	COMBINED TAILINGS	175mm
3	169	50/50 ROUGHER TAILINGS/ROCK	250mm
4	169	50/50 ROUGHER TAILINGS/ROCK	200mm
5	200	60/40 COMBINED TAILINGS/ROCK	200mm

Figure 5 – Gravity Distribution Cone

considerations are further influenced by whether one requires positive displacement pumps or not and whether surface disposal is used as well.

4.1 *Conventional thickening*

The routine dewatering step for most tailings is a combination of thickening followed by vacuum filtration.

Thickening efficiencies have improved over the last decades as a variety of flocculents and reliable dry flocculent mixing and addition systems became available. Flocculent supplied in bulk containers or bags makes the operation even more acceptable. Thickeners are now being marketed as high efficiency, high capacity and even paste thickeners. Thickener capacities have improved from the routine historical design rate of .45 ton per m2 of thickener area per hour to some conditions that allow 2.7 t/m2/hr to be used.

A large variety of flocculents are now available. Most flocculent suppliers will be more than willing to screen their reagents to optimize the particular requirements as to ionic specificity and the molecular weight of the polymer. Unless there are special circumstances, such as a small consumption level, it usually pays to purchase a dry polymer mixing system. Reagent screening is normally followed by graduate cylinder testing of the preferred polymer to allow settling rates to be determined at several addition levels, as well as determine the terminal density of the particular tailings. Thickener suppliers may use this data for design purposes or expand upon this testing.

Most manufacturers have their own preferred "feed well dilution, mixing or eduction system". These may be marketed under various trade names. Once the polymer has been mixed with the slurry feed stream at the right density in the feed well, gravity has to do the rest. Sometimes extreme dilution may be necessary before effective flocculation and settling occurs. For one particular application it was necessary to dilute the incoming feed with thickener overflow water before introducing it into the feed well for further dilution.

Thickeners are capable of "storing" some concentrate, however, operations are much improved and the danger of bogging a thickener down are much reduced by pulling the thickener continuously to target set points and storing densified material in agitated storage tanks for subsequent feeding to the next dewatering step.

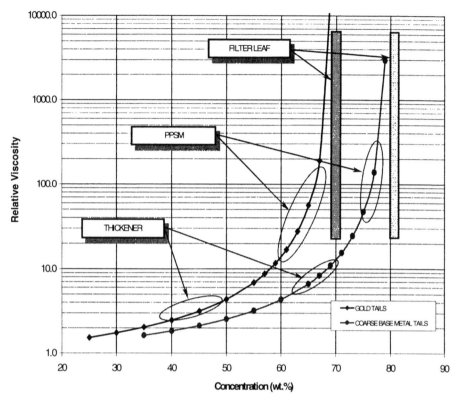

Figure 6 – Slurry Viscosity vs. Weight % Solids

With the interest in tailings paste, many thickener suppliers claim that their thickeners will produce paste. Most thickeners and especially pilot units can produce paste, the questions that are generally more difficult to answer is how long did it take to produce paste and can it be produced continuously and reliably and what type of control over the slump does one have? The developments over the last decade have clearly shown the limitations of conventional thickener designs in producing quality paste reliably. Figure 6 shows that relative slurry viscosity increases rapidly as the weight percent solids of the densified material increases for a relatively coarse base metal tailings. For this particular tailings the viscosity values are exceptionally high. However, the same type of behavior exists for other tailings. This behavior explains why "conventional" thickeners have a hard time producing quality paste reliably.

4.2 Deep tank dewatering

Surface paste disposal of deep tank thickened red muds has been practiced for many years in the aluminum industry. Deep tank dewatering for coarser mill tailings is relatively new.

Several manufacturers now market deep tank thickeners to allow a higher mud bed to de-water the tailings through gravitational / compression forces with the assistance of a specialized "rake and/or helix systems".

Deep tank dewatering units have a very distinctive niche to fill in surface paste disposal and underground paste fill. For surface disposal it may be quite acceptable to produce a product with a higher slump. For underground backfill it might be the answer when the deep tank thickener underflow is mixed with aggregate to assist in obtaining the structural strength of the fill or when the required paste strength can be obtained readily or is not of utmost importance.

One continuous paste production system is marketed by GL&V of Orillia, Ontario. This unit is marketed as a PPSM, a Paste Production and Storage Mechanism. This system was jointly developed with Inco, where there was a need to produce paste for backfill purposes, without having to resort to filtration or being limited to what could be produced from settled storage tanks. This system uses a tall tank, typically 13-m high and up to 10 m in diameter, with a hemispherical bottom and equipped with a rake system at the bottom and a helix around the center shaft to maintain fluidity of the paste and allow for storage. Paste is withdrawn or extruded from the bottom by gravitational forces.

Another continuous system is that marketed by Baker Process of Salt Lake City, Utah. Its development is based upon a liaison with Alcan, which has extensive experience in the disposal of red muds. These units have a typical height of 15 m and a diameter of some 10m. The bottom has a steeper angle of 30 degrees as compared to conventional thickeners and there is a specialized raking system. As well, as in the red mud industry, an underflow recycle system is incorporated to keep the settled material mobile in the discharge pod under the deep tank thickener. The gallonage that is recycled with a centrifugal pump is several times that of the continuous withdrawal rate.

All the above systems use flocculent to assist in rapid settling and increase thickener capacity. The impact of heavy flocculation is to decrease the slump and produce a "stiffer" product with a higher water content. This may result in an increased water release after material transport and deposition underground. The tanks are limited in diameter, because of torque limitations in the drives and rake mechanisms. The torque requirements, especially with some coarser tailings, can increase unacceptably as the paste stiffens.

The % solids of the material produced by the deep tanks is largely material specific and can vary from 45% solids by weight for some fine, lower specific gravity solids material that is fine and heavily flocculated to the 80% solids by weight level for coarser, higher specific gravity material.

There is a deep tank batch system that is marketed by Mag International Ltd. of Sudbury, Ontario. The number of tanks in use would depend upon the tonnage requirements and the turnover requirements. In a three-tank system, one tank would be in the filling and settling mode, another in the compaction and decantation mode and the third in the pouring mode. A patented aeration system is used to mobilize and mix the settled material prior to gravity discharge.

The author has made several project specific economic comparisons between deep tank dewatering systems and conventional thickener and filtration systems. For underground disposal for backfill purposes, the filtration option invariably has the highest capital cost, however, when the binder addition level is added into the equation, the NPV of the project can shift so drastically that general comments can be very misleading. Binder addition levels can readily vary from the 3% for stiffer pastes from filtration systems to the 5% addition level for more fluid pastes from deep tank dewatering units. With a cost of $100/ton for cement in most of the western world and as high as $200/ton for isolated areas, the binder cost can vary between $3 and $5/ton for $100/ton cement. Each project has to be assessed individually.

4.3 Vacuum filtration

Vacuum filtration is generally a routine operation and quite suitable for a variety of conditions. There are a variety of equipment available such as disk, drum and belt filters, as well as more specialised equipment such as belt presses and the ceramic media disk filters. Results are very much dependent on the specific material that has to be de watered. Of the filter types the cloth media disk filters have been used most frequently.

4.4 Pressure filtration

Various types of pressure filtration equipment are available. The most common types are the vertical chamber filter presses, the continuous belt horizontal chamber type and the type where

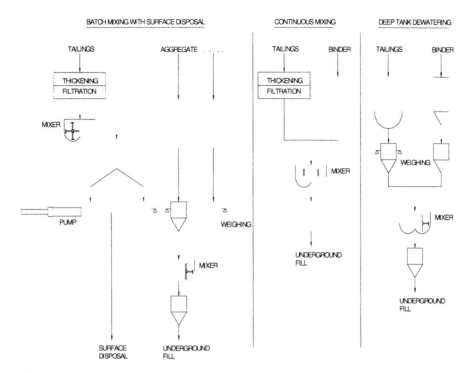

Figure 7 – Simplified Paste Flowsheet Options

a disk or ceramic disk filter is enclosed in a pressure vessel. Some pressure filters, of the vertical chamber type are in use for tailings dewatering for dry stacking purposes. Generally pressure filtration would be overkill for paste fill dewatering. There is a higher capital cost and unless there is another requirement such as cake washing of a cyanide containing gold tailings there normally would be no incentive to go beyond thickening and vacuum filtration or deep tank dewatering.

4.5 *Centrifuge dewatering*

For most base metal mining wastes centrifuge dewatering would not be considered. However, more and more work is being done developing surface and underground disposal options for insoluble mineral wastes from processing plants that recover soluble minerals. These very fine low specific gravity solids are "suspended" in a brine solution and will not readily densify. Under these circumstances it may be necessary to use centrifuge dewatering.

5.0 PASTE PLANT DESIGN

If conceptual evaluation of various fill methods deems it advantageous to consider paste then there are a series of steps that are typically followed to quickly zero in on the optimum system and facility for the particular mine site. The work plan to implement paste typically consist of the following steps:
- A visit to the mine site to review requirements, existing facilities or a meeting with the client and/or client's engineering design firm.
- Laboratory rheological testing of tailings to review the suitability of the tailings as a paste for underground or surface disposal

- Uni axial cylinder strength testing to define the paste recipe
- Pump loop testing to assess paste pipeline friction losses for design purposes of the underground/surface pipeline system and pumping requirements
- Paste plant flowsheet development. At this point major design criteria will be assessed and agreed upon, such as:
  - The tailings dewatering method
  - Requirements for binder and aggregate addition
  - Requirements for pumping for surface or underground disposal
  - Batch mixing or continuous mixing of the paste recipe components

The paste plant circuit design largely depends upon the mineral make up of the tailings, its size distribution, paste fill strength requirements, binder cost, and paste quality control requirements. Figure 7 shows three flowsheets, each tailored to local conditions. One option shows a batch mixing system with a surface paste disposal option.

Conventional thickening and disk filtration is followed by paste mixing to a required slump to allow either surface or underground disposal. For underground disposal, batch quantities of paste, aggregate and binder can be weighed to provide flexibility in fill design. The weighed components and water are mixed to the required slump. Through the use of a surge hopper a continuous paste fill flow is sent underground. Another option shows a continuous mixing system where filter cake and binder are proportioned prior to mixing. The third option depicts a deep tank dewatering system with batch weighing for deep tank paste and binder. The components are mixed and send underground.

6.0 SUMMARY

Paste fill operations are now becoming more common and can offer several advantages over traditional methods for backfilling and surface disposal. Each application has to be thoroughly evaluated to suit local conditions since the paste recipe, plant design and distribution system are very much dependent on the tailings/paste characteristics and mine requirements.

Tailings and Mine Waste'00 © 2000 Balkema, Rotterdam, ISBN 90 5809 126 0

Tailings management framework at QCM

Claude Y. Bédard & John E. Lemieux

Journeaux, Bédard and Associates Incorporated, Dorval, Qué., Canada

ABSTRACT: In the process of extracting and concentrating iron ore from the Mont-Wright mine in Northern Québec Canada the Québec Cartier Mining (QCM) company has successfully stored and controlled over 18-20 million cubic meters of tailings and 36 million cubic meters of process water annually since 1976.

The tailings impoundment area extends over 8 square kilometres and process water basins have a total surface area of 12 square kilometres. Tailings deposition is done parallel to a 6.5 kilometres long permeable dam reaching 100 meters high.

The success of the impoundment area is largely due to the use of a particular Tailings Management Framework different from that of traditional mining operations.

This article will present the tools developed to control the impoundment area including: water balance, mass balance, design, operation, management, inspection and emergency response manuals. It will also highlight some particular challenges at the Mont-Wright tailings impoundment area.

Previously presented at the 1999 Annual General Meeting of CIM

1 INTRODUCTION & MINE DESCRIPTION

All mining companies produce waste material that must be contained, as a concentration process by-product of a particular product required by society. The Québec Cartier Mining (QCM) operation employs over 1000 people from the neighbouring mining town of Fermont, located in northern Québec, Canada with an annual production rate of sixteen (16) million metric tons of iron concentrate which will generate eighteen (18) million cubic metres of tailings.

The transport, treatment and storage of this waste represents an important challenge for the mining company. In most cases, large sums of capital are spent on the construction of appropriate facilities. However, the efficiency or effectiveness of these facilities and structures cannot be measured by profit margin as with other investments. The tailing impoundment area represents a risk for the mining company and it must be planned and managed with this in mind.

QCM and Journeaux, Bédard & Associates Inc. (JBA) have developed an

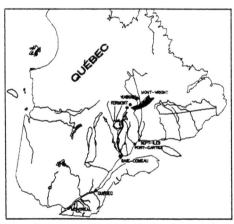

Figure 1: Location of the Mont-Wright mine, owned and operated by Québec Cartier Mining.

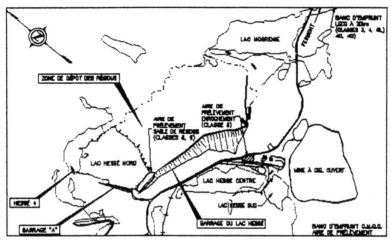

Figure 2: Schematic plan of tailings impoundment at OCM

administrative framework along with the appropriate tools to allow the use of this type of approach. Over the past four years this system has been implemented and improved upon at QCM.

2 DESCRIPTION OF THE QCM TAILINGS IMPOUNDMENT AREA

The impoundment area is comprised of a solids storage area and three water storage basins. The solids storage area of the impoundment includes separate areas for coarse, fine and slimes tailings. It is bordered to the North by a natural low mountain range, to the East and South by a 6.5 km long permeable tailings dam, with a maximum height of approximately 100 metres, and to the West by two impermeable compacted glacial till core dams, one of which measures 750 m in length and 30 m in height. During the summer, the coarse and fine portions of the pulp are separated with the coarse particles being used to raise the tailings dams. In winter the tailings are deposited from a single discharge point and flow to the West, parallel to the tailings dam.

The deposition slope runs for a distance of 4.5 km and drops 80 m before it reaches the sedimentation pond.

The three water storage areas comprise of two settling basins and one polishing basin. The first basin captures water directly adjacent to the solids storage area. Its discharge flow is controlled by a decant structure in one of the impermeable dams. The water is then directed down a 5 km long canal to the second reservoir where the water is recycled to the mill and any surplus water is treated. The third reservoir serves as a polishing basin for treated surplus water before being released to the environment.

The total average annual precipitation is approximately 980 mm. The precipitation falls as rain from May to September with significant flow during the spring melt when seven months of precipitation is liberated. All of the precipitation for a typical year flows during the five months of non-freezing conditions and enough water must be stored to ensure process water supply to the mill during the seven winter months.

On average annual precipitation in the region provides 15 000 000 m^3 more water than is required to mill and concentrate the ore by gravity spiral methods. This surplus water must be treated before its release to the environment. The treatment process is initiated only during the summer months at it's full capacity of 5 600 000 m^3/month to avoid any freezing problems of the treatment plant and to minimise operating costs. Fortunately this water must only be treated for suspended solids and excess iron concentration.

In total, 18 million m^3 of tailings solids and 17 million m^3 of water must be stored in the

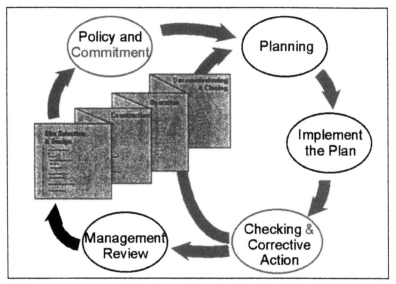

Figure 3: Application of the Framework through the Life Cycle, taken from The Mining Association of Canada, "A Guide To The Management Of Tailings Facilities" presented at Sudbury '99.

impoundment area annually. The impoundment area, including the three water storage basins, is approximately 20 km². Forty percent (40%) of this surface area is allocated to solids storage and the remaining area used for water storage is split between the 3 storage basins.

During winter, an average thickness of 5 m of transport water will freeze in the solids storage area and must be stored temporarily. Ice may build-up anywhere along the entire 5 km length of the impoundment area depending on the temperatures during the winter.

The description of the site outlined above depicts some of the constraints by which the tailings impoundment area at QCM must be managed. Unlike many smaller mines it is more practical to build the retaining structures in annual stages rather than all at once. Therefore design, construction, operation and surveillance aspects of the impoundment area are evolving all at the same time. New or changing data can be incorporated into the short term planning process such as weather, mining grade and tonnage, milling process changes, breakdowns and work force constraints.

3 MANAGEMENT ASPECTS

The work cycle starts from a series of reference documents that are not modified routinely. These documents form part of the core of the planning process and include: general long-term design strategy, environmental impact assessment (and other related documents), closure plan and any related government authorisation permits.

With these documents and a regular supply of up-to-date information the short-term planning cycle can begin. The product of the planning process is to determine parameters for four (4) principal management aspects that include design, construction (by mine personnel or by a contractor), operation and maintenance.

Two important working documents that the short-term planning process will create and maintain are:

Impoundment area fill plan

- Aerial photos - to produce topographic maps, measure volumes and deposition slopes.

87

- Mill production reports - to calculate the solids density.
- Tailing physical and chemical characteristics.

<u>Water balance</u>

- Bathymetric surveys - to produce under water contours, measure volumes, underwater deposition slopes, life span of each basin.
- Tailings pulp characteristics.
- Capacity and limits of the treatment plant.

Many tools have been created to aid in analysing and interpretation of the data and to locate trends and tendencies that can be used to form a solid understanding of the tailings behaviour inside the impoundment area.

As mentioned earlier the maintenance of these documents provides a clear view of what type of new structures and what modifications are required to existing structures to maintain the capacity of the tailings area. These working documents also provide design, construction, operating, maintenance and surveillance criteria for the future of the tailings impoundment.

Data from the short-term planning cycle is used to locate new structures and develop geotechnical reconnaissance programs, stability, seepage and risk analysis, cost evaluation and government permitting documents.

Based on construction plans and specifications the construction phase requires quality/cost control and surveillance guidelines. These documents are required whether the work is completed by mine personnel or by an outside contractor.

At QCM, the operation of the tailings management area is divided into winter deposition and summer construction modes with the mill personnel supervising the bulk of the operation. The mill operates the pumps and produces the pulp that is stored in the impoundment area. These employees have access to all the information required to perform pump start-up, shut-down and control over the fresh and recycled water basin levels. The mine department participates in the operation of the impoundment by advancing the tailings pipelines as required (little in winter but more during summer construction) and opening access roads. The mine also provides equipment and personnel for part of the summer construction.

Most structures in the tailings management area are subject to a minimum amount of maintenance except for the tailings pumps, pipelines and the treatment plant. The maintenance of these systems is split; the mine maintains the pipelines while pumps and treatment plant are maintained by the mill.

Surveillance is a very important aspect of management and ensuring on-site continuous or partial surveillance, as required, are provided by both the mine and the mill. Mill personnel provide continuous inspection of pressure and flow readings whereas the mine provides on-site surveillance. In addition, an outside consultant (JBA) also provides surveillance of the design parameters of the impoundment area.

4 ORGANISATION OF MANAGEMENT GROUP

The site description section was intended to provide information on the tailings impoundment area at QCM and some of the working characteristics of the tailings impoundment. The short term planning cycle section provides insight into the management philosophy of this tailings management area and a brief description of the method used at QCM to continually take into consideration new information from all over the mine site during the continuous evolution of the fill plan and water balance.

In this section a presentation of the actual administrative organisation that is used to ensure that all of the work outlined above is actually executed based on the QCM company objectives and then later verified and reviewed as necessary.

The impoundment area must be managed at minimum cost but also taking into account the risks associated with it's operation. QCM chose to have the tailings impoundment managed

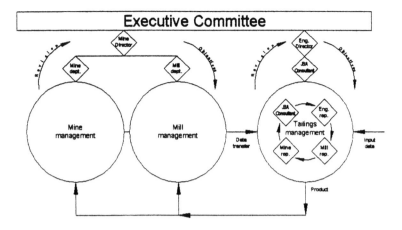

Figure 4: Organization of management group at QCM

directly under the executive committee. In this way, information is transmitted quickly, mine operation and tailings impoundment budgets are kept separate and it also allows hands-on control of the risk associated with the operation of the impoundment. Organised in this manner, specialised personnel can be better utilised in critical stages of the management process at minimum cost.

A representative from the mining department, from the mill, from direction (engineering) and an outside consultant (JBA) form a group that works under the engineering department director (see Figure 4). This group performs all of the tasks and assumes all of the responsibilities required by the short term planning cycle.

The mining department assumes all responsibilities associated with on-site surveillance, operation of the tailings lines (during summer construction and winter deposition) and maintenance of the tailings lines outside the concentrator.

The mill department oversees partial surveillance of the impoundment, treatment plant, pipeline usage and maintenance of pumps and tailings lines inside the concentrator.

QCM chose to implement an outside consulting firm (JBA) directly with the short-term planning cycle outlined earlier as project manager to co-ordinate the objectives and activities of the group. The principal reason for this choice was to ensure continuity as members of the team change over the years. All planning and associated expenses must be justified and approved by the executive committee.

As new information becomes available it is transferred to the group by the mine and mill representatives thus allowing quick action and revision of the fill plan, water balance, design and construction considerations and the surveillance program.

On a yearly basis the group meets with the executive committee to verify if global objectives have been met. If required, the process or the objectives are revised or confirmed.

5 CONCLUSION

The tailings impoundment is necessary for the operation of the mine. It must be managed on a minimum cost basis as with most other operations of the mine. Tailings impoundments expose companies to a level of risk. Because of this risk, QCM has chosen to manage the impoundment directly from the executive committee and with the direct participation as project manager of an outside consultant. In this way, the number of interventions is reduced thus allowing quicker transfer of information and facilitating the decision making process.

The system outlined in this paper has been implemented with success at QCM over the past four years.

Tailings and Mine Waste'00 © 2000 Balkema, Rotterdam, ISBN 90 5809 126 0

Instrumentation and monitoring behaviour of uranium tailings deposited in a deep pit

Y. Sheng
CSIRO Exploration and Mining, Nedlands, W.A., Australia

P. Peter
CSIRO Land and Water, Urrbrae, S.A., Australia

A. R. Milnes
Earth-Water-Life Sciences Pty Limited, Winnellie, N.T., Australia

ABSTRACT: The Ranger Uranium Mine in the Northern Territory, Australia has been depositing tailings into its 170 m deep No 1 pit since 1996. Potential slow consolidation of tailings deposited in such a deep pit has been a primary concern in terms of tailings management at the mine. In order to develop an optimum management plan for the operation, a program involving in situ instrumentation and monitoring was implemented to investigate the behaviour of tailings deposited in the pit. The present paper reports on the behaviour of tailings deposited in the pit based on the monitoring data at the early stage of the operation.

1 INTRODUCTION

In-pit tailings disposal is referred to backfill tailings into a previously mined out open pit. The concept of backfilling of tailings into a pit can be attractive. The significant environmental benefit is that there will be no risk of tailings spillage caused by breaches of the containment structure, and therefore significant environmental liability, as can often happen in above-ground tailings storage facilities (TSFs). Economic benefit may also be obtained from cost savings in the capping and decommissioning phases since smaller surface areas are generally associated with deeper pits. Due to these advantages, there is an increasing use of in-pit tailings disposal in the Australian mining industry.

Geotechnical challenges to the operation exist in the areas of selecting appropriate disposal methods, maximising the density of the tailings deposits, and minimising residual settlement and differential settlement to ensure long-term stability of capping structures. The primary concern associated with the operation is the creation of a thick slurry deposit with low density resulting from lack of drainage in pits. As a result of slow consolidation, extended periods ranging from several decades to several hundred years may be required for consolidation to advance to completion. Large settlements, resulting from incomplete consolidation under self-weight, and additional capping loads following decommissioning, would in turn have a dramatic impact on the long-term effectiveness of capping structures and hence on the behaviour and stability of the final landform.

In-pit disposal of tailings is relatively new to the Australian mining industry. At present, there is insufficient data available for a rational assessment of in-pit disposal by either operators or regulators and the need to obtain such information is an imperative (ACG 1996). Previous experiences from some mines that have adopted the in-pit disposal of tailings have demonstrated that the low density of tailings deposits could cause serious problems in decommissioning (Sheng and Peter 1999). In a typical case reported by the Australian Centre for Geomechanics (ACG 1996), the tailings surface at the centre of a pit at a gold mine was still not accessible on foot six years after the filling operation ceased and the surface level had dropped 7 m. It was estimated that further consolidation amounting an additional settlement of 18 m or so at the centre of the pit is yet to occur over many years.

ERA Ranger Mine has been operating in-pit disposal of tailings in its No 1 pit since 1996. To achieve the high standard environmental commitments set by the company, a research project was established to investigate the behaviour of tailings deposited in the pit before the commencement of the operation. The program included instrumentation (settlement measuring stations and vibrating wire piezometers), in situ testing, sampling and laboratory testing. The present paper provides a detailed description of the program and discusses the monitored results. An interpretation of the behaviour of tailings deposited in the deep pit environment is also provided.

2 DESIGN OF IN-PIT TAILINGS DISPOSAL SYSTEM AT RANGER MIME

2.1 Configuration of the Ranger #1 pit

The No #1 pit has a final depth of 170 m (between RL -148 m and RL 22 m) with respect to the surrounding land surface. The surface area of the pit is 420,000 m^2 (ERA – Ranger Mine, 1995) and the storage capacity is about 21 Mm^3 to RL 19 m. If the pit can be filled with tailings above the base of the weathered zone at RL 0.0 m, and this possibility is yet to be explored with the regulatory authorities, the final surface of tailings at the cessation of deposition would be at approximately RL 15 m and the relevant depth of tailings 157 m (from RL 15 m to RL –142 m).

The geological condition of the pit has been summarised by ERA Range Mine (1995). The Hanging Wall Sequence consists of a coarse grained quartz muscovite schist with minor magnetite. The unit is strongly foliated and is represented as micaceous clays in the uppermost 20 m but is moderately oxidised below that level. The Upper Mine Sequence consists of a fine grained quartz chlorite schist which is foliated and strongly jointed. This unit is represented as lateritic clays to 15 m depth, is variously oxidised to 40 m depth and is generally fresh and competent at greater depth.

The Footwall Sequence consists of a banded, weakly foliated granitic gneiss. The contact with the mine sequence rocks may represent a shear zone up to 10 m wide with little structural strength. The Lower Mine Sequence consists of two major individual units, namely massive chlorite rock and schists.

To ensure pit wall stability, face slopes in the pit were designed at 45° and 55° for the laterite and weathered material respectively. Final faces were scaled with a bulldozer to ensure that loose material was removed and that specified face angles were achieved to ensure continuation of wall stability. The pore water pressure in the pit wall did not create any stability problems.

2.2 In-pit tailings disposal system

The current operation of in-pit tailings disposal is based on the design by ERA Ranger Mine (1995). A schematic diagram for the disposal system is shown in Figure 1. The pit has an under-bed drainage system consisting of an underdrain and an partial side drain extended towards the northern pit wall, as shown in Figure 1. The 6 m thick underdrain consists of three layers of material: waste rock, coarse rock gravel and sieved fine sand-sized rock which is used as a filter layer. Two sets of in situ infiltration tests gave the permeability of the filter material as 4.3×10^{-6} and 5.6×10^{-6} m/s. Water seeping through the underdrain is collected in a horizontal rock-filled adit extending into the southern wall of the pit and then is pumped to the surface through a vertical bore.

Assessment of sedimentation and consolidation in the design was undertaken using 1-D large strain finite element model (Knight-Piesold 1992). It was assumed that the tailings would be deposited into the pit continuously for 15 years and that the underdrainage system would function effectively throughout the life of the operation. Consolidation properties of tailings used in the design were obtained in the laboratory on disturbed samples of tailings collected from the Ranger tailings dam in which tailings had been stored. According to the calculation, much of the space in the pit (from RL –115 m upwards) would be occupied by low density tailings, being around 1000 kg/m^3 of dry density at the end of deposition. This implies that consolidation of over a 150 m thick deposit of tailings in the pit will continue in the post deposition period. It was anticipated, according to the design, consolidation of tailings at the initial stage will benefit from the underdrain installed at the bottom of the pit, an average dry density of 1200 kg/m^3 would be achieved 8 months after the commencement of tailings deposition. It was also predicted that tailings would have an average dry density of 1210 kg/m^3 at the cessation of deposition (15 years from the commencement of deposition). The average dry density of tailings was predicted to increase to 1650 kg/m^3 5 years after the cessation of deposition.

Specific goals, in relation to tailings management and rehabilitation of tailings storage facilities at Ranger Mine, have been set up in the relevant regulation - the Uranium Mining (Environmental Controls) Act 1979 - and related authorisations to operate (Milnes 1998). The long term behaviour and stability of the capping system against residual settlement on decommissioning of the storage facility is a primary concern in tailings management at the Ranger mine. To ensure compliance with the relevant regulations, the mine has committed to a management target that the tailings in the pit will achieve an average dry density of 1200 kg/m^3 for every 20 m tailings deposited during the operational phase.

3 MONITORING BEHABIOUR OF TAILINGS DEPOSITED IN THE RANGER PIT

There has been little data available related to the physical properties of tailings deposited in deep pits. In early 1996 ERA committed to establish a research program to investigate behaviour of tailings to be deposited in its No 1 pit. As a result, a collaborative research project was established. CSIRO's Minesite Rehabilitation Research Program was requested to undertake the research project.

3.1 *Instrumentation in the pit*

Pore water pressure and settlement data provide direct information on the consolidation behaviour of tailings. Instruments including piezometers and settlement stations were designed to obtain these data.

Vibrating wire piezometers

Vibrating wire piezometers (VWPs), Geokon 4500S, were selected for monitoring pore water pressure. Four VWPs, with a pressure range of 0-345 to 0-690 kPa, were installed in the underdrain filter at two locations. Each piezometer was attached to a 200 m length of cable. The cables were extended to a monitoring station initially located at the top of the side drain. The station was later relocated to a higher elevation when the water level in the pit was

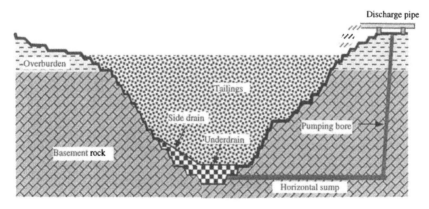

Figure 1 Designed tailings disposal system for the Ranger #1 pit.

increased. To avoid the cables being damaged when they were buried in the underdrain filter, each was individually pulled through heavy duty PVC electrical conduit with an internal diameter of 20 mm. The joints between each successive 4 m length conduit were cemented together for tensile strength using a high strength 'jointing cement'. The conduits were carefully placed at the bottom of a 0.5 m deep trench dug by a grader, as shown in Figure 2.

The plan for installing VWPs in tailings was to place an individual piezometer at different depths with progressive escalation of the tailings surface. A total of four piezometers was installed in the tailings during the early stages of the operation. The piezometers were tied to a vertical pipe, which was also used to monitor the vertical movement of tailings, installed with bottom concreted into the underdrain.

Settlement stations

To gain an appreciation of the consolidation behaviour of tailings deposited in the pit, settlement in tailings had to be effectively monitored. Two vertical movement stations were installed for this purpose. Each station consisted of a vertical shaft which was made from seamless stainless steel tube. Each extension tube was 1 metre in length and had a 1/2" BSP internal thread cut at each end to a depth 30 mm. Ring magnets were used to monitor the vertical movement of tailings. These were attached to 300 mm diameter and 5 mm thick perforated plastic plates. The use of the perforated plastic plates was intended to: a) minimise the effect of self-weight of the plate on the rate of consolidation of the tailings; and b) allow water to pass the perforations when the underlying tailings were consolidating.

When the tailings surface in the pit reached an appropriate level, a cell plate was carefully lowered over the top of the vertical shaft until it was in contact with the surface of the tailings sediments. Each plate was free to move up or down the shaft. It was anticipated that these plates would be embedded in tailings as additional tailings were deposited in the pit.

The position of each magnet was determined by lowering a torpedo shaped weight, attached to a graduated tape, inside the shaft. The graduated tape had two conductors embedded in it and these were attached to a reed switch located inside the weight. The reed switch was activated in the presence of a magnetic field indicated by an audible alarm attached to the end of the graduated tape. The accuracy of determining the position of each magnet so tested in the laboratory was +/- 3.0 mm.

3.2 *Measured pore water pressure responses*

Data provided in Figure 3 indicates that the pore pressure in the underdrain increased with continued deposition of tailings. The measured pore pressures were almost the same as those calculated for the static water pressure, which indicates that there was no excess pore water pressure in the underdrain. The phenomenon could be attributed to the fact that the side drain,

Figure 2 Installation of piezometers in the underdrain filter.

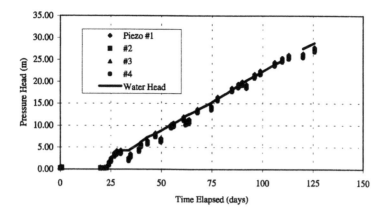

Figure 3 Measured pore water pressure in the underdrain.

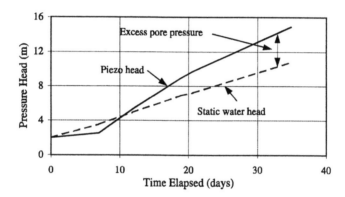

Figure 4 Measured pore pressure in the tailings.

which was connected to the main underdrain, had yet not been covered by tailings. Water entered the underdrain without passing through tailings, ie. the side drain acted as a water bypass channel during this period.

Figure 4 shows the measured pore pressure and static water heads from one of the piezometers installed in tailings. It can be clearly seen that the measured pore pressure was diverging from the static water pressure with time. This indicated that the pore pressure in the tailings was increasing with time i.e. building-up with increasing amounts of tailings deposited. The excess pore pressure head was about 4 m after 30 days of the piezometer being positioned.

3.3 *Measured settlement*

Figure 5 shows settlement of tailings at different depths within the profile for the first 15 m of tailings deposited at the location. It can be seen that greater amounts of settlement occurred with an increase in the depth of the tailings. The #1 settlement magnet was installed at the surface of the underdrain as a reference mark. Four other magnets were later placed in tailings with escalation of the tailings surface. The #2 settlement magnet was placed when the tailings depth was 3.45 m. It had settled about 0.8 m over a period of 30 days indicating a 23.2% compression of the deposit. Furthermore, greater settlement occurred as additional tailings were deposited, such as that shown for the #3 settlement magnet which settled 0.83 m over a period of 21 days. The fast rate of settlement that occurred in tailings indicated that the underdrain was performing well during the period.

95

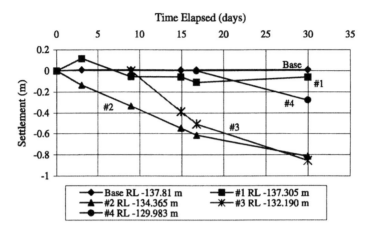

Figure 5 Measured settlement of tailings.

3.4 *Density of tailings*

To examine whether the targeted dry density of 1200 kg/m^3 was achieved, tailings were sampled from the pit twice between October and December 1996 using a portable rig developed by CSIRO. The tested density values were initially promising and ranged from 800 to 1600 kg/m^3 as shown in Figure 6. It can be seen that the dry density of tailings showed a linear increase with depth. This indicates that consolidation had proceeded well in tailings near the underdrain. The data can be divided into two categories: above and below the 1200 kg/m^3 threshold. The samples relevant to those data in the lower category were taken from the top 5 m of tailings deposited. If the dry density of these samples is plotted against their moisture content, as shown in Figure 7, the tailings should have a moisture content less than 55% in order to have a dry density of 1200 kg/m^3 or higher. The tested particle density of 2800 kg/m^3 was used in the above calculation. The average density determined from measured values was about 1280 kg/m^3. The above measured values are in agreement with the estimated value based on the bulk material which was processed and deposited into the pit. The average dry density derived by this method was about 1260 kg/m^3 in December 1997 (Woods, 1998).

3.5 *Undrained strength of tailings*

The progress of consolidation can also be assessed by determining the undrained strength of the tailings in a profile from the tailings at a particular position. The undrained strengths of tailings

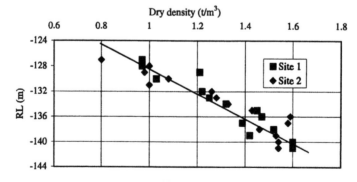

Figure 6 Density of tailings versus the Reduced Levels.

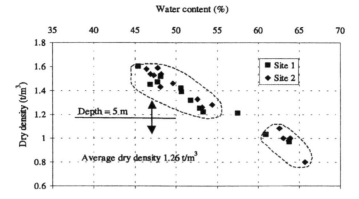

Figure 7 Dry density versus water content.

from the two sites in the #1 pit are shown in Figure 8. For comparison, an undrained strength profile for tailings from the tailings dam, which had been the prime tailings storage facility at Ranger Mine before implementation on in-pit tailings storage, is also provided. These data were derived from cone penetration resistances divided by an empirical parameter of 20 for the Ranger tailings obtained from a previous study (Sheng *et al.* 1998). Tailings in the pit showed virtually no strength in the upper 7 m of the deposits at the time when the tests were undertaken. If it is assumed that consolidation of tailings under self-weight was completed for the profile in the tailings dam, a period over 5 years could be required for the profile in the pit to achieve a similar degree of consolidation to the dam profile. The results were agreed well with those tested for density in the profiles where dry density is less than 900 kg/m^3. Therefore, we conclude that the tailings in the top 7 m of the deposit were mainly fine particles or slimes that experienced little consolidation during the period.

3.6 Tailings profile detected using geophysical acoustic techniques

Since early 1997, tailings in the pit have been consistently inundated with over 20 m water due to successive rainfall in the region. The water accumulating in parts of the mine where process water is stored has to be retained within the process water system (tailings storage facility and Pit #1) and may only be reduced in volume by evaporation. It was extremely difficult to undertake conventional geotechnical testing in the inundated pit. In an effort to map tailings deposits, marine geophysical survey techniques (sub-bottom profiling) were employed (Fugro Survey P/L 1997) to try to determine the density profile of tailings. The echo soundings were undertaken by using Pinger and Boomer instruments. With the assistance of the above-mentioned density data from geotechnical testing, a schematic density profile of tailings in the

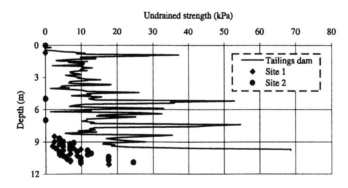

Figure 8 Undrained strength of tailings correlated from cone penetration testing.

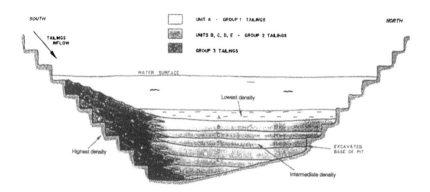

Figure 9 Schematic map of tailings deposits in Pit #1 using marine geophysics survey techniques — sub-bottom profiling (Fugro Survey P/L 1997).

pit was produced (Fugro Survey P/L 1997) and the tailings deposits were divided into three density categories: high density, moderate density and low density zones. The low density tailings were located in the shallow layers whereas the higher density zone was in the deeper layers. Zone A, shown in Figure 9, was mainly occupied by tailings that had a low density (<1.2 t/m^3). Tailings settled in the delta zone show the highest density due to the strong segregating effect.

The higher density values for tailings obtained during early deposition were largely attributed to the presence of the under drainage system. With an increasing volume of tailings in the pit, the dewatering rate decreases due to an increase in the length of the drainage path. Consequently, the volume of lower density tailings is expected to increase with time. It was found that the average dry density in the pit in January 1999 decreased to a level around 1200 kg/m^3 from around 1400 kg/m^3 at the early stage of the disposal.

3.7   Discussions on in-pit tailings disposal operation at Ranger Mine.

Beaching and settlement behaviour

Based on Boomer records (Fugro Survey P/L 1997), a submerged fan formed by coarser particles below a discharging point had a slope of 0.105 dipping towards the centre of the pit. The length of the slope was about 145 m with about 58 m being covered by low density slimes, similar to that shown in Figure 9. This submerged beaching behaviour is significantly different from that for sub-aerially deposited tailings in above-ground tailings dams. In the latter case, the beach slopes are normally less than 0.03 for hard rock tailings and 0.015 for bauxite tailings (Blight 1987; Fell *et al.* 1992).

The observed beaching slope in Ranger Pit #1 indicates that there exists significant segregation between coarser and finer particles. This is due mainly to the influence of kinetic energy due to the stream of tailings cascading down the pit walls. As a relatively small area existed in the pit, particularly at the early stage of the operation, finer particles tended to settle very slowly due to disturbance of the slime current. The above-observed behaviour is consistent with the monitored data described in the previous sections. The thick, mobile slime current also made instrumentation and monitoring extremely difficult in tailings because instrument stations were later damaged as the increasing mass of tailings moved.

Disposal methods

With continuing deposition of tailings, more and more fine particles accumulated with a potential to build up a thick deposit of suspended slimes moving around within the pit. These slimes would eventually settle in the centre of the pit. The possible worst scenario, consisting

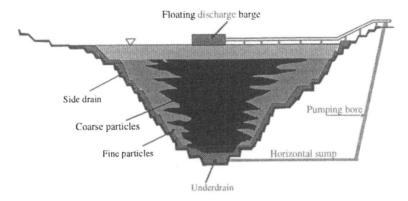

Figure 10 A conceptual segregating central disposal method.

of a very thick mass of slimes in the central parts of the pit, would have a great impact on decommissioning and rehabilitation of the pit after the cessation of operation by significantly extending the time of consolidation. Therefore, the peripheral disposal method is clearly not suitable for disposal of tailings into deep pits from the geotechnical point of view. An attempt to create a tailings deposit with minimum segregation was made in the design of the in-pit disposal system in Deilmann pit at the Key Lake uranium project in Northern Saskatchewan, Canada, where a central disposal system was designed (Mittal and Holl 1996). This approach has been extensively discussed at Ranger Mine and several tentative designs have been proposed. The conceptual central disposal system as shown schematically in Figure 10 (Sheng and Peter 1999) was among others. The proposed system is somewhat similar to that used in the Deilmann pit. Instead of creating a uniform deposit of tailings, segregation between fine and coarse particles is encouraged to allow the central part of the pit to be occupied by coarser tailings. Hydrocycloning techniques can also be integrated to enhance segregation. In this way, it is anticipated that a thick deposit formed by coarser particles would be built up in the central area of the pit, while fine tailings settle in areas near the pit walls where the deposits are built up on benches and are corresponding thinner.

An alternative approach, also being discussed, is to use paste technologies to produce a high density material that could be deposited uniformly across the pit, even under water. The possibility of using paste technologies is being evaluated at the mine.

4 SUMMARY

Based on the outcomes of the research presented in this paper, the instrumentation and monitoring of the behaviour of tailings deposited in a deep pit can be very difficult. Reliable instrumentation techniques are yet to be developed to allow instruments to sustain large displacements by mobile, thixotropic tailings currents. From the limited data monitored, it is concluded that the settlement and consolidation behaviour of tailings deposited in a deep pit would be significantly different from those deposited in conventional above-ground tailings dams. Strong segregation has occurred in Pit #1 at Ranger Mine. The tendency to form a low density deposit in the central part of the pit could result in large residual settlement problem and require a unacceptable time period for consolidation to be completed. We conclude that conventional hydraulic disposal along the periphery of the pit is not suitable for the deep pit situation from a consolidation point of view.

ACKNOWLEDGMENTS

The work reported forms part of a joint research project between Energy Resources of Australia Ltd (ERA) and the CSIRO Minesite Rehabilitation Research Program. Thanks are due to ERA for permission to publish the paper. Special thanks go to Mr Andrew Nefiodovas, Mr Martin Wright and Mr Geff Cramb for their effort in field and laboratory testing, instrumentation and monitoring.

REFERENCES

ACG. 1996. In-pit fill research news, *Australian Centre for Geomechanics*. Vol. 5.

Blight, G. E. 1987. The concept of the master profile for tailings dams. *Proc. of the Int. Conf. on Mining and Industrial Waste Management*, Johannesburg, 95-100.

ERA – Ranger Mine. 1995. Application for Approval – Deposition of neutralised tailings in Ranger #1 pit.

Fell, R., MacGregor, P. & Stapledon, D. 1992. *Geotechnical Engineering of Embankment Dams*. Blkema, Rotterdam.

Fugro Survey P/L. 1997. Echo sounder, pinger and boomer surveys in Fanger No 1 pit and tailings dam, Report to ERA Environmental Services Pty Ltd, Fugro Survey Pty Ltd. HY13261.

Knight-Piesold. 1992. Preliminary design for in-pit tailings disposal. Report to ERA Ranger Uranium Mines Pty Ltd, December 1992. Ref. 518/8.

Milnes, A. R. 1998. Ranger mine closure strategies for tailings disposal, *Seminar on Tailings Management for Decision Makers Organised by Australian Centre for Geomechanics*, 30 April – 1 May, Perth, Australia.

Mittal, H. K. & Holl, N. 1996. Management of fine tailings. *Proc. of 3rd Int. Conf. on Tailings and Mine Waste '96*, Balkema, 91-100.

Sheng, Y., Peter, P., Wei, Y. and Milnes, A. 1998. Capping of an extremely soft neutralised uranium tailings - A case history in environmental geomechanics, *Proc. of 4th Int. Conf. on Case Histories in Geotechnical Engineering*, March, 1998, Missouri, USA.

Sheng, Y. & Peter, P. 1999. Research priorities for management of in-pit tailings disposal at Ranger Mine. Research Report to ERA Environmental Services Pty Ltd. CSIRO Exploration and Mining, Report No. 608C.

Woods, P. 1998. Personal communication.

Tailings and Mine Waste'00 © 2000 Balkema, Rotterdam, ISBN 90 5809 126 0

Three-dimensional modeling for waste minimization and mine reclamation

C.H. Dunning & J.R. Arnold
Knight Piésold and Company, Denver, Colo., USA

ABSTRACT: Understanding the physical geometry of mine workings, tailings, waste dumps, and their relationship with the physical environment is key to beginning the process of waste minimization and mine reclamation. Building three-dimensional (3-D) computer models that accurately represent the components of surface and underground mine workings is a logical step toward creating comprehensive waste management and reclamation plans.

Visualization also is a very powerful tool for analysis, communication, and public education. Without the use of 3-D modeling, it can be very difficult to convey a sequence of events to laypersons. These types of presentations are used with clients, regulators, the public, and in court cases with great success. Various examples of recent modeling efforts are presented to illustrate how this process can form an invaluable tool to the overall design and remediation processes.

1 REPORT

To build an understanding of relationships that exist in a 3-D environment using computer tools, a model must be constructed. Components that contribute to a model can include topography, underground surveys, analytical sample points, geology, and hydrology. All components can vary through time and may be sampled more than once.

Constructing animations to assist in visualization is not new. By using recent advances in computer hardware and software combined with better survey data, the precision and usefulness of visualization has increased to the point where it is an engineering tool as much as it is a communications tool.

Construction of 3-D computer models can be fairly simple or quite arduous depending on the quality of the 3-D surveys available. Construction techniques include using Computer Aided Design (CAD) software which is 3-D capable. Data must then be compiled into 3-D vectors and points. CAD can then be used to extrude solid bodies or 3-D faces along the vectors and point data. Solids and 3-D faces are required to render scenes.

Once model construction is complete, the analysis of relative geometries can begin. Using a true scale model rather than a cartoon or relative representation yields insights and correlations that otherwise are not apparent. Supplemental designs and calculations can be incorporated in precise positions when true scale is maintained.

Figure 1 shows one frame in an animated example of the model used during the reclamation and treatment of an abandoned underground mine's wastewater. The 3-D correlations of old workings could not be made without the use of the model. Some old workings could not be entered safely, and hand drawn figures (circa 1904) were used to generate their geometry. Unmapped

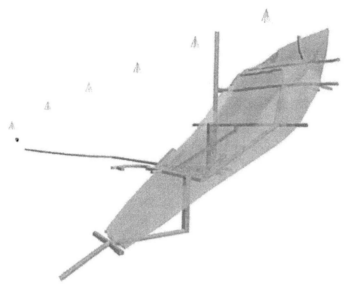

Figure 1.

Figure 2.

connections between levels which contain discharge points were interpolated directly by using the 3-D model. The isolation of discharge points was integral in helping design a passive treatment system that could handle the composite outflow that varies seasonally.

In Figure 2, the 3-D model and animated visualization were used as evidence in litigation showing a reclamation project by a contractor. Excavation of an old tailings impoundment was to

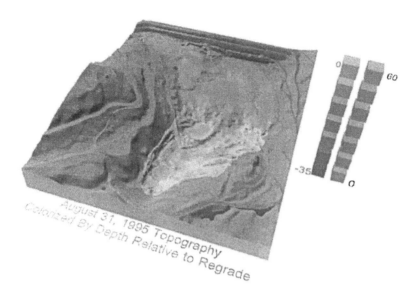

Figure 3.

convert the impoundment to water storage. The contractor over excavated 30 feet into preexisting topsoil. Demonstrating the sequence of events and the relative elevations of original topography and excavations through time to a judicial audience requires a self-explanatory graphic that reflects changes through time. An engineered graphic that is geometrically and technically accurate greatly enhances communications to a lay jury or judge. The judgment was a favorable one.

Figure 3 depicts a surface mine design where the waste is being used to reclaim the mine highwall concurrent with mining operations. This design is best understood and analyzed for practicality using a computer model for exemplification. In this case, visibility from certain populated areas was a key concern for permitting, and the mining method was design to both minimize waste outside the pit area but also to limit the visibility of the operations. Mine planners benefit from the opportunity to see the mine develop through time, and regulatory agency concerns can be addressed much more directly.

The model can give a view of the mine from multiple perspectives, which allows the audience to determine what the visual impacts will be from their homes before any surface operations start.

CONCLUSION

Mine planning and mine reclamation can become more efficient when first modeled on a computer. Fully integrated computer models can be analyzed through the life of a facility and can be used for practical analysis and communications. When attempting to convey a complex and technical scenario to a non-technical audience, visualization of the model then becomes paramount. By utilizing visualization techniques including transparency, material mapping, and animation, we can determine and convey the best methods for remediation and waste minimization to all parties concerned.

Geotechnical considerations

Tailings and Mine Waste'00 © *2000 Balkema, Rotterdam, ISBN 90 5809 126 0*

Settling of tailings

G.Ter-Stepanian
Armenian Academy of Sciences, Armenia

ABSTRACT: Numerous economic and environmental problems of mine tailings management follow from impossibility to settle tailings in form of a sediment able to carry a load. Discovery of suspension force made it possible to settle solid particles in suspensions by use of the related and acting in opposite direction seepage force. Economic and environmental advantages of settling of tailings are shown and its effect on future development of mine industry is briefly outlined. The first fiddle will play the protection of the environment from chemical and radioactive contamination.

1 INTRODUCTION

Numerous economic and environmental problems of mine tailings management, construction of tailings dams, restoration and rehabilitation of tailings impoundment and protection of the environment from the chemical and radioactive contamination follow from the impossibility to settle tailings in form of a sediment able to carry a load. Its economic and ecological consequences are well known.

This peculiarity is considered usually as an immanent property of tailings. However such a statement was never proved scientifically. This conclusion is based on ignorance of mechanism of tailings, and arose from the failure of attempts to settle solid particles by gravity or separation from liquid by filtration, centrifuging, etc. Discovery of suspension force (Ter-Stepanian 1998) made it possible to explain the mechanism of tailings and to propose a method for their settling .

2 SUSPENSION FORCE

Suspending is holding of solid particles in a liquid without attachment. Solid particle sink in suspensions. Suspension force arises as a result of friction formed between sinking solid particles and immovable water. This volumetric force is acting upward that is in opposite direction to movement observed. The suspension force J equals to the difference between unit weights of the suspension γ_m and of the water γ_w, $J = \gamma_m - \gamma_w$. Thus, the magnitude of the suspension force depends on the concentration of suspension.

The solid particles in a liquid experience the Archimedean or hydrostatic uplift. The static (submerged) unit weight of solid particles in water is equal to $\gamma_{st} = \gamma_s - \gamma_w$, where γ_s is the unit weight of solid particles. The hydrostatic uplift has a constant value.

Being directed upward the suspension force J decreases the static unit weight γ_{st} of solid particles, and makes it equal to the dynamic unit weight γ_{dy}, $\gamma_{dy} = \gamma_{st} - J$. Thus the hydrodynamic uplift is formed. Under braking action of the suspension force the solid particles sink slowly. The sinking of solid particles decreases the concentration in upper layers and increases that in the lower ones. Supernatant is formed in upper layers and a more concentrated suspension in the lower ones. The magnitude and the distribution of the hydrodynamic uplift changes correspondingly. At sufficiently high concentration of suspensions the dynamic unit weight of solid particles may decrease to zero. The solid particles become weightless, their sinking ceases, the suspension transits into the critical state, and remains in this state for indeterminate long time.

The suspension force J may be destroyed by the seepage force j acting in the opposite direction. On destroying the suspension force the dynamic unit weight of solid particles increases and settling of the suspension occurs rather quickly.

Depending on physical properties of suspensions that is on type of rocks, degree of grinding, and method of benefaction, the mine tailings are distinguished by their mineralogical content, size (sand-, silt-, or clay-size) and shape (angular, scale-like, or rounded) of particles, concentration, etc. Correspondingly the basic method of application of the seepage force for destroying the suspension force is supplemented with a number of auxiliary methods aimed at decreasing of the suspension force, increasing the seepage force, destroying the loose structure of sediments formed, removing this sediment, and so on. (Ter-Stepanian, in press). The patent is pending.

3 EFFECT OF SETTLING THE TAILINGS

The invention of the method for settling of mine tailings by use of seepage force without application of chemicals and special treatment will affect considerably the mining industry.

Economic effect of new technology for settling the tailings will be enormous. Settling plants may be located at concentration mills and form a common complex. The sediment obtained may be used as construction, road, and backfill material. The water and soluble reagents used for concentration of ores will be recycled.

Demand for new territories for tailings ponds, construction of tailings dams, their management, construction of pipelines, thickening of tailings, and so on will be eliminated. The threat of bursting of tailings dams, casualties and damages from flooding will disappear.

Still greater will be the ecological effect of settling the tailings. Each tailing pond forms a long tail of constant contamination along the seepage way. The menace of growing tailings ponds and thus of the sources of chemical and radioactive contamination concerns the public opinion today and will be intolerable tomorrow. The recent doubtful tolerance toward the existing situation is only apparent, and is explained by the alternative existing at present: either tailings ponds or life without metals. Naturally this circumstance suppresses the criticism addressed to mining industry.

4 FUTURE DEVELOPMENTS

The public opinion is prepared psychologically to changes and long-expected information about new technology in mine tailings management will be welcomed. Ecological community will be more active and demands for realization of possible changes will grow. Some of future developments can be predicted easily.

First will be the recession in orders for obsolete equipment, as thickeners of tailings and transportation facilities and many orders will be canceled. Second will be demand for new equipment for settling of tailings and utilization of sediments. But the equipment for the new technology does not exist even in drawings, only principal schemes are shown in publications.

Those prescient firms that will start earlier to work out and manufacture the new equipment for settling of the mine tailings can gain the highest profit because many constructive solutions will be found that can be patented. This field of innovations is quite vacant at present.

5 CHOICE OF METHOD FOR PARTICULAR SUSPENSIONS

As shown above depending on physical properties of particular tailings the basic method of use the seepage force should be supplemented by diverse auxiliary methods. This work may be likened to proportioning of concrete or determination of optimum water content of material for compacted fills. Special organizations as university laboratories must determine the optimum regime of settling, that is the use of open or closed type settlers, required concentration of the suspension, need for vibration or vacuuming, thickness of protective layer of sediments to be left on filters and their permissible height, and so on.

Besides the mode of utilization of sediment obtained should be determined. This is a vast field of activity. Probably the sediment obtained will be used mainly as backfill material in the underground mining, while it will find application as building and road material in the open cast mining.

6 UTILIZATION OF EXISTING TAILING PONDS

The existing tailing ponds being a serious source of contamination of the surface and groundwater should be liquidated. The most correct way will be the reverse transfer of tailings from ponds to settlers using the existing pipelines and installing pumps. Contaminated earth formed on the bottom of tailings ponds and material of the tailings dams should be also extracted and directed for purification and utilization. These operations will reduce the contamination of the surface and ground water.

The water of tailings ponds can be recycled and used as circulating water. That will decrease considerably the need in fresh water. The soluble mill reagents of this water may find use after recycling too.

When dealing with abandoned, stagnant tailing ponds due account must be taken to the possibility that a natural separation of minerals by gravity could have been occurred in ponds. The following explains the possible mechanism of this separation of tailings. Since the suspension force acts equally on all solid particles, light minerals at quartz, clay minerals, calcite, etc. could became weightless while heavy minerals as hematite, magnetite and other combinations of heavy metals could have kept sufficient dynamic unit weight and continue to settle. Then their content in lower layers should be higher that that in the upper ones. Deriving carefully tailings from lower layers and reworking them with use of new technology, it will be possible to use advantageously the natural process of separation, and extract additional quantities of concentrates.

Thus management of tailings from a major debit item will turn unto a major credit item. However the main advantage of settling of tailings is its environmental aspect.

7 ENVIRONMENTAL PROBLEMS

The stimulating reason for concern is the constantly increasing pollution and contamination of environment. There is no difference in this respect between the acting, still accreting tailings ponds and the abandoned, stagnant ones. Contamination occurs not in vicinity of tailings ponds only, but along the whole seepage way of groundwater until it enters into the streams, and continues further enveloping the biota. Even if we succeed to eliminate all tailings, the contaminated groundwater will continue a long time to enter into streams. Therefore the work should be done without any delay. A comprehensive review of the situation in the mining industry was done by Nancarrow et al. (1989).

The mankind is in great debt as regards Nature, the environment. The pollution of the Earth increases everywhere. Only a few developed countries, USA included carry on a real struggle against pollution. The successes achieved in this country as in cleaning of the Boston Bay area are well known. Forced pollution and contamination of the environment by mine tailings disposal made this field lagging behind. It must be assumed that the situation will be changed in the near future.

Benevolent organizations spreading information and knowledge concerning ecological problems can render help to the Environmental Protection Agency in her noble work. This information is important especially for the reason that at the present mainly individuals are interested in the problems of protection of natural environment, although it should be of interest for everybody. Protection of and benevolence to Nature are no less important than that of the people. Such a work should be started immediately and exactly in the United States of America, the most advanced country of the world.

REFERENCES

Nancarrow, R. & Amrault, J. 1989. Environmental audits for mining industry. *Proc.,Intern. Symp. on Tailings and Effluent Management*, Halifax, Canada, 1989, pp. 119-128
Ter-Stepanian, G. 1998. Suspension force induced landslides. *Proc., 8th Intern. Congress, Intern. Assoc. for Eng. Geol. and the Environment,* Vancouver, Canada, 1998, v. 3, pp. 1905-1912
Ter-Stepanian, G. (in press). Enigmatic features and mechanism of mine tailings.

Tailings and Mine Waste'00 © 2000 Balkema, Rotterdam, ISBN 90 5809 126 0

Prediction of long-term settlement for uranium tailings impoundments, Gas Hills, Wyoming

D. B. Durkee
Colorado State University, Fort Collins, Colo., USA

D. D. Overton & K. C. Chao
Shepherd Miller Incorporated, Fort Collins, Colo., USA

T. E. Geick
Umetco Minerals Corporation, Grand Junction, Colo., USA

ABSTRACT: Settlement analyses are typically performed on tailings impoundments for the purpose of reclamation design. The predicted settlements are used to analyze cover performance such as clay cover cracking potential or the development of flow concentration zones in riprap erosion protection covers. This paper presents the results of settlement analyses performed for Umetco Minerals Corporation (Umetco) on two separate uranium tailings impoundments at their Gas Hills Wyoming location. These impoundments were constructed using different methods and are at different stages of reclamation construction.

Settlement at the previously reclaimed Above-Grade Inactive tailings impoundment at Umetco's Gas Hills, Wyoming uranium mine and millsite was analyzed using a one-dimensional finite difference technique for predicting time-rate consolidation. Settlement was predicted at various points of known soil conditions and was used to develop settlement contours for evaluating differential settlement. A two-dimensional finite element seepage model (SEEP/W) was also used to predict time rate of pore pressure dissipation.

Settlement at the A-9 Below-Grade Repository at Umetco's Gas Hills uranium mine and millsite was analyzed using coupled two-dimensional finite element seepage and stress-strain models (SEEP/W-SIGMA/W). The model was used to predict long-term settlement due to construction of the proposed cover system. Cover cracking potential was evaluated directly from the two-dimensional profile.

1 INTRODUCTION

This paper presents the results of long-term settlement analyses that were conducted on two uranium tailings impoundments located in Gas Hills, Wyoming, and are operated by Umetco Minerals Corporation of Grand Junction, Colorado. Each of the impoundments has been nonoperational for several years and are at different stages of reclamation. The purpose of the settlement analyses presented in this paper was to evaluate the effects of placing riprap on the previously constructed cover. Specifically the cover cracking potential and the development of surface water concentration zones due to differential settlement were evaluated. If surface water concentration zones were to develop, it would result in larger riprap erosion protection.

2 ABOVE-GRADE INACTIVE TAILINGS IMPOUNDMENT

The above-grade inactive tailings impoundment at Umetco's Gas Hills uranium mine was operated between 1960 and 1979. The original impoundment was built in 1960 by constructing an earth dam across a gully that drains through the site. The dam was constructed from locally available silty clayey sands. The 1960 impoundment was enlarged between 1969 and 1974 by the construction of several earth-filled dams (1969, 1972, and 1974 impoundments) in a terraced configuration downstream (north) and east of the original impoundment. The 1969 impoundment was constructed east of and adjoining the 1960 impoundment. The 1960 and 1969 impoundments reached capacity by 1972 at which time the 1972 impoundment was constructed north of the 1960 impoundment. The third and final addition to the tailings impoundment was completed in 1974 east of the 1972 and 1969 impoundments. The tailings pond configuration before closure is shown in Figure 1.

The tailings impoundments reached capacity in 1979 and were reclaimed between 1985 and 1992. Reclamation was performed in accordance with the NRC approved reclamation plan (D'Appolonia, 1980). The constructed reclamation plan consisted of regrading the pile to a 10:1 slope or flatter and constructing an earth fill cover. The cover consists of 1 foot of compacted clay radon barrier, 1 foot of filter material on top of the compacted clay, and 7.5 feet of cover material to provide erosion protection and frost protection. The surface of the reclaimed impoundment was to be vegetated for erosion protection.

Figure 2 shows the final surface of the reclaimed impoundment. Zones 1 through 4 in Figure 2 represent the areas of thickest slimes deposits, determined from piezocone tests previously conducted by Water Waste and Land (1984). Settlement monitoring points, SP-1 through SP-4 were installed in Zones 1 through 4, respectively. Settlement data has been collected at the site since June 1989. In addition, a number of standpipe and pneumatic piezometers shown in Figure 2 have been installed in the impoundment since 1984. Shepherd Miller installed strings of pneumatic piezometers at locations BH-1/P-1 through BH-6/P-6.

NRC and Umetco reviewed the previously approved reclamation plan (D'Appolonia, 1980) for effectiveness in 1996. The NRC identified surface erosion on the cover and determined that the approved vegetative cover lacked adequate vegetation due to the climatic conditions at the Gas Hills site and was not providing the necessary erosion protection. Thus, Umetco proposed an enhanced reclamation plan that included armoring the cover with riprap to provide erosion protection throughout the design life.

2.1 *Settlement Analysis*

The settlement analysis was divided into two parts: Part 1 from the end of tailings deposition to January 1997 (time of analyses); and Part 2 from January 1997 to the end of settlement. Part 1 consisted of settlement predictions on soft saturated tailings. The analyses were conducted using a finite difference solution to Terzaghi's classical consolidation equation. Due to potential

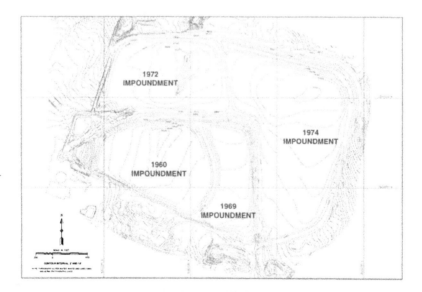

Figure 1 Tailings Pond Configuration Prior to Closure

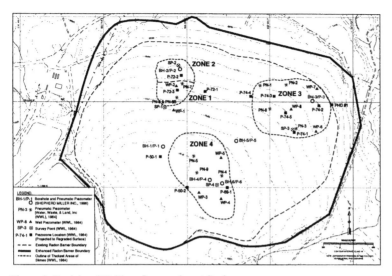

Figure 2 Reclaimed Tailings Impoundment Surface.

large strains it was necessary to account for permeability changes that occur during consolidation. The large strains were accounted for by updating the coefficient of compressibility using hydraulic conductivity-void ratio functions developed from laboratory tests on remolded tailings and compressibility-stress functions developed from laboratory consolidation tests. In addition, the Part 1 modeling results allowed calibration of the model to the recorded settlement and piezometer data for predicting future settlement.

Part 2 of the analyses was conducted by continuing the analyses until the pore pressure dissipated to zero under the final loading conditions. In addition, an independent calculation of future settlement was performed using data from field sampling and pore pressure data from pneumatic piezometer strings, P-1 through P-6. The settlement analyses were performed to develop settlement contours in the areas of thickest slimes deposits and determine the potential for

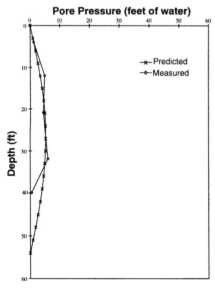

Figure 3 Measured and Predicted Pore
Pressures, December 1997, Zone 4

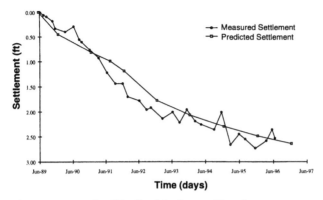

Figure 4 Measured and Predicted Settlement, Zone 4

cover cracking in the radon barrier, and to determine if flow concentration zones would develop that would affect riprap sizing.

2.2 *Settlement Analysis Results*

Typical results from Part 1 of the analyses are presented in Figures 3 and 4. Predicted and measured pore pressure distributions in Zone 4 (the thickest tailings deposit) from December 1997 are shown in Figure 3. The predicted and measured settlement in Zone 4 is shown in Figure 4. The close agreement between the predicted and measured pore pressure and settlement indicates that the model is calibrated to the field soils and that the future settlement can be predicted with confidence.

The future settlement based on the model is shown in Figure 5. Figure 5 also shows the measured and predicted settlement from June 1989 to present. These results indicate that the tailings in Zone 4 had consolidated approximately 93 percent by January 1994 and that less than

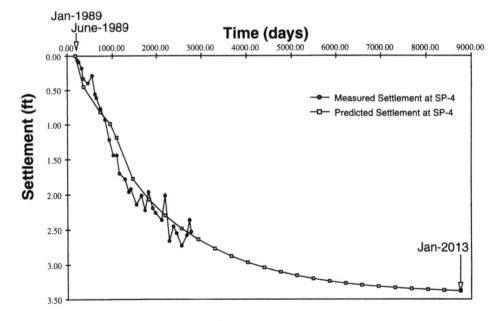

Figure 5 Measured Settlement and Predicted Future Settlement, Zone 4

114

Table 1 Summary of Predicted Settlement, Above-Grade Inactive Impoundment

Zone	Total Settlement Based on the Model (ft)	Future Settlement Based on Model (ft)	Future Settlement Based on Current Conditions (ft)	Percentage of Total Settlement Completed	
				Based on Model (%)	Based on Current Conditions (%)
1	9.3	0.79	0.003	91	100
2	4.2	0.08	0.005	98	100
3	2.5	0.53	0.47	80	88
4	6.7	0.58	0.29	93	96

1 foot of additional settlement will occur at that location. The results from Zones 1, 2, and 3 showed similar agreement with measured data in those areas.

The results of future settlement calculations using field measured pore pressures and consolidation parameters from recent laboratory testing agree with the model predictions. Results of the future settlement predictions from both methods are presented in Table 1.

The difference in Zone 1 between the settlement predicted by the model and that based on measured data most likely results because of the soil profile that was used. The profile used for the model was obtained from piezocone log P-72-3 and the soil profile for the measured data was obtained from B-1 outside the area of thick slimes deposits. The predicted settlements from the other three zones show reasonably close agreement based on the two methods.

The calculated settlement was used to develop a settlement versus tailings thickness relationship for various soil types. These relationships were then applied to an isopach of compressible tailings thickness to determine contours of settlement of the slime areas. The predicted settlement contours are shown on Figure 6.

2.3 *Seepage Analysis*

An independent seepage analysis was performed on the Above-Grade Inactive Impoundment. The analysis was performed using a finite element seepage model (SEEP/W) and initial condi-

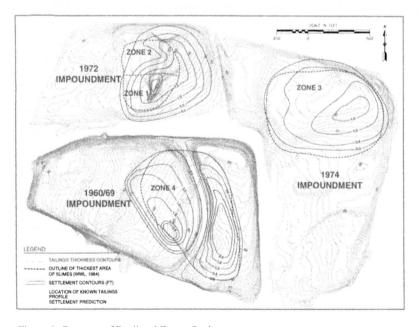

Figure 6 Contours of Predicted Future Settlement

tions based on the current pore pressure measurements. The results of the seepage model are consistent with the results of the settlement and consolidation modeling.

3.0   A-9 REPOSITORY

Between 1950 and 1984 Union Carbide developed several open pits, both inside and outside of the restricted area at their Gas Hills, Wyoming uranium mining and mill site. In 1979 after the above-grade tailings impoundment had reached capacity, the A-9 pit was converted for use as a below grade tailings repository.

The bottom of the A-9 pit was graded to a 3 percent slope toward the south end and a 1 percent slope toward the centerline. A compacted clay liner approximately three feet thick was constructed on the bottom of the pit and an underdrain system was constructed on top of the liner.

The underdrain system consists of a perforated drainage pipe and sand drainage blanket along the centerline of the repository that connects to a solid pipe and continues south to a gravel collection sump. The sand drainage blanket provided a level surface for operation of the spigot line. A filter blanket and perforated pipe were also constructed on the south end of the repository and extend approximately 300 feet westward from the gravel collection sump. The second filter blanket was constructed to provide a beachhead for initial tailings disposal and an operating area large enough for future lifts of coarse sand material around the gravel collection sump. In addition, it provided control of the phreatic surface on the south end, against the pit wall where seepage was of concern.

Operations were to consist of placing tailings through a spigot from the top of the constructed sand drainage blanket, toward the south end and the east and west pit walls of the repository. The original design planned for the slimes to settle out along the unlined pit walls and to provide a low permeability layer.

Approximately 1,580,000 cubic yards of tailings from the mill operations were deposited into the repository between December 1979 and December 1984 at which time operations were discontinued. In addition to the Umetco tailings, imported wastes from sites in the surrounding area were deposited in the A-9 repository after 1984. The additional waste consists primarily of 1,793,801 cubic yards of Title I tailings that were transported from the former Susquehanna Western Mill site in Riverton, Wyoming. Approximately 110,000 cubic yards of material from various other sites were also placed in the repository since 1984. The additional materials were placed and compacted in accordance with standard engineering procedures.

As was the case with the above-grade impoundment the approved reclamation plan for the A-9 repository included a vegetated cover for erosion protection. Since it was determined by the NRC that vegetated covers do not provide adequate erosion protection at the Gas Hills site, Umetco proposed to enhance the previously approved plan with a riprap cover. The previously approved plan also included the criterion that reclamation construction at the A-9 repository was not to begin if the tailings were still settling. The settlement analysis was performed to demonstrate that additional settlement from the current loads and settlement resulting from cover construction would not result in cracking of the radon barrier, or result in flow concentration zones that would require larger riprap.

3.1 *Settlement Analysis*

Three modes of settlement were evaluated for this investigation, immediate settlement, consolidation settlement, and long-term settlement. The analyses were conducted on the cross sections shown in Figure 7. The differential settlement, which is used to evaluate cover cracking potential, was evaluated along each cross section.

3.1.1 *Immediate Settlement*
Immediate settlement due to placement of the cover was evaluated using the two-dimensional finite element program SIGMA/W. The soils were assumed to be elastic and a two-dimensional plane strain model was used for the analyses.

3.1.2 *Consolidation Settlement*

The evaluation of consolidation settlement was performed by coupling computer programs SEEP/W and SIGMA/W. SIGMA/W computes displacements and stresses while SEEP/W computes the changes in pore-water pressure with time in a coupled consolidation analysis. Considering the soil properties as well as the stress range applied in the field, a nonlinear elastic-plastic soil model was used to predict the consolidation settlement due to the dissipation of pore pressure in the saturated tailings. A linear elastic soil model was used to model the other materials.

Initial pore pressure conditions in the repository were determined from existing pneumatic piezometers installed by Umetco, strings of pneumatic piezometers installed by Shepherd Miller, and areas of ponded water determined from aerial photographs. Figure 7 shows a plan view of the repository and the locations of field instrumentation. Aerial photographs were also used with the boring logs from this and previous investigations to determine the distribution of soils in the repository. Based on the soil profiles and the pore pressure distributions in the repository the critical area in terms of future settlement was determined to be the southwestern area. The northern half of the repository consists of sand and sandy slimes that are fully drained. The southeastern area contains slimes but is not as deep as the southwestern area. The soil profile and pore pressure data from the southwestern area are shown in Figure 8. The pore pressure distribution measured from the pneumatic piezometers is shown in Figure 9 This pore pressure distribution was used as the initial condition for the seepage-deformation analysis.

Stress strain parameters and hydraulic parameters were obtained from laboratory triaxial tests, flexible walled permeability tests, and water retention tests conducted on samples collected during the installation of pneumatic piezometers. The deformation model was calibrated using the results from the laboratory triaxial tests.

3.1.3 *Long-Term Settlement*

In order to extrapolate the strain data to predict the maximum creep strain it was assumed that the creep strain data will fit a hyperbola of the form:

$$\varepsilon_c = \frac{t}{a + bt}$$

where:
   t = time since creep began, and
   a and b = hyperbolic parameters.

From the hyperbolic creep strain equation a plot of t/ε_c as a function of t would be of the form:

$$\frac{t}{\varepsilon_c} = a + bt$$

This equation shows that t/ε_c is a linear function of t if the data fits a hyperbolic function. From consolidation tests the strain data after 90 percent of consolidation occurred were plotted in the above form, and the hyperbolic parameters a and b were determined as the intercept and the slope, respectively, of the linear function. In this way, the creep strain as a function of time was predicted by the hyperbolic creep strain equation. The maximum creep strain approaches a value of $1/b$ as time approaches infinity.

3.2 *Results of Settlement Analyses*

The immediate settlement, the consolidation settlement, and the long-term settlement along each of the four cross sections analyzed are shown in Table 2. The immediate settlement due to placement of the cover at the location with the maximum total settlement is 0.19 feet. The consolidation settlement was modeled by coupled pore-pressure and deformation models along each of the four cross-sections. The consolidation settlement at the location with the maximum total settlement was 1.60 feet, and the results of all four cross sections are shown in Table 2. Long-term settlement was calculated using a hyperbolic model of creep data. The long-term

117

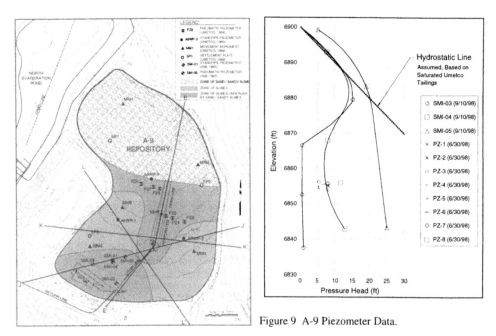

Figure 9 A-9 Piezometer Data.

Figure 7 Site Plan of A-9 Repository

settlement at the location with the maximum total settlement was 0.83 feet, and the results for all four sections are shown on Table 2. The total settlement profiles from each cross-section were used to develop contours of total settlement for the A-9 repository as shown on Figure 10.

4.0 SUMMARY

Long-term settlement analyses were conducted on two uranium tailings impoundments to verify stability of the cover systems. These two tailings impoundments, the Above-Grade Inactive Tailings Impoundment and the A-9 Repository are located in Gas Hills, Wyoming, and are op-

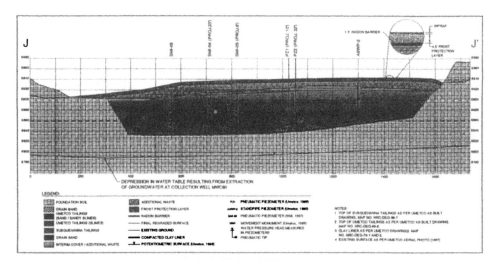

Figure 8 Cross-Section J-J´ on A-9 Impoundment

118

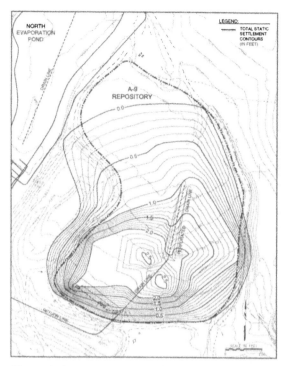

Figure 10 Contours of Predicted Total Settlement for A-9 Repository

Table 2 Summary of Predicted Settlement, A-9 Repository

Cross Section	Immediate Settlement (ft)	Consolidation Settlement (ft)	Long-Term Settlement (ft)	Maximum Total Settlement (ft.)
E-E'	0.21	1.32	0.86	2.39
J-J'	0.19	1.38	0.83	2.40
K-K'	0.20	1.46	0.84	2.50
L-L'	0.19	1.60	0.83	2.62

erated by Umetco Minerals Corporation of Grand Junction, Colorado. Each of the impoundments is currently at different stages of reclamation. The long-term settlement analyses were performed to determine the effects of completing reclamation and the potential for cover cracking, and to determine riprap sizing.

Settlement analyses at the Above-Grade Inactive Tailings Impoundment was performed using a one-dimensional finite difference technique for predicting time-rate consolidation. The settlement analyses were divided into two parts: Part 1 from the end of tailings deposition to January 1997 (time of analyses); and Part 2 from January 1997 to the end of settlement. Part 1 of the settlement analyses was performed in the areas of thickest slimes deposits. Additionally, the results allowed calibration of the model for predicting future settlement. The analyses were conducted using a finite difference solution to Terzaghi's classical consolidation equation, because the settlement predictions were conducted on soft saturated tailings. Part 2 of the analyses were also performed using a finite difference solution and was conducted by continuing the analyses until the pore pressure dissipated to zero under the final loading conditions.

The model was initially calibrated to the field conditions (see Figures 3 and 4). The close agreement between the predicted and measured pore water pressure and settlement indicates that the model is calibrated to the soil conditions at the site and the future settlement can be analyzed with confidence. The results of the future settlement and the measured and predicted settlement indicate that the tailings has reached approximately 93% consolidation by January

1997 and that less than 1 foot of additional settlement will occur. The results of future settlement calculations using field measured pore pressures and consolidation parameters from recent laboratory testing agree with the model predictions (refer to Table 1). The predicted settlement contours for the impoundment are presented on Figure 6, and were used to compute the cracking potential of the radar barrier, and to determine if flow concentration zones would develop.

Settlement analyses at the A-9 repository were performed with the finite element computer programs SEEP/W and SIGMA/W. Three modes of settlement including immediate settlement, consolidation settlement, and long-term settlement were evaluated in this study. Immediate settlement due to placement of the cover was evaluated using the two-dimensional finite element program SIGMA/W. The increase in pore pressure and stress and the associated deformation due to the placement of the proposed additional waste and cover was obtained from the SIGMA/W model. Consolidation settlement due to dissipation of the pore pressure in the saturated tailings was evaluated by coupling computer programs SEEP/W and SIGMA/W. Long-term settlement due to creep of the impoundment materials was predicted by assuming that the creep strain data will fit the hyperbolic form. From consolidation tests the strain data after 90 percent of consolidation occurred were plotted in the hyperbolic form.

The settlement analyses were conducted on typical cross sections (see Figure 7). Table 2 shows the results of the immediate settlement, the consolidation settlement, and the long-term settlement along each of the typical cross sections analyzed. The maximum total settlement predicted for the repository is 2.62 feet. The predicted total settlement contours for the A-9 Repository are shown in Figure 10 and were used to compute the cracking potential of the radon barrier, and to determine if flow concentration zones would develop.

5.0 REFERENCES

D'Appolonia Consulting Engineers, Inc., 1980. "Reclamation Plan: Inactive Tailings Areas and Heap Leach Site: East Gas Hills Uranium Mill, Wyoming," December.

GEO-SLOPE International, Ltd. 1998a. SEEP/W Software Package for Seepage Analysis, Version 4.20. Calgary, Alberta, Canada.

GEO-SLOPE International, Ltd. 1998b. SIGMA/W Software Package for Deformation Analysis, Version 4.20. Calgary, Alberta, Canada.

Nuclear Regulatory Commission, 1999, "Technical Evaluation Report, Umetco Minerals Corporation Design for Enhancement of the Previously Approved Reclamation Plan for the Above-Grade Inactive Tailings Impoundment" June 21.

Shepherd Miller, Inc. (SMI), 1997. Design for Enhancement of the Previously Approved Reclamation Plan for the Above-Grade Inactive Tailings Impoundment, Gas Hills, Wyoming, October, Fort Collins, Colorado.

Shepherd Miller, Inc. (SMI), 1998. "Design for Enhancement of the Previously Approved Reclamation Plan for the A-9 Repository, Gas Hills, Wyoming," Fort Collins, Colorado, October.

Shepherd Miller, Inc. (SMI), 1998. "Response to NRC Comments, Design for Enhancement of the Previously Approved Reclamation Plan, Above-Grade Inactive Tailings Impoundment, Gas Hills, Wyoming," Fort Collins, Colorado, September.

Water, Waste, and Land, Inc. (WWL), 1984. Stabilization and Reclamation of an Inactive Tailings Impoundment, Volumes I and II, November, Fort Collins, Colorado.

Tailings and Mine Waste'00 © 2000 Balkema, Rotterdam, ISBN 90 5809 126 0

Consolidation behaviour and modelling of oil sands composite tailings in the Syncrude CT prototype

G.W. Pollock & E.C. McRoberts
AGRA Earth and Environmental Limited, Edmonton, Alb., Canada

G. Livingstone & G.T. McKenna
Syncrude Canada Limited, Fort McMurray, Alb., Canada

J.G. Matthews
Syncrude Canada Research, Edmonton, Alb., Canada

ABSTRACT: Syncrude Canada Limited has been developing a technology known as Composite Tailings or CT as part of their tailings reclamation. CT is a combination of fine tailings, tailings sand, and a chemical amendment (eg. gypsum) to prevent segregation. A benefit of CT is that the deposit on the mine site will consolidate enough to provide a reclaimable surface.

During 1997 and 1998 Syncrude Canada Limited conducted a field scale experiment (CT Prototype) to assess the behaviour of Composite Tailings (CT). The paper discusses the influence that depositional technique had on the deposit. In addition, the finite strain consolidation theory was used to model the CT Prototype. It was determined that there are differences between the behaviour of the CT in the field and in the laboratory – the field deposit consolidates much faster and to a greater density than lab deposits. When the differences are accounted for, the deposit can be modelled with reasonable accuracy.

1 INTRODUCTION

As part of developing a dry landscape reclamation scenario at Syncrude Canada Limited's oil sands mine site near Fort McMurray Alberta, Syncrude has been pursuing a technology known as Composite Tailings or CT. CT is essentially a combination of fine tailings, tailings sand, and a chemical amendment (eg. Gypsum) to prevent segregation. A potential benefit of CT is that the deposit on the mine site will consolidate enough to provide a reclaimable surface.

In the summers of 1997 and 1998, Syncrude conducted a field scale experiment on CT at the north end of the Mildred Lake Settling Basin dyke. The experiment is referred to as the CT Prototype.

AGRA Earth & Environmental was commissioned to model the consolidation of the CT Prototype. The purpose of the commission was to determine the ability to accurately model the consolidation behaviour of CT based on available data and theories as an indication of the ability to predict future performance of proposed tailings ponds. This paper summarizes the analyses and conclusions from this work.

2 SYNCRUDE'S CT PROTOTYPE

Figure 1 shows the layout of the CT Prototype and the approximate locations of instrumentation and sampling locations. During the 1997 stage of the CT Prototype, CT was discharged from a pipe and deposited subaerially into the pit at the south west end of the pit. A slope developed on the CT and release water ponded at the distal end of the pit. The upstream edge of the ponded water ranged between Stations 3 and 4. Daily monitoring of the CT from the discharge pipe provided data on average daily solids loading rate and Sand to Fines Ratio (SFR)[1]. The SFR averaged 4 with a range of 2.8 to 4.7.

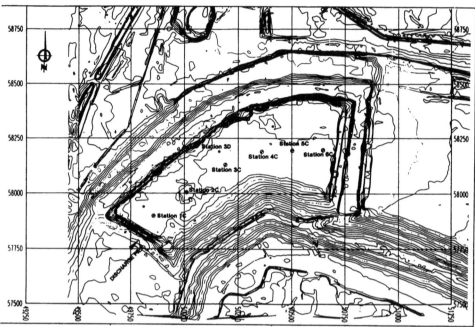

Figure 1. Layout of Syncrude's CT Prototype.

In 1998, discharge into the CT Prototype was undertaken by discharging at several locations around the pond instead of at one end. As will be discussed, this change in depositional procedure had an impact on the contents of the pond. Subsequent to the filling of the CT Prototype pond, a sand cap was placed hydraulically on the CT surface on August 12, 1998. Discussion in this paper considers the consolidation of the pond up to the start of the sand cap placement.

3 FIELD BEHAVIOUR - ZONATION

A sampling program was undertaken on December 3 and 4, 1997. Solids contents and grain size distributions were done on samples taken from the different stations. Figures 2 and 3 show the solids content and fines content (less than 44 microns) profiles obtained for each of the stations.

From Figure 3 it is evident that there is an increasing amount of fines in the deposit the greater the distance from the discharge pipe. Table 1 shows the average fines content and Sand to Fines Ratio (SFR) for each of the stations (Station 1 is closest to the pipe). The target SFR was 4 (20%fines), and the average was approximately 3.9 (20.4%fines).

It should be noted that towards the end of the field experiment, fine tailings was discharged into the pit for a short period of time resulting in a layer of fine tails on top of the CT. Table 1 does not average in the results from the samples at the very top of Stations 4 to 6, because they were taken in the fine tails layer on top of the CT.

The CT remained non-segregating, however the fines content varies significantly above and below the average and target fines content. In addition, there is a consistent trend to the variation, that is, the CT is coarser near the inlet pipe and becomes finer with distance from the inlet pipe.

---

[1] Throughout this paper the term fines content (F) refers to the ratio of the mass of material finer than 44 microns (#325 sieve) to the mass of all the mineral grains. Sand to Fines Ratio (SFR) refers to the ratio of the mass of the material coarser than 44 microns to the mass of the material finer than 44 microns. The two terms are related as follows: SFR = (1-F) / F.

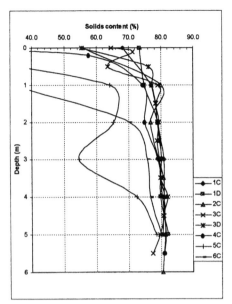

Figure 2. Measured solids content with depth (1997 data).

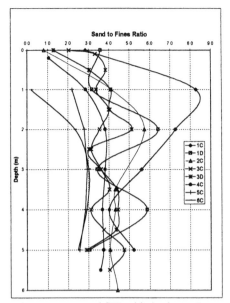

Figure 3. Measured SFR with depth (1997 data).

Table 1 - Average Solids and Fines Contents from Dec 3, 1997 samples

	Station 1C	Station 2C	Station 3C	Station 4C	Station 5C	Station 6C
Ave F.C.%	14.8	18.5	21.4	22	26.4	27.3
Ave. SFR	5.7	4.4	3.7	3.5	2.8	2.7

Several mechanisms could be postulated to account for the variation and the zonation trend. First, it could be speculated that the feed was variable enough to result in a CT which ranged from SFR of 6 to 2.4. The coarser CT (i.e. higher SFR) would then deposit close to the discharge pipe and the finer CT would deposit further downstream. However, when examining the SFR data, there does not appear to be enough high SFR or low SFR material to account for the volume of the material in those ranges in the test pit. It could be further argued that the measured values represent daily averages and that hourly spikes could account for the extreme SFR's. Although this is possible, it was concluded to be unlikely based on field observations and discussion with the plant operators.

Another possible mechanism to explain the fines content results is the redistribution of the fines within the CT. That is, if the combination of water, gypsum and fines is such that the material is close to the non-segregation boundary, the fines may be able to move with the CT river away from the discharge point. The mechanism is consistent with some findings from static segregation tests that show that there is a zone between segregation and non-segregation as opposed to a distinct line. This mechanism would also explain the SFR's which are higher and lower than the range which was deposited into the pit.

It is also possible that CT release water strips some of the fines from previously deposited CT. These fines are then carried downstream and deposited and eventually mixed in with newly deposited CT.

In 1998, more discharge locations were used and CT was discharged using a tremie pipe instead of subaerially. This lower energy regime resulted in much less zonation and suggests that the zonation is related to some form of "segregation" at the discharge point.

The zonation of fines, by whatever mechanism or combination of mechanisms, has implications in the area of pond management and pond reclamation. In the area of pond management, if the CT is discharged subaerially from the perimeter of the pond, it can be anticipated that zonation will occur and that beach slopes will develop that will be consistent with the coarser CT. In terms of pond reclamation, if the zoning process occurs, the long term surface settlement could be more pronounced in the finer zone region (possibly the centre of the pond) which would push reclamation to a wet land concept or require additional work to develop the dry land concept.

4 MODELLING THE CT PROTOTYPE

The modelling of the prototype was done in two stages. First, the 1997 period was modelled using laboratory data as input. The input parameters were then calibrated to the field measurements. Second, the calibrated parameters were then used to model the combined 1997 and 1998 filling.

4.1 Consolidation Model

The computer program FSConsol (GWP Software Inc., 1999) was used to model the one dimensional consolidation of the CT. The program is based on the finite strain consolidation theory (Gibson et al., 1967). The finite strain theory is required for this case since it accounts for:
- large deformations;
- non-linear compressibility and permeability relationships; and
- self weight consolidation.

The finite strain theory also considers the consolidation of the deposit during the filling operation.

The required input parameters are material properties and analysis conditions. The required material properties include compressibility and permeability-void ratio relationships and appropriate sand to fines ratio. Analysis conditions which are required include such as solids loading rate, solids contents and pond area rating curve.

4.2 Material Properties

4.2.1 Compressibility

Laboratory experiments were previously done to determine the relationship between effective stress and void ratio for CT (Caughill, 1992 and Suthaker and Scott, 1996 for example). It was necessary to develop a "family of curves" for Syncrude's CT based on fines content since the CT fines content varied along the deposit.

The permeability input into the FSConsol model is typically of the form:

$$e = A \, \sigma'^{\,B} + M \tag{4.1}$$

where e is void ratio, σ' is effective stress, and A, B, and M are constants unique to the deposit being modeled (in this case CT).

4.2.2 Permeability

As with compressibility, previous laboratory testing on Syncrude CT had provided permeability data for CT at various void ratios for various fines contents.

Pollock (1988) showed that for oil sands tailings, a unique relationship could be developed between permeability and the fines void ratio of the material as illustrated on Figures 4 and 5. Fines void ratio (e_f) is defined as

$$e_f = e \, / \, F \tag{4.2}$$

124

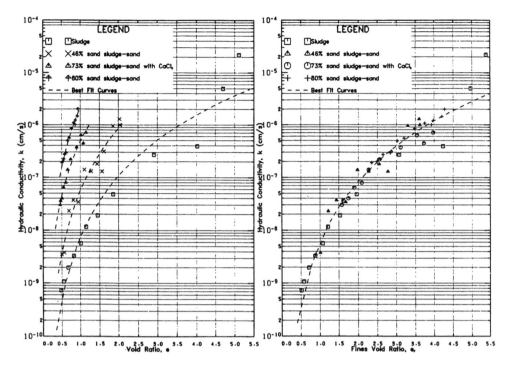

Figure 4. Permeability of Oil sands Fine Tailings/Sand Mixtures (Pollock 1988)

Figure 5. Permeability Compared to Fines Void Ratio (Pollock 1988)

where F is the fines content (in decimal).

The use of the fines void ratio allows the comparison of permeability results from materials of the same type (i.e. oil sands tailings) but differing in grain size distribution. It should be noted that this relationship masks the influence of clay content. That is, if the fines content to clay content ratio varies for the materials being compared, additional scatter will be shown in the fines void ratio – permeability relationship.

4.3 Analysis Conditions

The elevation / volume curve (pond rating curve) provided by Syncrude is shown on Figure 6. As shown on Figures 1 and 6, the pond widens with increasing elevation, which is typical of

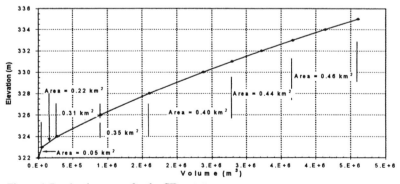

Figure 6. Pond rating curve for the CT prototype .

125

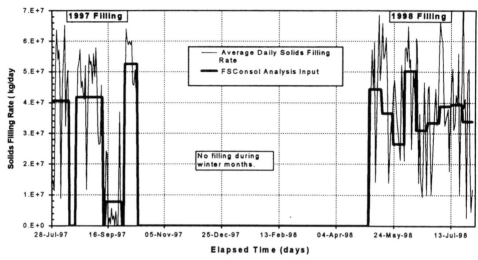

Figure 7. Solids loading rate used in consolidation model compared to actual.

virtually all tailings ponds. In order to represent the changing shape of the pond in the model, the pond was divided into a series of 7 layers and an area was determined for each layer. The layers and respective areas used in the model are shown on Figure 6.

To model the filling of the prototype pond, the filling schedule was broken down into several filling periods (in addition to periods of no filling) for each year. The average values that were used in the model for each filling period are compared to the actual filling schedule in Figure 7. Note the gap in filling during the winter of 1997/1998.

5 MODELLING RESULTS

5.1 Phase I- 1997 Test Period

Upon reviewing the field data (solids content with depth and pore pressure measurements) it became apparent that the laboratory compressibility relationship did not describe the compressibility encountered in the field as shown on Figure 8. It should be noted that this finding is a function of the material properties and independent of any consolidation theory.

It is evident from the figure that the CT becomes denser in the field, at a given effective stress, than in the laboratory. It is possible that this is due to shear effects and particle arrangement caused by the flow type of deposition which occurs in the field as opposed to the homogenous vertical deposition in laboratory standpipes. Additional measurements were taken at several locations with different techniques to confirm the solids content data. The results were within +/- 1 percentage point lending support to the original data. Therefore it was necessary to adjust the compressibility relationships used in the analyses as shown on Figure 8 to the field compressibility curve.

As discussed in above, it was apparent from the field data that there was a zonation with respect to the fines content. Therefore a series of analyses were conducted for each specific station along the prototype (Figure 1). Instead of varying the fines content over the analysis, the average fine content for the specific station (as shown in the Table 1) was used in the analysis for that station.

Once the field fitted compressibility relationships were used, the 1997 phase of the CT Prototype was modeled. The results indicated that the deposit is consolidating at a rate much faster than indicated by the analyses even using the field fitted compressibility data. This is illustrated on Figure 9 which shows the solids content profile at 137 days at station 3D and on Figure 10 which shows the pore pressure profile at the same time and station.

126

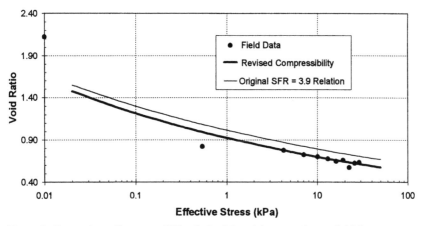

Figure 8. Comparison of compressibility derived from laboratory data to field data.

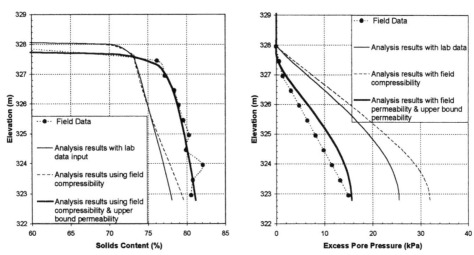

Figure 9. Solids content profile on Dec 4, 1997
(137 days) at location 3D (SFR = 3.9)

Figure 10. Pore pressure profile on Dec 4, 1997
(137 days) at location 3D (SFR = 3.9)

The permeability was adjusted to the upper bound of the laboratory data range and subsequent analyses were conducted. The results of the analyses for station 3D using the increased permeability in addition to the field fitted compressibility are also shown on Figures 9 and 10. Figures 11 to 14 show the results of the analysis compared to field data for other stations. As shown on the figures, both the solids contents and excess pore pressures are modelled with reasonable accuracy. (Not all stations had pore pressure measurements.)

It should be noted that the change in permeability to achieve the good results was less than an order of magnitude. In terms of permeability, differences between laboratory permeability and field permeability are often an order of magnitude or more. This is generally attributed to the structure of the material in the field at a scale larger than tested in the lab. In addition, it was previously speculated that the laboratory settling tests (which provide permeability data in the high void ratio range) were influenced by the narrow aspect ratio of the settling columns. This has been supported by some recent laboratory data (Boratynec et al. 1998).

5.2 *Phase II - 1998 Test Period*

The filling during the 1998 summer period differed from the 1997 filling period in that several

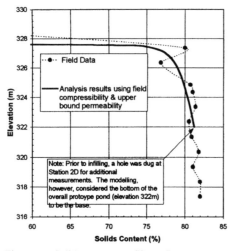

Figure 11. Solids content profile on Dec 4, 1997 (137 days) at location 2C (SFR = 4.4)

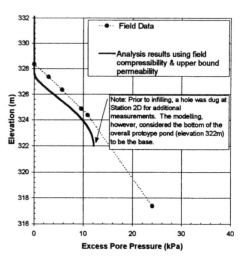

Figure 12. Pore pressure profile on Dec 4, 1997 (137 days) at location 2C (SFR = 4.4)

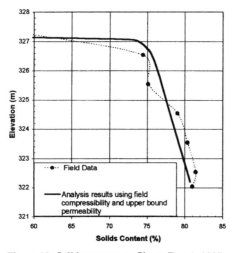

Figure 13. Solids content profile on Dec 4, 1997 (137 days) at location 4C (SFR = 3.5)

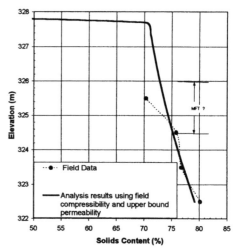

Figure 14. Solids content profile on Dec 4, 1997 (137 days) at location 6C (SFR = 2.8)

discharge pipes around the prototype pond were used instead of one pipe at one end. Therefore the zonation that was encountered after the 1997 filling was not present in the 1998 deposit. Also, there was less field data available for comparison for the 1998 phase than for the 1997 phase.

Figures 15 and 16 show the height versus time comparison for the two available stations. As shown on the figures, the height of the deposit over time is modeled with good accuracy.

Figures 17 and 18 show the pore pressure comparison for the same two stations The pore pressure is modeled quite well at station 3D but is over predicted at station 2C. The over prediction of excess pore pressure may be attributed to downward drainage at this location, which was indicated by the field data. This condition deviated from the assumption of an impermeable bottom boundary used in the analyses.

There was no solids content data available for the 1998 phase at the time the analyses were conducted.

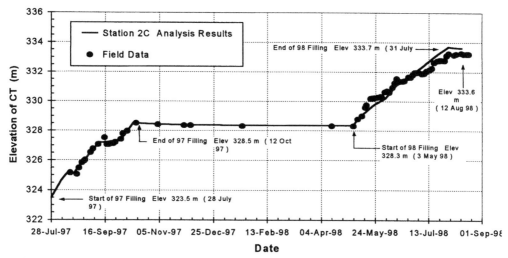

Figure 15. Elevation at Station 2C.

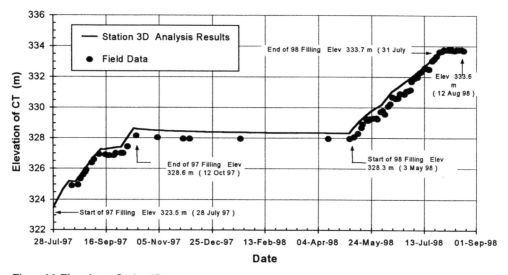

Figure 16. Elevation at Station 3D.

6 SUMMARY AND CONCLUSIONS

The depositional technique has been found to influence the composition of the deposit. At the CT Prototype, discharging solely from one end of the pond resulted in a deposit that was zoned with respect to fines. That is, the CT had more sand near the discharge and more fines further from the discharge point. This zonation was eliminated in the subsequent year when additional discharge locations and a tremie-style discharge were used.

In terms of consolidation, the field behavior is typically better than that predicted when using data solely from the laboratory. In the field, it appears that the material is more compressible than in the laboratory. This is especially so at low stresses where the material in the field is measured to be much denser than in the laboratory. Possible reasons could be related to the dynamic effects during placement in the field.

Results indicated that the permeability in the field is greater than obtained from laboratory

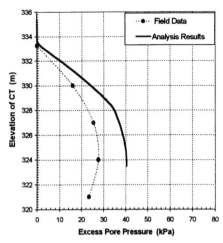 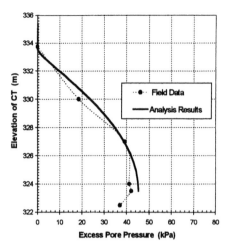

Figure 17. Pore pressure profile at Station 2C on August 12, 1998.

Figure 18. Pore pressure profile at Station 3D on August 12, 1998.

testing. This is a reasonably common finding in geotechnical practice and is likely attributable to macro structure or channeling.

Once the field compressibility relationships and increased permeability parameters are used, the height of the deposit, solids contents, and pore pressures can be modelled with reasonable accuracy.

The reliability of using laboratory data for predicting consolidation rates and ultimate densities of composite tailings was found to be somewhat lacking. However, using field results to calibrate the finite strain model proved to provide a good fit with data from later CT deposition and has increased the confidence in predicting the behaviour of future CT deposits at Syncrude.

REFERENCES

Boratynec, D.J., Chalaturnyk, R.J. & Scott, J.D. 1998. Experimental and Fundamental Factors Affecting the Water Release Rates of CT. *Proceedings of the 51st Canadian Geotechnical Conference.* pp 607-614.

Caughill, D.L. 1992. *Geotechnics Of Non-Segregating Oil sands Tailings.* University of Alberta. M.Sc. Thesis. 243p.

Gibson, R.E., England, G.L. & Hussey, M.J.L. 1967. The Theory of One-Dimensional consolidation of Saturated Clays, I, Finite Non-Linear Consolidation of Thin Homogeneous Layers. *Geotechnique.* 17. pp 261-273.

GWP Software Inc. 1999. FSConsol Version 2. Canada.

Pollock, G.W. 1988. *Large Strain Consolidation of Oil Sand Tailings Sludge.* University of Alberta. M.Sc. Thesis. 276p.

Suthaker, N.N. & Scott, J.D. 1996. Measurement of Hydraulic Conductivity in Oil Sands Tailings Slurries. *Canadian Geotechnical Journal.* 33. pp642-653.

Tailings and Mine Waste'00 © *2000 Balkema, Rotterdam, ISBN 90 5809 126 0*

Tailings as a construction and foundation material

J.J.Crowder & M.W.F.Grabinsky
Department of Civil Engineering, University of Toronto, Ont., Canada

D.E.Welch
Golder Associates Limited, Mississauga, Ont., Canada

H.S.Ollila
Kinross Gold Corporation, Kirkland Lake, Ont., Canada

ABSTRACT: Throughout northern Ontario and Quebec there exist dozens of successfully operating upstream and stacked tailings facilities, as well as dykes built directly on tailings. Tailings dams, however, have been increasingly subject to intense public scrutiny due mainly to several significant failures. Although the properties of tailings vary considerably from site to site, as well as at different locations across one site, tailings can sometimes be used successfully as a construction material in dykes, or as a foundation soil. The Macassa tailings facility demonstrates how a dyke can be constructed on top of a tailings deposit and subsequently raised by the upstream method. The Quirke tailings management area has performed to expectations having used tailings as a construction material in dykes and upstream shells of dams. Using proper engineering design, construction, monitoring, and maintenance approaches, both case studies have satisfied technical concerns often cited by the public and technical communities.

1 INTRODUCTION

The failures of tailings dams have created a clear picture of public alarm due to past failures that resulted in severe environmental damage and loss of life. Indeed these recent experiences suggest an underestimation of the environmental risk associated with the design of mine waste management facilities (East 1999). The perception of tailings facilities' environmental performance has been negatively impacted by a relatively few number of failures. This negative overview seen in the public and technical communities override the many positive contributions made by many mining companies. A recent study has illustrated that upstream tailings dam design, which is often targeted as a poor construction technique, may not be the culprit of most failures.

In fact, there are dozens of successful tailings operations in northern Ontario and Quebec (Canada), which is a seismically benign region. Gold and base metal tailings found in this area, although consisting of variable grain size and composition, can be used quite successfully either as foundation soil for dykes, or as an excavated borrow and construction material.

The upstream raise construction technique, often employed in northern Ontario and Quebec, has received much public and technical attention. The main technical concerns stressed are about stability and seepage of effluent. Stability concerns focus on pond location, phreatic surface control, induced pore pressures, and seismic liquefaction.

A case study of the Macassa mine tailings area demonstrates that these concerns are generally dealt with appropriately, resulting in safe, long-lasting, environmentally sound tailings facilities. As well, a case study of the Quirke tailings basin shows how tailings can be used successfully as a borrow and construction material, as several dykes and the upstream shells of some dams were constructed of compacted tailings.

Failures of tailings impoundments continue to occur. Recent failures such as Los Fraises, Spain (1998), Marcopper, Philippines (1996), and Omai, Guyana (1995) have attracted much publicity. This attention, along with other sources of public information, criticises the mining industry, as a whole, for its practises.

After the failure of Los Fraises in 1998, an article appeared in The Globe and Mail newspaper in Canada, titled "Mining's dam problem" (Robinson & Freeman 1998). This article, while attempting to be well rounded, came across with some statements that could very well create negative perceptions of the tailings storage process. "'Tailings dams are supposed to last forever,' said one exasperated mine consultant. But they don't." The article also describes tailings, not as crushed rock particles that are either produced or deposited in slurry form (Vick 1990), but as "toxic black sludge", and "toxic soup" that "oozed". Tailings dams are described as "leftover dams" and "by their very nature and location ... are vulnerable [to failure]." The article quoted the president of the Center for Science in Public Participation, a grassroots group in Montana, as saying, "If there is seepage through the dam or around the dam, that weakens the dam structurally." In fact most earth-fill embankments and tailings dams are designed with the knowledge that seepage does and will occur. Some designs rely on seepage to allow for pore water dissipation. This is achieved often by a rock-fill starter dam followed by course rock-fill upstream raises.

Tailings are also widely discussed through Internet sites. Project Underground is an organisation that focuses on "campaigns against abusive extractive resource activity." Their goal is to serve "those communities threatened by the mining and oil industries." The web site for this organisation portrays a pessimistic perception of the mining industry as a whole: "We expose the myths perpetrated by powerful resource industry interests" (Project Underground, 1998a). In the description of one of their projects, the site says that a civil war, which has cost "as many as ten thousand lives ... stems in large part from the impacts of tailings dumping" (Project Underground 1998b).

The World Information Service on Energy (WISE), Uranium Project also portrays a negative outlook on tailings disposal. The web site, which gives a bibliography of tailings information sources and lists failures, says that the construction of tailings dams "in most cases ... are not built as engineered structures." (WISE 1999) The message the site stresses is all about the potential problems. The ideas are stressed, however, using poorly defined statements such as: "tailings slurries may liquefy" and as a consequence "may be released." As well, "in this case ... slurries would flow ... and the slurry wave would threaten inhabitants of the area," and surrounding land then "would be devastated." Furthermore, it is described that water level rise in an impoundment "can be caused by heavy precipitation events. ... If the exposed beach becomes to [sic] small ... the whole dam can collapse." While the site seems to have a reasonable grasp of the technical challenges involved in tailings dam design and operation, the statements as shown above, again try to create a perception of tailings dams as poorly-engineered, unstable structures.

While all of the above cases may just be perspectives, these types of information are widespread across all media concerning tailings facilities, and serves to develop a negative view of tailings disposal and the mining industry. Other recent studies show a different perspective.

As reported by Strachan (1999), the United States Commission on Large Dams (USCOLD) Committee on Tailings Dams conducted in 1989 a review of tailings dam incidents. The survey consisted of a compilation of 185 incidents from 1917 through 1989 (USCOLD 1994). As well, in 1996, the United Nations Environment Programme (UNEP) published a survey (UNEP 1996) of 26 incidents which occurred from 1980 to 1996, that were independent of those reported by the USCOLD. The results of both surveys were compiled and analysed by Strachan (1999). The surveys included most types of tailings dams (upstream, downstream, centreline, and conventional water-retention). One key finding within the combined incident survey is that there is no overriding cause or mechanism for all tailings dam incidents. As well, there exists no particular type of tailings dam or operational condition that is more susceptible to incidents. The key recommendation was that tailings dams should be located, designed, operated, and reclaimed with a thorough knowledge of site conditions, mill operation, site monitoring, and tailings characteristics.

While about 200 tailings dam incidents were reported in these surveys, East (1999) is quick to point out that since the 1985 Stava tailings dam failure in Italy, which caused 269 deaths, there have been only 14 reported failures of tailings dams. The reader must also be aware that a failure does not always mean breach of a dam and release of tailings and effluent water. Another type of failure is a loss of serviceability by settlement of the dam, which does not allow passage of vehicles, but has still not released any tailings. Failure could also be defined as loss of wind blown tailings, or piping through a dam – both of which are not necessarily disastrous and are generally possible to remediate.

3 TAILINGS IN NORTHERN ONTARIO AND QUEBEC

Gold and base metals constitute the majority of mining in northern Ontario and Quebec. Although many milling operations employ similar processing and deposition strategies, the tailings from any particular operation will have its own unique characteristics. One major difference in and tailings deposition plan depends on whether the tailings are acid generating. Acidic tailings, which are often found in the Canadian Shield, must be kept saturated to avoid oxidation and acid drainage. This has implications on how high a facility can be raised – the higher it is, the more tailings are exposed to air. Grain sizes of these tailings vary from sand to silt depending on sample location.

Properties of tailings at a specific site can vary considerably across the impoundment. If deposition is by spigotting, the coarser fraction is generally deposited near the spigot and grades finer toward the centre of the pond. Permeability also varies as distance from the beach increases. This phenomenon is advantageous to using the tailings as a construction or foundation material. If the most coarse, and most permeable, material exists at the starter dam, then this zone will have the highest strength and the best tendencies to drain seepage or excess pore pressures. These properties exist also for end discharged tailings. Both of these methods are incorporated in northern Ontario and Quebec.

Although the tailings deposited and stored by these methods mostly remain saturated, seismic stability concerns are minimised because this region is associated with a low seismic risk.

4 UPSTREAM CONSTRUCTION ISSUES

On-surface tailings impoundments are most commonly constructed by the stage-raise method. The advantages of such a system are well known. These include:
– embankments usually begin with a starter dyke built of borrow materials to accommodate two to three years of deposition, and is raised as needed as the height of the tailings surface in the impoundment increases;
– tailings themselves can be used as the construction material, hence lowering the cost of borrow materials;
– construction expenditures are distributed over the lifetime of tailings deposition, rather than as an initial capital cost incurred using a water-retention type of tailings dam;
– low initial capital cost and associated cash-flow benefits;
– flexibility in materials selection, e.g. mine wastes, natural soils, cycloned or spigotted tailings, waste from other nearby mines, etc.; and
– design changes may be employed throughout the lifetime of the impoundment.

The upstream method of raising tailings impoundments is common in northern Ontario, Canada. Construction begins with a starter dyke, commonly composed of course rock fill or mine waste to allow for free drainage of seepage. As tailings are deposited, they fill up against the starter dyke, eventually reaching the initial crest height. At this point, this newly created beach becomes the foundation for the raised embankment. Such a system of disposal and construction carries its own benefits. In general, it is the most cost effective method of raised tailings impoundments. Only minimal amounts of mechanically placed materials are required for the construction of the dams. The method is also one of simplicity. Construction is an ongoing process that can be performed with minimal equipment and personnel.

Vick (1990) and MAC (1998) report, however, that the method has several technical concerns: phreatic surface control and pond location; susceptibility to seismic liquefaction; and seepage and development of excess pore pressures.

4.1 Pond Location and Phreatic Surface Control

The location of the phreatic surface is critical in determination of stability of the tailings dam (Vick 1990). A high phreatic surface can lead to increased pore water pressures within the dam, as well as piping and eventual failure. For a tailings dam created by spigotting, there are few structural measures to control the location of the phreatic surface within the embankment. The main factor that influences the phreatic surface is the location of the pond. A high pond (i.e. close to the crest of the dam) will certainly lead to a high phreatic surface and the potential for seepage out of the downstream dam face. Another critical factor, as described in Stauffer & Obermeyer (1988), is the permeability of the foundation. In the course of a detailed monitoring program on upstream tailings dams, the authors found that less than hydrostatic pore water pressure conditions existed in the dam, leading to the conclusion that the foundation material was permeable enough to dissipate pore pressures significantly.

One method to help reduce the failure potential due to a high phreatic surface, as discussed in Minns (1988), is to raise the upstream dam using an outer rock-fill shell. This is accomplished by creating a starter dam out of waste rock-fill, from which tailings are spigotted. Each successive raise is also constructed from rock-fill. The starter dam becomes a toe drain for the embankment, in this case. This design also reduces any pore pressures at the face and allows the tailings near the face to dry.

To avoid a high-pond condition, spigotting of tailings can be used to divert the pond toward the decant or spillway structure. If the pond begins to encroach against a dam, tailings can be spigotted in that area, increasing the tailings elevation, and hence driving the pond away. An increase in the decanting rate can also lower the pond away from the crest.

As Vick (1990) points out, near total diversion of runoff and flood water is essential in upstream construction. This means that any decant structures, often prone to clogging, must be well maintained. Alternatively, a spillway could be used to divert water to a settlement or conditioning pond.

4.2 Susceptibility to Seismic Liquefaction

Since upstream tailings dams generally consist of saturated soil of an initially low relative density, they may be susceptible to liquefaction under severe seismic ground motion (Vick 1990). The consequence of seismic activity is liquefaction-induced flow of tailings. If there is a breach of a dam during the event, there is a possibility of the tailings to be released and flow some distance downstream. Breach is often caused by the pond overtopping the dam and subsequent erosion of the shell material.

One method of reducing the potential of dam rupture and loss of containment is to flatten the downstream slope of the dam. In Ontario, closure regulations generally call for a slope of no steeper than 3 (horizontal) : 1 (vertical). This slope angle helps to ensure slope stability even during seismic events.

Even if there was no failure during a seismic event, excess pore pressures developed due to shaking may remain well after. This is another reason to be sure there is adequate opportunity for drainage from the impoundment by means of a well-maintained, functioning decant structure or spillway.

Any design must consider the consequences of liquefaction. One major consideration involves deformation. Dam deformation may occur due to seismic activity. The crest of the dam, therefore, must have sufficient freeboard to account for any potential settlement of the crest.

4.3 Seepage and Excess Pore Pressure Development

The control of seepage is a major issue in the design and construction of tailings dams. Mittal and Morgenstern (1976) suggests that seepage control is needed to maintain a low phreatic surface for stability concerns, to control pollution of the downstream environment, and for reasons of water supply.

Seepage through any earth fill embankment or tailings dam is to be expected. That seepage, however, must be controlled for reasons of environmental performance and dam stability. This means that all seepage water must be collected and treated to reduce particulate matter and toxins before release to the outside environment. Alternatively, water may be re-circulated to the mill.

A lower phreatic surface can be achieved by allowing seepage to occur by placing the most pervious material at the embankment face. The general principle is that permeability should increase in the direction of flow (Vick 1990). This is achieved through spigotting, as the most course material is deposited at the dam face, while the finer fraction flows toward the centre of the impoundment. A waste rock shell and toe drain, which may exist in the form of a waste rock starter dam, act well to conduct seepage from the dam.

Excess pore water pressures can be created by rapid loading of the tailings. This often occurs by rapid rates of pond level rise and deposition of tailings. Excess pore pressures may occur for raising rates in excess of five to ten metres per year (Mittal & Morgenstern 1976). Slower rates of raising will allow for any excess pore pressures to dissipate as more tailings are deposited.

5 CASE STUDY: MACASSA MINE, KIRKLAND LAKE, ONTARIO

5.1 Tailings Deposition

The Macassa Mine, located west of the town of Kirkland Lake in northern Ontario, Canada, has been in operation since 1926, and started milling in 1933. The tailings management facility is comprised of two cells for the primary solids impoundment, a conditioning pond, and a system of small treatment ponds downstream of the treatment plant.

The Macassa tailings area, which and contains about 10 million tonnes of tailings, was initially contained topographically, with dams in the lows. After decades of dam raises, however, the facility essentially operates as a stacked system. That is, the major tailings retaining dams have been raised by the upstream method and now nearly contain the entire basin. The general arrangement plan can be viewed in Figure 1. This shows that the major tailings retaining dams are Dams B, E, F, and G.

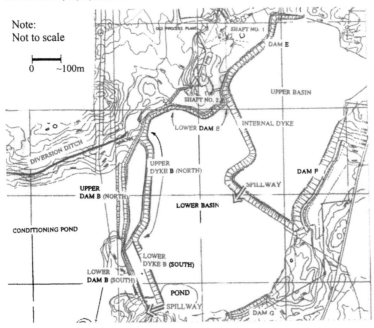

Figure 1. General arrangement of the Macassa tailings area.

135

Dam B, which was constructed with a mine waste rock shell raised over tailings by the upstream method (slope angle ~ 37 deg.), separates the tailings basin from the conditioning pond. Just upstream of the dam, lies a deposit of mine tailings about 13 m (40 ft.) thick, overlying natural varved silty clay and glacial till deposits. After a drilling investigation revealed the undrained shear strength of the silty clay to be about 38 kPa (800 psf), a stability study suggested that for the chosen factor of safety, the height of the dam should not exceed 13 m. Since Dam B was nearing that height, an alternative deposition plan was required.

Dyke B was proposed to be built greater than 33 m (100 ft.) upstream of Dam B, with its foundation on top of the deposited tailings. The distance upstream was chosen based on an estimate that the zone of influence of increased vertical stress was far enough away as to not affect Dam B. The construction of Dyke B effectively flattens the overall slope of the tailings stack, from toe to the point of deposition, to about 4H : 1V, creating an inherently safer structure.

To facilitate construction of Dyke B, biaxial geogrid was laid down on top of the tailings before the first lift of fill was placed. The aim was to introduce the new load on the tailings gradually, to allow for the dissipation of excess pore water pressures in the tailings. A typical cross section and construction details can be found in Figure 2. The dyke consists of a thin layer of sand on top of the geogrid to facilitate construction, on top of which is coarse rock-fill. The upstream face of the rock-fill has been covered with a blanket of compacted tailings to act as a filter between the rock and deposited tailings. Construction of the dyke was completed without any difficulties and has experienced no problems due to settlement. The dyke has now been raised up to the 5th bench as shown in Figure 3. A similar use of geogrid in an upstream raise tailings dam was described in Burwash et al. (1993).

5.2 Tailings Properties

The tailings deposited in the Macassa Mine tailings area are consistent across the site. They can be described as wet, grey, angular, medium SILT, with trace sand and rock flour. A grain size distribution can be seen in Figure 4. The water content is about 30%, but ranges depending on whether the sample is above or below the water table. The specific gravity is 2.81. The tailings are not acid generating. Results of drained and undrained, consolidated triaxial testing at The

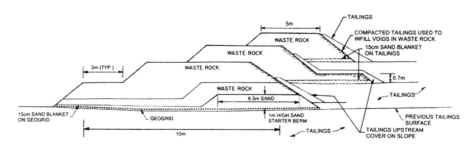

Figure 2. Typical cross-section of Dyke B.

Figure 3. Photograph of Dyke B from May 1999.

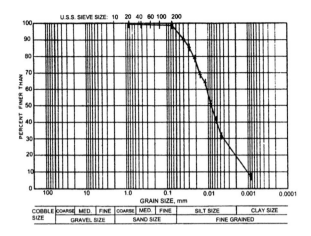

Figure 4. Typical grain size distribution of Macassa tailings.

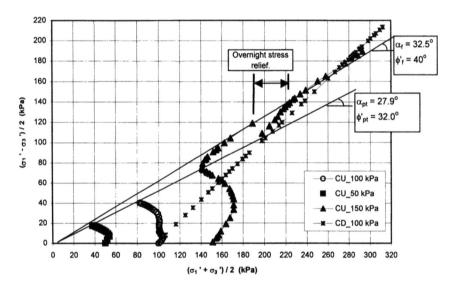

Figure 5. Results of triaxial testing on Macassa tailings.

University of Toronto, shown in Figure 5, indicate that the angle of internal friction is about 40 degrees. These results are consistent with friction angles obtained by direct shear tests conducted at the University of Toronto in 1999 and by Coy (1998).

5.3 Dealing with Technical Concerns

5.3.1 Pond Location and Phreatic Surface Control
The tailings basin is divided by an internal dyke built from tailings. During operation, tailings were deposited alternately from the east and west sides of the basin to maintain a drainage course down the centre of the basin. This procedure has allowed the tailings beaches upstream of the embankments to be drained and to facilitate raising by the upstream method. Furthermore, discharging by spigotting from the dyke deposits relatively coarse tailings suitable for dyke construction. Dyke construction generally took place concurrently with tailings deposition at different areas of the basin during the summer months.

These practices of maintaining drainage down the centre of the basin, keeping water away from the dykes, and alternating between deposition in one part of the basin and construction in the other part, has led to the creation of dryer, stronger tailings at the time of raising. A drilling investigation in 1990 of Dam B revealed a drawn-down phreatic surface, indicated by water table levels, that exited the dam through the starter waste rock dyke.

Drainage from the tailings facility used to be decanted through a structure located in the south end of Dam B. This decant structure has now been decommissioned and replaced by a spillway located at the south abutment of Dyke B and spills into the conditioning pond. The spillway has been designed to handle the routed probable maximum flood. Perimeter spigotting facilitates closure with the formation of a natural drainage valley directing drainage to the final spillway. This practice keeps water away from the dam and dyke crests, allowing tailings to dry and consolidate prior to upstream raises.

5.3.2 Susceptibility to Seismic Liquefaction

Kirkland Lake is located in a region of low seismic risk. A seismic stability study, prior to the construction of Dyke B, used input seismic ground motion from the Saguenay, Quebec earthquake of 1987. Amplitudes were scaled back to a peak horizontal acceleration of 0.078 g. The output of a computer model showed that the induced peak horizontal acceleration at the top of the tailings deposit at Dam B was 0.121 g. The analysis indicated that a potential zone of liquefaction was found in the top half of the tailings deposit.

The recognition of this also led to the decision not to raise Dam B any further. To continue deposition, Dyke B was constructed, as already mentioned, some 30 m (100 ft.) or more upstream of the dam. In effect this creates an overall slope angle of less than 4:1, which constitutes a more inherently safe structure. Freeboard is maintained adequately enough on both Dam B and Dyke B that even if some deformation did occur due to seismic liquefaction, there would not be overtopping of either structure. A toe berm has been added to Dam B as a further measure to ensure stability.

5.3.3 Seepage and Pore Pressures

Some seepage does occur through Dam B. Monthly and annual inspections have always found that the seepage water is free of sediment that indicates internal erosion is not occurring. Any water that does seep through Dam B, or is conducted by the spillway is collected in the conditioning pond. This pond allows for the settling of sediment and heavy metals, as well as retention time for natural degradation of other chemicals by means of ultraviolet breakdown. Treated water flows through a series of polishing ponds before release into the environment. A series of ditches have been designed to divert runoff and seepage flow into the polishing pond. The facility is consistent with the regulatory guidelines of the Ministry of Natural Resources, Ministry of Northern Development and Mines, and the Ministry of the Environment. The operation has also developed a "Strategic Tailings Operating Plan" modelled after MAC (1998), which details the operational procedures, inspections and quality control program for the tailings area.

6   CASE STUDY: QUIRKE TAILINGS BASIN, ELLIOT LAKE, ONTARIO

The Quirke tailings basin, near the town of Elliot Lake in northern Ontario, started tailings deposition in 1956 and ended in the early 1990's. The impoundment is found in a confined lake basin with retaining dams in the topographic lows. The capacity of the site has been increased in the past by constructing dykes on the surface of previously deposited tailings and employing the stage-raise method. Internal cells were essentially created by these dykes, which were then infilled with tailings, as shown schematically in Figure 6. Discharge was often by a single point, which was moved frequently. Spigotting was performed occasionally to optimise the filling of the cells (Matyas et al. 1984).

The dykes constructed at Quirke are similar to other dykes found in the Elliot Lake region. The foundations for these dykes are often previously deposited tailings. Typically constructed in stages over many years, some of the dykes reach heights of 10 m. Often these structures are used as haulage and maintenance roads.

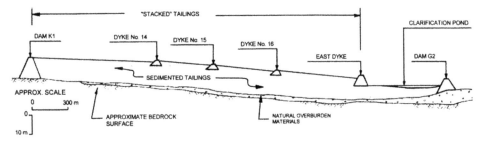

Figure 6. Stacked tailings scheme at Quirke tailings management area.

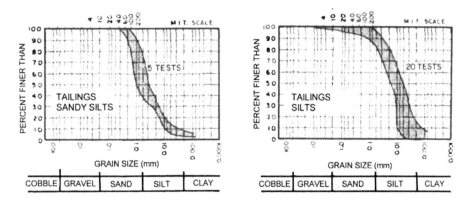

Figure 7. Typical grain size curves for tailings sandy silts (left) and tailings silts (right).

A test embankment at the site was reported by Matyas et al. (1984). Test data were used to assess the performance of dykes built on tailings and the tailings as a foundation material. Results showed that rapid rates of pore pressure dissipation and tolerable settlement, coupled with high friction angles, ensured no stability problems during construction.

The Quirke tailings were also used as a borrow and construction material. Course tailings were readily excavated, using front-end loaders or scrapers, from well drained areas and were used for the construction of dykes, causeways and the upstream shells of several dykes and dams. The tailings were often compacted with ease, requiring an average of only four passes of smooth-wheeled or vibratory rollers to achieve the required degree of compaction. The average optimum dry unit weight, based on about 100 standard Proctor tests, was 16.9 kN/m^3. Using this value, the compaction yielded an average in-situ dry density of 97.2% of the standard Proctor dry density. Tailings were found to be fairly uniform, as shown by the typical grain size curves in Figure 7.

Other results of the study found that trafficability of the tailings sands in the borrow areas and on fills was very good when drained and compacted. Matyas et al. (1984) suggests that this is due to the angularity of the particles. This angularity and needle-shaped particles make for a high angle of internal friction for the silt tailings (Conlin 1980).

7 CONCLUSION

Failures of tailings facilities, and the resulting media coverage, has led to a negative outlook on tailings management practice and the mining industry. Such cases seem to override the positive contributions of mining to society.

The case study of the Macassa tailings facility has demonstrated that, despite the concerns and critics among the public and technical communities, a successful upstream-raised tailings facility can be constructed and maintained in a region of low seismic risk. The ability to construct a dyke on tailings shows that proper engineering design and good construction and management practices can produce a safe, reliable tailings structure. It must be noted, however that in regions of high seismic activity, upstream construction may pose too much risk, given liquefaction potential.

The Quirke case study has shown again that tailings can be used as foundation material, as dykes were built right on tailings. The study also demonstrated that tailings can be excavated and used as construction materials to build dykes or as a fill material like the upstream shell of a dam. Due to the angularity and needle-shaped quality of the tailings, as well as good pore pressure dissipation characteristics, a high degree of compaction can be achieved during construction.

Although failure will continue to delay and impede the permitting and construction of well designed tailings facilities, the art and science of geotechnical engineering, as stated by East (1999), "is well capable of designing structures which minimise the risk of failure given the appropriate fiscal and technical management support."

REFERENCES

Burwash, W.J., Arnall, P.G., and Kerr, J.R. 1993. Geogrid stabilization for raising the elevation of a tailings dam, British Columbia, Canada. *Geosynthetics Case Histories,* ISSMFE.

Conlin, B. 1980. *Design considerations for earthquake-resistant mine tailings impoundment structures in northern Ontario.* M.A.Sc. thesis, University of Toronto, Canada.

Coy, G.A.G. 1998. *Basic Geotechnical Properties of Macassa Mine Fine Tailings.* Undergraduate Thesis, University of Toronto, Canada.

East, D.R. 1999. Tailing dam failures – why do they continue to occur? *SME Annual Meeting,* Pre-print 99-102, Denver, CO.

Matyas, E.L., Welch, D.E., & Reades, D.W. 1984. Geotechnical parameters and behaviour of uranium tailings. *Canadian Geotechnical Journal,* 21:489-504.

Mining Association of Canada (MAC). 1998. *A guide to the management of tailings facilities.*

Minns, A. 1988. A review of tailings disposal practices in North America and Australia. *Hydraulic Fill Structures,* ASCE Geotechnical Special Publication No. 21, pp. 52-68.

Mittal, H.K., & Morgenstern, N.R. 1976. Seepage control in tailings dams. *Canadian Geotechnical Journal,* 13:277-293.

Project Underground. 1998a. Mission Statement, *Biennial Report,* www.portal.org/ProjectUnderground /info.html

Project Underground. 1998b. Freeport McMoRan: Environmental Issues, www.moles.org /ProjectUnderground/index1.html

Robinson, A., and Freeman, A. 1998. Mining's dam problem. *The Globe and Mail Newspaper,* Toronto, Canada, May 16.

Stauffer, P.A., and Obermeyer, J.R. 1988. Pore water pressure conditions in tailing dams. *Hydraulic Fill Structures,* ASCE Geotechnical Special Publication No. 21, pp. 924-939.

Strachan, C. 1999. Tailings dam performance from USCOLD incident survey data. *SME Annual Meeting,* Pre-print 99-40, Denver, CO.

United Nations Environment Programme (UNEP). 1996. *Environmental and safety incidents concerning tailings dams at mines,* based on survey conduced by Mining Journal Research Services for UNEP. (as cited in Strachan, 1999).

U.S. Committee On Large Dams (USCOLD), Committee on Tailings Dams. 1994. *Tailings dam incidents,* USCOLD.

Vick, S.G. (2nd ed.) 1990. *Planning, Design, and Analysis of Tailings Dams.,* Vancouver: Bitech Publishers Ltd.

World Information Service On Energy (WISE). 1999. Uranium project: *Safety of tailings dams.* www.antenna.nl/wise-database/uranium/mdas.html

Tailings and Mine Waste'00 © 2000 Balkema, Rotterdam, ISBN 90 5809 126 0

Disposal of industrial waste liquids by evaporation and capillary storage in waste dumps

G. E. Blight
Witwatersrand University, Johannesburg, South Africa

A. Kreuiter
Eskom, Johannesburg, South Africa

ABSTRACT: In semi-arid and arid regions the atmosphere has a considerable capacity to evaporate liquid industrial wastes such as brines. The evaporation capacity may be as high as 2000 mm/year, which means that 20 megalitres of liquid can potentially be evaporated per year per hectare. In the past, lined evaporation ponds were used to evaporate waste liquids. Recently, the tendency has been to spray-irrigate waste liquids over the surface of deposits of solid waste, eg landfills, ash dumps and mine waste dumps. To be environmentally acceptable, the rate of irrigation must be related to the overall water balance of the waste body, so that all waste liquid that infiltrates the waste is either re-evaporated or permanently held within the pores of the waste by capillarity. No liquid can be allowed to escape from the usually unlined base of the waste body to become a potential source of groundwater pollution.

The paper will describe how to establish acceptable rates of irrigation as well as irrigation cycles. Also, how to establish the quantity of water it is possible safely to absorb in the waste body so that nothing will leach into the ground water even under the most adverse weather conditions.

1 INTRODUCTION: THE SOIL WATER BALANCE

The 'water balance' principle has been widely used to predict changes in moisture in a waste body (e g Fenn et al, 1975). It depends on the application of the principle of conservation of mass to water movement within the soil. The balance can be most simply stated as:

$$\text{water input} = \text{water output} + \text{water stored} \tag{1}$$

Considering the components of each term of equation (1), the water balance for a waste deposit may be written as:

$$R = ET + LE + RO + \Delta ST. \tag{2}$$

Where R is rainfall
$\quad ET$ is water lost by evapotranspiration
$\quad LE$ is water lost as leachate from the base of the waste body

RO is water lost as runoff

Δ*ST* is the change in water stored in the waste by capillarity.

RO is small for most rainfall events, and hence can usually be disregarded (Blight and Blight, 1993). This leads to a simplified equation for the water balance, from which the leachate recharge may be calculated if the loss to evapotranspiration (ET) and the change in storage ΔST can be estimated:

$$LE = R - ET - \Delta ST \tag{3}$$

The larger ET, and the smaller R, the smaller will be the chance of leachate being produced. The value of ΔST will fluctuate, the waste acting as a capillary sponge, storing moisture which may later be released as leachate (if the limiting moisture storage capacity is exceeded), or lost by evapotranspiration. In arid and semi-arid areas, where ET is perennially larger than R, ΔST may remain small enough that the limiting moisture storage capacity of the waste is never exceeded and leaching never occurs. Over a reasonably long period (i.e. a couple of years) ET cannot exceed R even if the annual pan evaporation far exceeds R. In this situation, ΔST = 0 overall, although it will fluctuate seasonally. In a situation like this, there may be sufficient spare moisture storage capacity to enable the waste body to be used deliberately as a temporary storage and source for evapotranspiration of waste liquids. It is this situation that the paper will address.

2 OPERATING A WASTE DEPOSIT TO DISPOSE OF SURPLUS LIQUIDS

To operate such an evaporative waste water disposal system optimally, the application rates for the water should be matched to the evaporation rates. If too much water is applied, the limiting moisture storage capacity of the waste in the body of the deposit will be exceeded and waste water will start leaching out at the base of the deposit.

Over any 24 hour period, part of the waste surface will be sprayed and part will not. For the part that is sprayed, suppose that the spraying takes place for h hours in 24. If the total surface area being irrigated is A and the area that is sprayed on a typical day is A_s, the water balance for the area that is being sprayed will be:

$$S.h/24 \, A_s = (E_A.h/24 + E_{ss} + I + P) \, A_s - RA_s$$

Hence $(I + P) = (S - E_A) \, h/24 - E_{ss} + R$ (4)

In equation (4) all parameters are measured in mm per 24 hour day, i.e. mm/d

S = application of waste water by spraying;

E_A = evaporation rate of water into the air above the waste surface over the area that is being sprayed;

E_{ss} = evaporation rate from the saturated surface that is being sprayed (E_{ss} is assumed to apply throughout the day on which A_s is sprayed);

I = infiltration rate of waste water into the waste;

P = ponding depth of waste water on the waste surface;

E_s = evaporation rate of water from the waste surface that has not been sprayed on a particular day;

R = rainfall for the day.

(I + P) forms the input moisture for the area (A - A$_S$) of waste that is not being sprayed during the (A/A$_s$ - 1) days left in the spraying cycle. For this area the water balance is

$$(I + P) = (E_s - \check{R})(A/A_s - 1) \tag{5}$$

where (A/A$_S$ - 1) is the number of days in each rotation when water is not being sprayed over an area, and $\check{R}$ is the average daily rainfall for days on which no spraying took place. (Note that R is the rainfall for the day on which spraying took place).

Thus for an exact water balance for the entire surface being spray-irrigated:

$$(S - E_A) h/24 = E_{SS} - R + (E_S - \check{R})(A/A_S - 1) \tag{6}$$

For example, suppose A/A$_S$ = 5, E$_A$ = 3 mm/d, E$_{SS}$ = 4 mm/d, E$_S$ = 2 mm/d, R = 0, $\check{R}$ = 0 and h = 10 hours
Then   S = 31.8 mm/d
In other words, if 20% of the waste surface area is sprayed every day on a 5 day rotation, with no rain, 31.8 x 10/24 = 13.25 mm of water could be applied each day without causing an overall increase of the water content of the waste body.

Clearly, if it rains by more than a value given by

$$R + \check{R} (A/A_s - 1) = E_{SS} + E_S(A/A_S - 1) \tag{7}$$

spraying will not be possible without upsetting the water balance.

3 MEASURING WATER BALANCE PARAMETERS

The water balance parameters that need to be measured are E$_A$, E$_{SS}$ and E$_S$. These parameters all vary seasonally and with the weather and considerable effort is required to assess them and their variation and limits. S, the rate of application of waste water by sprinkler, also needs to be established in terms of the performance characteristics of the sprinklers and pumping system being used.

3.1 Aerial Evaporation, E$_A$

This can be measured by comparing the rate of application measured by means of a flowmeter in the hose feeding a sprinkler, with measurements made by a number of rain gauges placed just above the surface, within the area covered by the sprinkler. The difference between water supplied and water actually reaching the waste will be a measure of E$_A$, while the water reaching the waste will be S.

3.2 Surface Evaporation, E$_S$ and Saturated Surface Evaporation, E$_{SS}$

These can be estimated by means of radiation balance measurements or by means of microlysimeters (Bowen, 1926 Blight, 1997, Blight and Blight, 1999). The radiation balance at the surface of the waste deposit will be given by:

143

$$R_n = R_I - R_O = G + H + L_e \qquad (8)$$

Where R_n is the net radiation, incoming (R_I) less outgoing (R_O);

    G   is the energy consumed in heating the waste surface and near-surface waste;

    H   is the energy consumed in heating the air over the waste surface; and

    L_e is the latent heat of evaporation consumed in evaporating water from the waste surface.

Knowing L_e and the latent heat of vaporization of water, the evaporation rate E_s can be calculated in mm/d.

To find L_e, R_n is measured by means of a solar radiometer, G is derived by measuring the changing near-surface temperature gradient below the waste surface, and H is measured from temperature and humidity gradients in the air just above the waste surface. L_e is then found by means of the "Bowen's Ratio" calculation (see Blight, 1997).

To measure E_{SS}, the measurements would be made above an extensive area that had recently been sprayed, while to measure E_S, the measurements would be made above an extensive area that had last been sprayed a few days previously. The measurements need to be made at different times of the year and in various weather conditions so that different seasonal evaporation rates and the effects of variable weather can be assessed.

Microlysimeter measurements are a cheap and effective way of measuring evaporation from a soil or waste surface, providing that only a short-term estimate is required. The method was originally suggested by Boast and Robertson (1982) and was recently evaluated by Blight and Blight (1999). The comparative measurements described by Blight and Blight (1999) show that measurements can be made over periods of up to 2 weeks, with results that compare well with radiation balance measurements.

A microlysimeter consists of a short section of rigid plastic pipe open at both ends. Sections of 150 mm to 200 mm diameter and of a similar length are convenient, but in principle, any suitable size can be used. If the waste is soft enough and of suitable consistency, the lysimeter can be driven into the surface, and an undisturbed core taken of the near surface waste including any surface vegetation. The core, in the plastic pipe, is carefully extracted and the lower end of the extracted waste is sealed by means of a plastic sheet taped over the outside of the lysimeter pipe, using adhesive plastic tape. The core, in its lysimeter pipe, is weighed and then re-inserted into the hole from which it was extracted. Losses of moisture from the microlysimeter are then measured over a period of few days by re-extracting the lysimeter each day and weighing it.

If the waste is fibrous or too hard to drive in the lysimeter tube, it will be necessary to trim the core to fit the lysimeter tube by excavating all round the cylinder, trimming away the excess and easing the tube down over the waste core. Once the core has been extracted, sealed at its lower end and weighed, it is reinserted into the hole. The hole is then backfilled, tamping the loose waste around the walls of the microlysimeter to achieve a similar condition to that of a driven lysimeter. The aim is to produce a removable undisturbed core that has the same surface condition as the undisturbed surface that surrounds it.

3.3 Limiting Moisture Storage Capacity

The overall limiting moisture storage capacity depends on the height of the waste deposit, the elevation of the local ground water table, and the moisture content-suction relationship of the waste (see Blight and Roussev, 1995).

For the present purpose, however, the limiting moisture storage capacity can be defined as the maximum moisture content the near-surface waste can hold in its pores by capillarity against the force of gravity. It is conventionally measured by compacting a specimen of the waste into a CBR mould with a perforated base plate and ponding the surface of the waste until the water starts dripping out of the base. The surface of the waste is then sealed with a plastic sheet to prevent water loss by evaporation and the specimen left until it stops draining. The drained moisture content of the waste is then taken as the limiting moisture storage capacity. As moisture can easily be drawn to the surface by evaporation gradients from a depth of 500 mm, or more, this limiting moisture capacity, less the initial water content over the top 500 mm of waste can be used as a guide to deciding on the maximum value of S, the permissible application of water by spraying.

4 TYPICAL MEASUREMENTS

The following measurements are illustrative of measurements made on the surface of a power station ash dump:

4.1 Radiation balance to estimate E_S and E_{SS}:

Figure 1 shows radiation balance measurements made on a chilly, overcast day (air temperature varying from 2°C at 06h30 to 15° C at 11h00). These measurements were made to assess a lower limit to daily E_S and E_{SS}. The figure shows the recorded variation of the net incoming radiation R_n, the energy absorbed by the near-surface ash and the sensible heat absorbed by the air, H. The latent heat of evaporation, L_e, was then found by subtracting $(G + H)$ from R_n. Except for a bright period between 10h00 and 11h00, the level of net radiation was unusually low for the area in which the ash dump is situated.

The net evaporation (E_S) corresponding to Figure 1 was 1.1 mm for the 9 hours of observations. The corresponding microlysimeter evaporation for a 24 hour period (from 16h00 on the previous day) was 1.2 mm. Figure 2 shows radiation balance measurements for a cool sunny day (air temperature varying from 4°C at 07h00 to 16°C at 14h00). In this case general levels of incoming net radiation were higher, and the net evaporation corresponding to Figure 2 was 2.8 mm. Corresponding microlysimeter evaporation for 24 hours from 17h00 on the previous day was 2.6 mm (ranging from 2.3 to 2.9 mm).

Experience so far has shown that there is little difference between E_s, the evaporation from a "drying" surface, and E_{SS}, that from a saturated surface.

4.2   Spray-irrigation to estimate E_A

Figure 3 shows the results of a spray irrigation test to measure E_A, the aerial evaporation. The figure shows measured contours of irrigation water reaching the surface of the ash, expressed in mm (ℓ/m^2) of water over the 6.5h duration of the test. The total volume of irrigation was found by calculating the volume under the contoured surface (m^2 x $\ell/m^2 =$ ℓ) and then subtracting this from the total volume of water pumped through the sprayer in

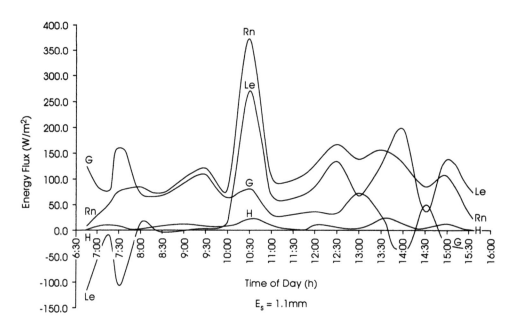

Figure 1: Radiation balance measurements on a chilly, overcast day

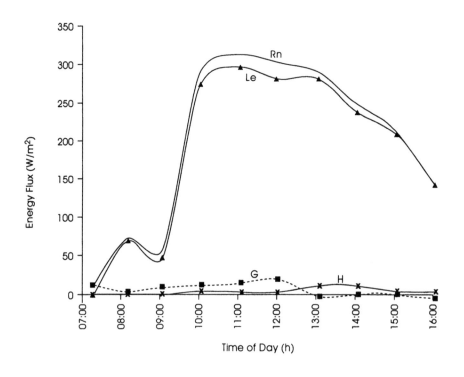

Figure 2: Radiation balance measurements on a cool sunny day

the test period. The difference represented E_A, the aerial evaporation. The average volume of E_A was found by dividing the volume of evaporation by the area enclosed by the zero irrigation contour in Figure 3. This gave an average value for E_A of 3.2 mm for 6.5h or about 3.5 mm for the period covered by Figure 2 (pro rata to the incoming net radiation). Hence E_A was 25% more than E_s for this day. The total evaporation for the day was thus

2.8 + 3.5 = 6.3 mm

or 63kℓ per ha of ash surface.
In period of hot dry weather, it is expected that these figures could double.

5 CONCLUSION

Evaporation by controlled spray-irrigation is a promising way disposing of waste water without causing pollution. There are, however, a number of aspects that need to be investigated before such a system can be designed with confidence. These include the behaviour of the wetting front produced by a period of a spray irrigation and the drying front that will develop during the ensuing period of drying. Provision must also be made to store water during periods of rain when spray irrigation has to be stopped. The possible problem of pollution of surrounding land by saline spray blown off the waste deposit by high wind also needs investigation.

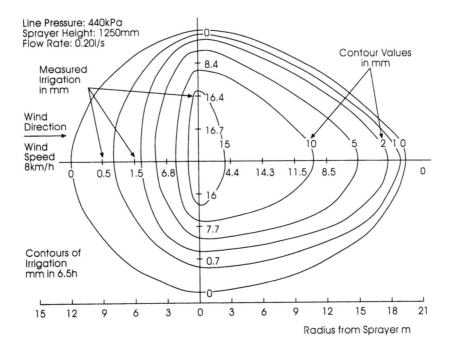

Figure 3   Contours of spray irrigation water reaching waste surface

ACKNOWLEDGEMENT

Eskom, South Africa's national power generating authority that sponsored the research, is thanked for permission to publish this paper.

REFERENCES

Blight, G.E. (1997) Interactions between the atmosphere and the earth. *Geotechnique*, Vol 42; p715-766

Blight, G.E, and Blight, J.J. (1993) Runoff from landfill isolation layers under simulated rainfall. In *Geology and Confinement of Toxic Wastes*. Balkema, Rotterdam: p313-318

Boast, C W and Robertson, T M 1982. A Microlysimeter method for determining evaporation from bare soil: description and laboratory evaluation. J. Soil Science Society of America, 46; 689-696.

Blight, J J and Blight, G E 1999. The microlysimeter technique for measuring evapotranspiration losses from a soil surface. *12th African Regional Conf*, Durban, South Africa.

Blight, G E and Roussev, K 1995. The water balance for an ash dump in a water-deficient climate. *1st Int. Conf. on Unsaturated Soils*, Paris, France, Vol 2, 833-840.

Bowen, I S. 1926. The ratio of heat losses by conduction and by evaporation from any water surface. *Physical Review*, 27, 779-787.

Fenn, D G, Harley, K J, De Geare, TV. (1975). *Use of the water balance method for predicting leachate generation from solid waste disposal sites*. US. EPA Report No EPA/530/S168, Washington, USA.

Tailings and Mine Waste'00 © *2000 Balkema, Rotterdam, ISBN 90 5809 126 0*

Static liquefaction as an explanation for two catastrophic tailings dam failures in South Africa

A. B. Fourie, G. Papageorgiou & G. E. Blight
University of the Witwatersrand, Johannesburg, South Africa

ABSTRACT: The Bafokeng platinum tailings dam and the Merriespruit gold tailings dam failures both resulted in loss of life and enormous financial and environmental damage. Although investigations of both disasters provided detailed explanations of the sequence of events that led to the failures and likely mechanisms of failure, why these dams were susceptible to flow failures has not been examined in any detail. The results of a laboratory testing programme to define the Steady State Line for both tailings and how the position of the line varies with particle size distribution are described. The results confirm that if the tailings in-situ density is low enough, static liquefaction is highly likely. Results of field investigations at the Merriespruit tailings dam are evaluated in terms of liquefaction susceptibility. Piezocone data show the anomaly of extremely low ratios of undrained shear strength to effective overburden pressure, which is consistent with the argument that zones of the tailings were in a meta-stable state.

1 INTRODUCTION

Destructive mudflows as a consequence of the breaching of a tailings dam impoundment are fortunately relatively rare. The two incidents in South Africa that attracted the most attention were the failures of the Bafokeng platinum tailings dam in 1974 and the Merriespruit tailings dam in 1994, principally because of the loss of life that occurred. Many important lessons were learned from the post-failure investigations of these occurrences, hopefully resulting in improvements to the design and operation of subsequent tailings deposits. However, there is also usually a number of contradictory opinions as to the mechanism of these failures and as to why mudflows ensued. Often hypotheses are put forward that are difficult to either prove or refute because the tailings that was part of the mudflow is obviously in a highly disturbed state, completely different from its in-situ pre-failure state.

In this paper we describe the results of laboratory tests that were carried out to determine the Steady State Line for both the Merriespruit and the Bafokeng tailings. By comparing these results with values of in-situ void ratios of tailings immediately adjacent to the Merriespruit failure scar, it is shown that static liquefaction may well have been the cause of the mudflow that resulted at Merriespruit. Results of piezocone testing carried out as part of the post-failure investigation are re-evaluated and it is shown that in one case the tailings still appeared to be in a meta-stable state in-situ.

2 DESCRIPTION OF FAILURES

Detailed descriptions of the Bafokeng and Merriespruit failures and the sequence of events prior to the failures are given by Jennings (1979), Wagener et al (1998) and Blight (1998) and the discussion here will accordingly be kept brief.

The Bafokeng platinum ring-dyke tailings dam failed as a result of overtopping after a large storm (Jennings 1979). The overtopping caused a breach in the wall of the impoundment, through which about 3×10^6 m^3 of tailings slurry flowed. This slurry was extremely mobile and at a distance of 4km from the breach had spread to a width of 800m and was 10m deep (Blight 1998). It continued down the course of a nearby river bed, with about 2×10^6 m^3 of tailings eventually flowing into a water storage reservoir 45km downstream of the Bafokeng mine.

Failure of the Merriespruit gold ring-dyke tailings dam in 1994 was similar in many respects to the Bafokeng failure. It again occurred after a significant rainfall event, although in this instance it appears to have been exacerbated by the fact that a large volume of water was being stored on the top of the dam at the time of the failure. Overtopping of the tailings dyke occurred, causing a breach to develop. Approximately 500 000 m^3 of tailings flowed through this breach, wreaking havoc in the adjacent village and flowing for a distance of about 2km.

In both cases there had been unmistakable signs (particularly in hindsight) in the weeks and months prior to the disasters that the dams were not in acceptable states. There had been evidence of limited toe sloughing, which in the case of Merriespruit led to the construction of a slimed buttress and excessive seepage on the face of the dams occurred. However, there are many instances on South African tailings dams (mostly unreported) where similar instances of local instability or excessive seepage have been attended to using a variety of palliative measures, and the dam has continued to be operated successfully. As mentioned above, it is only with hindsight that conclusions can be drawn with certainty about the inevitability of these dams failing catastrophically. What is clearly needed is the ability to make such deductions before the event, so it can be avoided. This paper describes the results of some laboratory and field testing that may go some way to achieving these goals.

3 POSTULATED CAUSES OF FAILURES

In both cases the failures were initiated by overtopping of the perimeter tailings dyke, resulting in erosion of the dyke and exposure of tailings that had previously been confined. The exposed tailings then flowed through the breached dyke, like a viscous liquid, coming to rest at some distance from the tailings dam. In both cases the mudflow that resulted was preceded by a series of loud bangs (in the case of Merriespruit a single, loud bang was reported by eyewitnesses).

The trigger mechanism for both failures thus appeared to be inadequate management of tailings water on the top of the tailings dam. For a mudflow to occur, however, there must be both a trigger to initiate the mudflow and the tailings must be in a state such that it is susceptible to flow. There are many reported cases where failures of the perimeter dyke have not resulted in catastrophic mudflows and have been remediated fairly easily. As an example, earlier in 1974 (the year of the Bafokeng tragedy) a slope failure occurred at the same dam, but tailings only moved about 20m from the original toe line of the dam before coming to rest. Blight (1998) reports on the failure of the perimeter dyke of a gold tailings dam at the Simmergo mine in South Africa, where the tailings only moved a very short distance.

Why then do some trigger mechanisms not result in mudflows whilst others do? Clearly, one of the main contributing factors to a mudflow is the amount of water stored on top of the dam (or indeed inside the dam, eg a very elevated phreatic surface). If there is a large tailings pond present at the time of a breach occurring, then this water will obviously flow through the breach, eroding tailings as it flows and perhaps being a major contributor to a mudflow developing. There are, however, instances where relatively dry deposits of waste material have also flowed great distances. In the case of the failures of coal waste dumps that flowed, no water was stored on top of the dump, which itself was relatively dry (Dawson et al 1998). Furthermore, Sasivarathan et al (1994) have demonstrated collapse behaviour of completely dry specimens in laboratory triaxial tests.

Much of the discussion in South Africa of the Bafokeng and Merriespruit failures refers to erosion mechanisms when describing the mudflows that occurred. There has not been much discussion of the concept of static liquefaction, in contrast to much of the literature around failures such as the Nerlerk artificial islands (Been et al 1991). However, from the above discussion, it seems that the state of the tailings in-situ is inextricably linked to whether or not it will flow when a particular trigger occurs. This concept is embodied in the definition of 'state' that was so clearly described by Been et al (1985).

4 STATIC LIQUEFACTION

The term liquefaction is often associated with the consequences of earthquakes or other forms of dynamic loading. Our concern when discussing the Bafokeng and Merriespruit failures is not with dynamic loading, but rather the phenomenon that has become known as static liquefaction. Trigger mechanisms for this type of failure are usually slope instability, a rise in the phreatic surface, toe erosion or rapid application of surface loading. When considering the potential liquefaction of a soil deposit, it has been found useful to use as a frame of reference the concept of the steady state line (SSL), as discussed by Yamamuro and Lade (1998) amongst many others. The concept will not be reviewed here. The SSL defines the relationship between the ultimate void ratio and the mean effective stress p' (where $p' = \frac{1}{3}(\sigma_1' + 2\sigma_3')$ in a triaxial test. The ultimate condition is defined as that for which deformation takes place under constant stress and at constant volume. As shown by Been et al (1991), the steady state and the critical state are the same condition. In-situ soil at a void ratio above the SSL will strain soften (contractive behaviour) when loaded undrained, whilst a void ratio below this line results in strain-hardening (or dilative) behaviour.

5 LABORATORY TESTING TO DETERMINE POTENTIAL FOR STATIC LIQUEFACTION

Four different particle size distributions of the Merriespruit gold tailings were used in this investigation. A large bulk sample of tailings was recovered from a number of locations within the failure scar of the dam and immediately adjacent walls. Aside from the as-recovered sample, which had 60% finer than 75μm, three other particle size distributions were created by selectively sieving this material, the objective being to investigate the influence of particle size distribution on the location of the Merriespruit SSL.

The resulting four curves are shown in Figure 1a, together with the upper and lower grading limits of samples recovered from the Merriespruit tailings dam during the post-failure investigation. Although the two coarser 'manufactured' curves fall outside this envelope, it was conceivable that tailings that had been part of the flow slide may have fallen outside this envelope. Furthermore, we wanted to investigate a wide range of particle size distributions, including 'clean' specimens.

Two different gradings of Bafokeng tailings were used. The particle size distribution curves for these samples are shown in Figure 1b, together with the upper and lower limits of curves obtained during the post-failure investigation. The 60% fines material was a bulk sample obtained from the tailings dam beach, while the 30% sample was created by sieving out fine material. It was created to coincide with the coarse limit previously obtained, as shown in Figure 1b.

The maximum and minimum void ratios for the different specimens were determined according to ASTM D2049 (1983) and are summarised in Table 1.

5.1 Specimen preparation

Previous work has shown that a very high degree of specimen homogeneity is achievable with the moist tamping method (Sladen et al, 1993). Great care however needs to be taken to account for the volume changes that occur during specimen set-up and consolidation. The procedures described by Garga and Zhang (1997) were carefully followed and changes in volume monitored throughout the pre-shear stage of testing. Specimens were isotropically consolidated to the required confining effective stress prior to shearing, which was carried out undrained.

5.2 Results

An example of a stress-strain curve and excess pore pressure versus axial strain curve for the

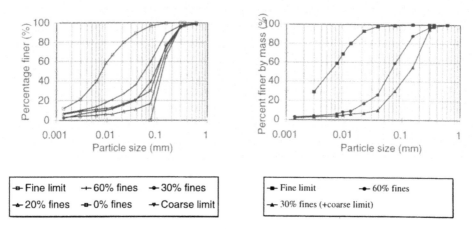

Figure 1: Particle size distribution curves for (a) Merriespruit and (b) Bafokeng tailings

Table 1: Maximum and minimum void ratios for Merriespruit gold and Bafokeng platinum tailings

Sample	Maximum Void Ratio	Minimum Void Ratio
M1- 0 % fines	1.221	0.738
M2- 20 % fines	1.326	0.696
M3- 30 % fines	1.331	0.5772
M4- 60% fines	1.827	0.655
B1- 30% fines	1.493	0.375
B2- 60% fines	1.593	0.406

(where percent fines is defined as the percentage finer than 75μm)

Merriespruit tailings is given in Figures 2 and 3. A characteristic of the tests carried out on both Merriespruit and Bafokeng tailings is the rapid increase in excess pore pressure, which then remains relatively constant. Little or no dilation occurs and the specimen does therefore not strain harden. The opposite in fact is most common. As shown in Figure 4, which is a stress path plot for the same Merriespruit specimen, the stress path shows a decreasing mean stress which eventually precipitates a drop in shear stress. The specimen strain softens and would fail suddenly and very rapidly after passing the peak shear stress value, were it not for the strain-controlled nature of the applied loading.

The end points of stress paths such as those shown in Figure 4 were used to construct Steady State Lines for the various tailings and the results are summarised in Figure 5. It was clear that the percentage of fines in a tailings sample had a marked effect on the location of the resulting SSL and it would be inappropriate to define a single SSL for a particular tailings deposit if there was any chance of particle sorting or segregation on the tailings beach leading to variations in particle size distribution across the dam.

Reference to post-failure investigations at Merriespruit showed that many of the undisturbed specimens had void ratios that placed them above the relevant SSL shown in Figure 5. Undrained loading of such specimens would cause a decrease in mean effective stress, thus causing strain softening and collapse. Based on reports of the sequence of events that occurred at Merriespruit

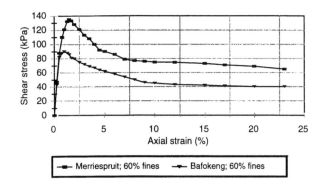

Figure 2: Typical shear stress vs axial strain curves for Merriespruit and Bafokeng tailings

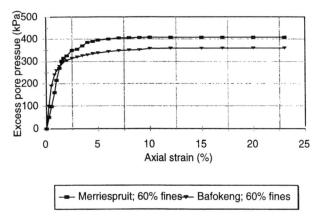

Figure 3: Typical excess pore pressure versus axial strain plots for Merriespruit and Bafokeng tailings

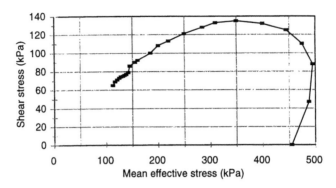

Figure 4: Example of stress paths for Merriespruit tailings displaying contractive characteristics

it appears that the tailings immediately inside the perimeter of the dam was exposed very rapidly because of the erosion that occurred when the perimeter of the 30m high tailings dam was breached and the dam emptied. This material was thus likely loaded in an undrained manner.

The fact that it liquefied and flowed is, we suggest, because it was in a meta-stable state in-situ and once exposed it was unable to resist the shear stress imposed by its own self-weight. Papageorgiou et al (1999) collected results of void ratio measurements made on undisturbed samples of Merriespruit tailings that were collected during the post-failure investigation. These results showed that more than 50% of these samples had in-situ void ratios greater than 1.0, which would place them above all but the 'clean' Merriespruit Steady State Line.

6 EVIDENCE OF LIQUEFACTION SUSCEPTIBILITY FROM FIELD TESTS

As part of the investigation of the Merriespruit failure a number of piezocone tests were carried out. Most of the tests that were carried out around the perimeter of the dam or adjacent to the failure scar showed fairly dense material, with predominantly dilative characteristics. One exception was a test carried out immediately west of the failure on the middle berm of the dam. The results of this test are shown in Figure 6 (after Wagener et al 1998). The type of piezocone commonly used in South Africa is not fitted with a friction sleeve and thus results are only shown for cone end resistance and excess pore water pressure. As can be seen from the figure, between depths of about 2m and 11m the end resistance is consistently low, usually being below 1MPa. The material in this region was thus still relatively soft (and probably relatively loose) some weeks

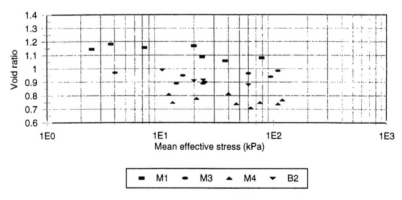

Figure 5: Steady State Lines for Merriespruit and Bafokeng tailings

154

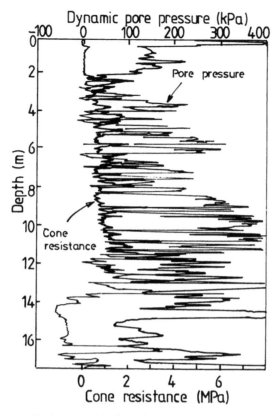

Figure 6: Piezocone profile for test at Merriespruit (after Wagener et al, 1998)

after the actual failure had occurred. The in-situ undrained shear strength of the tailings was calculated from the piezocone data and compared with the likely in-situ shear stress in the tailings. As described by Robertson and Fear (1995), if the in-situ shear stress exceeds the large-strain undrained shear strength of a deposit, the material is inherently unstable.

The determination of the undrained shear strength, C_u, using a piezocone has been relatively successful in fine-grained, cohesive soils. Using correlations with vane shear test results and the back-analysis of embankment failures, the following equation is widely used:

$$C_u = \frac{q_c - \sigma_{v0}}{N_{kt}} \qquad (1)$$

where N_{kt} is an empirical cone factor and is analogous to the bearing capacity factor N_γ. From tests on normally consolidated clay deposits, the value of N_{kt} varies between 11 and 19, with a typical average value of 15, (Senneset et al, 1989).

The above equation was obtained for normally consolidated clay deposits. Its suitability to a mine tailings, which will usually consist of fine sand and silts, is unproven. Tailings is likely to drain much more rapidly than a normally consolidated clay and it may thus be difficult to measure a truly undrained shear strength. If anything, the piezocone is likely to overestimate the undrained shear strength, but in the absence of field correlations it seems acceptable to use the above equation, with $N_{kt}=15$.

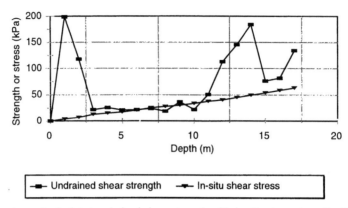

Figure 7: Comparison of calculated undrained shear strength with in-situ shear stress for Merriespruit piezocone profile

6.1 *Application to Merriespruit piezocone data*

The undrained shear strength profile for the piezocone data shown in Figure 6 was calculated using the above procedure. The results are plotted in Figure 7 which shows the undrained shear strength compared with the estimated in-situ shear stress. It is clear that between depths of about 6m and 10m, the in-situ shear stress may be greater than the large-strain undrained shear strength. At this particular location there therefore appears to be a zone of confined tailings that only remains stable by virtue of the fact that it is confined by other denser, dilative tailings, ie tailings that have an in-situ void ratio that is less than the equivalent steady state value. While there are many factors that impact on the value of the undrained shear strength inferred from piezocone data, such as rate of loading and possible anisotropy of the deposit, the fact that not only does the undrained shear strength approach the in-situ stress, but even drops below it at some depths, indicates that the tailings is very much in a meta-stable state.

The history of the Merriespruit tailings dam that was described in detail by Wagener et al (1998) indicates that sub-aqeous deposition of tailings frequently occurred in the vicinity of where the flow failure eventually occurred in 1994. Such procedures would almost certainly have encouraged the development of tailings with a very high void ratio, quite possibly higher than the equivalent steady state value.

Unfortunately no data on the post failure in-situ void ratios for the Bafokeng tailings dam are available, nor were any piezocone tests carried out at any time. It is therefore not possible to make such a compelling case for static liquefaction at Bafokeng as it is for Merriespruit, but the simple fact that so much of the material at Bafokeng flowed and for such a great distance, coupled with the demonstration in this paper that contractive behaviour of Bafokeng tailings is easily achieved, makes this a highly likely scenario.

The results shown in Figure 7 are surprising and worth further consideration. The data was reworked in terms of the variation of the ratio C_u/σ_v' with depth. This ratio, termed the C_u ratio in this paper, is rarely less than 0.2 for normally consolidated natural clays and is usually between 0.2 and 0.4. This ratio was calculated for the results shown in Figure 7 and the resulting variation with depth is shown in Figure 8. The values drop below 0.2 and remain low over much of the profile. The obvious question is why such low C_u ratios occur? Surely the overburden stress must be sufficient to compress and consolidate the tailings so that the void ratio decreases sufficiently to produce a reasonably large undrained strength? The answer to this apparent anomaly may be

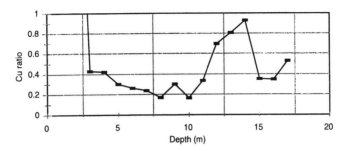

Figure 8: Variation of the ratio C_u/σ_v' with depth for Merriespruit piezocone profile

found in the results of the laboratory tests published in this paper, and indeed in many other papers dealing with contractive behaviour under undrained conditions.

The stress path for the Merriespruit specimen shown in Figure 4 starts at an initial mean effective stress of about 450kPa. The steady state undrained shear strength of this specimen was about 65kPa, giving a ratio of approximately 0.14 which is below the typical lower limit for natural clays. It is likely therefore that if the tailings is placed at an initially high void ratio, it is possible that even under the large imposed overburden stress of an accreting tailings deposit, the void ratio at depth may still be very high.

7 CONCLUSIONS

It is argued that the reason the Bafokeng failure of 1974 and the Merriespruit failure of 1994 resulted in catastrophic flowslides was because a sufficiently large volume of the tailings was ina meta-stable state prior to the failures. Laboratory triaxial compression tests were used to define the Steady State condition for both tailings. In the case of the Merriespruit failure it was likely that tailings existed at higher void ratios than the critical value and were thus highly susceptible to liquefaction under undrained loading. Piezocone tests carried out during the post-failure investigation were re-evaluated and in one case the tailings at depth still appeared to be in a meta-stable condition. Very low values of the ratio C_u/σ_v' were calculated for this particular piezocone profile and this apparently anomalous result is worth further investigation in other tailings deposits.

REFERENCES

Been, K. and Jefferies, M.G. 1985. A state parameter for sands. *Geotechnique*, Vol.35, No.2, pp99-112.

Been, K.., Jefferies, M..G. and Hachey, J. 1991. The critical states of sands. *Géotechnique*, 41(3): 365-381.

Blight G .E .1998. Destructive mudflows as a consequence of tailings dyke failures. *Proc. Instn Civ. Engrs.* Vol.125, pp 9-18.

Dawson R.F., Morgenstren N.R. & Sokes A.W. 1998. Liquefaction flowslides in Rocky Mountain coal mine waste dumps *Canadian Geotechnical Journal*, Vol 35, pp 328-343.

Garga, V.K., and Zhang, H. 1997. Volume changes in undrained triaxial tests on sands. *Canadian Geotechnical Journal*, Vol.34, pp762-772.

Jennings, J.E. 1979. The failure of a slimes dam at Bafokeng. Mechanisms of failure and associated design considerations. *Civil Engineer in South Africa*, Vol.21, No.6, pp135-141.

Papageorgiou, G., Fourie, A.B. and Blight, G.E. 1999. Static liquefaction of Merriespruit gold tailings. *Proc 12th African Regional Conference: Geotechnics for Developing Africa*. Durban, South Africa, October 1999.

Robertson, P.K. and Fear, C.E. 1995. Application of CPT to evaluate liquefaction potential. *Proceedings of the International Symposium on Cone Penetration Testing*, CPT'95, Linköping, Sweden, Vol.3, pp 57-79.

Sasitharan, S., Robertson, P.K., Sego, D.C. and Morgenstern, N.R. 1994. State-boundary surface for very loose sand and its practical implications. *Canadian Geotechnical Jnl*, Vol.31, pp321-334.

Senneset, K, Sandven, R. and Janbu, N. 1989. Evaluation of soil parameters from piezocone tests. *Transportation Research Board*, National Research Council, 1989, Washington, DC, pp24-37.

Sladen, J.A. and Handford, G. 1987. A potential systematic error in laboratory testing of very loose sands. *Canadian Geotechnical Journal*, Vol.24, pp462-466.

Wagener, F., Craig, H.J., Blight, G.E., McPhail, G., Williams, A.A.B. and Strydom, J.H. 1998 The Merriespruit tailings dam failure - A review. *Proceedings Tailings and Mine Waste '98*. Colorado State University, January 1998, pp 925-952.

Yamamuro, J.A. and Lade, P.V. 1998. Steady-State concepts and static liquefaction of silty sands. *ASCE Journal of Geotechnical and Geoenvironmental Engineering*, Vol.124, No.9, pp 868-877.

Tailings and Mine Waste'00 © 2000 Balkema, Rotterdam, ISBN 90 5809 126 0

A study of granulation distribution in a beach of tailings dam

M.J.Lipiński & A.Gołębiewska

Department of Geotechnics, Warsaw Agricultural University, Poland

ABSTRACT: Stability evaluation of a tailings dam constructed by upstream method must consider possibility of liquefaction of the sediments. Such an analysis requires undrained shear strength, which is strictly related to soil kind. Due to large non uniformity of material, resulting from a discharged method, precise identification of granulation distribution in a beach of tailings dam is a difficult task and is rarely addressed in geotechnical literature.

In the proposed paper, approach to a quantitative assessment of sediment granulation distribution in the beach of Żelazny Most tailings pond is described. Large body of index properties data obtained by three laboratories were elaborated. In order to quantify soil kind, a several parameters were introduced to substitute a grain size curve. Mutual correlation among these parameters made possible to distinguish those of them, which most closely represent grain size distribution and can be used for reconstruction of a granulation curve in any point of the cross section of the beach.

1 INTRODUCTION

Stability evaluation of tailings dams constructed by upstream method should consider the most critical conceivable performance of the construction i.e. liquefaction of soil, which usually leads to a flow failure of the dam. Such a analysis requires undrained shear strength of tailings material as an input parameter in stability calculation. However, this parameter is very sensitive not only to small change in state but also to a change in granulation (Lipiński 1999). Precise identification of undrained shear strength value is usually determined in a laboratory for known material (rarely with considerable amount of fines), void ratio and state of stresses. In analysis of tailings dam performance such a tediously determined parameter has very limited applicability, because of large nonuniformity of tailings material resulting from the placement method. This nonuniformity - in practice - makes impossible to distinguish any geotechnical layers of significant thickness. Farthermore, quantitative interpretation of shear wave velocity measurement in tailings have to be referred to certain volume of specified granulation. Therefore in large projects, any projection of laboratory and field test results on in situ conditions should be preceded by evaluation of grain size distribution of tailings material in the beach of a dam.

The paper presents results of experimental work focused on analysis of grain size distribution in the beach of post-flotation copper tailings pond Żelazny Most located in Poland. The pond has been constructed by upstream method since 1975. The present capacity is more than 270 mln cubic meters and the designed one is around 1bln cubic meters. The total circumferential length of the dam exceeds 14 km and present height of the dam ranges between 22 and 44 m. More details on the pond description can be found in monograph of KGHM (1996).

Classification tests were carried out by five laboratories in the period 1992 - 1997. Tailings material was sampled in 31 profiles located in 5 cross sections. The analysis of grain size distribution of tailings was focused on beach area in three cross sections as shown in Figure 1. Test profiles in each cross section had the following distances from the crest of the dam: 40, 80, 120,

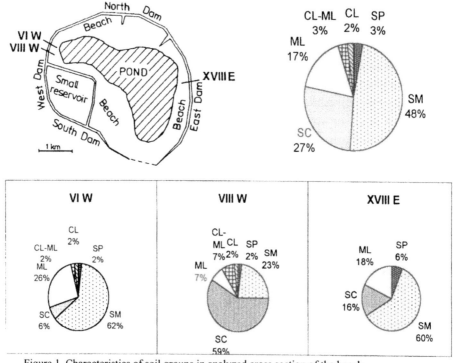

Figure 1. Characteristics of soil groups in analyzed cross section of the beach.

160, 200 and 240m. In each profile cylindrical cores were taken (by MOSTAP sampler produced by AP van den Berg) usually every third meter. For refined analysis of granulation distribution 145 granulation curves were considered. Each curve represented granulation of mixture obtained from sampled 1 m long tailings core. Before mixing tailings material each sample was carefully examined and cohesive soils of clay fraction (< 0.002 mm) content exceeded 10 % were removed. The rationale for that rests on the assumption that sand/silt mixtures is a matrix of tailings skeleton and control its mechanical behavior. Presence of small amount of clay fraction in a mixture could have shifted the obtained granulation curve towards fines, thus making the resulting curve not representative for the deposit.

2 OBJECTIVE OF THE STUDY

The objective of the study was to work out a procedure, which makes possible reconstruction of grain size distribution of tailings in any point of the analyzed cross section of the beach area. This study is a first step which is to provide data for correlation with static penetration in tailings, which hopefully will make possible to convert sounding parameters from CPTU to granulation of material represented by a number.

At start of work three hypotheses were assumed to hold true:

1. During placement of tailings segregation process takes place, what results in decreasing diameter of grain with increasing distance from a discharge point
2. If a granulation curve can be expressed by a single parameter, there is a relationship between this parameter and distance from a discharge point.
3. Most suitable parameter for numerical representation of granulation curve is *SFR*, which is defined as the ratio of dry weight of sand fraction (> 0.074 mm) to fine fraction(< 0.074 mm).

SFR is a parameter defined for a ternary diagram developed by Scott and Cymerman (1984).

160

This diagram is very useful for distinguishing between sandy behavior and clayey behavior and can be used for evaluation of soil susceptibility to liquefaction (Carrier 1991).

The first hypothesis is actually confirmed by Stokes law, while genuineness of the second and the third hypotheses has been proved on the basis of analysis of granulation curves and index properties of tailings.

3 TAILINGS MATERIAL DESCRIPTION

As shown in Figure 1, according to USCS, in beach tailings prevail silty sands, sand-silt mixtures (SM) and clayey sands, sand-clay mixtures (SC). *SFR* for tested tailings is in the range 0.3 - 10. Around 90 % of granulation curves starts at diameter 0.25 – 0.30 mm and show less than 5% of clay content.

Index properties of majority of the data are in the following ranges:

Bulk Density	ρ	$\in$	1400 - 2200 kg/m^3
Dry Density	ρ_d	$\in$	1300 - 1850 kg/m^3
Density of Solids	ρ_s	$\in$	2650 - 2790 kg/m^3
Water Content	w	$\in$	20 - 30 %
Void Ratio	e	$\in$	0.4 - 1.05

In cross section VIII W, SC soils create majority of tailings. This cross section was selected as an example and therefore all data presented later in this paper originate from VIIIW.

4 THE METHOD OF DATA EVALUATION

In order to verify assumed hypotheses partial analyses of granulation curves, index properties and granulation indexes were carried out and they are summarized as follows:
- Granulation curves were carefully examined in each profile and also at the same depths along a direction perpendicular to the crest of the dam. An example of granulation curves at various depths (representing one of 28 profiles) is shown in Figure 2, while an example of curves gathered in the other configuration (40 sets) is presented in Figure 3.
  The influence of each variable on the grain-size distribution has been assessed as the ratio of number of combinations confirming the existence of the influence to the number of possible combinations.
- Index properties of tailings were analyzed in profiles. An examples of distribution of void ratio, bulk density, dry density and density of solids is shown in Figure 4, while example of *SFR* profile is shown in Figure 5.

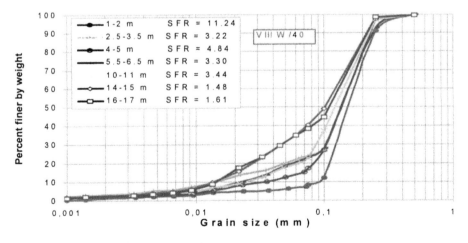

Figure 2. Grain size distribution of tailings at various depths

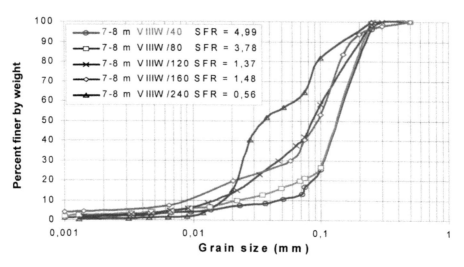

Figure 3. Grain size distribution of tailings at the same depth but various distance from the crest of the dam

− For verification of the hypothesis 3, relationship between *SFR* and several granulation indexes was analyzed. At first, the following indexes used in soil mechanics have been applied:
c_u = coefficient of uniformity,
c_c = coefficient of curvature,
c_d = coefficient of domination.
Having stated that not all of the above indexes are satisfactory for characterization of grain-size distribution of tested tailings new indexes have been introduced:
U_x = curve slope index,
tg α = curve slope tg α,
ASFR - areal sand to fines ratio.
Definitions of all grain size distribution measures used in the analysis are depicted in Figure 6 and specified below:
Coefficient of Uniformity

$$c_u = \frac{d_{60}}{d_{10}} \tag{1}$$

characterizes the degree of grain-size distribution uniformity in the diameter range of d_{60} - d_{10}
Coefficient of Curvature

$$c_c = \frac{d_{30}^2}{d_{10}d_{60}} \tag{2}$$

characterizes the degree of smoothness of granulation curve in the diameter range of d_{10}, d_{30}, d_{60}, used as suitability criterion for soil compaction
Coefficient of Domination

$$c_d = \frac{d_{10}d_{90}}{d_{50}^2} \tag{3}$$

characterizes degree of domination of coarse and fine grain-size in relation to diameter d_{50} (Kollis et al. 1968)
$c_d > 1 \Rightarrow$ coarse fraction prevails
$c_d < 1 \Rightarrow$ fine fraction prevails

Sand to Fines Ratio *SFR*

$$SFR = \frac{S}{F} \qquad (4)$$

S - sand fraction content (≥ 0,074 mm)
F - fine fraction content (< 0,074 mm)
S + *F* = 100%
used also for identification of soils susceptible to liquefaction.

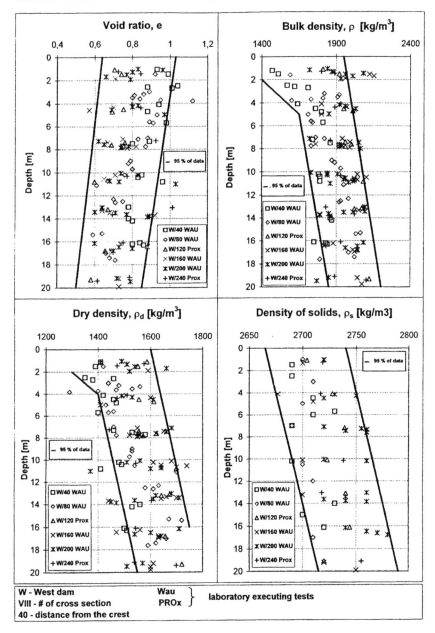

Figure 4. Example of index properties distribution in the beach profiles.

163

Curve Slope Index

$$U_x = \frac{d_{80}}{d_{30}} \qquad (5)$$

characterizes degree of diversification of grain-size distribution within the diameter range of d_{80}, d_{30}.

It has been found that the central part of granulation curve within the diameter range of d_{80} and d_{30} is close to a straight line.

Curve Slope tg α

$$\text{tg}\,\alpha = \frac{0.5}{\log \dfrac{d_{80}}{d_{30}}} K \qquad (6)$$

K = scale factor

$$K = \frac{vertical\ scale}{horizontal\ scale} \qquad (7)$$

example: vertical scale 0.1 = 13 mm
horizontal scale lg_{10}-lg_1 = 6.5 cm
$K = 2$

tgα was calculated assuming $K = 1$
tangent of inclination angle of part of granulation curve within the diameter range of d_{80} - d_{30}.
Distribution of all the above indexes against *SFR* (with the exception of *ASFR*) are shown in Figure 7.

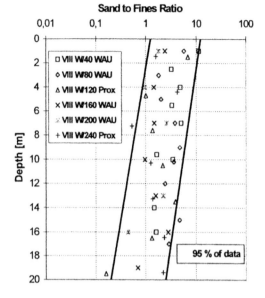

Figure 5. Example of *SFR* distribution in the cross section

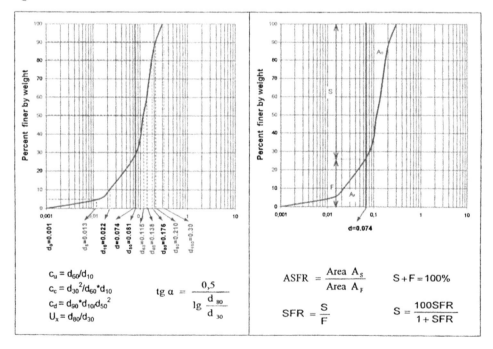

Figure 6. Definition of grain size distribution measures used in the analysis.

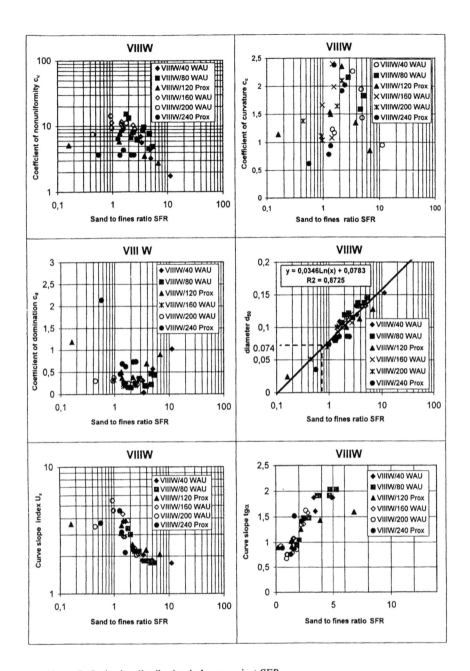

Figure 7. Grain size distribution indexes against *SFR*.

A special kind of parameter is areal sand to fines ratio *ASFR*, which characterizes the entire granulation curve (contrary to *SFR* ratio that informs about location of only one point on granulation curve at diameter 0.074 mm).

$$ASFR = \frac{A_S}{A_F} \tag{8}$$

A_S = area over a granulation curve from its intersection point with diameter 0.074 mm line,
A_F = area under a granulation curve from its intersection point with diameter 0.074 mm line.

However, determination of *ASFR* in practice is troublesome and time consuming, therefore it was necessary to obtain relationship between *SFR* and *ASFR.*. As shown in Figure 8, these two parameters correlate very well what proves genuineness of the hypothesis 3.

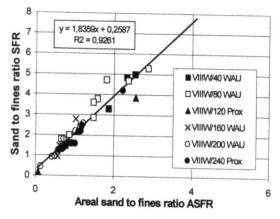

Figure 8. Sand to fines ratio SFR versus ASFR

The hypothesis 1 assumes that grain size distribution of tailings sample depends on its distance from a discharge point L'. For particular geometry of the dam, the true value of L' can be calculated on the basis of sampling depth and current distance from a discharge point (see procedure shown in Figure 9). Having estimated distance from a discharge point for each sample, it have made possible to obtain desired relationship between *SFR* and L'. An example of such relationship for cross section VIII W is shown in Figure 10.

All the above relationship made possible to elaborate a procedure for reconstruction of grain size distribution of tailings in any point of specified cross section on the basis of its location. The procedure can be briefly characterized in the following steps:
– determination of distance (L) between the profile of selected point and the upstream slope toe of the last completed stage of the dam corresponding to a crest elevation (R_z)
– determination of selected point elevation (R_o)
– having estimated tgα for the pond (see Figure 9) the distance from the discharge point (L') for selected point is determined , according to the formulas:

$$L' = L + \Delta l \tag{9}$$

$$\Delta l = (R_z - R_o)\text{tg}\alpha \tag{10}$$

– on the basis of L' value from the plot of *SFR* versus L' (Figure 10) the relevant *SFR* value is determined and thus fine fraction (< 0.074 mm) content is calculated

$$F = \frac{100}{1 + SFR} \tag{11}$$

– having got *SFR* value the diameter d_{50} from the plot $d_{50} = f(SFR)$ in Figure 7 is determined
– having F & d_{50} value the central part of a granulation curve my be drawn
– diameters d_0 , d_{100} and percent of particles <0.002 mm are assumed according to *SFR* values

$SFR \geq 9$	$d_{100} = 0.35$ mm	$d_0 = 0.020$ mm	0 % of particles 0.002 mm
$SFR = 3 - 9$	$d_{100} = 0.30$ mm	$d_0 = 0.005$ mm	0 %
$SFR = 1 - 3$	$d_{100} = 0.25$ mm	$d_0 = -$	3 %
$SFR = 0.5 - 1$	$d_{100} = 0.25$ mm	$d_0 = -$	5 %
$SFR = 0.18 - 0.5$	$d_{100} = 0.25$ mm	$d_0 = -$	8 %
$SFR < 0.18$	$d_{100} = 0.1 - 0.15$ mm	$d_0 = -$	15 %

5 ANALYSIS OF THE RESULTS

In majority of the beach profiles, granulation curves moves towards fine fraction with increasing depth (Figure 2). It is particularly evident in place, where a granulation curve cuts the line of diameter 0.074 mm. On average, in all cross sections, from among 521 possible combinations, 63 % confirm the above assumption.

From among 155 possible combinations of granulation curves comparison, 70 % confirmed that grain-size distribution becomes finer with increasing distance from the crest of the dam.

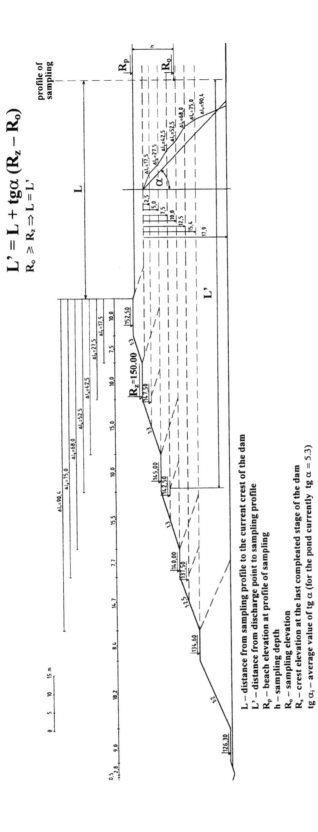

$$L' = L + tg\alpha \, (R_z - R_0)$$
$$R_0 \geq R_z \Rightarrow L = L'$$

L – distance from sampling profile to the current crest of the dam
L' – distance from discharge point to sampling profile
R_p – beach elevation at profile of sampling
h – sampling depth
R_0 – sampling elevation
R_z – crest elevation at the last compleated stage of the dam
tg α_i – average value of tg α (for the pond currently tg α = 5.3)

Fig. 9 Determination of distance from a discharge point

167

In spite of apparent scatter of the index properties distribution, there are certain trends in profiles of each examined parameter. Bulk density ρ, dry density ρ_d and density of solids ρ_s increase with increasing depth h and distance from the current crest of the dam L, while void ratio e and SFR decrease when h and L increase. These results confirm genuineness of the hypothesis 1.

The analysis of relationships between indexes describing grain size distribution and SFR showed that the following parameters describe in the best way the granulation curves of tested tailings:
- curve slope index U_X,
- tangent of curve slope tgα,
- diameter d_{50},
- $ASFR$.

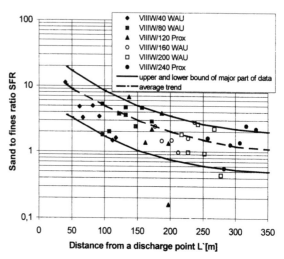

Figure 10. Envelopes of sand to fines ratio SFR in analyzed cross section

Direct correlation of $ASFR$ to SFR confirmed that the later parameter can be used for representation of a granulation curve in tested tailings, thus the hypothesis 3 was proved to hold true.

6 CONCLUSION

Presented in the paper data and analyses entirely justify assumption expressed in the form of three hypotheses set as the objective of the work, thus formulating possibility for determination of procedure of reconstruction of granulation curve in particular cross section in any point without auxiliary test or data. Such a result is a good starting point for further analyses combined with CPTU results, giving a chance for extending the conclusion from selected cross sections to the whole pond. It was also proved that in some applications, grain size distribution represented usually by granulation curve, can be substituted by parameter SFR.

7 ACKNOWLEDGEMENT

The authors thank KGHM Polska Miedż SA - the owner of the pond, for the opportunity to participate in the Żelazny Most project and for the opportunity to present this paper.

REFERENCES

Carrier, W.D.III, 1991. Stability of Tailings Dams. *Conferenze di Geotecnica di Torino, XV ciclo.Torino.*
Kollis W.et.al. 1968. *Gruntoznawstwo techniczne.* Warszawa.
Geoteko Ltd. 1997. Grain Size Distribution of Deposits. *Unpubl. report prepared for Hydraulic Engineering Division KGHM Polska Miedż S.A.*
Geoteko Ltd. 1997. Evaluation of Granulation Distribution of Tailings Material in the Beach. *Unpubl. report prepared for Hydraulic Engineering Division KGHM Polska Miedż S.A.*
Lipiński, M.J. 1999 Undrained response of cohesionless soils to monotonic loading. Ph.D. thesis submitted to Technical University of Gdańsk.
Monografia KGHM Polska Miedż S A 1996. Wrocław, CBPM "Cuprum".
Scott, J.D. & Cymerman, G.J. 1984 Prediction of Viable Tailings Disposal Methods, *Proceedings of a Symposium sponsored by ASCE, Sedimentation Consolidation Models Prediction and Validation,* San Francisco: 522-544.

Tailings and Mine Waste'00 © *2000 Balkema, Rotterdam, ISBN 90 5809 126 0*

Exploiting the rheology of mine tailings for dry disposal

Fiona Sofra & David V. Boger
Advanced Mineral Products Centre, Department of Chemical Engineering, The University of Melbourne, Parkville, Vic., Australia

ABSTRACT: Disposal of highly concentrated mineral tailings requires a thorough understanding of the effects of dewatering on the flow characteristics of the tailings. For thickened tailings disposal, dry stacking and paste fill, the rheology must be manipulated and exploited to ensure maximum efficiency of the entire disposal operation. The paper presents examples involving the dewatering and disposal of bauxite residue (red mud), nickel tailings and coal tailings and outlines alternative and simplified methods of rheological characterisation. The effect of both shear and compression history is examined in order to identify the most favourable processing schemes, in addition to the effect of other factors such as mineralogy, particle size and the use of additives.

1 INTRODUCTION

Technological advancements in many mineral-processing industries have increased the feasibility of refining low-grade ores. As a result of this progression and increasing environmental legislation, larger volumes of waste material are produced which must be disposed of in a manner that is both environmentally acceptable and economically viable. In the past, mine tailings have typically been deposited as dilute slurries so the disposal area required has been large and the associated costs high. In addition to land costs, the construction and lining of large dams is expensive. Reclamation and rehabilitation is a difficult and lengthy process as the dam may take many years to fully dry and consolidate.

To reduce the environmental impact of mine tailings, various techniques that involve dewatering the waste prior to disposal have been identified. Schemes such as thickened tailings disposal, dry stacking and paste fill all involve the deposition of dewatered waste material for improved water/reagent recovery, decreased tailings volume and 'footprint' (surface area required) and ease of rehabilitation.

Thickened tailings disposal involves deposition of the dewatered waste material from a central disposal point to form a tailings stack of conical geometry (Robinsky, 1975, 1978). The formation of such a cone is advantageous as it eliminates the need for high dam walls and stability is enhanced through improved consolidation. In addition, both steep tailings slopes and the surface ponding of slimes are avoided. Land utilisation is improved as higher tailings densities are achieved compared with conventional disposal methods. The deposit is fully drained and consolidated at decommissioning so the reclamation procedure is simplified.

Dry stacking involves the progressive deposition of thickened tailings onto sloped and underdrained drying beds. Once a given drying bed is covered to the desired depth, the discharge point is moved to an adjacent bed to allow the freshly covered area to dry via evaporation and underdrainage. When all drying areas are covered, the process is repeated by returning to the original (now dry) bed and depositing a fresh layer of tailings. The dry stacking method allows the timing and depth of successive depositions to be varied to accommodate changes in climate and material characteristics.

Paste fill is used for backfilling underground mines in order to ensure good support properties for the overlying land and to reduce the surface placement of tailings. The thickened tailings that form the paste are often mixed with small amounts (3-5%) of cement to increase their strength characteristics. The paste usually flows in a pipe by gravity down a mineshaft and then horizontally to the area being filled. Once deposited, the material must spread sufficiently to totally fill the void whilst quickly consolidating to provide the required strength.

The dry disposal methods briefly outlined above all require the flow properties of the tailings to lie within a narrow range to ensure adequate spreading of the material after discharge. If the solids concentration of the tailings is too high, the material will not spread and will accumulate directly below the discharge point. If the concentration is too low, the material will flow too far and the drying area will not retain the required geometry. Drying times will be greatly increased due to the higher water content.

In addition to the deposition behaviour, other essential considerations are the ability to economically pump the material from the thickener to the disposal area and the ability to dewater the tailings to the high concentrations required.

Concentrated suspensions usually exhibit non-Newtonian behaviour, so determination of the disposal plant operating conditions requires a thorough understanding of the rheological characteristics of the material. The implementation and optimisation of dry disposal methods involves three concurrent and interdependent rheological studies to determine i) the concentration required to achieve the optimum spreading and drying characteristics of the tailings once deposited ii) the optimum conditions for pipeline transport and iii) the feasibility of dewatering the slurry to the required concentration.

2 SHEAR RHEOLOGY

2.1 Measurement

Characterisation of the shear rheology allows determination of the spreading characteristics, the requirements for pipeline start-up and the conditions for minimal energy expenditure during pipeline transport from the thickener to the prepared disposal area.

The rheological characterisation of concentrated mineral suspensions requires specialised equipment and techniques. Mineral suspensions are generally non-Newtonian fluids at high solids loadings, exhibiting a yield stress, which is the minimum stress required for material deformation and flow to occur. Furthermore, the rheology of many suspensions is time dependent (thixotropic) and shear rate sensitive (pseudoplastic).

Measurement of fundamental flow behaviour is undertaken using a capillary rheometer (Want et al., 1982, Nguyen and Boger, 1983, Leong, 1987). The capillary rheometer generates shear stress-shear rate data for the determination of pumping energy requirements in addition to describing the influence of thixotropy and pseudoplasticity.

A significant amount of work on the measurement of the yield stress of mineral suspensions has been completed at the University of Melbourne. From this work novel and simplified measurement techniques have resulted. The vane-shear instrument and technique allows direct and accurate determination of the yield stress from a single point measurement (Nguyen, 1983, Nguyen and Boger, 1983,1985). Many workers worldwide have adopted the vane-shear method and confirmed its applicability for all types of yield stress materials (Yoshimura et al., 1987, James et al., 1987, Avramidis and Turian, 1991, Liddell and Boger, 1996, Pashias, 1997).

In an attempt to further simplify yield stress measurement, the 'slump test' has been modified to accurately evaluate the yield stress of mineral suspensions (Showalter and Christensen, 1998, Pashias et al., 1996). This technique has been typically used to determine the flow characteristics of fresh concrete. The slump test is conducted using only a cylinder and a ruler, eliminating the need for sophisticated equipment and allowing easy, on-site yield stress measurement by plant operators.

2.2 Shear rheology results

The flow properties of concentrated mineral suspensions vary significantly with solids

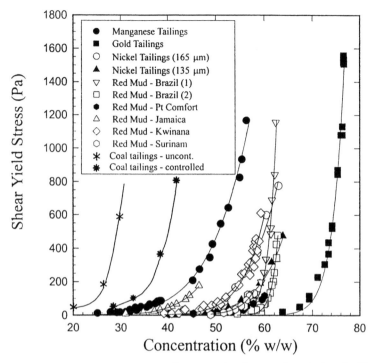

Figure 1. Yield stress as a function of concentration for a number of mineral tailings

concentration and type; however, a number of common characteristics have been observed for concentrated suspensions in general. The strong dependence of the rheology on concentration is exemplified in figure 1. This figure shows the yield stress as a function of concentration for a number of industrial slurries. Although the relationships vary for the different materials, all materials exhibit an exponential rise in the yield stress with concentration. Furthermore, for all materials the yield stress begins to rise rapidly beyond 80 to 100 Pa, regardless of the concentration.

The concentration at which the yield stress begins to rise rapidly is significant when optimising pumping energy requirements; further explained in section 2.3. The yield stress must be sufficient to allow laminar pipeline transport to prevent solids deposition, but not so high that start-up problems will be encountered.

The presence of a yield stress is essential for dry stacking and thickened tailings disposal to ensure that the material comes to rest at the required angle for stability and maximum storage capacity. An adequate yield stress also ensures that particle size segregation does not occur and that the final tailings stack will be homogeneous.

Figures 2 and 3 show the decrease in viscosity with increasing shear rate (shear thinning) for a number of concentrations of red mud and coal tailings respectively. The shear thinning often evident in mineral suspensions is attributed to the alignment of particles or flocs. An increase in the shear rate from rest results in the alignment of particles in the direction of shear, therefore providing a lower resistance to flow.

To determine the effect of shear history on the flow properties, the suspension is sheared (by mixing) in between measurement of the shear stress-shear rate behaviour by capillary rheometry. Typical flow data for 47-wt % bauxite tailings samples (red mud) from alumina production are shown in figure 4. The effect of shear history on the yield stress is determined concurrently using the vane technique or the slump test (figure 5).

Capillary rheometry data in figure 4 illustrate a yield stress material that is strongly thixotropic and shear thinning. Although knowledge of the shear thinning behaviour is imperative, for many materials the time dependent nature, which results in thixotropic behaviour, has a more significant influence on the flow properties.

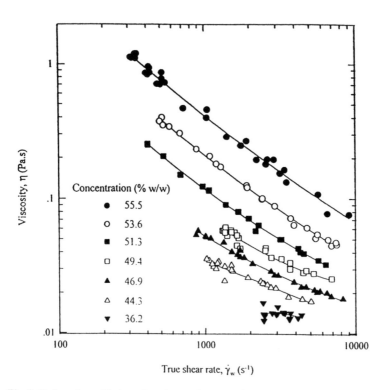

Fig. 2. Red mud; equilibrium viscosity vs. shear rate for various solids concentrations.

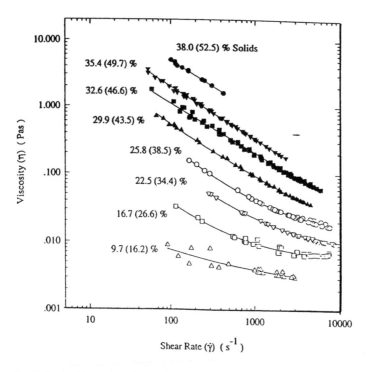

Fig. 3. Coal tailings; viscosity vs. shear rate for various concentrations (wt%)

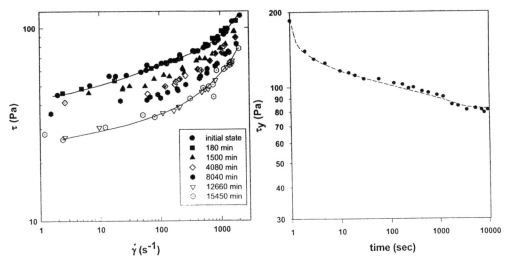

Fig. 4 Shear stress vs shear rate variation
with shear history

Fig 5. Decrease in the vane yield stress
due to thixotropic breakdown

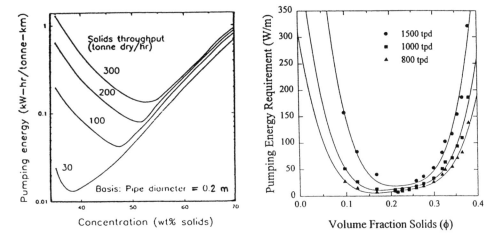

Fig 6 Red mud; pumping energy requirements as a
function of solids throughput and conc.

Fig 7. Coal tailings; pumping energy requirements
as a function of solids throughput and conc.

Thixotropy is the result of structural breakdown under shear and manifests itself as a decrease in the viscosity and yield stress with time for a given, constant shear rate. As time of shear elapses, the rate of breakdown will decrease, as fewer structural bonds are available for breakdown. Structural reformation may take place and the rate of this process will increase with time of shear due to the increasing number of bonding sites available.

2.3 Pumping energy determination

For laminar flow, the data in figure 2 can be used directly to determine the pipeline pressure drop and energy requirements for given pipe sizes and flow velocities. In the turbulent regime, the rheological are is used in conjunction with an empirical correlation. An exhaustive list of correlations were reported in Munn, (1988). Munn concluded that the method proposed by

Dodge and Metzner (1959) was suitable for plastic materials irrespective of particle size. Other authors favour different methods (Hanks,1986, Darby et al, 1992).

For a given solids throughput and pipe diameter, figures 6 and 7 highlight the presence of an optimum concentration where the pumping energy is minimised. The optimum concentration for all throughputs is found to correspond to the laminar-turbulent transition (Nguyen, 1983). Where the solids loading is higher than the optimum, flow will be laminar but the large increase in the viscosity and yield stress results in an increase in the pressure drop. For concentrations much less than the optimum, the total volume of material to be transported increases resulting in high flow velocities, turbulent flow, higher friction factors and therefore higher pressure drops. For thixotropic materials such as red mud and some nickel tailings, an additional complication exists due to the changing nature of the rheology with shear history. A state of dynamic equilibrium, where the rate of breakdown of the structure and the structural reformation rate is equal is possible. However, this state is not always achieved in industrial applications due to the extended times required for equilibrium to be reached. Generally, the material in the pipe will be in a partially sheared state, where the shear stress-shear rate behaviour is still changing with the time of shear.

The design of pipelines and pumping energy determination requires knowledge of the structural state of the material. For highly thixotropic materials, problems may arise when using flow curves generated in a laboratory environment due to the difficulties in ensuring that the material is in the same structural state as in the pipeline.

The only region in which one can be sure of the structural state of the material is the equilibrium state; thus, the equilibrium state provides a good point of reference. Flow curves are shown to be reliable in this state, with the viscosity and yield stress being reproducible. For fully sheared results to be applied to a partially sheared slurry, the effect of shear history on the generated flow curves must be determined.

In order to determine the variation in flow curve shape with shear history, the curves shown in figure 4 were superimposed by vertically shifting so that the intercept was the yield stress for the equilibrium case (figure 8).

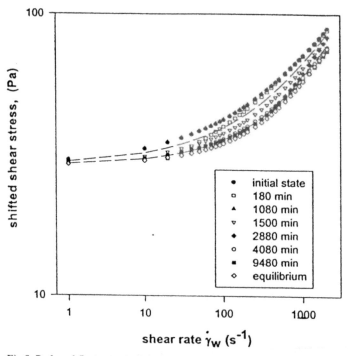

Fig 8. Red mud flow curves vertically shifted to a common yield stress. 95% confidence limits shown.

It can be seen that thixotropy manifests itself primarily as a variation of the yield stress; the actual shape of the curve is relatively consistent (Sofra`and Boger, 1999). As such, if both the yield stress at the pipeline conditions and any flow curve for the material are available, the shape of the curve can simply be shifted along the y-axis to coincide with that yield stress. The resulting curve can then be used directly to calculate the shear stress, hence the pumping energy required at the desired shear rate for laminar flow, or in conjunction with the Dodge and Metzner correlation for turbulent conditions.

To ensure that the yield stress used is representative of the structural state of the material in the pipe, it is recommended that the slump test (Pashias et al., 1996) be used on site at the time of sample collection. This method eliminates further thixotropic breakdown during transport and resuspension or thixotropic recovery due to time delay. The slump test has the added advantage of being inexpensive and simple to perform and analyse.

2.4 Modification of shear rheology

As shown above, a knowledge of the effects of concentration, shear rate and shear history allows the rheology of a given mineral suspension to be modified and exploited. Although knowledge of these 'macro' effects is imperative, it is also possible to manipulate the rheology by altering the physico-chemical properties of the suspension.

2.4.1 Surface chemistry of fine particles.

The rheology of most mineral tailings is governed primarily by the surface chemistry of the fine particles in the suspension (Fangaray, 1997, Doucet, 1999). Modification of the rheology may be accomplished by varying the surface chemistry of the fine particles using pH, the concentration of dissolved salts or by the type and concentration of flocculant or dispersant. Alternatively, since the rheology is governed by fine particles, altering the particle size distribution will also change the flow properties (Nguyen, 1983, Pashias, 1997).

2.4.2 Tailings containing swelling clays

For tailings containing swelling clays, for example coal tailings containing sodium-montmorillonite, the dewatering process and the rheology of the dewatered material is dominated by the surface chemistry of swelling clay platelets rather than fine particles.

When dry, clay particles exist as compact platelet stacks, or tactoids. Montmorillonite platelets have a net negative charge on the platelet faces, which is compensated by exchangeable sodium cations between the platelets. When hydrated, intracrystalline swelling occurs as water molecules penetrate the layers and sodium ions enter the water phase. The platelet face is left with a negative charge which results in electrostatic double layer repulsion of the platelets and the clay swells to form a structured material containing large amounts of entrapped water. The swelled clay is thixotropic, difficult to dewater and has unfavourable rheological properties making pipeline transport difficult at the high concentrations required for dry disposal methods.

An increase in the ionic strength (salt concentration) of the hydrating medium suppresses the electric double layer, reducing double layer repulsion and therefore reducing the extent of swelling and structure formation. Combined with ion exchange through a method known as 'controlled dispersion', it is possible to achieve a concentrated suspension containing larger particles with fewer interparticle interactions and a less structured platelet network is formed.

Controlled dispersion involves addition of calcium ions for ion exchange with sodium whilst suppressing the swelling of the clay via elevated ionic strength (de Kretser et al, 1997). Exchanging the monovalent sodium cations for divalent calcium ions results in improved neutralisation of the negatively charged platelet faces. The enhanced neutralisation leads to coagulation at lower concentrations and promotes ordered aggregation in favour of the random, space filling structure observed in the sodium form. The resulting tailings contain less interfloc water, leading to faster settling and easier dewatering to a higher concentration.

Figures 9 and 10 show the improved rheology (specifically, yield stress) and pumping energy

175

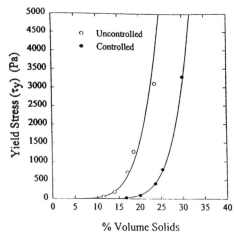

Fig. 9. Yield stress vs. conc for uncontrolled and controlled dispersed coal tailings

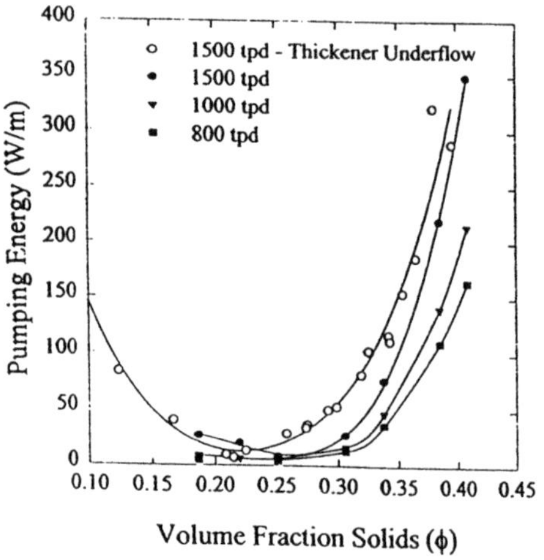

Fig. 10 Pumping energy requirements for controlled dispersed coal tailings (thickener underflow is uncontrolled)

requirements for controlled dispersed clay tailings over tailings dispersed in an uncontrolled manner.

Controlled dispersion of coal tailings leads to a lower effective volume of the clay network, so the rheological properties at high concentrations are improved and consequently dewatering and pipeline transport are simplified.

3 COMPRESSION RHEOLOGY

Prior to deposition, the slurry must be thickened to the concentration found in shear rheology tests to provide the desired deposition characteristics whilst ensuring that the material is still easily pumpable. Thickening also ensures that little or no particle size segregation occurs during deposition (Robinsky, 1978; Wood and McDonald, 1986; Williams, 1992). Segregation is inhibited due to an increase in the viscosity and yield stress with increased solids concentration and results in the uniform gradient required for dry disposal methods. The observed uniformity is desirable as it leads to improved stability while minimising tailings storage volume, which facilitates easy site rehabilitation.

Thickening is usually achieved using gravity thickeners with an appropriate polymeric flocculant. Choice of the flocculant and operating conditions within the thickener is determined by examining the shear, compression and permeability characteristics of the suspension.

The shear yield stress (the shear stress required for irreversible flow to occur) is important in the determination of piping and deposition behaviour. Likewise, the compressive yield stress coupled with the permeability of the material provide information regarding the feasibility of dewatering the residue to the concentration required for dry disposal. The compressive stress is the stress required for irreversible compression of the network structure and the permeability dictates the rate of dewatering and accounts for the hydrodynamic interactions between falling particles in a suspension.

In a conventional gravity thickener, a continuous network structure is formed through the aggregation of particles or flocs containing interparticle water. The compressive yield stress, beyond which the transmitted network pressure of the overlying structure will cause the collapse of the structure and syneresis of liquor, is concentration dependant (Buscall and White, 1987). By increasing the compression zone depth, the material at the bottom of the thickener will be subjected to a greater applied pressure and will consolidate to a higher concentration until the compressive yield stress is equal to the increased stress.

176

An understanding of the compression zone depth required to overcome the compressive yield stress at a given concentration, and the variation in this relationship with factors such as flocculant dosage and shear history will facilitate optimisation of thickener performance.

Figures 11 and 12 show typical results for red mud. In the unflocculated state, a conventional thickener with negligible compression zone will yield an underflow concentration of 38 wt%. By increasing the sediment height to 5m, an underflow concentration of over 52 wt% is possible and flocculating with 145ppm flocculant can further increase this to 58 wt%. Figure 12 indicates that the shear history experienced by the mud effects the compressive yield stress in a similar manner to the shear yield stress. For a given concentration, the yield stress will decrease as the network structure of the material is progressively disrupted and a larger mean particle size will result in a lower yield stress at a given concentration.

Figures 13 and 14 illustrate the benefits of flocculation and controlled dispersion of coal tailings in addition to the effect of shear history. The effect of flocculation is evident in both the settling rates and the compressive yield stress for clay tailings. The unflocculated thickener feed settled slowly and required a greater sediment height to overcome the compressive yield stress than the flocculated thickener underflow.

Controlled dispersed tailings achieved equivalent underflow concentrations for given sediment heights as the flocculated tailings, but markedly improved settling rates. Both observations are remarkable given the absence of flocculant in the controlled dispersed case.

The effect of a compression zone is greater for coal tailings than for red mud. A compression zone of only 1m can increase the underflow concentration from 17 wt% for a conventional thickener with no compression zone to about 55 wt% (de Kretser, 1995).

While the compressive yield stress gives and indication of the underflow concentration possible, the permeability determines the rate of the dewatering process. Permeability is influenced by the liquor viscosity, the flocculation conditions and the solids concentration. The liquor viscosity varies with temperature and dissolved solids concentration and affects the rate at which the liquid can move upward through the particulate matter. Flocculation conditions (type, dose and addition method) affect the suspension structure formed, the amount of entrapped interfloc liquid and the ease of liquid permeating through the structure. For a given set of process conditions, ie temperature, dissolved solids concentration and flocculation conditions, permeability is a function of the solids volume fraction.

A portable apparatus that is capable of determining the compressive yield stress and the permeability has been developed and built at the university of Melbourne to allow optimisation of the thickening process (de Kretser et al, 1999)

4 APPLICATIONS

Characterisation of the shear and compression rheology of mineral tailings using the aforementioned techniques has been instrumental in the development and implementation of many dry disposal schemes in Australia and abroad.

Alcoa of Australia first implemented the dry stacking scheme in 1985, has since adopted the method for all of its Australian and Suriname plants. Flow and compression data were initially used for the pipeline design, prediction of optimum pumping conditions, determination of bed slope and thickener operation. Presently, rheological information combined with knowledge of how to manipulate the rheology continues to be used for the optimisation of the entire disposal process.

The dry stacking disposal method has proved to be very successful in minimising many environmental, technical and economic problems inherent in the previous wet disposal schemes (Cooling and Glennister, 1991, Pashias, 1997, Nguyen and Boger, 1998).

Paste fill at the BHP Cannington mine is presently being implemented (Skeeles, 1998) with a concurrent rheological study being undertaken in the Advanced Mineral Products Centre (Clayton, 1999). Yield stress measurement via the slump test is used extensively to ensure that the material to be deposited underground will have the required spreading and strength properties.

The thickened tailings disposal at WMC's Mt Kieth nickel operation uses the thickener as a type of rheometer (Bentel, 1999). The yield stress of the thickener underflow is kept relatively

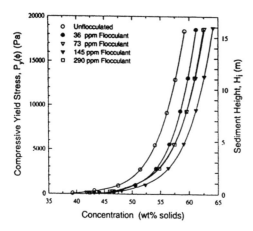

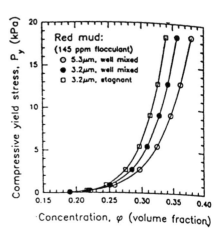

Fig 11 Compressive yield stress vs concentration for flocculated and unflocculated red mud

Fig 12 Effect of particle size and shear history on the compressive yield stress

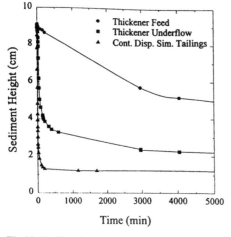

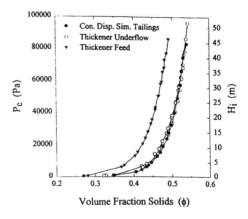

Fig 13. Settling data for thickener feed, underflow and controlled dispersed coal tailings

Fig 14 Compressive yield stress for thickener feed, underflow and cont. disp. tailings

constant (leading to a relatively constant beach slope) using the torque reading of the rake in the thickener. The concentration of the underflow will vary with ore type as the relationship between concentration and yield stress will differ. If the underflow was kept at a constant concentration, the yield stress of the material being deposited would fluctuate and the constant beach slope required for the formation of a uniform cone would not be obtained.

In the coal industry, the impact of general operating practices coupled with the modification of the rheology via surface chemistry (controlled dispersion) will increase the tailings processing rate and decrease the water content of the residue (de Kretser et al, 1997). Although full-scale commercial operation has not been implemented, significant improvements are anticipated in terms of plant economics, environmental and practical considerations.

5 CONCLUSION

Due to the complex rheology of mineral tailings under shear and compressive forces, knowledge of the rheological properties allows the design of a disposal scheme that takes best advantage of the material behaviour without affecting upstream plant performance. Furthermore, an insight into how to change or manipulate the rheology may facilitate waste disposal in a manner that is more environmentally or economically favourable than conventional methods.

The implementation and optimisation of dry disposal methods involves three concurrent and interdependent rheological studies to determine i) the concentration required to achieve the optimum spreading and drying characteristics of the tailings once deposited ii) the optimum conditions for pipeline transport and iii) the feasibility of dewatering the slurry to the required concentration. The technical methods outlined in this paper provide the rheological characterisation required to complete these studies.

As environmental factors translate into economic issues, the push for minimising waste production using dry disposal methods is gaining popularity. The use of rheological information is of high importance in evaluating the spreading characteristics, the transportation and the dewatering of thickened mineral tailings. The principles outlined in the examples given in this paper may be applied to many industries encompassing a wide range of waste materials.

REFERENCES

Avramidis,K.S,Turian,R.M 1991, Yield stress of Laterite suspensions. *J.Colloid Interface Sci.* 143;54-68

Bentel G. Pers. Comm. June 1999.

Buscall,R.,White,L.R.,1987.On the consolidation of concentrated suspensions, I. The theory of sedimentation.*J.chem.Soc.Faraday Trans.* 1(83);873-891

Clayton, S. 1999, Pers. Comm. August 1999.

Cooling,D.J.,Glennister,D.J. 1992 Practical aspects of dry disposal. *Light Metals 1992.* TMS;25-31

Darby, R., Boger, D.B., Munn,R.,1992. Predicting friction loss in slurry pipes.Chemical Engineering, September:116-119

de Kretser,R.G.1995 *The rheological properties and dewatering of slurried coal tailings.* PhD Thesis, The University of Melbourne

de Kretser,R.G.Scales,P.J.,Boger,D.V.1997. Clay-based tailings disposal;a case study on coal tailings.*AIChEJ* 43(7);1894-1903

de Kretser,R.G.Scales,P.J, Aziz,A.A.A., 1999, Optimising the surface cxhemistry of suspensions for dewatering in mineral processing. *Rheology in the mineral Industry II.*

Doucet,J. 1999.Effect of sand addition on the rheological properties of red mud slurries. *Proc. 5th International Alumina Quality Workshop* Bunbury Aust ;478-488

Fangary,Y.S,AbdelGhani,A.S.,ElHaggar,S.M.,Williams,R.A. 1997. The effect of fine particles on slurry transport processes. *Minerals Engineering,*10(4);427-439

Green, M.D., de Guingand, N.J., Boger, D.V.,1994. 'Exploitation of Shear and Compression Rheology in Disposal of Bauxite Residue', *Proceedings of Hydrometallurgy*, p971-982.

Hanks,R.W.1986. Principles of slurry pipeline hydraulics. *Encyclopedia of Fluid Mechanics, vol. 5 Slurry Flow Technology* ch. 6. Ed Cheremisinoff, N.P.

James,A.E., Williams,D.J.A, Williams,P.R. 1987. Direct measurement of static yield properties of cohesive suspensions. *Rheol. Acta* 26;6489-6492.

Loeng Y.K. 1988. *Rheology of modified and unmodified Victorian brown coal suspensions.* PhD. Thesis. The University of Melbourne.

Liddel,P.,Boger, D.V. 1996, Yield stress measurement with the vane. *J Non-Newtonian Fluid Mech.* 63:235-261

Munn,R.1988 The pipeline transportation of suspensions with a yield stress. Masters thesis. Univ. Melb.

Dodge,D.W.,Metzner,A.B. 1959.Turbulent flow of non-Newtonian systems. *AIChE J.*5(2);189-204

Nguyen, Q.D., 1983. *Rheology of Concentrated Bauxite Residue*, PhD Thesis, University of Melbourne.

Nguyen, Q.D., Boger, D.V.1983 Yield stress measurement in concentrated suspensions. *Journal of Rheology*, 27(4); 321-349.

Nguyen, Q.D., Boger, D.V. 1985.Direct Yield Stress Measurement with the Vane Method. *Journal of Rheology*, 29(3);335-347.

Nguyen, Q.D., Boger, D.V. 1998, Application of rheology to solving tailings disposal problems, *Int. J. Miner.Process*, 54;217-233

Pashias, N; Boger, D.V; Summers, J; Glennister, D.J.1996. A Fifty Cent Rheometer for Yield Stress Measurement. *Journal of Rheology*, 40(6); 1179-1189.

Pashias, N.1997, *The Characterisation of Bauxite Suspensions in Shear and Compression*, PhD Thesis, The University of Melbourne.

Robinsky, E.I,1975. 'Thickened Discharge - A New Approach to Tailings Disposal', *Canadian Mining and Metallurgical Bulletin*, 68;47-53.

Robinsky, E.I, 1978, 'Tailings Disposal by the Thickened Discharge method for Improved Economy and Environmental Control', *Tailings Disposal Today, Proceedings 2nd Int Tailings Symp.*:75-95

Skeeles, B.E.J. 1998. Design of paste backfill plant and distribution system for the Cannington project. *Australasian Institute of Mining and Metallurgy Publication Series n1*: 59-63

Showalter, W.R., Christensen,G. 1998. Toward a rationalization of the slump test for fresh concrete: comparisons of calculations and experiments. *J.Rheol*, 42(4):865-870

Sofra`,F.,Boger,D.V.1999. Application of equilibrium state rheological data in the red mud disposal process. *Proc. 5th International Alumina Quality Workshop* Bunbury Aust ;550-558

Want, F.M., Colombera,P.M., Nguyen, Q.D., Boger, D.V 1982. Pipeline design for the transport of high density bauxite residue slurries. *Proc. 8th Int Conf. Hydraulic Transport of Solids in Pipes*, Johannesburg;242-262

Williams, M.P.A. 1992 'Australian Experience With The Central Thickened Tailings Discharge Method for Tailings Disposal', *Environmental Issues and Waste Management in Energy and Minerals Production*;567-577, Balkema, Rotterdam.

Wood, K.R., McDonald, G.W.1986 'Design and Operation of Thickened Tailings Disposal System at Les Mines Selbaie', *CIM Bulletin*,79(895);47-51.

Yoshimura,A.S. Prud'homme,R.K. Princen,H.M. Kiss,A.D,1987. A comparison of techniques for measuring yield stresses. *J. Rheol.* 31;699-710

Tailings and Mine Waste'00 © 2000 Balkema, Rotterdam, ISBN 90 5809 126 0

Dynamic run-out analysis methodology (DRUM)

P. Tan & Y. Moriwaki
URS Greiner Woodward Clyde, Santa Ana, Calif., USA

R. B. Seed
University of California, Berkeley, Calif., USA

R. R. Davidson
Woodward Clyde, Sydney, N.S.W., Australia

ABSTRACT: Flow slides have been induced by liquefaction in embankments or in the foundation layer beneath embankments caused by earthquakes and other loading. One key aspect of a flow slide is its run-out distance, which inherently involves a dynamic process. The assessment of whether a flow slide is triggered is relatively straightforward, but the estimation of the run-out distance is a far more difficult task. Current methods used to estimate the run-out distance have modeled a complex dynamic process by evaluating static "equivalent" force equilibrium of the assumed final at-rest geometry. Because of these simplifications, subjective judgements and model uncertainty are considerable in these methods. To partially overcome these limitations, a new methodology based on a dynamic process is proposed to estimate the run-out distance of flow failure. The proposed approach can be used with the more common procedure to compute residual shear strength of liquefied material based on case histories. This paper presents the proposed methodology to calculate the run-out distance that provides insight into the actual dynamic nature of the run-out process. The methodology satisfies the dynamic equilibrium at each step of the run-out process. This methodology, referred to as "Dynamic RUn-out Method" (DRUM), can be implemented in a computer program with a modest effort. Calibration of the methodology using two published case histories of flow slides is also presented in this paper. Although the new methodology has limitations, it offers significant advantages over the current method. It also can be used to back-calculate residual shear strength from case histories in a consistent manner.

1 INTRODUCTION

Current methods by which run-out of failed embankments is evaluated, in general, model a complex dynamic process by static force equilibrium. Because of the practical simplifications required to produce meaningful results, subjective judgements and parametric variability are inherent in these methods. Such methods have been widely used in the past to evaluate the run-out distance of failed embankments due to liquefaction. One such approach was presented by Lucia et al. (1981).

To overcome some of these limitations, a new dynamic methodology has been developed to evaluate the amount of run-out distance of failed embankments. The new methodology was tested on two flow slide cases histories.

2 DYNAMIC RUN-OUT ANALYSIS METHODOLOGY

2.1 General approach

The method that is currently used most often to evaluate the amount of run-out from a flow failure is the method developed by Lucia et al. (1981). This approach is illustrated on Figure 1. Essentially, the "Lucia" method is a simplified two-dimensional limit equilibrium analysis of the liq-

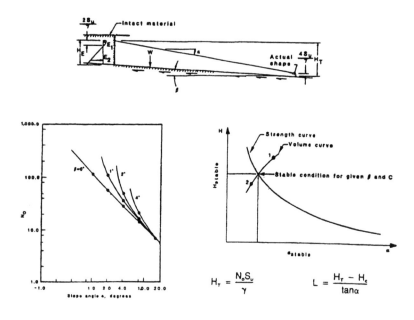

$$H_\tau = \frac{N_o S_u}{\gamma} \qquad L = \frac{H_\tau - H_c}{\tan\alpha}$$

CALCULATION METHOD (AFTER LUCIA ET.AL. 1981)

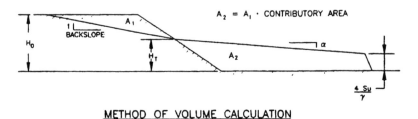

METHOD OF VOLUME CALCULATION

Figure 1. Lucia method of flow slide analysis (After Lucia, et al. (1981)).

uefied material downstream from an impoundment. Forces taken into consideration include the resisting force along the base provided by the residual shear strength of the liquefied material, weight of the liquefied mass, and lateral pressure of the liquefied material acting on the back of the wedge. Two curves are developed as shown on Figure 1, one based on strength and the other based on volume. The intersection of the two curves yields the residual height of tailings at the embankment breach. Knowing the residual height and the slope of the run-out material, the run-out distance can be calculated.

The Lucia method estimates the end configuration of a flow slide using only the equilibrium of the static forces: gravity force and boundary forces. However, an actual flow slide is a very complicated dynamic process and requires considerable simplification before a practical analysis can be performed using the Lucia method. One key limitation of the Lucia method is in modeling inherently dynamic run-out processes using only static forces. Thus, in theory, it is desirable to calibrate the method for each case by comparing with similar case history results in order to obtain reasonable results.

The dynamic nature of the actual run-out process is schematically shown on Figure 2.

1. In the initial configuration (1), liquefaction of the materials composing at least part of the section induces slope instability along the postulated failure surface AC because liquefaction reduces the shear strength of the materials. The associated "static" factor of

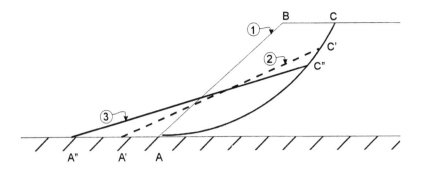

① (Initial Configuration)	$FS_{static} < 1$	$F_{driving} > F_{resisting}$	Velocity increases from 0.
② ("Statically Stable" Configuration)	$FS_{static} = 1$	$F_{driving} = F_{resisting}$	Velocity is greater than 0.
③ ("Dynamically Stable" Configuration)	$FS_{static} > 1$	$F_{driving} < F_{resisting}$	Velocity equals 0.

Figure 2. Configurations and conditions during a flow slide.

safety (FS) is less than one, and the driving force is greater than the resisting force. Although the initial velocity of the slide mass (ABC) is zero, it will quickly increase, resulting in increasing momentum of the slide mass.

2. In the "statically stable" configuration (2) shown on Figure 2, the slide mass has moved enough to reach a configuration (A'C') such that the static slope stability factor of safety would be one. This means that the driving force is equal to the resisting force. However, because the slide mass has been moving, it would have some velocity with associated momentum, and the slide mass would not stop.

3. In the "dynamically stable" configuration (3) shown on Figure 2, the slide mass has finally stopped. In this configuration the computed static stability factor of safety would be greater than one, with the resisting force being greater than the driving force, and the velocity of the slide mass finally becomes zero.

The dynamic process described above does not include seismic forces as part of the driving force (or the resisting force, depending on the direction of the seismic force). This is because the level of shaking typically becomes much less by the time the materials liquefy. In addition, the effects of seismic forces, which act both as the driving and resisting forces, are generally relatively minor in affecting the amount of flow slide. Unlike the case of limited seismically induced movements of earth structures, where the seismically induced forces tend to dominate in causing ground movements, flow slides are typically dominated by the gravity induced driving force. Nevertheless, this is one limitation of this method.

The approach adopted in this method is to use a procedure that follows the flow slide process summarized in Figure 2 in an approximate way to estimate the amount of run-out. The procedure thus reflects the dynamic nature of flow slides and is referred to as the Dynamic RUn-out Method or DRUM herein. The procedure, calibrated using two case histories, is consistent with approaches used to back calculate residual shear strength from case histories of flow slide in recent years.

2.2 Analysis methodology

First, the potential for flow slide and the shape and size of the slide mass are estimated using post-earthquake two-dimensional limit equilibrium method of slices, typically using Spencer's method. The initial configuration such as ABC shown on Figure 2 is incrementally changed in shape to eventually reach the shape A"C" on Figure 2. At each increment, the volume (or the area in two dimensions) of the slide mass is kept the same. Each increment corresponds to a small time increment determined as part of the analysis. As the failure mass changes in shape, the crest corner B shown on Figure 2 is incrementally smoothed out to a straight line such as A'C' on Figure 2. At each of these incremental shapes, the slide mass is considered to be a "rigid" body for the purposes of calculating forces acting on it and the consequence of unbalanced forces. However, this sliding block is considered capable of "deforming" in such a way that it moves on the shear surface without losing contact. The driving force acting on the sliding block is its weight, and the resisting force is the shear strength resistance along the slide surface.

On Figure 3(a), wedge EBCF shows an incremental shape of the sliding block at time t_i. The center of gravity of this sliding block is identified as CG_i. The base of the sliding block, represented by segment EBCF, provides shear resistance vector $\vec{S}_{EBCF}$. In the next increment of the sliding block shown on Figure 3(b), the top of the sliding block has moved from point F to point H, and the toe of the sliding block has moved from E to G. Wedge GBCH shown on Figure 3(b) is an incremental shape of the sliding block at time t_{i+1}. The center of gravity for this case is at CG_{i+1}, and the shear resistance vector for this sliding block is $\vec{S}_{GBCH}$.

For this time increment from t_i to t_{i+1}, the dynamic run-out calculations are performed as follows:

1. The shear resisting force during this increment is set equal to the average of those associated with the two slide mass increments using the following equation:

$$\vec{S} = (\vec{S}_{EBCF} + \vec{S}_{GBCH})/2 \tag{1}$$

Figure 3(c) shows the displacement vector of the center of gravity between the two increments as $\vec{d}_{i+1}$. This is the postulated direction of the sliding block movement from t_i to t_{i+1}. The driving force for this movement is provided by the component of the sliding block weight in the direction of $\vec{d}_{i+1}$. The resisting force for this movement is provided by the component of the above average shear force in the direction of $\vec{d}_{i+1}$. Thus, letting θ_{i+1} be the angle of this displacement vector with horizontal and δ_{i+1} be the angle between the shear resistance force vector and the direction of $\vec{d}_{i+1}$, the following equation of motion can be formulated:

$$-Mg\sin(\theta_{i+1}) + S\cos(\delta_{i+1}) = M \cdot a_{i+1} \tag{2}$$

where a_{i+1} is the acceleration of sliding mass.
3. Given a_{i+1} from the above equation and a_i and d_i from the previous increment, the following equations can be used to calculate Δt and v_{i+1}:

$$v_{i+1} = \frac{(a_i + a_{i+1}) \cdot \Delta t}{2}$$
$$d_{i+1} = d_i + v_{i+1} \cdot \Delta t \tag{3}$$

4. This process is repeated until the calculated velocity reaches zero.

3 CALIBRATION ANALYSIS OF TWO CASE HISTORIES

The DRUM procedure was calibrated using the following two case histories:
1- Lower San Fernando Dam: During the 1971 San Fernando earthquake (Mw 6.6), the upstream section of the Lower San Fernando dam (Seed, et al., 1973, 1989), located northwest of Los Angeles, broke into blocks which slid into the reservoir on top of the liquefied hydraulic fill material. The dam was about 120 feet high. At the time of the slide, the reservoir level was approximately 35 feet below the crest of the dam. The run-out distance beyond the upstream toe was about 150 feet.

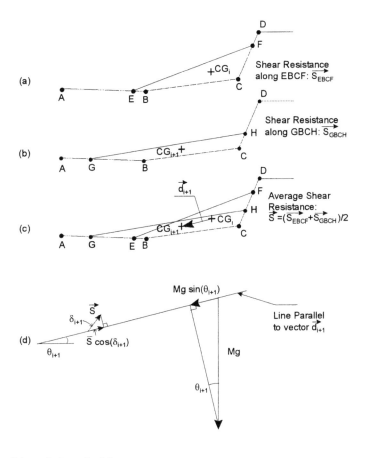

Figure 3. Flow slide analysis methodology.

2- La Marquesa Dam: During the 1985 Central Chile earthquake (Mw 8.0), the upstream and downstream sections of this dam located southwest of Santiago, Chile, moved by about 11 and 6.5 meters, respectively, because of liquefaction (De Alba et al., 1987).

3.1 Analysis of the Lower San Fernando Dam slide

The upper drawing on Figure 4 shows the post-failure geometry of the Lower San Fernando dam. Based on the blocks that slid during the flow slide, Seed et al. (1973) had reconstructed the pre-failure geometry using the post-failure components. A layer below the upstream slope liquefied. As a consequence of this liquefaction, blocks 1 through 11 slid into the reservoir. Some of these blocks moved mainly toward upstream while others moved mainly downward. From Figure 4, the run-out distance of the upstream slide is about 150 feet.

Subsequent study performed by Seed et al. (1989) indicated that the residual shear strength of the liquefied material underneath the upstream slope was about 400 psf. This value of residual shear strength was used in the present run-out calibration analysis. The portion of the slide used in the analysis involves blocks 5 through 11 riding on the liquefied material below. Because a significant portion of these blocks was submerged at the time of the slide, a submerged unit weight was used for the entire portion of the slide.

Figure 5 shows the idealized pre-failure geometry of blocks 5 through 11 together with the static factor of safety of the slide mass. The strength of the material above the liquefied zone is assumed to have no shear resistance due to failure. The liquefied zone was assigned a uniform residual shear strength of 400 psf. The resulting post liquefaction factor of safety based on the initial geometry was 0.6.

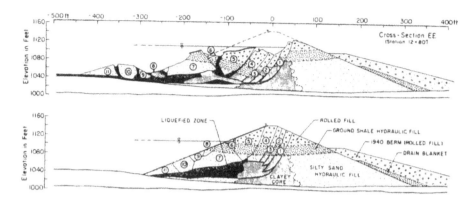

Figure 4. Post-failure geometry of the Lower San Fernando Dam and its reconstructed pre-failure geometry (After Seed et al., 1973).

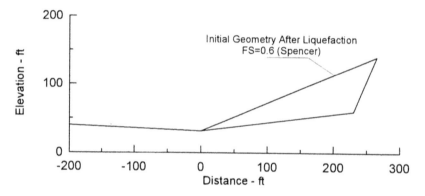

Figure 5. Idealized pre-failure geometry used in the DRUM analysis and computed factor of safety of the initial geometry after liquefaction.

Figure 6 shows the post-failure idealized geometry calculated using the dynamic run-out analysis. Along the sliding plane, shear strength of 400 psf was assigned, but zero shear strength was assigned to the back scarp area corresponding to area between blocks 5 and 4 (Figure 4). Furthermore, a submerged unit weight of 58 pcf was assigned to the entire slide mass. Based on this geometry, and material property characterization, a run-out analysis was performed.

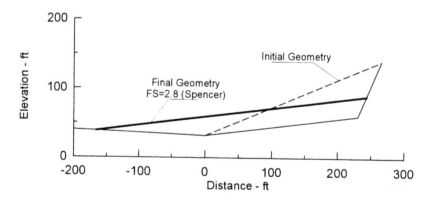

Figure 6. Run-out analysis results of the Lower San Fernando Dam: initial and final geometry.

Figure 7 shows the calculated time histories of the computed factor of safety, the acceleration, velocity, and upstream toe horizontal displacement of the sliding block from one set of analysis. At about 4 seconds after the initiation of the slide, the acceleration of the block reaches zero. At this point the computed factor of safety also reaches one. This stage corresponds to the "statically stable" configuration. However, the sliding block has not stopped. As the sliding block continues to move upstream, the shear resistance continues to increase because the shear surface increases. This increasing resisting force and the decreasing driving force then results in slowing down and eventually stopping of the sliding block. Once the velocity reaches zero, all sliding motion stops, and the sliding block has come to a "dynamically stable" configuration. At the "dynamically stable" position, the upstream toe of the slide has moved approximately 166 feet upstream. This computed run-out distance value compares favorably with the observed run-out distance of 150 feet. The computed distance did not take into account the additional push generated by block 1 through 5 and the expected mounding of materials at the toe of the slide.

A slope stability evaluation was also performed for the "dynamically stable" geometry as shown on Figure 6. The resulting factor of safety obtained was 2.8 indicating that indeed a stable position has been reached. Note that the initial factors of safety are 0.6 and 0.4, respectively, using the Spencer's method and DRUM. The final factors of safety are 2.8 and 2.3, respectively, using the Spencer's method and DRUM. These small differences caused by the simplifications made in the computation of the factor of safety using DRUM highlight the importance of calibration study. The computed factors of safety also identify the difficulty of back-calculating residual strength from static slope stability analysis alone. How do you, a priori, select a pre-failure factor of safety such as 0.6 or a post-failure factor of safety such as 2.8?

In the run-out analysis performed above, the shear strength along the bottom of the reservoir was assumed to be 400 psf. Because of the presence of water in the reservoir at the time of the slide, it can also be argued that the shear strength along the bottom of the reservoir could be lower than 400 psf. A sensitivity analysis addressing this issue by assigning lower shear strength values along the bottom of reservoir resulted in the calculated run-out values shown on Table 1. In general, the computed run-out distances do not appear to be highly sensitive to the shear strength reduction.

3.2 Analysis of La Marquesa Dam slide

Thirty small storage reservoirs, retained by earth embankments, were located within 50 miles of the epicenter of the Central Chilean earthquake (Mw 8.0) of March 3, 1985. The reservoir embankments were typically built of silty clayey sands weathered from the granodioritic Coast Batholith. During the earthquake, 14 of the embankments suffered cracking, while two of them essentially failed. De Alba et al. (1987) studied these two embankment failures to estimate the residual shear strength of these embankments. One of the two embankments was La Marquesa dam.

Although the embankment suffered major slides both in the upstream and downstream slopes, we focused only on the downstream slope. Figure 8 shows the pre-earthquake geometry of the downstream slope of the dam, which had a maximum height of 10 m. The loss of freeboard caused by slides was about 2 m in the middle one-third of the embankment. The observed horizontal displacement at the downstream toe of the dam was 6.5 m.

Based on a postulated failure mechanism shown on Figure 8, De Alba et al. (1987) performed a back calculation of the residual shear strength along the base of the postulated failure mechanism which ranged between 266 to 580 psf. The lower bound of the residual shear strength was

Table 1. Sensitivity analysis results for the Lower San Fernando Dam.

Shear strength along bottom of reservoir psf	Percentage of shear strength reduction %	Run-out distance of upstream toe feet
400	0	166
300	25	183
240	40	195

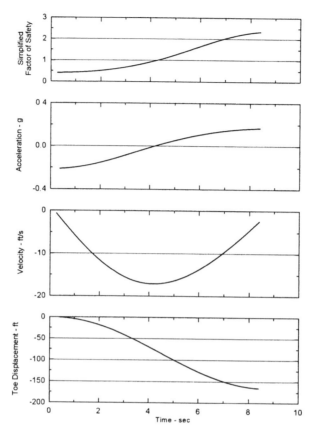

Figure 7. Time histories of computed simplified factor of safety, acceleration, velocity, and toe displacement.

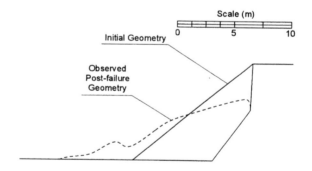

Figure 8. Initial and post-failure geometry of La Marquesa Dam (After De Alba, 1987).

obtained using the final configuration of the slopes after the earthquake, whereas the upper bound value was obtained using the pre-earthquake geometry of the embankment.

For the calibration analysis, the same range of residual shear strength (266 to 580 psf) was uniformly assigned to the base of one of the postulated sliding mechanism proposed by De Alba, et al. (1987). Figure 9 shows the computed post-failure geometry of the slope for a postulated residual shear strength equal to 420 psf superimposed on the observed post-failure geometry.

The calculated values of run-out distance for the range of residual shear strength used are summarized on Figure 10. The computed run-out distance ranges from about 1.5 to 12.5 m for

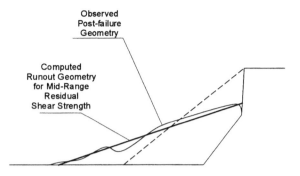

Figure 9. Comparison of computed and observed run-out geometry for La Marquesa Dam.

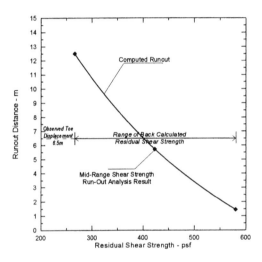

Figure 10. Run-out analysis results for La Marquesa Dam – Run-out distance vs. residual shear strength.

the residual shear strength of 580 to 266 psf, respectively. This is consistent with the observed run-out distance of 6.5 m. The higher estimate of the residual shear strength at 580 psf was reflecting the pre-failure geometry, and would be expected to be higher than that mobilized during the failure and underpredicts run-out distance. The residual shear strength value estimated based on post-failure geometry would probably be more appropriate, depending on the factor of safety assumed.

4 CONCLUSIONS

Based on the limited analysis of the calibration case histories presented herein, the proposed DRUM (Dynamic Run-out Method) for dynamic run-out analysis provides reasonable calculated run-out distances when compared to the observed distances. Although some simplifying assumptions were made in the proposed methodology, in comparison to the current analysis approach, Lucia et al. (1981), the proposed methodology is a useful improvement, capable of a more site-specific evaluation of the run-out distance reflecting the dynamic nature of the process. One advantage of the proposed method over more sophisticated non-linear analysis is that it is easy to perform parametric analysis to obtain the "feel" of the problem. Further, many non-linear analysis programs will have re-meshing problem because of large deformation.

The calibration analysis process used is somewhat "incestuous" in the sense that the values of residual shear strength were obtained by others relying largely on static methods that are inherently similar to the proposed method. However, a practical alternative does not appear to exist at this time and the results of calibration analysis do provide a level of comfort in using the proposed procedure. Since the method was first developed, it has been applied to predict the observed run-out distance and time of a number of tailings dam flow failures and road embankment landslides with excellent success.

One possible improvement to the proposed method would be to perform slope stability analysis using, for example, Spencer's method at each incremental stage and evaluate the unbalanced forces from the slope stability analysis results. Although it is not clear whether such refinement is warranted, the proposed method is considered an appropriate way to back-calculate the values of residual shear strength from available case histories in a consistent manner.

Reflection on Figure 4 clearly indicates the complexity of flow slide phenomena. Any methods to be used in run-out calculation should be used with caution, critical thinking, and plenty of engineering judgment.

5 REFERENCES

De Alba, P., Seed, H.B., Retamal, E. & Seed, R.B. 1987. Residual strength of sand from dam failures in the Chilean earthquake of March 3, 1983. University of California, Berkeley. Report No. UCB/EERC 87-11.

Lucia, C.P., Duncan, J.M. & Seed, H.B. 1981. Summary of research on case histories of flow failures of mine tailings impoundments. Mine Waste Disposal Technology, Proc. Bureau of Mines Technology Transfer Workshop, Denver, July 16, 1981.

Lucia, C.P. 1981. Review of experiences with flow failures of tailings dams and waste impoundments. Dissertation submitted in partial satisfaction of the requirements for the degree of Doctor of Philosophy, University of California, Berkeley.

Seed, H.B., Lee, K.L., Idriss, I.M. & Makdisi, F. 1973. Analysis of the slides in the San Fernando Dams in the earthquake of Feb. 9, 1971. University of California, Berkeley. Report No. UCB/EERC 73-2.

Seed, H.B., Seed, R.B., Harder, L.F. & Jong, H.L. 1989. Re-evaluation of the slide in the Lower San Fernando Dam in the 1971 San Fernando earthquake. University of California, Berkeley. Report No. UCB/EERC 88-04.

Liners, covers and barriers

Tailings and Mine Waste'00 © 2000 Balkema, Rotterdam, ISBN 90 5809 126 0

Geomembrane/soil interface strength relationships for heap leach facility design

J.E.Valera & B.F.Ulrich

Knight Piésold and Company, Denver, Colo., USA

ABSTRACT: It is well documented in the literature that the Mohr-Coulomb failure envelope for soils is not a straight line but, rather, is somewhat curved. This curvature is especially evident when the failure envelope represents a wide range of stresses.

Since the failure of the Kettleman Hills Landfill, much has been written about the shear strength behavior of geomembrane-to-soil interfaces, and a wealth of information has been gained in this area. Recently, some researchers have presented findings that indicate such interfaces may also exhibit curved failure envelopes.

The results of extensive laboratory interface shear tests using 8- and 12-inch (200- and 300-mm) square shear boxes are summarized herein. These tests were carried out to characterize the curvature of the failure envelope for various fine-grained soils using both smooth and textured geomembranes for stress ranges up to the equivalent pressure imposed by a 300-foot-high heap (approximately 36,000 pounds per square foot [psf]. Test data indicate that the curvature of the failure envelope is dependent upon the soil liner properties including its moisture content, geomembrane type, and pressure range. The effects of this curvature on the stability of a lined heap leach facility, waste rock pile, or (in some cases) a tailings storage facility can be significant. Examples of stability analyses considering the effects of a curved failure envelope are presented.

1 INTRODUCTION

The stability assessment of a heap leach facility comprises a significant portion of the geotechnical analyses pertaining to the heap. Each aspect examined during the geotechnical investigation is represented in the stability analysis including the foundation characteristics and zonation, ore and geomenbrane-to-soil liner interface material properties, and location of the phreatic surface. Together with design and operational aspects such as pad topography, heap geometry, loading sequence, and leachate application schedule, these parameters contribute to the overall assessment of the pad stability.

As mining companies study the potential benefits of constructing heaps to greater heights or on unfavorable terrain, the role of the stability assessment in the overall design of a heap leach facility becomes of greater importance. In order that a proper stability analysis be conducted, it is critical that each aspect of the heap design be properly evaluated.

2 HISTORICAL BACKGROUND

It has long been understood by the geotechnical engineering community that although a shear strength relationship (failure envelope) may be approximated by a straight line within a given stress range, the true failure envelope for essentially all soil types is a nonlinear function of the normal effective stress (several references are cited in Maksimovic, 1996). This curvature of the failure envelope for soils is attributed to several factors including dilatancy, grain crushing, and particle rearrangement (Mitchell, 1993).

The failure envelopes of various rockfills have also been shown to be nonlinear (e.g., Leps, 1970, Barton and Kjaernsli, 1981, Indrartna et al., 1998). Failure envelopes of rock discontinuities as well as rock/rockfill interfaces have been reported to be nonlinear by Barton and Kjaernsli (1981).

Since the failure of the Kettleman Hills landfill in 1988, the shear strength relationships of soil/liner interfaces have been studied and reported by numerous investigators. This incident has probably been responsible for more advances in the characterization of soil/liner interfaces than any other single event. Recently published information (Stark and Poeppel, 1994, Gilbert et al., 1998) regarding the strength of the soil/liner interfaces at Kettleman Hills indicates that these interfaces also exhibited nonlinear failure envelopes.

3 BACKGROUND FOR HEAP LEACH FACILITIES

Since the advent of the heap leaching process for which underlying composite liners at the heap pad-foundation interface include clay and geomembrane layers in intimate contact, the shear strength of these interfaces has been of considerable interest. Typically, the soil/liner interface is the overriding feature in the assessment of the stability of a heap leach facility. This interface may be termed the "weak link" in the stability of a heap due to its almost inherently low strength properties. The phenomenon of a nonlinear failure envelope for a soil/liner interface has many important implications with consideration to the slope stability of a heap leach facility.

In conjunction with the various aspects of designing a heap leach facility, it is standard practice to conduct limit equilibrium slope stability analyses to provide guidance on the overall heap layout and material selection, including geomembrane and clay liner types. In order to conduct a proper stability analysis, which includes the examination of potential wedge-type failures with the failure plane passing along the soil/liner interface, it is vital that direct shear tests be carried out to assess the appropriate shear strength relationship to be adopted for the analysis.

Based on the results of extensive direct shear tests conducted on various soil types and geomembranes under the supervision of the authors, a number of key issues developed which deserve special consideration. These considerations are discussed in the following sections.

3.1 Curvature of the Failure Envelope

As additional reserves of low grade ores are identified, it has become common practice to load existing heaps to greater and greater heights. This practice has illuminated a phenomenon that is not always apparent. As evidenced in other materials, the degree of curvature in a failure envelope for a soil/liner interface is often more pronounced when data are available for a wide range of stresses. Typical curvature of a failure envelope is shown in Figure 1. This failure envelope was established for a sandy lean clay (Liquid Limit (LL)=45, Plasticity Index (PI)=19, percent passing the #200 sieve (% fines)=51%) at a moisture content of 27 percent tested against a 60-mil smooth very flexible polyethylene (VFPE) geomembrane. It can be seen that the friction angle (secant) of the material decreases from a value of 22 degrees at a normal stress of 1,000 psf to 11 degrees at 15,000 psf. It can also be noted from Figure 1 that the shear strength of this soil/liner interface is essentially the same regardless of whether the peak stress or some other failure criteria is used. As will be discussed subsequently, this is not always the case.

194

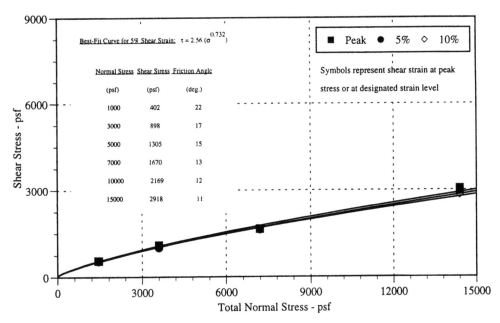

Figure 1. Typical Failure Envelope for a Clay Liner/Geomebrane Interface

(Sandy Clay & 60 mil VFPE Smooth Geomenbrane - 12 inch shear box)

Curvature of the failure envelope similar to that in Figure 1 has been observed for a variety of soil and geomembrane combinations. Typically, however, this curvature is less pronounced for granular, non-plastic soils.

3.2 *Extrapolation of Data*

The maximum normal stress applied during the testing shown in Figure 1 corresponds to the maximum load that can be applied to the specimen by this particular 12-inch by 12-inch shear box test apparatus with a maximum normal load capacity of 14,000 pounds (equivalent to a maximum normal stress of 14,400 psf). For a heap with an average total density of 110 pounds per cubic feet (pcf), this maximum normal load corresponds to a heap height of about 130 feet. This is well below the heights at which heaps are currently being designed. In order to overcome this deficiency, the authors had a rigid insert built for the shear box that reduces the sample dimensions to 8 inches by 8 inches. Given this reduced sample size, the maximum normal stress is increased by a factor of 2.25 to 32,400 psf (equivalent to a heap height of nearly 300 feet, based on an average total density of 110 pcf).

One of the dangers of extrapolating laboratory shear test data is in establishing a best-fit line through linear regression. While selection of for example a power function may appear to be appropriate for the stress range represented by the test data, there is no guarantee that such a function will represent the failure envelope at higher stress levels; perhaps a hyperbolic function would be more appropriate. In many cases, this 8-inch insert for the shear box device alleviates the requirement of extrapolating the failure envelope. An example of this is shown in Figure 2.

Figure 2 represents test results conducted on a clayey sand with gravel (LL=43, PI=20, 46% fines, 23% gravel) tested against a 60-mil smooth VFPE geomembrane. Based on these test results, the 8- and 12-inch shear boxes were observed to yield essentially the same results for stresses up to 14, 400 psf. However, it can be seen that extrapolation of the failure envelope for data obtained

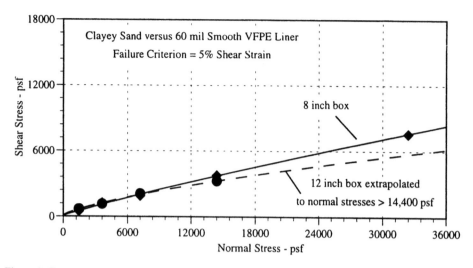

Figure 2. Comparison of Interface Test Results Using 8 and 12 inch Shear Boxes

using the 12-inch box to normal stresses greater than 14,400 psf (in this case using a power function determined through linear regression analysis) under predicts the shear strength at higher levels of normal stress when compared to the results of the 8-inch box at a normal stress of 32,400 psf. The opposite effect has also been observed where extrapolation of the failure envelope over predicts the shear strength at higher levels of normal stress. Thus, it may be observed that extrapolation of a curved failure envelope should be avoided whenever possible.

3.3 Failure Criterion: Peak vs. Post-peak Strength

Quite frequently the development of a peak shearing resistance is often mobilized at relatively low strain levels and may be followed by a rapid loss of shear strength. An example of this is shown at the bottom of Figure 3, which represents stress-strain plots obtained using a textured high-density polyethylene (HDPE) geomembrane against a highly plastic clay material (LL=104, PI=67, 76.6% fines). It may be noted from these plots that the peak shearing resistance was mobilized at less than 2 percent strain for each of the tests. As these tests were conducted in a 12-inch shear box, this strain is equivalent to less than 0.25 inches (6 mm) of deformation. As illustrated in Figure 4, the shear strength envelope for the tests shown at the bottom of Figure 3 suggests that the interface shear strength at peak shearing resistance is approximately twice that available at a post-peak condition taken at 5 to 10 percent shear strain.

Stress-strain plots obtained from soil/liner interface tests using a smooth HDPE geomenbrane on the same highly plastic clay material noted above are shown in the upper portion of Figure 3. In this case, there is no sudden loss of strength after the peak strength is reached. Instead, the strength remains approximately constant with increasing strain levels.

Based on the above example and similar findings from numerous other tests conducted, use of the peak shearing resistance in stability analysis of a heap for soil/liner interfaces may not be prudent engineering and is not recommended. It is quite likely that minor strains induced in the interface during installation and initial loading may be sufficient to bring about a residual strength condition. As suggested by Mitchell et al. (1990), Seed and Boulanger (1991), and Eigenbrod and Locker (1987), it is evident from test results such as those shown in Figure 3 that a post-peak strength should be adopted for use in a slope stability analysis.

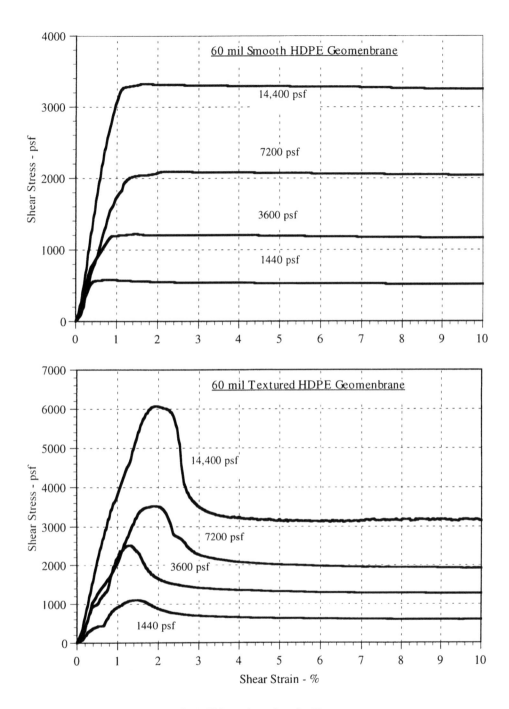

Figure 3. Stress - Strain Curves for Soil/Geomebrane Interface Tests

4 EFFECT OF NONLINEAR FAILURE ENVELOPE ON SLOPE STABILITY

In order to illustrate the importance of considering the curved failure envelope, a series of slope stability analyses were carried out. The analysis was conducted using the computer program

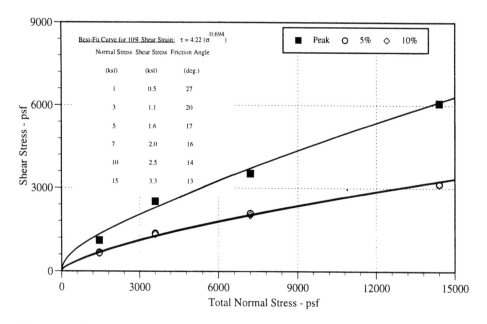

Figure 4. Interface Test Results for a Clay Liner and a 60 mil Textured HDPE Geomenbrane

(12 inch shear box)

XSTABL (Sharma, 1998) to analyze a typical heap slope. The modeled heap is illustrated in Figure 5 and consists of a series of 50-foot-high (total height of 430 feet) angle of repose lifts with 25-foot benches provided between each lift. The overall heap slope is approximately 2:1 (H:V). The soil/liner interface slopes at a 5 percent grade, and the results of the direct shear interface testing are shown in Figure 6. These results were obtained for a clayey sand with gravel (LL=34,

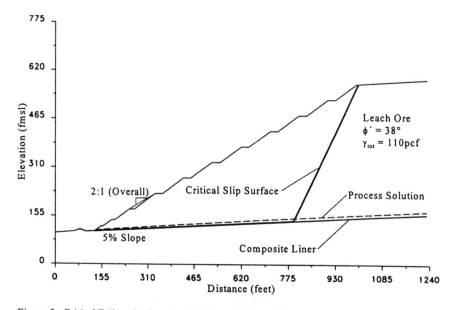

Figure 5. Critical Failure Surface for Stability of Ultimate Heap

PI=18, 46% fines, 26% gravel) tested against a 60-mil smooth HDPE geomembrane using the 8-inch insert for the shear test apparatus (highest normal stress). The best-fit power function to the actual data points is plotted on the figure together with values of normal stress, shear stress, and secant friction angle corresponding to the various normal stress levels. The values of the secant friction angle range from 28 degrees al low stress levels (1,000 psf) to 13 degrees at a stress level of 25,000 psf.

Also shown on Figure 6 is a conventional linear interpretation of the failure envelope which results in a friction angle of 17 degrees. For this envelope, the test result obtained using the 8-inch insert was not used for the linear interpretation since, conventionally, the 8-inch insert would not be available, and a straight line would be extrapolated using the available results.

Results of the analysis are shown on Figures 5 and 7 for analyses of the overall heap and the initial lift, respectively, and factors of safety resulting from the analyses are summarized below.

	Failure Envelope	
Case	Straight Line	Curved
Ultimate Heap	1.3	1.2
Initial Lift	1.0	1.3

It is readily apparent that the two failure envelopes yield different results. If, as is conventionally assumed for non-water-retaining heaps, the desired factor of safety is 1.3, neither of the analyses fully satisfies the requirements. While the results obtained for the straight line failure envelope over predicts the factor of safety for the overall heap, the factor of safety of the

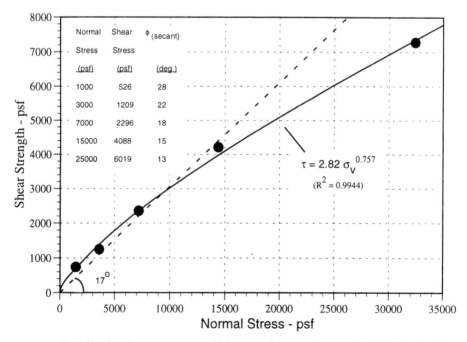

Note: X and y axis not to same scale; friction angle of straight line obtaained from a separate plot.

Figure 6. Nonlinear and Straight Line Interface Failure Envelopes Used in Stability Analysis

199

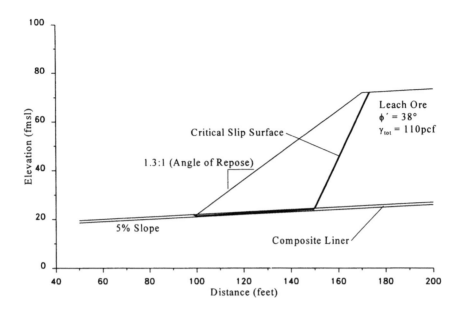

Figure 7. Critical Failure Surface for Stability of Initial Lift

initial lift is under predicted. This effect can be explained by examination on the information shown on Figure 6. At low ranges of normal stress (such as the initial lift), the available shear strength is greater than is estimated by using the straight line failure envelope. Conversely, at high ranges of normal stress (such as the ultimate heap), the available shear strength is less than is estimated by using the straight line failure envelope. Thus, it may be observed that it is not necessarily conservative to adopt a straight line interpretation of the failure envelope for use in slope stability analyses.

5   CONCLUSIONS

Based on extensive testing of soil/liner interfaces and stability analyses of numerous heap pads ranging from less than 100 feet to more than 400 feet in height, the following conclusions are provided:

1. The interface strength of a clay liner/geomenbrane depends on a number of factors. These include the soil characteristics, geomenbrane type (smooth versus textured), field compaction effort and moisture content, and applied normal stress.

2. The failure envelope for clay liner/geomenbrane interface tests is generally nonlinear. This nonlinearity becomes more pronounced at higher stress levels. Since heaps are now being designed to greater and greater heights, it is important to include the curvature of the failure envelope in stability analyses. Existing stability computer programs such as XSTABL and SLOPE/W can readily incorporate nonlinear failure envelopes.

3. Review of shear strain-shear stress plots for various types of interface tests indicates that in many cases, the peak stress occurs at relatively low strain levels and is followed by a signification reduction in strength to a residual strength value. It is our recommendation that the residual strength value be used in stability analyses. For cases where there is not a significant reduction in strength with increasing strain level, we recommend that the shear stress corresponding to a shear strain on the order of 10 percent be used in stability analyses.

4. We strongly recommend that site-specific interface testing be conducted for all heap pad

designs using the appropriate clay liner characteristics and geomenbrane types. In addition, the range of normal stresses used in the testing should correspond to the maximum heap height being designed.

REFERENCES

Barton, N. and B. Kjaernsli. 1981. Shear strength of rockfill. *ASCE Geot. Jour., Vol. 107, No. GT7*, July: 873-891.

Eigenbrod, K.D. and J.G. Locker. 1987. Determination of friction values for the design of side slopes lined or protected with geosynthetics. *Canadian Geotech Jour., Vol. 24*: 509-519.

Gilbert, R.B., S.G. Wright, and E. Liedtke. 1998. Uncertainty in back analysis of slopes: Kettleman Hills case history. *ASCE Jour. of Geot. and Geoenv. Eng., Vol. 124, No. 12*, December: 1167-1176.

Indrartna, B., D. Ionescu, and H.D. Christie. 1998. Shear behavior of railway ballast based on large-scale triaxial tests. *ASCE Jour. of Geot. and Geoenv. Eng., Vol. 124, No. 5*, May: 439-449.

Leps, T.M. 1970. Review of shearing strength of rockfill. *ASCE Jour. of Soil Mech. and Found. Div., Vol. 96, No. SM4*, July: 1159-1170.

Maksimovic, M. 1996. A family of nonlinear failure envelopes for non-cemented soils and rock discontinuities. *Electronic Journal of Geotechnical Engineering*, Oklahoma State University.

Mitchell, J.K. 1993. *Fundamentals of Soil Behavior, Second Edition*, Wiley Interscience, 437 p.

Mitchell, J.K., R.B. Seed, and H.B. Seed. 1990. Kettleman Hills waste landfill slope failure. I: liner-system properties. *ASCE Jour. of Geot. Eng., Vol. 116, No. 4*, April: 647-668.

Seed, R.B. and R.W. Boulanger. 1991. Smooth HPDE-clay liner interface shear strengths: compaction effects. *ASCE Jour. of Geot. Eng., Vol. 117, No. 4*, April: 686-693.

Sharma, S. 1998. XSTABL Version 5.202, Interactive Software Designs, Inc., Moscow, Idaho.

Stark, T.D. and A.R. Poeppel. 1994. Landfill liner interface strengths from torsional-ring-shear tests. *ASCE Jour. of Geot. and Geoenv. Eng., Vol. 120, No. 3*, March: 597-615.

Tailings and Mine Waste'00 © *2000 Balkema, Rotterdam, ISBN 90 5809 126 0*

Use of a chemical cap to remediate acid rock conditions at Homestake's Santa Fe Mine

D.A. Shay
Resource Investigation and Management Consultants, Spearfish, S. Dak., USA

R.R. Cellan
Homestake Mining Company, Grants, N. Mex., USA

ABSTRACT: The Santa Fe Mine, located in west central Nevada, began leaching ore in 1988 and completed ore processing in 1995. Homestake Mining Company has completed final reclamation of the site, including remediation of waste dump acid conditions resulting from the oxidation of near-surface sulfide minerals. The production of acidity and the migration of acidity into coversoil became a significant detriment to previous revegetation success at the site. Acid geochemical conditions at the site indicated that the actual production and concentrations of acidity were relatively low, the result of low moisture regime, low biological activity, and limited oxidation of ferric ions. The oxidation of sulfides was a near-surface condition and the spread of acidity into coversoil materials was largely the result of chemical diffusion processes. After examining several remediation options, Homestake elected to construct a chemical cap on the waste dumps to prevent the diffusion of acidity into new coversoil materials. Monitoring data from the site has indicated that the chemical cap has been successful in preventing the diffusion of acidity into coversoil materials and the long-term goal of revegetation success is likely.

1 INTRODUCTION

Hardrock mine plans in the early 1980's did not generally include provisions for managing waste rock containing pyrite or other sulfidic minerals. By the late 1980's, mine operators and regulators began to understand the potential problems associated with mining sulfidic rocks. As has been well documented, the oxidation of sulfides will result in the generation of significant acidity and the potential for the release of acid mine drainage. The sulfide oxidation process, and resultant acid production, has plagued many hardrock mines in recent years. The lack of planning for management of sulfidic rock during mine-planning and permitting phases resulted from the fact that most cyanide leach facilities developed in the early 1980's mined primarily oxide deposits. That is, the minerals in the deposits, including pyrite, were highly oxidized resulting in a deposit that is amenable to cyanide leaching. In addition, the majority of rock characterization was aimed at defining the ore reserve and less attention was paid to waste rock characterization. Waste rock was placed randomly in waste dumps and it was generally thought that the rainfall amounts in arid and semi-arid climates were not sufficient to cause environmental or reclamation concerns.

The Santa Fe Mine was developed in the late 1980's with ore processing completed in late 1995. Homestake Mining Company acquired the Santa Fe Mine in 1992 as part of a merger with another mining company. The operator began to reclaim the waste dumps in 1991 and 1992. The waste dumps were characterized by long angle-of-repose side-slopes and relatively flat tops, configurations common at the time for dumps constructed by end-dumping trucks. The original reclamation plans included the placement of 4 to 6 inches of coversoil on the tops and sides of the dumps, straw mulching, and seeding with a native seed mix. Initial revegetation growth was excellent with the establishment of a native grass, shrub, or forb community.

In 1997, revegetation monitoring and site observations indicated that the re-seeded vegetation was dying, leaving bare areas on the tops and side-slopes of the dumps. Examination of these areas confirmed that the oxidation of sulfides and the subsequent migration of acidity were responsible for the decline of vegetation. Homestake contracted with RIMCON to develop an acid rock remediation plan for the site and to assist Homestake in implementing that plan. Details of the remediation and results of first year monitoring are presented in this paper.

2 GEOLOGIC AND MINING BACKGROUND

The geologic setting and the mining practices at the Santa Fe Mine were typical of the mining operations and operational practices of the 1980's. The Santa Fe gold-silver deposit is located in the foothills of the eastern margin of Todd Mountain, within the south-central portion of the northwest trending Gabbs Valley Range in Nevada. The surface elevation of the mine site ranges from 5975 feet to 6160 feet above sea level. Three major rock units are present at the site. Figure 1 depicts a geologic cross-section of the Santa Fe pit area. The oldest unit is the Triassic Luning Formation, which is intruded by a Mesozoic pluton. These rocks are overlain by, or in fault contact with, Tertiary volcanics and Quaternary to Recent alluvium (Fiannaca, 1987).

The Luning Formation comprises blue-grey, medium-bedded to massive brecciated limestone, micrite, and rare amounts of siltstone. These rocks are usually carbonaceous and occasionally pyritic. Near intrusive margins, dull-green hornfels, whitish brown marble pods and calc-silicate skarns occur with pyrite as a common accessory mineral. The Tertiary volcanic rock unconformably overlies, or is in fault contact with the Luning Formation. This ash flow has a densely welded basal rhyodacite which grades upward into a partially welded rhyolite and both are generally non-pyritic (Fiannaca, 1987).

As is common, the gold mineralization at this site was controlled by a normal fault system, in this case the Santa Fe fault zone. This structure has experienced repeated movement during the Tertiary. The distribution of oxidized rocks, alteration, and mineralization has been controlled by the structure. Near the hanging walls of the structure, oxidation of sulfides has occurred to extended depths. Often, the oxidation has been incomplete and minor concentrations of pyrite may exist in the oxidized rock. Unoxidized rocks are commonly carbonaceous, contain minor to moderate amounts of pyrite, and occur in areas where low permeabilities existed, generally away from the fault zone. The low-grade gold-silver mineralization occurred as sheeted and pipe-like bodies found principally within the breccia in carbonate rocks (Fiannaca, 1987).

The gold and silver mined at the Santa Fe Mine came largely from the highly oxidized zones within the pit. As such, the rock placed on the heap leach pads did not contain appreciable concentrations of pyrite or sulfides and did not produce acidity upon leaching. Drill cuttings from the in-pit blast-hole drilling were analyzed for ore grade and oxidation state in relation to sulfides. Ore rock containing appreciable concentrations of pyrite (sulfide ore) was selectively handled and placed in a stockpile. This ore was generally not considered readily amenable to cyanide leaching and was stockpiled as a contingency for possible future processing.

Waste rock consisted of a combination of oxidized, mixed and sulfidic rock that was not mineralized. Waste rock dumps at the Santa Fe Mine were typical end-dump repositories. That is, the rock was dumped by trucks in piles and the piles were dozed flat to form a layer in the dump. This process resulted in long, angle-of-repose sideslopes and relatively flat tops. The mining process moved in and out of rock that contained pyrite and this process resulted in the sulfidic rock being placed randomly in the dumps. As stated previously, during this mining period, there was little concern by mining companies or regulators that the sulfides could create environmental or reclamation problems in the future. The Santa Fe Mine provides a case study of the types of remedial problems that can occur in an arid climate when pre-planning for management of sulfides has not occurred.

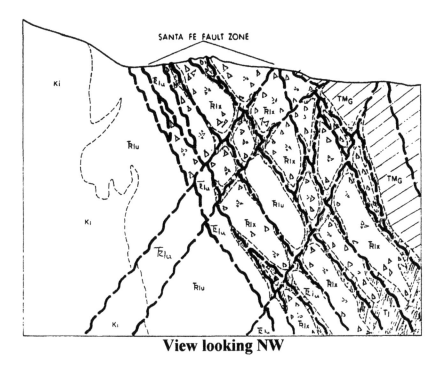

SANTA FE FAULT ZONE

View looking NW

EXPLANATION

TERTIARY

| | jasperoid |

| TMG | Guild Mine Member, Mickey Pass tuff |

CRETACEOUS

| Ki | quartz monzonite of Todd Mountain stock |

TRIASSIC

| Rlx | breccia of Luning Formation PRINCIPAL ORE HOST |

| Rlu | undifferentiated limestone & hornfels of Luning Formation |

fault

breccia

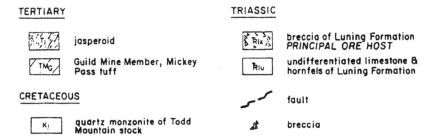

Figure 1. Santa Fe pit geologic cross-section.

3 SITE GEOCHEMICAL CHARACTERIZATION

In 1997, Homestake personnel observed that vegetative cover was declining, significant bare areas were developing, and the bare areas appeared to be growing in number and size. Revegetation specialists examined the site and evaluated numerous agronomic characteristics to determine the cause of the dying vegetation. Based on their review and the subsequent field examination by RIMCON, it was determined that the majority of bare areas were the direct result of the oxidation of sulfides in the waste rock.

Since the waste rock was placed randomly in the dumps, the occurrence of the acid areas was also random and the location of other potential acid generating areas was difficult to predict. Numerous areas of the waste dumps exhibited classic symptoms of the sulfide oxidation process. That is, oxidation by-products, such as iron and manganese hydroxides and gypsum were present. Surfaces were crusted and exhibited telltale red, yellow-orange, and black colors. Where some of the acidity had been neutralized by adjacent limestone, clayey residues resulted, further enabling surface crusting as the materials dried. In some areas, the oxidation and weathering processes allowed for significant expansion or swelling of the waste materials creating a hummocky landscape.

The development of a remediation plan for the site depended on numerous factors including the sulfide oxidation mechanisms. The sulfide oxidation process is generally complicated and involves both chemical and biological oxidation components. The rate of chemical oxidation is influenced by factors including pH, oxygen concentrations, temperature, moisture content and the ferrous/ferric iron equilibria. The chemical oxidation of pyrite is quite sensitive to the pH of the reaction media, particularly on the influence of pH on the ferrous/ferric iron equilibria. That is, at very low pH, ferric iron serves as an oxidant of ferrous iron resulting in a significant increase in both the rate and quantity of acid generation. If the oxidation process is limited and pH levels remain generally above 4.5, the oxidation of ferric iron will be limited (Sobek, etal., 1997).

Microorganisms also serve as catalysts to the oxidation process and are most active in the pH range of 2.0 to 4.5. The organisms have optimum temperature requirements and must have moisture to be active. Generally the organisms obtain their energy from oxidation of elemental sulfur and can actually regenerate ferric iron in the process. When bacteria are active, the generation of acidity may be increased 50-fold by increasing the rate of ferrous iron oxidation (Sobek, etal., 1987).

Table 1 summarizes some of the site geochemical characteristics from selected areas. The table has been divided into two sections, one summarizing data from areas that did not contain oxidizing sulfides and the other from areas that had oxidizing sulfides. As is typical of most acid generating materials, the sulfidic areas exhibited elevated iron, manganese, and sulfate levels. The sulfate concentrations for the sulfidic areas ranged from 100 to 880 ppm while the sulfate concentrations for the non-sulfide areas ranged from 6.6 to 64 ppm. Manganese concentrations for the sulfide areas ranged from 98 to 820 ppm as compared to the concentration range for non-sulfide areas of 3.3 to 46 ppm. The iron concentrations, often the most identifiable element associated with acid generation, ranged from 190 to 4600 ppm for the sulfide samples and ranged from 6.7 to 95 ppm for the non-sulfide samples. With the exception of one sample (5C), pH values for the sulfidic areas were greater than 4.5 and ranged up to 5.9. The sampled identified as 5C exhibited a very low pH value of 2.6.

While the generated acidity was sufficient to degrade coversoil, the actual concentration of acidity was somewhat limited. Laboratory weathering tests of samples 5C and 23C-1 show that total titrateable acidity generated in the samples was only 380 and 420 mg/l, respectively. These results can be compared to numerous other mine sites where total generated acidity often exceeds several thousand mg/l. This data, along with the observations described above and field observations, indicated that the relatively low acid generation was the result of several rate limiting factors, including low moisture regime, limited biological activity, and limited oxidation of ferric ions. This conclusion was very important for the selection of a cost-effective remediation plan.

Table 1. Selected waste dump geochemical characteristics.

Sample Site	pH	Iron mg/l	Manganese mg/l	Sulfate Mg/l
Non-sulfides				
4A	7.6	13	7.4	64
15	7.0	79	46	39
16	6.9	95	24	41
17	7.9	8.2	3.8	6.6
21A	7.9	6.7	3.3	12
Sulfides				
13B	5.5	1000	320	180
4D	5.9	190	98	140
5C	2.6	4600	820	880
23C-1	4.5	2800	410	320
23C-2	5.6	1900	110	100

Several factors resulted in the degradation of coversoil and vegetation at the Santa Fe Mine. Because precipitation is limited at the site (7 inch average), the generation of acidity was strictly a near-surface condition. Limited grading of the dump and dump settlement resulted in brief ponding and collection of rainfall runoff in low areas. This moisture was sufficient to allow for oxidation of near-surface sulfides and the production of low-level acidity. While the sulfide rock was often deposited in the vicinity of other waste rock derived from the Luning limestone, geochemical kinetics were limited and little neutralization of the acidity occurred. Generally, the acid generation process was very rapid and the limestone surface area was too large to allow for significant neutralization of the acidity. Additional field observations of acid areas showed typical iron hydroxide coating of the limestone, further restricting the ability of the limestone to neutralize the acidity. In areas where rainfall did not collect for extended periods of time, the sulfide waste rock had not yet began to oxidize and generate acid.

The migration and spread of acidity into the coversoil was the result of chemical diffusion rather than movement by convective flow. That is, the acid solute moved from areas of high concentrations to areas of lower concentrations without significant water movement. Because of this process, the final remediation design had to consider ways to prevent the diffusion of the acidity into new coversoil. Since many areas of the dump had sulfide waste that had not yet significantly oxidized, the plan had to be inclusive of these areas also. Covering the dumps with non-sulfide waste rock was not an option since mining had ceased, all waste rock was randomly mixed with sulfidic rock, and coversoil availability was also limited.

Since the oxidation was accelerated in low areas on the dumps, the final remediation plan involved grading the surface of the dumps to prevent ponding of the limited precipitation. The crests of the dumps were graded to prevent rainfall from flowing directly over the steep sideslopes creating rilling and erosion. In addition, drainage channels were designed into the plan to route precipitation from potential high rainfall events away from the dumps. While the grading plan may create a less diverse vegetative stand over time, the prevention of ponding was considered very important to preventing isolated high oxidation areas.

The final, and most important aspect of the remediation plan, was the construction of a chemical cap that was placed on the dump tops and sideslopes before coversoil application. Chermak and Runnells (1997) discussed the use of a chemical cap at an unnamed Nevada mine site. They attempted to create a permanent chemical cap, or hardpan, by neutralizing surface acidity with lime. The chemical reactions produced gypsum and amorphous iron oxyhydroxides, which were produced in an attempt to limit water infiltration into the dump. Conversely, the Santa Fe Mine chemical cap function was to neutralize existing near surface acidity and to serve as a barrier for diffusion of acid solutes. The chemical cap was placed over the entire surface of the dumps and selected areas of the sideslopes since the location of potentially new acid generating areas was difficult to locate even with intensive sampling.

The chemical cap consisted of the placement of 20 tons per acre of magnesite (magnesium carbonate) and brucite (magnesium hydroxide) mined in nearby Gabbs, Nevada. The application rate was significantly higher than was necessary to neutralize any potential acidity. The application rate was finally estimated by the ability of available equipment to apply the chemical uniformly over the dump surfaces. The chemical was not disked into the surface as is common but was left as a concentrated amendment, maximizing chemical gradient differences. A total of 1850 tons of magnesite and brucite were applied to the dumps. Following placement of the cap, 8 to 12 inches of coversoil, borrowed from a nearby native draw, was placed carefully over the cap and seeded with a native seed mix.

While used at times for water treatment, the use of magnesite and brucite as a chemical amendment for remediation of acid rock conditions is somewhat rarer. Theses chemicals had a total alkalinity of 1920 mg/kg and a neutralizing potential (NP) of 600 tons per 1000 tons $CaCO3$ equivalent. Laboratory observations indicated that the rate of chemical neutralization was slightly slower than for calcium carbonate due to a slower dissolution rate. However, the magnesite and brucite had significant buffering capacities and the fact that the materials were mined very close to the site made them an extremely cost-effective and technically sound amendment.

5 REMEDIATION RESULTS

Construction of the chemical cap and coversoil placement was finished in late 1997 and revegetation efforts on the remediated dumps were completed by spring 1998. The site was visited in late April 1999 to monitor the effects of the chemical cap and to make general observations of the site conditions. Vegetation monitoring has been conducted separately from this study.

Monitoring procedures consisted of hand-dug holes to a depth below the chemical cap. Monitoring sites were selected randomly throughout the tops and sideslopes of the waste dumps. A total of 21 monitoring holes were dug and observations of coversoil depth, coversoil pH, waste pH, chemical cap conditions, and other general observations were recorded. Photographic documentation of the monitoring holes was also performed.

Table 2 and Table 3 summarize the results of those field observations for the east and west waste dumps. In all cases, where sulfidic waste was noted below the cap, no diffusion of acid solutes has occurred. One site, identified as site W-9 showed evidence that the coversoil, chemical cap, and sulfidic waste had been disked and thoroughly mixed. While oxidation by-products were clearly visible in the materials, field pH measurements indicated that neutralization of the acidity had occurred as evidenced by a 6.9 pH value. The mean pH of all coversoil immediately above the chemical cap was 7.2 with a range of 6.9 to 7.6. The mean pH of the sulfidic waste present immediately below the chemical cap was 5.5 with a range of 4.7 to 5.9.

At this point in time, rainfall quantities have not been sufficient to reach the chemical cap. The 8 to 12 inches of coversoil is sufficient to store all of the moisture received during a normal rainfall year. In the absence of sufficient moisture to allow for convective movement of acid solutes, conditions are ideal for chemical diffusion to occur. The presence of the chemical cap has prevented the diffusion of acidity into the coversoil and it is likely that the chemical gradient created by the cap will result in long-term mitigation of any potential diffusion.

Table 2. West waste dump chemical cap monitoring observations.

Location	Coversoil Depth - in.	Coversoil pH	Waste pH	Visual Observations
W-1	11	7.6	5.9	No visible acid by-products in coversoil; amendments visible
W-2	13	7.5	4.9	No visible acid by-products in coversoil; amendments visible
W-3	8			Non-sulfidic waste; amendments visible
W-4	9			Non-sulfidic waste; amendments visible
W-5	10			Non-sulfidic waste; amendments visible
W-6	8	7.2	5.9	No visible acid by-products in coversoil; amendments visible
W-7	12			No visible acid by-products in coversoil; amendments visible
W-8	8			No visible acid by-products in coversoil; amendments visible
W-9*	0	6.9	6.9	Coversoil/waste mixed; small area; acidity neutralized
W-10	8			No visible acid by-products in coversoil; amendments visible
W-11	10			No visible acid by-products in coversoil; amendments visible
W-12	10			Non-sulfidic waste; amendments visible
W-13	8	7.1	5.9	No visible acid by-products in coversoil; amendments visible
W-14	8			No visible acid by-products in coversoil; amendments visible
W-15	8			Non-sulfidic waste; amendments visible
W-16**	12	7.4	4.7	No visible acid by-products in coversoil; amendments visible
W-17	8	7.1	5.8	Side slope site; no visible acid by-products in coversoil
W-18	8			Side slope site; no visible acid by-products in coversoil

*   Coversoil, waste, and amendments have been disked together at this location. Acidity has been neutralized.
** Location in area of intense weathering and oxidation before remediation. Area was very wet and generating steam on cold days. Amendments were applied at a rate of 40 tons/acre.

Table 3. East waste dump chemical cap monitoring observations.

Location	Coversoil Depth - in.	Coversoil pH	Waste PH	Visual observations
E-1*	24			No visible acid by-products; amendments not reached
E-2	9	7.3	5.9	No visible acid by-products in coversoil; amendments visible
E-3	11			No visible acid by-products in coversoil; amendments visible
E-4**	12	7.1	2.7	No visible acid by-products in coversoil; amendments visible
E-5	8			Non-sulfidic waste; amendments visible
E-6	9			Non-sulfidic waste; amendments visible
E-7	8			No visible acid by-products in coversoil; amendments visible
E-8	8	7.0	5.9	No visible acid by-products in coversoil; amendments visible
E-9	8			No visible acid by-products in coversoil; amendments visible
E-10	10	7.2	5.8	No visible acid by-products in coversoil; amendments visible
E-11	10			No visible acid by-products in coversoil; amendments visible
E-12	8	7.0	5.2	No visible acid by-products in coversoil; amendments visible
E-13	8			

* For unexplained reasons, this sampling site is very deep and amendments were not reached.
** This location is the site of significant acid generation compared to other areas. The chemical cap is functioning at this severe site.

6 ECONOMIC CONSIDERATIONS

Table 4 summarizes the costs associated with the acid remediation plan and does not include the internal management expenses encountered by Homestake. In addition to the expenses noted in Table 4, Homestake also had internal site review costs, remediation plan development and review costs, on-site construction management costs, and site maintenance costs.

As can be seen from Table 4, Homestake's voluntary waste dump remediation efforts cost the company in excess of $775,000 not including their internal costs. These costs are a case study of a relatively uncomplicated remediation plan, the result of low precipitation levels. Many sites that have developed acid problems, including acid mine drainage, are faced with remediation costs that often exceed several million dollars. In fact, acid remediation construction costs at two mines in South Dakota that are of similar size to the Santa Fe Mine are estimated in excess of $12,000,000 each, not including post-remediation water treatment and management costs (SDDENR, 1999).

7 CONCLUSION

The selection of the chemical cap technology for remediation of acid conditions at the Santa Fe Mine was cost-effective. Other alternative options, such as a combined physical and chemical cap, would have cost significantly more than the chemical cap. The voluntary effort by Homestake to improve the revegetation success of the site appears to have been very successful. This project is indicative of Homestake's desire to be proactive in the reclamation and management of all of their mine sites.

Table 4. Santa Fe Mine remediation and reclamation costs.

Expense category	Cost
Mobilization, demobilization, site grading, and drainage construction	$146,000
Chemical amendment purchase, haulage, and application	$139,000
Coversoil haulage, application, and revegetation	$442,000
Engineering and construction management	$ 50,000
Total	$776,000

8 REFERENCES

Fiannaca, M. 1987. Geology of the Santa Fe gold-silver deposit, Mineral County, Nevada, in Johnson, J.L. ed., Bulk Metal Deposits of the Western United States, Geological Society of Nevada. 1987 Symposium, p. 233-239

Chermak, J.A. and D.D. Runnells. 1997. Development of chemical caps in acid rock drainage environments. Mining Engineering. P. 93-97.

Sobek, A. A., Hossner, L.R., Sorenson, D.L., Sullivan, P.J. and Fransway, D.F.. 1987. Acid-base potential and sulfur forms. In Reclaiming mine soils and overburden in the western United States, R. Dean Williams and Gerald E. Schuman, editors, Soil Conservation Society of America, Ankeny, Iowa, pp. 223-258.

South Dakota Department of Environment and Natural Resources, Minerals and Mining Division, personnel communication, Pierre, South Dakota, 1999.

Groundwater and geochemistry

Tailings and Mine Waste'00 © 2000 Balkema, Rotterdam, ISBN 90 5809 126 0

Application of 3D groundwater models to the assessment of ARD source loadings in surface water bodies

M. M. E. Uwiera & M. J. Reeves
M. D. Haug and Associates Limited, Saskatoon, Sask., Canada

ABSTRACT: An approach to acid rock drainage (ARD) source term modelling involving sources (waste rock piles, open pits and underground mines) and receptors (streams and lakes) is illustrated. The approach convolves a mass source term for numerous advective particle paths with computed travel times from source to receptor using groundwater flow modelling software. The result is a temporal mass loading function for each receptor. A case history of a mine containing acid generating rock, deposited in subaerial waste rock piles and used as subaqueous backfill in open pits is presented. The results illustrate the flexibility and utility of the source-term/particle path modelling approach for estimating mass loads to surface water receptors. The use of high vertical resolution 3D models is critical to the successful estimation of flows between sources and receptors. The computational effort required is much less than a conventional transport simulation. The approach presented successfully captures "travel time dispersion".

1 INTRODUCTION

Computer modelling of contaminant transport in groundwater is a standard tool in the design of waste disposal systems. Models provide the only effective method of extrapolating the behavior of complex three-dimensional (3D) flow systems over extended time periods and are of particular use in the comparative evaluation of alternative decommissioning strategies.

Mine sites typically involve some or all of the following features; underground workings, surface pits, waste rock piles and tailings management areas; often in close proximity and interacting with one another. The major difficulties of applying 3D models to such problems are the large meshes and small time steps needed to adequately resolve the spatial and temporal complexities and the consequent high demands placed on computational resources.

The compromises needed to simplify real-world problems typically involve either 2D analysis or pseudo-3D analysis of advection and dispersion using modelling packages such as MODFLOW and MT3D. This allows the modeller to predict the development of contaminant plumes in the subsurface but, because of the time-step and mesh-size limitations imposed by attempting to model dispersion, vertical resolution of the flow-domain is usually poor and the time period over which predictions can be made is limited.

In practical terms, it is only necessary to simulate dispersion if the model is aimed at predicting the distribution of solutes in the subsurface. If the focus of the model is the prediction of mass loadings in surface water bodies, the subsurface plume geometry is not necessarily required. Tracing of advective pathlines from source to receptor is often sufficient to estimate surface water mass loadings. Because flow line tracing is much less demanding on computational resources, true 3D models are feasible for long-term simulations with excellent vertical and horizontal resolution.

2 MODELLING CONSIDERATIONS

The most important factors in determining mass loadings to surface water bodies are the mass source term, the advective path of the mass and the travel time. A true 3D mesh is essential for accurate estimation of pathline trajectories and travel times.

2.1 *Modelling Software*

The numerical codes MODFLOW, MODPATH and ZONEBUDGET can be used to simulate groundwater flow (MODFLOW), determine travel times for solute transport to the nearest receptor (MODPATH) and to calculate water balance within specified regions of the numerical grid (ZONEBUDGET). Both MODFLOW and MODPATH are supported by the Groundwater Modeling System (GMS) graphical interface (ECGL, 1996).

MODFLOW is a 3D, finite difference, saturated groundwater flow code developed by the United States Geological Survey (USGS; McDonald and Harbaugh, 1988) and is recognized as a standard code for groundwater flow modelling. MODFLOW can be used as a true 3D-simulation package by assigning appropriate values of vertical conductance and block thickness and building the topography into the layer structure. Figure 1 shows a schematic representation of a true 3D MODFLOW mesh.

MODPATH is a particle tracking code that calculates an "imaginary" path and advective travel times of particles within the groundwater flow field (ECGL, 1996; Pollock, 1994). Once the steady-state regional groundwater flows are calculated using MODFLOW, the particle travel times can be determined with MODPATH.

The ZONEBUDGET code calculates subregional water budgets using the results from MODFLOW simulations (Harbaugh, 1990). Areas of interest are assigned a particular zone number and ZONEBUDGET then calculates the volumetric flow of water into and out of a region of interest.

2.2 *Source Term Estimation*

Perhaps the most difficult part of any mass transport model in the context of ARD is the generation of source terms. A source term is simply a mass or concentration value expressed as a func-

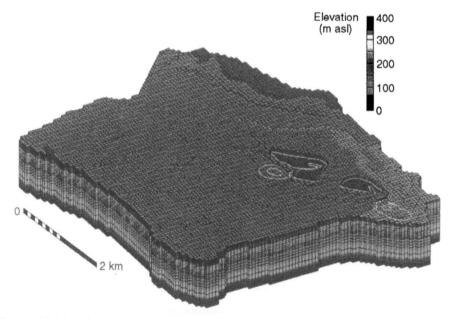

Figure 1. Model mesh.

tion of time. For example, simple mathematical forms include: $M(t) = M_o$ (constant mass source term), $M(t) = M_o (1-\beta t)$ or $M(t) = M_o \exp(-\beta t)$ (declining or finite mass source terms).

To assign source terms, the available data may include: the total mass of a solute species present in the waste rock (from whole rock analyses), estimates of the amount of leachable solute species (from acid or water extraction experiments), estimates of initial and long term leach rates (from kinetic column tests or humidity cell tests), levels of the solute species in waste rock leachate and background levels present in the groundwater (from monitoring wells).

2.3 Modelling Approach

Groundwater modelling using MODFLOW and MODPATH provides travel times from the sources to the receptors. Finite mass source terms can estimate using the leachate composition of groundwater monitoring wells and estimates of the bulk waste rock chemistry.

To generate mass fluxes for the receptors, the time-dependent source terms are convolved with the steady-state flows and travel times produced from MODFLOW, MODPATH and ZONEBUDGET. The convolution algorithm is simple. Source terms are assigned to each particle path, offset by the travel time and then added to generate a total mass loading versus time function.

Dilution between source and receptor is modelled by extracting the total flows to receptors from the MODFLOW output using the ZONEBUDGET program. The flows and masses leaving the waste rock piles are estimated from the infiltration rates and the source term models respectively.

To model retardation, travel times are adjusted by a retardation factor. A simple analysis involves the application of a single retardation factor for every flow path. For a more rigorous analysis, each flow path can be broken down by geological material and estimates made of the travel times in each material with different retardation factors.

3 CASE HISTORY

The case history chosen to illustrate the application of the methodology is a site where both underground cut and fill stoping and surface open pit mining has taken place. Waste rock has been deposited in two subaerial waste rock piles and used as subaqueous backfill in two open pits. The waste is acid generating and the low pH environment allows leaching of additional metals from marginal ore deposited in the waste rock piles.

3.1 Topography and Surface Water Drainage

The topography in the study area is gently undulating and gradually sloping to the south and west towards a major lake. The topographic elevations above sea level range from a high of approximately 368 m to about 317 m at lake level. The landscape has been significantly affected by continental glaciation and is dominated by low relief drumulinized till.

Surface water drainage is poor due to the gently sloping topography. Poorly drained wetlands are common in the region. Coniferous with subordinate deciduous trees cover most of the area.

3.2 Geology and Hydrogeology

The regional geology comprises low permeability basement gneisses overlain by a thin (5-10m) layer of more permeable sandy till. The upper 10m of the gneisses are weathered to produce a near surface zone of enhanced permeability. Extensive faulting occurs in the study area and the major lakebeds and river valleys are fault-controlled. Fault zones are assumed to represent areas of enhanced bedrock permeability.

The major hydrostratigraphic units are: 1) overburden (surficial till, sand and peat); 2) weathered bedrock; and 3) intact bedrock. There are three basement lithologies of importance in the study area: 1) sandstone; 2) aluminous gneisses; and 3) felspathic gneisses and granitoids. For each lithology, three weathering zones were defined based on depth. In addition to the lithologic subdivisions of the basement rocks, fracture zones were designated along major fault trends. The fault zone material was further subdivided based on depth from surface.

The three basement lithologies and fault zones combined with the weathering classification gives twelve (4x3) hydrogeologically distinct materials. The overburden, waste rock and backfill material for the pits and underground mines provide three additional units for a total of fifteen materials in the 3D model.

3.2.1 Aquifer Properties
The hydraulic properties of the hydrostratigraphic units encountered in the study area were estimated from laboratory testing, Environmental Impact Statement (EIS) documents, consultant reports and the literature. The estimated ranges of the saturated hydraulic conductivity and porosity are summarized in Table 1.

Table 1. Estimated ranges of saturated hydraulic conductivity and porosity.

Hydrostratigraphic Unit	Saturated Hydraulic Conductivity (m/s)	Porosity (%)
Waste Rock	$1 \times 10^{-2} - 4 \times 10^{-7}$	26 – 40
Backfill	$1 \times 10^{-2} - 4 \times 10^{-7}$	26 – 40
Till	$3 \times 10^{-5} - 3 \times 10^{-7}$	25 – 50
Fault Zone	$1 \times 10^{-5} - 5 \times 10^{-8}$	20 – 40
Sandstone	$1 \times 10^{-6} - 4 \times 10^{-9}$	5 – 30
Aluminous Gneiss	$5 \times 10^{-6} - 1 \times 10^{-8}$	1 – 10
Feldspathic Gneiss and Granitoids	$5 \times 10^{-6} - 1 \times 10^{-8}$	1 – 10

3.2.2 Infiltration Rates
It is estimated from net infiltration measurements for the regional surface water catchment that the mean net infiltration for the area is between 35 to 45% of the total precipitation. Measurements conducted on site indicate a net infiltration rate of 40% of the precipitation. An earlier modelling study deduced a calibrated net infiltration value of 93 mm/yr or approximately 23% based on a review of data for the site indicating net infiltration in the 10 to 40 % range of annual precipitation. For the present study, 31% of precipitation, or 125 mm/yr, was adopted.

3.3 Groundwater Flow Model
The development of the groundwater model made the following assumptions:
- the bedrock hydrostratigraphic units were considered homogeneous and isotropic;
- net infiltration applied spatially throughout the model was a constant average annual value;
- incorporation of fracture and weathered zones took in account the variability in fracture porosity throughout the hydrostratigraphic units; and
- particle paths were developed for a steady-state flow regime.

3.3.1 Model Discretization
The model discretization in the horizontal plane consists of 90 x 112 grid blocks. Each grid block is a 50 m x 50 m square. In the vertical plane, topography is modelled by 5 m high grid blocks from 385 m down to 300 m asl (17 layers), then 10 m high blocks from 300 to 210 m asl (9 layers), then 20 m high blocks down to 150 m asl (3 layers) and finally 50 m high blocks from 150 m to sea level (3 layers). No blocks have aspect ratios greater than 10. The complete mesh is 90 x 112 x 32 for a total of 322,560 grid blocks (more than 250,000 active grid blocks).

The model represents a true vertical section of the regional geology with a horizontal resolution of 50 m and a vertical resolution of 5 to 50 m. The "cut-away" cross-section of Figure 2 shows the model representation of the geology, underground mines open pits and waste piles. The model is a true 3D MODFLOW representation, rather than the more familiar pseudo-3D use of the USGS package.

3.3.2 Boundary Conditions
The boundary conditions used to represent the regional groundwater flow regime included con-

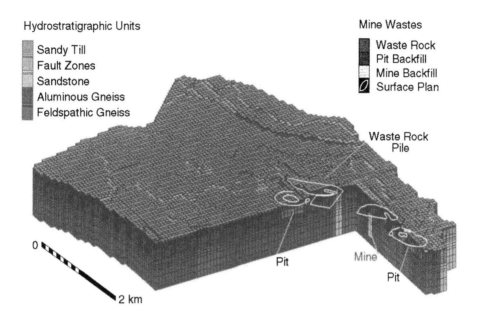

Hydrostratigraphic Units

■ Sandy Till
▨ Fault Zones
▨ Sandstone
■ Aluminous Gneiss
■ Feldspathic Gneiss

Mine Wastes

■ Waste Rock
■ Pit Backfill
▨ Mine Backfill
▨ Surface Plan

Waste Rock
Pile

0

2 km

Pit

Mine

Pit

Figure 2. Model hydrostratigraphy.

stant head, no flow, recharge and drain boundaries. Most boundary conditions used are shown in Figure 3.

The limits of the regional flow domain were selected from natural hydrological and hydro-geological boundaries, topographic highs and from the numerical results of an earlier study. The base of the regional flow model was located at sea level and designated as a no flow boundary condition. It was judged that truncating the model at this elevation would not affect the shallower flow conditions.

The small lakes located within the interior of the model were assigned as constant head boundary conditions that were equivalent to the water level within the lake. The water levels assigned to these lakes were determined from topographic maps. The boundaries formed by rivers and lakes were assigned to the surface layer only. Constant head boundary conditions used to represent the major deep lake were assigned to four layers of the regional model in order to simulate the 20m deep lake within the region of interest.

The low relief areas above the free water surfaces surrounding the lakes were mapped as bogs. Artesian groundwater conditions cause groundwater to discharge at these bogs. Drain boundary conditions were used to represent the release of water from the groundwater flow system. When the water table is below the ground surface the drain node remains dormant and does not act as a boundary.

3.3.3 Steady-State Calibration

Calibrating the steady-state groundwater flow model for the area involved a trial and error iterative process. During model calibration the hydraulic properties of the geologic media, infiltration rates, and initial and boundary conditions were adjusted so that the numerical model results match the data obtained in the field and the results of the pre-existing model.

Data from twenty monitoring wells completed in the vicinity of the mining operations were used during the model calibration process. For monitoring wells that were routinely monitored the average water level was calculated and used in the calibration process.

The modelled steady-state heads are shown in Figure 4. These heads were computed for a horizontal plane at the elevation of the major lake. All calibrated hydraulic properties are within the range of values estimated in Table 1 for the site. The calibrated infiltration was 125 mm/yr, approximately 31 % of the mean annual precipitation.

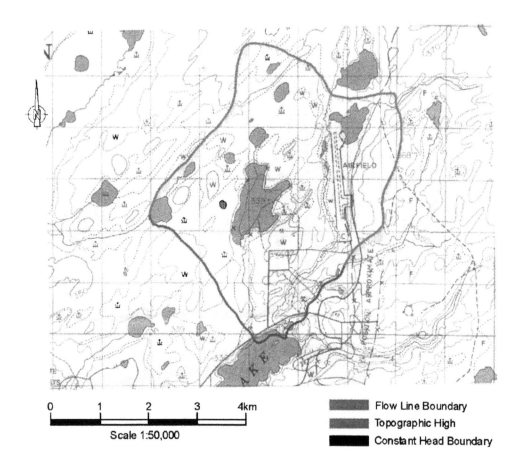

0	1	2	3	4km

Scale 1:50,000

- ▬ Flow Line Boundary
- ▬ Topographic High
- ▬ Constant Head Boundary

Figure 3. Model boundary conditions.

3.4 Particle Paths and Travel Times

The groundwater modelling package MODFLOW, through the use of the ZONEBUDGET program, can predict advective flows through any subregion or material. This capability was used to predict flows through specific pits, pit lakes and mines. MODPATH was used to predict travel times from the sources (two waste rock piles) to the backfilled open pits, pit lakes and mines and to the principal surface water receptors. Figure 5 shows the pattern of flow vectors in the vicinity of the waste rock piles, projected onto the 320 m plane (the approximate elevation of the major lake receptor).

Particles were placed in the first saturated model cells beneath the entire footprint of both waste piles. A total of 117 particles were used for the larger waste pile and 53 for the smaller pile. Each particle is assumed to represent $1/117^{th}$ of the larger pile total mass source term (0.85 %) and $1/53^{rd}$ (1.9 %) for the smaller waste pile.

3.4.1 Flows through Waste Piles

The amount of water passing through the waste piles was calculated from the infiltration rate and the area of the footprint of the pile. The volumes are 33,000 m^3/yr for the larger waste pile and 19,000 m^3/yr for the smaller waste pile.

3.4.2 Flows through Pits and Pit Lakes

Flows through the backfilled open pits and pit lakes were calculated using 125 mm/yr infiltration. Average travel times from the each of the waste piles to each of the pits were also com-

218

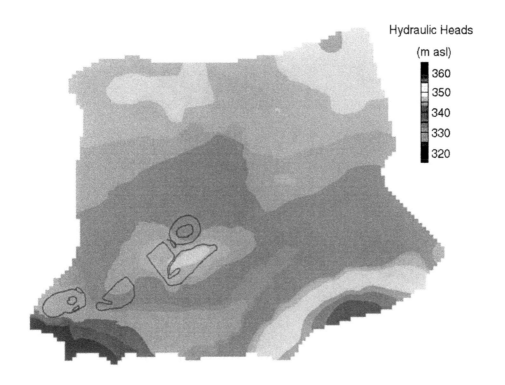

Figure 4. Modelled hydraulic heads.

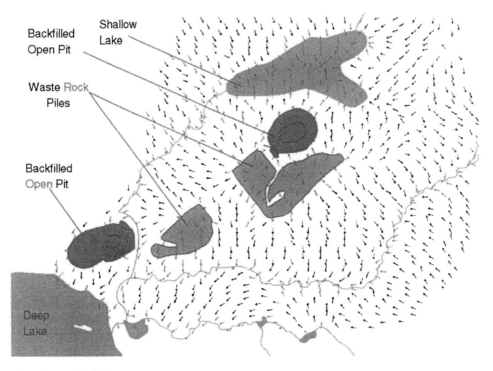

Figure 5. Modelled flow vectors.

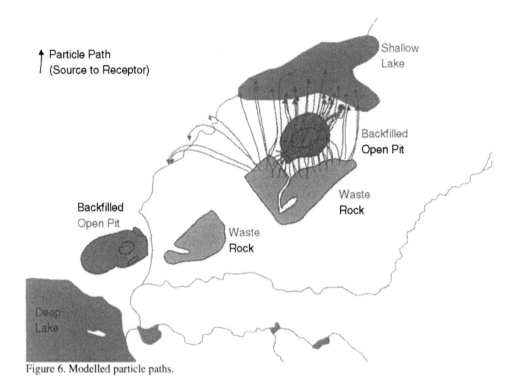

Figure 6. Modelled particle paths.

puted. The total flow modelled through the pit and pit lake adjacent to the larger pile is 61,000 m³/yr. Most of this flow (37,000 m³/yr) passes through the pit lake and decline significantly with depth. The modelled flow through the second open pit is also approximately 61,000 m³/yr.

3.4.3 Flows through Underground Mine Workings

Total flows through the two zones of underground mine workings and travel times to the receptors were modelled for an infiltration rate of 125 mm/yr. The mine flows are estimated at 8,800 m³/yr for the small lakeshore mine and 55,000 m³/yr the more extensive underground mine. The higher modelled values at the larger mine result from the high permeability assumed for the fault zone in which the mine is situated.

3.4.4 Flows to Receptors

Total flows and travel times from the waste piles to the major receptors were modelled for an infiltration rate of 125 mm/yr. The total flow at the receptors is estimated at 880,000 m³/yr. The total infiltration through the two piles was previously estimated at 33,000 m³/yr +19,000 m³/yr = 52,000 m³/yr. The average dilution of the discharged groundwater is thus about a factor of seventeen. The particle paths from a source to a receptor are plotted in Figure 6.

Flows and travel times to the receptors for the reduced infiltration rate of 93 mm/yr were also computed. These results show that significant increases in travel times from source to receptor are possible as infiltration is reduced, particularly for the critical shallow lake receptor adjacent to the large waste rock pile where the modelled median travel time increases by a factor of more than three.

This is a significant result since an engineered cover, possibly combined with a perimeter drain to minimize increased peripheral infiltration from runoff, could delay and attenuate the mass loading peaks at the critical shallow lake receptor.

4 CONCLUSIONS

The case history illustrates the flexibility and utility of the source-term/particle path modelling approach for estimating mass loads to surface water receptors. The use of high vertical resolution 3D models is critical to the successful estimation of flows between sources and receptors. The computational effort required is much less than a conventional transport simulation including mesoscale mechanical dispersion. The simpler particle-tracking models successfully capture "travel time dispersion" which is the dominant process in controlling the temporal mass flux to surface water receptors.

5 ACKNOWLEDGEMENTS

The authors wish to thank Brian Janser and Moir Haug for critically reviewing the manuscript.

REFERENCES

Engineering Computer Graphics Laboratory (ECGL), 1996. *The Department of Defense Groundwater Modeling System: Reference Manual.* Provo: Brigham Young University.

Harbaugh, A.W., 1990. A computer program for calculating subregional water budgets using results from the U.S. Geological Survey modular three-dimensional finite-difference ground-water flow model. Reston: U.S. Geological Survey.

McDonald, M.G., and Harbaugh, A.W., 1988. *A modular three-dimensional finite-difference ground-water flow model.* Reston: U.S. Geological Survey.

Pollock, D.W., 1994. *User's guide for MODPATH/MODPATH-PLOT, Version 3: a particle tracking post-processing package for MODFLOW, the U.S. Geological Survey finite-difference ground-water flow model.* Reston: U.S. Geological Survey.

Tailings and Mine Waste'00 © *2000 Balkema, Rotterdam, ISBN 90 5809 126 0*

Probe sampling and geophysics applied to ground water evaluation of mine dumps

F. Malen
Colorado School of Mines, Golden, Colo., USA

R. Wanty
US Geological Survey, Lakewood, Colo., USA

J. H. Viellenave & J. V. Fontana
TEG Rocky Mountain, Golden, Colo., USA

ABSTRACT: Metals migration into an alpine wetland setting prompted application of Direct Push sampling and installation of nested, multi-level piezometers to evaluate the ground water flow path through mine dumps near the abandoned Waldorf mine near Georgetown CO. Access to the site prohibited use of traditional large scale drill rigs, without significant road modification. A Direct Push Strataprobe™ was used to hydraulically advance either a push point or a core sampler to bedrock beneath the dumps, with maximum depths up to 40 feet. In several locations, nested piezometers, with thermister, were installed with short screens at multiple depths. This allowed discrete sampling of ground water and the vertical profiling of water temperature and chemistry. In addition, natural gamma logging was performed to evaluate the potential for using the tool to characterize the lithology while minimizing the number of soil samples taken.

1 INTRODUCTION

Acid mine drainage from the oxidation of sulfide minerals (mainly pyrite) depends on the sulfide content and the oxidation rate, which in turn is controlled by the availability of oxygen as the oxidizing agent. Thermal anomalies generated by the exothermic pyrite oxidation reaction may be observed in the ground in zones of active pyrite oxidation (Harries and Ritchie, 1980; Fielder, 1989)

The two general mechanisms of non-bacterial oxidation of pyrite in acidic solution are shown in reaction Eq.1 and Eq.2;

$$FeS_2+7/2O_2+H_2O \text{ ------> } Fe^{2+}+2SO_4^{2-}+H^+ \qquad \Delta H=1440 \text{ KJmol}^{-1} \qquad (\text{Eq.1})$$

$$FeS_2+14Fe^{3+}+8H_2O \text{ ------> } 15Fe^{2+}+2SO_4^{2-}+16H^+ \quad \Delta H=11 \quad \text{KJmol}^{-1} \qquad (\text{Eq.2})$$

Equation 1 involves dissolved molecular oxygen as oxidant, while Equation 2 involves ferric iron as the oxidant (Garrels and Thompson, 1960). Equation 2 is rate limited by availability of Fe^{3+}, which is generated by oxidation of Fe^{2+} with molecular oxygen (Singer and Stumm, 1970) as show in Equation 3.

$$Fe^{2+}+1/4O_2+ H^+ \text{ ------> } Fe^{3+}+1/2H_2O \qquad \Delta H=102 \quad \text{KJmol}^{-1} \qquad (\text{Eq. 3})$$

Equations 1, 2 and 3 show the amount of oxidant consumed per mole of pyrite, as well as the heat generated per mole of pyrite (the enthalpy of reactions). Equation 1 produces more heat, but requires the presence of oxygen, whereas the reaction in Equation 2 produces less heat, but can proceed in the absence of oxygen provided that there is a source of ferric ion.

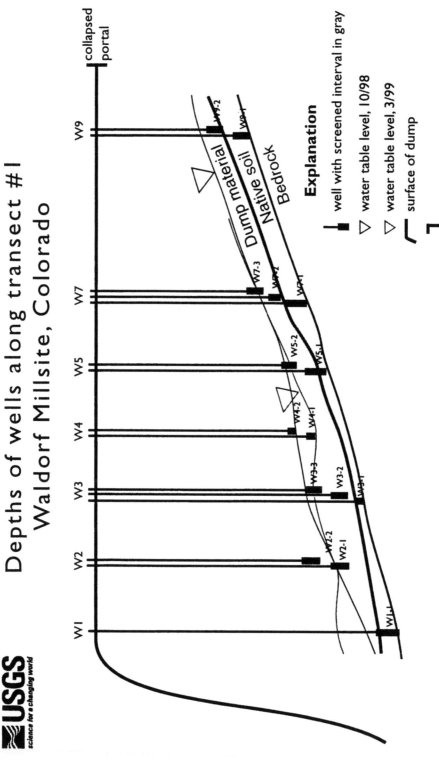

Figure 1. Water levels in the Waldorf mine dump.

Equations 1 and 2 produce acidity (H^+) which may be released to drainage water causing the problem that is known as acid mine drainage. Oxygen has been recognized as an important factor controlling the generation rates of acid mine drainage (Clark, 1965; Morth and Smith, 1966; Myerson, 1981; David and Ritchie, 1986; Nicholson et al., 1988; Nicholson et al., 1989; Guo et al., 1994; Elberling et al., 1994).

The temperature profiles within mine dumps are used to locate sites of pyrite oxidation. Several studies report elevated temperatures within dumps (Cathles and Apps, 1975; Harries and Ritchie, 1985; Harries and Ritchie, 1987), and in some instances the temperatures have exceeded 50 °C and have been observed as high as 80 °C (Cathles and Apps, 1973; Cathles and Schlitt, 1980; Murr, 1980). Temperatures within dumps depend on the composition of the waste rock and the dump properties (Janes et al., 1983). The temperature of the spoil environment is also critical in determining the oxidation rate of pyrite. The rate at which pyrite reacts directly with oxygen, and thus the rate of oxygen consumption, increases with temperature (Clark, 1965).

A field study of the Waldorf mine dump is being conducted to understand the distribution of oxygen in the saturated zone in the mine dump and its thermal profile, the nature of groundwater flow in the dump, and the distribution of sulfide minerals. Temperature profiles are being used to estimate the rate and location of pyrite oxidation. Understanding oxygen transport in mine waste requires an understanding of solid material properties, oxygen and sulfur isotopic fractionation, and reactive transport of oxidation products. This study will give some insight into the dependence and sensitivity of the oxidation rate on physical and compositional characteristics of the dump, and aid in developing assessment methods for future mine-waste remediation programs at the site, and at other similar mine-waste impoundments.

Critical to the field study was the use of a powerful Direct Push drill rig. Direct Push, started by Geoprobe, Inc., in the 1980s, has until recently been insufficiently powerful to penetrate rocky or cobbly formations. Further, the pipe size was too small to permit the use of borehole logging tools, even those classified as small bore or slim-line. TEG, starting in the early 1990s, developed a new class of Direct Push rigs, including the present Strataprobe, using large diameter drive pipe (up to 8.1 cm OD). Larger pipe admits the use of TEG's specially designed gamma and other loggers.

2 SITE DESCRIPTION

The Waldorf mine was established in about 1902 by Edward J. Wilcox, owner of the Imperial Mines of Waldorf and builder and owner of the Argentine Central Railroad. It is located about 7 miles southwest of Georgetown at an elevation of 11,660 feet. The Georgetown mining district has been a prominent metal producer since its discovery in 1867 (Southworth, 1997) and was once the richest silver region in the world (Draper, 1940). The Waldorf ore typically contains galena, sphalerite, argentite (silver plume), native silver, and silver-bearing antimony sulfides, such as tetrahedrite, pyrargyrite, and polybarite (Sims, 1988). Pyrite is common but not abundant, and the trace concentrations of gold in the silver-lead-zinc veins commonly are associated with minor chalcopyrite (Spurr et al., 1908; Lovering and Goddard, 1950; Grybeck, 1969, Sims, 1988). Studies of mine waste transport from the mining district have found that water flowing through abandoned mines and dump piles can have high concentrations of Cd, Cu, Pb, Zn and other metals (Kimball et al., 1995; Lewis and Clark, 1997).

3 METHOD

The overall approach in this field study incorporated the installation of multi-level piezometers to measure hydrology, temperature, and other factors within the dump. The Waldorf dump, like many in the Rocky Mountains, is inaccessible to large, conventional drilling rigs without significant road modification. Most light duty rigs cannot penetrate the rocky dumps. It was necessary to find a drill rig that could get to the site inexpensively and yet be heavy-duty enough to penetrate up to 12 m of run-of-mine waste rock. The rig chosen in this case was the

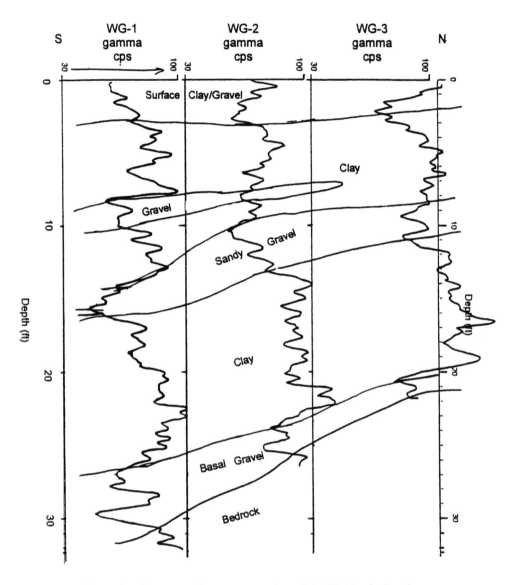

Figure 2. Gamma ray log cross section of the Waldorf mine dump.

Strataprobe™, a Direct Push rig operated by TEG Rocky Mountain of Golden CO. The Strataprobe™ is a hydraulic push and hammer rig mounted on a 1.35 metric ton, Ford F450 Super Duty truck. The total weight of about 6,800 kg gives the rig great penetrating power but without sacrificing maneuverability.

Soil Sampling and Piezometer Installation

To establish the piezometers, and acquire sufficient core samples to characterize lithology in the dump, TEG employed a dual tube sampling system. A 5.7 cm OD drive casing was advanced at the same time as a sampling tube and inner rod. The casing remained in place while 1.2 m long samples were retrieved. No examples of refusal, short of bedrock below the mine dump, were experienced in the study.

When the maximum depth was reached, small bore (1.25 cm ID) PVC screen and riser was inserted to the bottom of the hole inside the casing, which was then withdrawn as the well was completed. The deepest hole in each nest was equipped with a thermistor string so that a temperature profile could be established (note Figure 1).

At each nest of piezometers, surface-to-bedrock core samples of dump solid material were collected. The core samples were logged and oven-dried at approximately 60 °C in the laboratory. The samples were analyzed for sulfide contents. The piezometers are installed to locate the water table, which is used to estimate the direction of groundwater flow. The hydraulic heads were measured during October 1998 to August 1999. Piezometers also provide access to the subsurface for temperature measurement. The piezometer depths depend on depth to bedrock. Temperature was measured by thermistors, which provided temperature measurements accurate to within +/- 0.2 °C. The tip of the thermistor was coated with epoxy to protect the thermistors from damage during installation and later operation. The thermistor was connected by a wire lead through which the thermistor resistance is measured at the surface. The resistance is converted to temperature using calibration curves.

Borehole Logging

Borehole logging using a natural gamma ray logger was performed on a limited number of holes as a demonstration. Gamma logging, if successful, would be a powerful method of characterizing lithologies without continuous coring. This would minimize the amount of potentially contaminated cuttings and assure lithologic data in areas that experience poor recoveries of sample, common in dumps. TEG employed a SlimLine natural gamma logger (1.8 cm OD x 61 cm long probe) with MGX Data Logger to perform the logging. Logging could be conducted inside 2.5 cm PVC, open hole, or inside the drive casing. An example log cross section is shown in Figure 2.

4   RESULTS AND DISCUSSION

Lithology

The soil sampling and borehole logging procedures effectively characterized the lithology of the dump. In certain portions of the dump, the presence of coarse material made core recovery ineffective, particularly those parts with large (>5-7.5 cm) cobbles. In such areas, the gamma ray

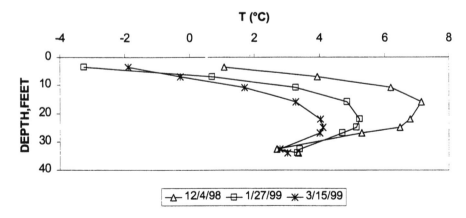

Figure 3. Winter temperature profile in Well 1-1, Waldorf mine dump.

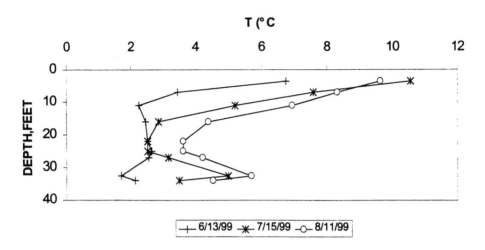

Figure 4. Summer temperature profile in Well 1-1, Waldorf mine dump.

12/98 TEMPERATURE TRANSECT.

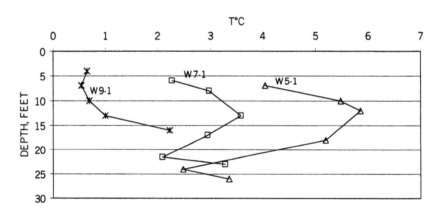

Figure 5. Temperature profile of Wells W5-1, W7-1 and W9-1,
December 1998 in Waldorf Mine Dump.

logs were very effective. An optimum application of gamma ray logging and coring would allow
rapid sampling and effective lithologic characterization. Figure 2 shows a typical gamma
ray/lithologic cross section from the upper part to the toe of the dump.

Hydraulic Head

The water table configurations were measured regularly during the investigation period using the
network of piezometers. Seasonal fluctuations were observed.   The water levels show the flow
direction to be from the adit portal near well W9-1 to the tip of the dump at well W1-1. The
water table level fluctuates seasonally, an example of which is shown in Figure 1.

Temperature Distribution

The temperature was measured from October 1998 to August 1999. The measured profiles of temperature distributions indicate a wide range of temperature. The change of temperature profiles indicates seasonal variation as shown in Figure 3.

Temperature profiles in a dump depend on the rate of heat production from pyrite oxidation, the heat transport by heat diffusion and gas and water flow, and on changes in ambient temperature at the dump surface (Ritchie, 1994). The heat generated was detected in most deep holes during the investigation period. The effects of environmental conditions on the temperature profiles in the wells varies as shown in Figure 3.

5 CONCLUSIONS

The drilling and sampling method used to generate data from the site was successful. The Strataprobe Direct Push rig was more than capable of sampling and emplacing the piezometers, and would be capable of penetrating up to 30 m or more in such dumps or waste piles. Piezometers were easily installed and completed. The gamma logging successfully identified zones of coarse and fine material with limited coring, enabling the investigators to proceed more rapidly than would have been the case otherwise. Other logging methods are also applicable with the Strataprobe, including nuclear density and radar.

Water levels and temperature profiles inside the Waldorf mine waste were measured from October 1998 to August 1999. The water levels in the mine waste represent the possibility of groundwater-flow direction, and temperature profiles show the evidence of heat generated from pyrite oxidation.

6 REFERENCES

Cathles, L. M., and Apps, J. A., 1975, A model of the dump leaching process that incorporates oxygen balance, heat balance and air convection: Metallurg. Trans., v. 6B, p. 617-624.

Clark, C. S., 1965, The oxidation of coal mine pyrite: Ph.D. Thesis, U. Johns Hopkins, Baltimore, MD.

David, G. B., and Ritchie, A. I. M., 1986, A model of oxidation in pyrite mine waste: Part 1: Equations and approximate solution: Appl. Math. Model, v. 10, p. 314-322.

Draper, B. P., 1940, Georgetown: high points in the story of the famous Colorado silver camp, Georgetown, Colorado: , 25 p.

Elberling, B., Nicholson, R. V., and Scharer, J. M., 1994, A combined kinetic and diffusion model for pyrite oxidation in tailing: a change in controls with time: J. Hydrology, v. 157, p. 47-60.

Fielder, D., 1989, Identification of acid producing zones within a reclaimed surface mine utilizing thermal surveying: Master thesis, U. Pennsylvania State, College Park, P.A.

Garrels, R. M., and Thompson, M. E., 1960, Oxidation of pyrite in ferric sulfate solution: Am. J. Sci., v. 258, p. 57-67.

Grybeck, D. J., 1969, Geology of the lead-zinc-silver deposits of the Silver Plume area, Clear Creek County, Colorado: Colorado school of Mines, Master thesis:, 147 p.

Guo, W., Parizek, R., and Rose, A. W., 1994, The role of thermal convection in resupplying O_2 to strip coal - mine spoil: Soil Sci., v. 158, p. 47-55.

Harries, J. R., and Ritchie, A. I. M., 1980, The use of temperature profiles to estimate the pyritic oxidation rate in a waste rock dump from an open cut mine: Water, Air, Soil Poll., v. 15, p. 405-423.

Harries, J. R., and Ritchie, A. I. M., 1985, Pore gas composition in waste rock dumps undergoing pyritic oxidation: Soil Sci., v. 140, p. 143-152.

Harries, J. R., and Ritchie, A. I. M., 1987, The effect of rehabilitation on the role of oxidation of pyrite in a waste rock dump: Environ. Geochem. Health, v. 9, no. 2, p. 27-36.

Jaynes, D. B., Rogoski, A. S. S., Pionke, H. B., and Jocoby, E. I., 1983, Atmosphere and temperature changes within a reclaimed coal strip mine: Soil Sci., v. 136, p. 164-177.

Kimball, B. A., Callender, E., and Axtmann, E. V., 1995, Effects of colloids on metal transport in a river receiving acid mine drainage, upper Arkansas river, Colorado, U.S.A.: App. Geochem., v. 10, no. 3, p. 285-306.

Lewis, M. E., and Clark, M. L., 1997, How does streamflow affect metals in the upper Arkansas river?. Fact sheet - U.S. Geological Survey: , FS0226-96.

Lovering, T. S., and Goddard, E. N., 1950, Geology and ore deposit of the Front Range, Colorado: U.S. Geological Survey Professional Paper 223, 319 p.

Miller, R.J., Culig, M.J., and Viellenave, J.H., 1999, Image is Everything: Tailings and Mine Waste '99, Balkema, rotterdam, p 9-11.

Morth, A. H., and Smith, E. E., 1966, Kinetics of the sulfide to sulfate reaction: Am.Che.Soc., Division of Fuel Chemistry, v. 10, p. 83-92.

Myerson, A. S., 1981, Oxygen mass transfer requirements during the growth of Thiobacillus Ferrooxidans on irn pyrite: Biotechnol. Bioeng., v. 23, p. 1413-1416.

Nicholson, R. V., Gillham, R., and Reardon, E., 1988, Pyrite oxidation in carbonate-buffered solution: Geochem. Cosmochem. Acta, v. 53, p. 1077-1085.

Nicholson, R. V., Gillham, R. W., Cherry, J. A., and Reardon, E. J., 1989, Reduction of acid generation in mine tailings through the use of moisture-retarding cover layers as oxygen barrier: Can. Geotech. J., v. 26, p. 1-8.

Ritchie, A. I. M., 1994, The waste-rock environment: in: Short course handbook on environmental geochemistry of sulfide mine-waste. Mineralogical Association of Canada, p. 133-162.

Sim, P. K., 1988, Ore deposit of the Central city - Idaho springs area: Geological Society of America 1888-1988, Field trip guidebook, Centennial meeting, Denver, Colorado, p. 82-121.

Singer, P. C., and Stumm, W., 1970, Acid mine drainage- the rate determining step: Science, p. 1121-1123.

Southworth, D., 1997, Colorado mining camp, Wild horse publishing, p. 311.

Spurr, J. E., Garrey, G. H., and Ball, S. H., 1908, Economic geology of Georgetown quadrangle: U.S. Geological Survey Professional Paper 63, 416 p.

Tailings and Mine Waste'00 © 2000 Balkema, Rotterdam, ISBN 90 5809 126 0

Analyzing flow through mine waste-dumps

Rizwanul Bari
Faculty of Computer Science, Dalhousie University, Halifax, N.S., Canada

David Hansen
Department of Civil Engineering at DalTech, Dalhousie University, Halifax, N.S., Canada

ABSTRACT: This paper reports the results of numerical and experimental investigations on steady flow through mine waste-dumps comprised of coarse rockfill. In such mine waste-dumps the former stream passes through very coarse porous media but the behavior of this flow does not follow Darcy's law. Rather, it behaves in a manner similar to that of ordinary open channel flow. The longitudinal variation in the depth of water along the dump is not however governed by the roughness of the stream-bed, but by the characteristics of the coarse porous media which now fills the channel. Such flow is theoretically governed by non-Darcy flow operating under the Dupuit assumptions. An integrated software package FABS was developed as part of this research effort. FABS simulates 1-D non-Darcy water surface profiles through mine waste-dumps. This paper describes the first module of FABS – the steady flow module.

1 INTRODUCTION

Open-pit mining operations in mountainous areas often result in the permanent infilling of some of the valley terrain with large volumes of coarse rockfill. Under such circumstances the most common practice for the continued conveyance of streamflow is to allow it to pass through the rockfill, instead of diverting it or passing the flow under the deposit via culverts (Ritcey, 1989). The pre-existing watercourses in such cases become buried streams which continue to flow through the same valleys, but under great depths of waste rock, and sometimes over considerable distances. There exist a number of such mine waste-dumps in coal mines of the Canadian Rockies, particularly in the Kootenays in south-eastern British Columbia. Although in such mine waste-dumps the former open channel flow passes through a porous media, it does not follow Darcy's law, the governing equation for most porous media flow. Rather, it behaves in a manner similar in some ways to natural open channel flow (Bari, 1997). This unique, relatively rapid flow (as compared to groundwater flow) through coarse porous media is often referred as *non-Darcy flow*. The formation of these mine waste-dumps causes permanent local changes in the hydraulic, hydrologic, and sediment regimes of the watershed in which they are located. The water surface elevation along the streams affects the design, planning, and operation of the coal mines that create them. Also, elevated water depths along these rockfill deposits are sometimes associated with large-scale slope failures, particularly at the downstream toe. It is therefore necessary to have an efficient tool for the hydraulic modeling of water surface profiles in such mine waste-dumps.

This paper presents FABS (Flow Analysis of Buried Streams), an integrated software package that simulates non-Darcy flow through mine waste-dumps. FABS is based on gradually-varied open channel flow algorithms. It is intended that FABS will have three modules: a steady profile simulation module, an unsteady profile simulation module, and a water quality module. FABS was developed as part of on-going research on non-Darcy flow at DalTech (see http://is.dal.ca/~hansend/non_darcy.html). This paper describes the first module of FABS – the *steady* non-Darcy flow profile simulation. FABS appears to be the first such system in its domain. It is a result of the authors' extensive academically-based research on non-Darcy flow and as such it is not a commercial software product. However, distribution copies of FABS and related literature can be obtained by writing to the second author and paying a nominal distribution cost to the university.

2 CHARACTERISTICS OF FLOW THROUGH MINE WASTE-DUMPS

For flow through porous media which follow Darcy's law, the velocity is very small ($< 10^{-2}$ m/s). Consequently, the momentum and kinetic energy of the flow are neglected. This is not the case for non-Darcy flow through mine waste-dumps. The flow in such cases behaves in a manner similar in some ways to open channel flow. However, the longitudinal variation in water is no longer governed by the roughness of the stream-bed, as is the case for open channel flow, but by the characteristics of the coarse porous media, which now fills the formerly open channel. Using the definitions shown in Figure 1, the one-dimensional dynamic equation of flow through mine waste-dumps under steady-state conditions can be shown to be:

$$\frac{dy_e}{dx} = \frac{S_o(x) - S_f(x, y_e)}{1 - \mathsf{Fr}_P^2(y_e)} = f(y_e, x)$$

[1]

where:

x : distance along the channel bed (L),
y_e : depth used in energy considerations (L), $= y_{nominal} \cdot \cos^2 \theta$,
θ : stream bed slope angle (degree),
S_o : channel bed slope ($= \sin \theta$, dimensionless),
S_f : friction slope (dimensionless),
Fr_P : pore Froude number $= \dfrac{U_V}{\sqrt{g \cdot D}}$ (dimensionless),

U_V : void velocity (bulk velocity divided by porosity, L/T),
D : hydraulic depth (area of flow divided by top width, L),
g : gravitational constant (L/T^2).

Eqn. [1] is analogous to the dynamic equation applicable to open channels for which the term Fr_P is replaced by the Froude number Fr associated with open channel flow, and the friction slope S_f is computed using a uniform-flow resistance equation (such as the Manning equation). For non-Darcy flow through mine waste-dumps, S_f can be evaluated using a non-Darcy flow equation. The following section provides a brief overview of some of the options available for the choice of non-Darcy flow equation.

3 OVERVIEW OF 1-D NON-DARCY FLOW EQUATIONS

Experimental studies suggest that, for non-Darcy flow, the relation between bulk velocity U and hydraulic gradient i becomes non-linear, taking either a power law form $i = aU^N$ (where a is an empirical constant determined by the properties of the fluid and of the porous medium, and N is an exponent between 1 and 2) or a quadratic form $i = sU + tU^2$ (s and t being empirical constants determined by the properties of the fluid and of the medium). Both the power and quadratic forms are used to describe one-dimensional non-Darcy flow phenomena and are inter changeable (George and Hansen, 1992). Some of the better-known non-Darcy flow equations are the Ergun equation (Ergun, 1952), the Ergun-Reichelt equation (see Fand and Thinakaran, 1990), the Martin equation (Martin, 1990), the McCorquodale equation (McCorquodale et al., 1978), the Stephenson equation (Stephenson, 1979), and the Wilkins equation (Wilkins, 1956). Hansen et al. (1995) have provided a review of these equations. The Ergun, McCorquodale, and Ergun-Reichelt equations represent generalizations of large sets of data into unified equations. These data-sets included the results of researchers other than those to whom the final equation is

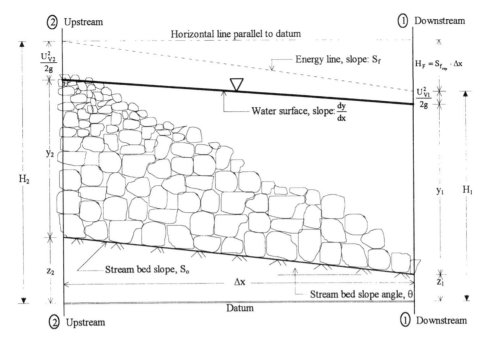

Figure 1. Energy considerations for non-Darcy flow through mine waste-dumps
(water surface and energy line are in fact completely buried).

attributed. The Martin equation is based on experiments performed on a moderate range of
porous media types but included little or no data from other sources. Both the Stephenson and
Wilkins equations are based on experiments on crushed rocks of a relatively narrow size-range
and of a given angularity.

Among the non-Darcy flow equations mentioned above, the Stephenson and the Wilkins
equations are the simplest in form and are also widely-used (although in differing parts of the
globe). Both equations were used in the numerical models developed in this study (described in
Section 4). Brief overviews of these two underlying equations are provided in the following
sections.

3.1 The Stephenson equation

By analogy to flow in conduits, Stephenson (1979) proposed that the hydraulic gradient for flow
through coarse porous media might be expressed as:

$$i = \frac{K_{st} \cdot U^2}{g \cdot d \cdot n^2}$$
[2]

where:
 d : particle diameter (L),
 n : porosity (dimensionless),
 K_{st}: Stephenson's friction factor (dimensionless).

Stephenson (1979) further suggested that the following relation could be used to evaluate K_{st}:

$$K_{st} = \frac{800}{Re} + K_t$$
[3]

233

where:

Re : pore Reynolds number $= \dfrac{Ud}{n\nu}$ (dimensionless),

ν : kinematic viscosity of water (L^2/T),

K_t : parameter to account for the angularity of the particles, ranging from 1 for polished sphere to 4 for rough and angular crushed stone (dimensionless).

3.2 The Wilkins equation

Based on experimental work done in a large packed-column, Wilkins (1956) proposed the following dimensionally-unbalanced equation for flow through coarse porous media:

$$U_V = W \cdot m^{0.50} \cdot i^{0.54} \tag{4}$$

where:

W : Wilkins' constant (equal to 52.4 if m in cm and U_V in cm/s),

m : hydraulic mean radius of the coarse porous media (cm).

The product $Wm^{0.50}$ in eqn. [4] can be thought of as a hydraulic conductivity of the porous media. The exponent 0.54 indicates that this equation is suited to the flow regime of nearly fully-developed turbulence. Knowing that $U_v = U/n$, eqn. [4] can be restated as:

$$i = \frac{1}{m^{0.93}} \left(\frac{U}{W \cdot n} \right)^{1.85} , \equiv S_f \tag{5}$$

which renders it useful for independently computing S_f along water surface profiles beneath mine waste-dumps.

4 DEVELOPMENT OF THE NUMERICAL MODEL

Since unconfined non-Darcy flow phenomenon is analogous to that of open channels, it was hypothesized that flow profiles for the former could be computed in the same general manner as is done for the latter. Two of the computational schemes for delineating steady Gradually Varied Flow (GVF) profiles are the method of Prasad (Prasad, 1970) and the Standard Step Method (the SSM; Chow, 1959). In this study both of these computational schemes were modified in order to simulate non-Darcy flow profiles through mine waste-dumps. Details of these schemes and their underlying assumptions/limitations are described elsewhere (see Hansen and Bari, 1999a); brief overviews follow.

The method proposed by Prasad numerically integrates the one-dimensional dynamic equation for open channel flow (the differential equation analogous to eqn. [1]) at successive cross-sections, starting at a known water level. Under this method the direction of computation does not depend on whether the flow is subcritical or supercritical. On the other hand, the SSM applies the energy equation successively across pairs of cross-sections where the depth at one of them is known. Under this method the direction of computation must be downstream-to-upstream for subcritical flow, and vice versa for supercritical flow.

5 IMPLEMENTATION OF THE NUMERICAL MODEL

The models described in Section 4 were used to develop a software package for non-Darcy flow profile computation through mine waste-dumps – FABS (Flow Analysis of Buried Streams). FABS was developed using Visual Basic (VB) 6.0. Since its first release VB has evolved into a major development environment that covers all aspects of programming, from educational applications to databases, and from financial applications to Internet components. VB is an

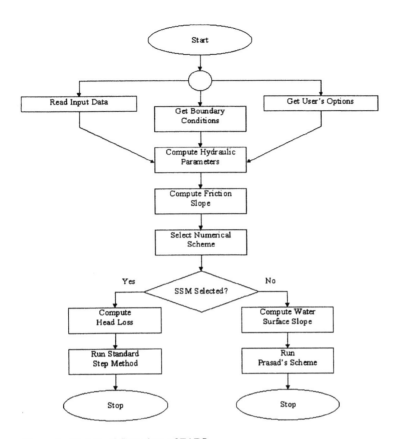

Figure 2. High level flow chart of FABS.

event-driven programming language. It includes a complete set of graphical tools and has high-level language constructs. While developing FABS, efforts were made to eliminate some of the physical constraints generally associated with most of the available commercial open channel flow computation software packages. For example, the system is implemented in such a way that it is not constrained by the number of cross-sections – FABS can accept as many cross-sections as the memory of the host machine can support. FABS can also accept an unlimited number of data points in each cross-section. These flexibilities were achieved by using dynamic arrays to store geometric and related data. This would normally make the program slightly slower because accessing dynamic arrays is slower than accessing static arrays, but this drawback was partially overcome by using a more efficient hydraulic algorithm.

A high-level flowchart for FABS is shown in Figure 2. Data input in FABS via keyboard was kept to a minimum. The system reads all geometrical data and some of the porous media data from an input file. Based on the input data and user options, FABS computes the related hydraulic parameters for the starting cross-section (the most downstream cross-section). The friction slope is then computed for this cross-section. Based on the user's choice, the system then executes the Prasad scheme or the SSM to compute the depth of flow at subsequent cross-sections in the upstream direction.

FABS is compiled in native code. The size of the executable file for the current version of FABS is only 328 KB. The executable file was optimized for speed by instructing the compiler to favor speed over size. The executable file also favors Pentium Pro processors. FABS will run on other Intel processors but will not do so as efficiently.

6 SYSTEM DESCRIPTION

The hardware requirements for FABS are: Pentium class processor, 16 MB of RAM, 50 MB minimum of hard disk space, a 17 inch monitor, and a mouse. FABS was developed for the Windows (95 or 98) operating system.

In order to provide screen displays that create an easy operating environment for the users and to form an explicit visual and functional context for the computer user's actions, the user interface for FABS was based on the prevailing windows, icons, menus, and pointers (WIMP) approach. An effort was made to design the interface to be as simple and as easy to use as possible. This was achieved by using graphics, in addition to text, in order to communicate with the user. Also, the user is always presented with a finite number of options rather than requiring the user to memorize and manually enter commands from a large number of options. This significantly improves FABS' learning curve and helps first-time users become productive almost immediately. In developing the user interface, most of the fundamental principles of human-computer interaction were considered, including: visibility, consistency, feedback and dialogue, user control, aesthetic integrity and see-and-point. Screen layouts were designed based on accepted principles of graphic design and usability (see http://www.acm.org/cacm/AUG96/-antimac.htm). For example, the menu-bars contain strategic choices and the feedback areas are close to the associated controls so as to keep the user informed about his or her interactions with the program.

A screen snapshot of the FABS data-input window is shown in Figure 3. The data input window provides the user with options for selecting the numerical scheme, the precision level, the maximum number of iterations, the friction-slope averaging technique, and friction slope equation. Figure 4 shows the FABS' simulation window. Steady-flow profile simulation is controlled by this window. This window also presents the depth of water at locations specified by the user, and presents all related hydraulic parameters at the different cross-sections. The graphical representation of cross-sections is viewable from this screen. Figure 5 shows FABS' summary output window. This window shows all the options previously selected by the user and has the printing and saving functions.

In addition to hard copies, FABS has on-line technical and user's manuals. Two 'WebBrowser' controls were programmed as part of the development of the system, which provide excellent web-browsing capabilities to FABS. The controls are basically simplified custom web-browsers which enable users to access the web from within FABS so as to view the html documents for the technical and user's manuals.

7 PERFORMANCE OF FABS

Since the mathematical foundation of FABS is based on arbitrary stream cross-sections, FABS had to be capable of simulating non-Darcy flow profiles through any natural mine waste-dump. However, the performance of FABS could not be evaluated for a field-scale mine waste-dump due to the fact that such data were not available. It was therefore decided to carry out the performance evaluation by physical model testing. A model mine waste-dump was built as part of the study. Figure 6 shows the model mine waste-dump built in a glass-walled flume located in DalTech's Hydraulics Laboratory. Figure 7 shows a schematic of the laboratory set-up for the model mine waste-dump. Crushed limestone was used in the experiment as the porous media (see Figure 8).

The simulations in general indicate that the nature of the porous media greatly affects the hydraulic properties of flow through mine waste-dumps. This necessitates an accurate characterization of the media in order to be able to independently predict accurate water surface profiles. In this study the models developed were calibrated to get optimum value of media properties. The details of the parameter optimization are presented elsewhere (see Hansen and Bari, 1999b).

The performance of FABS was evaluated by comparing its output to that collected from the experiments performed on the model mine waste-dump. Some of these comparisons are presented in Figures 9 and 10. Figure 9 presents the outcomes when FABS used Stephenson's equation to compute the friction slope and Figure 10 presents the outcomes when Wilkins' equation was used.

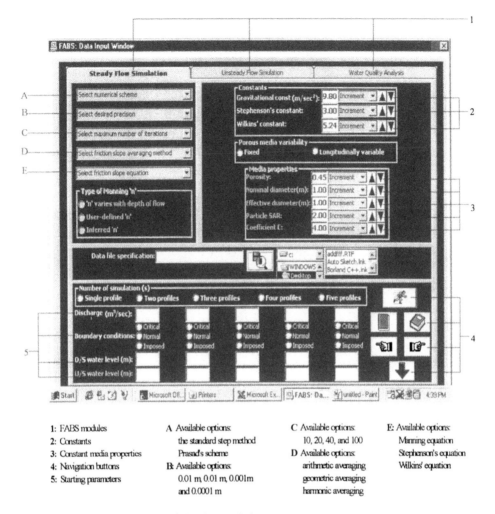

1: FABS modules	A Available options:	C Available options:	E: Available options:
2: Constants	the standard step method	10, 20, 40, and 100	Manning equation
3: Constant media properties	Prasad's scheme	D Available options:	Stephenson's equation
4: Navigation buttons	B: Available options:	arithmetic averaging	Wilkins' equation
5: Starting parameters	0.01 m, 0.01 m, 0.001m	geometric averaging	
	and 0.0001 m	harmonic averaging	

Figure 3. A screen snapshot of FABS data input window.

It can be seen from Figures 9 & 10 that the performance of FABS in simulating water surface profiles is quite satisfactory. It can also be readily seen that the Wilkins and Stephenson equations simulated similar water surface profiles for the three different discharges. It can therefore be concluded that the Wilkins and Stephenson equations performed equally well in simulating the experimental water surface profiles through the model mine waste-dump.

8 CONCLUSIONS

An integrated software system called FABS was developed and tested. FABS simulates non-Darcy flow through mine waste-dumps. It is intended that FABS will eventually have three modules: a steady profile simulation module, an unsteady profile simulation module, and a water quality analysis module. This paper describes the development of the first module of FABS – steady non-Darcy profile simulation.

Two numerical schemes for GVF of open channel were modified in this study in order to handle of non-Darcy flow. These modified algorithms were implemented in FABS.

FABS is capable of simulating non-Darcy water surface profiles through mine waste-dumps. The performance of FABS in simulating water surface profiles under laboratory conditions was found to be satisfactory. FABS uses Wilkins' or Stephenson's equation as the headloss

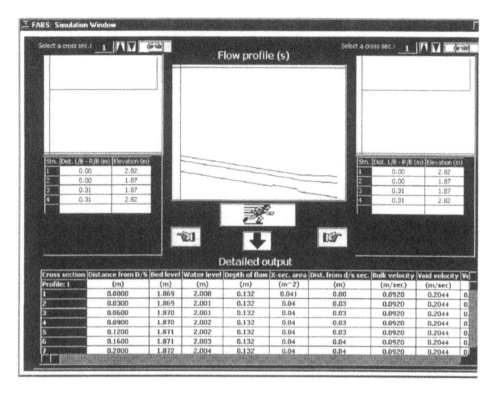

Figure 4. FABS simulation window.

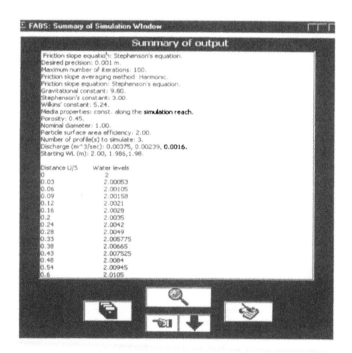

Figure 5. FABS summary of output window.

238

Figure 6. Model mine waste-dump in the glass-walled flume.

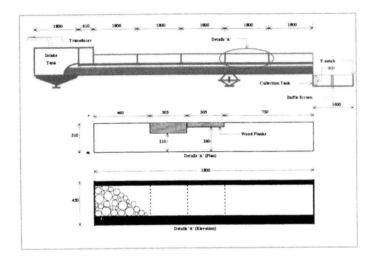

Figure 7. Schematic of the experimental setup (all dimensions in mm).

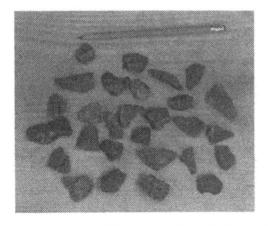

Figure 8. Sample of the porous media used in the experiment.

239

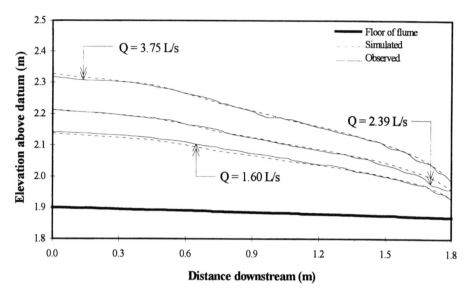

Figure 9. Comparison of observed and simulated water surface profiles
            (friction slope computed by Stephenson's equation).

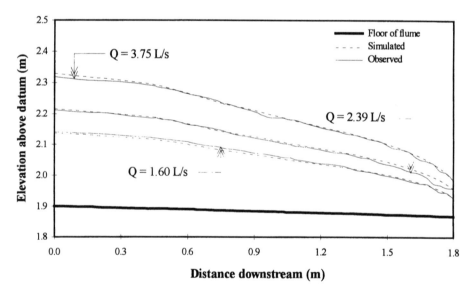

Figure 10. Comparison of observed and simulated water surface profiles
             (friction slope computed by Wilkins' equation).

equation. It was found that both the equations performed equally well in simulating the
experimental water surface profiles through model mine waste-dumps.
  It is believed that despite the limitations and possibilities for further improvement, FABS
represents a step toward more explicit assistance in the non-Darcy flow profile simulation
process, and will hopefully be welcomed by the concerned communities, such as managers of
waste rock at open-pit mines.

REFERENCES

Bari R. 1997. *The Hydraulics of Buried Streams*. MASc thesis, Department of Civil Engineering, Technical University of Nova Scotia, Halifax, NS.

Chow V.T. 1959. *Open Channel Hydraulics*. New York: McGraw-Hill Book Co. 217-296.

Ergun S. 1952. Fluid flow through packed columns. *Chemical Engineering Progress*. 48(2):89.

Fand R.M. & R.Thinakaran. 1990. The influence of the wall on flow through pipes packed with spheres. *ASME Journal of Fluid Engineering*. 112:84-88.

George & D. Hansen. 1992. Conversion between quadratic and power law for non-Darcy flow. *ASCE Journal of Hydraulic Engineering*. 118(5):792-797.

Hansen D. & R.Bari. 1999a. Application of gradually-varied flow algorithms to simulate buried streams. *ASCE Journal of Computing in Civil Engineering*. Submitted for publication.

Hansen D. & R.Bari. 1999b. In preparation.

Hansen D., V.K.Garga, & D.R.Townsend. 1995. Selection and application of one-dimensional non-Darcy flow equation for two-dimensional flow through rockfill embankments. *Canadian Geotechnical Journal*. 32:223-232.

Martins R. 1990. Turbulent seepage flow through rockfill structures. *Water Power and Dam Construction*. 90:41-45.

McCorquodale J. A., A.Hannoura, & M.S.Nasser. 1978. Hydraulic conductivity of rockfill. *Journal of Hydraulic Research*. 16(2):123-137.

Prasad R. 1970. Numerical method of computing flow profiles. *ASCE Journal of the Hydraulics Division*. 96(HY1):75-86.

Ritcey G.M. 1989. *Tailings Management: Problems and Solutions in the Mining Industry*. New York: Elsevier Science Publishing Co. 78-83.

Stephenson D. 1979. *Rockfill in Hydraulic Engineering*. Amsterdam: Elsevier Science Publishers B.V. 19-37.

Wilkins J.K. 1956. Flow of water through rockfill and its application to the design of dams. *Proceedings of the 2nd Australia-New Zealand Conference on Soil Mechanics and Foundation Engineering*, Canterbury University College, Christchurch, New Zealand. 141-149.

Tailings and Mine Waste'00 © 2000 Balkema, Rotterdam, ISBN 90 5809 126 0

In-situ and laboratory testing for predicting net infiltration through tailings

D.A. Swanson, G. Savci & G. Danziger

Savci Environmental Technologies, LLC, Golden, Colo., USA

ABSTRACT: A hydrologic assessment was conducted for an existing tailings impoundment in the southwest United States to predict net infiltration through the unsaturated tailings. Net infiltration was predicted using a combination of semi-empirical and numerical soil-atmosphere hydrologic models with model input parameters based on site-specific climate and hydrologic property data. The hydrologic properties of the tailings were measured using both in-situ and laboratory testing techniques. In-situ measurements were made for soil-water characteristic curves, hydraulic conductivity, density and runoff. Laboratory testing included texture, density, soil-water characteristics and hydraulic conductivity. The results of the in-situ and laboratory testing indicated that the tailing material was an efficient material for limiting net infiltration.

1 INTRODUCTION

A hydrologic assessment of a tailings impoundment in a dry climate (i.e., arid to semi-arid) of the southwest United States was conducted in support of a facility closure permit application. The assessment focused on predicting net infiltration through the existing tailing surface to determine if the existing surface was effective at limiting net infiltration.

The tailings impoundment averages 85 feet thickness and the regional groundwater table is situated approximately 500 feet below the ground surface. Annual lake evaporation (~72 inches) exceeds annual precipitation (~17.5 inches) by a factor of approximately four. A competent, hard surface crust has formed over much of the tailings impoundment, which promotes runoff that collects in a seasonal pond at the center of the facility.

Key components of the hydrologic assessment included identifying site-specific factors affecting net infiltration through the tailings and collecting accurate and defensible site-specific data from which to base the prediction of net infiltration. Net infiltration was predicted using a combination of semi-empirical and physically-based numerical soil-atmosphere hydrologic models. The following paragraphs outline the technical approach applied in the hydrologic assessment and detail the in-situ and laboratory testing methods used to measure the site-specific hydrologic properties. Key results of the testing are also presented and discussed.

2 TECHNICAL APPROACH

The technical approach developed to predict net infiltration through the unsaturated portions of the tailings impoundment included a site-specific hydrologic characterization and the combined use of a semi-empirical hydrologic model and a physically-based numerical model. These components of the technical approach are described in the sections that follow.

2.1 Hydrologic Characterization

A site-specific hydrologic characterization of the tailings impoundment was conducted through a review of historical climatic data and an assessment of the hydrologic properties and conditions of the tailings impoundment surface using in-situ measurement and laboratory testing.

Climatic data for a period of record of up to 60 years was compiled from site and regional climate stations. Climate data parameters collected to facilitate the prediction of net infiltration included evaporation, precipitation, relative humidity, solar radiation, temperature and wind speed.

A field investigation program was developed to collect the data necessary for the definition of input parameters for infiltration modeling of the tailings impoundment. The hydraulic properties and conditions of the active soil zone (i.e., the depth of soil affected by evaporation) were characterized using both in-situ field and laboratory test methods. The parameters of texture, density, runoff, soil-water characteristics, hydraulic conductivity, water content and matric suction were measured. As shown on Figure 1, ten sampling stations were identified including four shallow trench and six surface grab sampling stations.

2.2 Prediction of Net Infiltration

The technical approach used to predict net infiltration was sensitive to the dry environment of the southwest United States. In dry environments, net infiltration rates can be significantly overestimated if the predictive method is not climate-specific (Hutchison and Ellison 1992). Consequently, Hutchison and Ellison (1992) recommend the use of physically-based hydrologic modeling that is capable of accurately simulating upward unsaturated flow. The technical approach for predicting net infiltration used the U.S. Environmental Protection Agency's HELP model (Schroeder et al. 1994) to identify mean and extreme climate years combined with the physically-based soil-atmosphere hydrologic model SOILCOVER (USG 1997) to predict the net infiltration rate for the mean and extreme wet years.

3 IN-SITU AND LABORATORY TESTING METHODS

The following hydraulic properties and conditions of the unsaturated tailings were measured using in-situ and laboratory test methods.

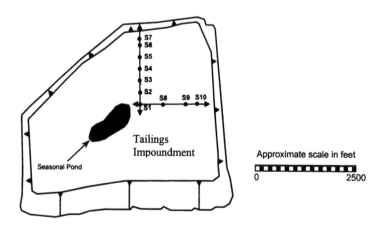

Figure 1 Sampling locations for shallow trench and surface grab samples.

- Texture;
- Density;
- Runoff;
- Soil-water characteristics (i.e., moisture retention);
- Hydraulic conductivity; and
- Water content and matric suction.

3.1 Texture

Material texture was determined for use in the development of conceptual hydrologic models of the tailings impoundment and for extrapolating the results of soil-water characteristic testing to samples that were not subjected to soil-water characteristic curve testing. The texture of tailings impoundment samples was determined using the sieve and hydrometer method in general accordance with ASTM D422.

3.2 Density

Density was measured to estimate the porosity of the tailings (an important property affecting the rate of infiltration) and to aid in interpreting soil-water characteristic and hydraulic conductivity data. Porosity was also measured in the laboratory on selected samples as part of soil-water characteristic curve and hydraulic conductivity testing. The in-situ density testing was conducted using a Troxler nuclear density gage (Troxler 1970). The density gage is commonly used for quality control in road and building foundation construction and has also been used to characterize mine soils (Pederson et al. 1980).

3.3 Runoff

Preliminary analysis of the tailings impoundment indicated that runoff would be a major component of the water balance. Runoff was measured through the installation and monitoring of runoff troughs (Figure 2) and the analysis of existing seasonal pond depth data. Each runoff trough was constructed from a fifty-five gallon steel drum, cut into two halves with an entrance cut into each front. As shown in Figure 2, the contributing area was defined using plastic edging and a precipitation gage was installed next to the trough to ensure accurate measurement of incident precipitation.

The presence of a seasonal pond near the center of tailings impoundment allowed for a unique opportunity to measure the runoff characteristics of the tailing surface. Pond water level data had been collected on a bi-weekly basis for approximately ten years. The tailings impoundment had a well-defined contributing area (i.e., area contributing runoff to the pond) that graded toward the pond situated at the center, which, combined with measurements of pond

Figure 2 Photograph of a runoff trough installed on the tailings surface.

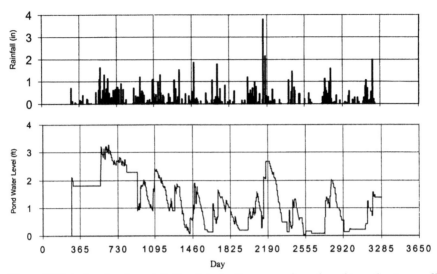

Figure 3 Water level and precipitation data for the seasonal pond used to estimate runoff

depth, pond evaporation and precipitation, allowed for an accurate estimate of runoff for the impoundment. Figure 3 presents the precipitation, evaporation and pond water level data used to estimate runoff.

3.4 Soil-water characteristics

The soil-water characteristic curve is defined by the relationship between volumetric water content and matric suction (infiltration estimates are sensitive to this parameter) and was determined to assess the soils ability to retain water. Two tailings samples were collected using a Shelby-tube driven into the tailings impoundment with a backhoe. The soil-water characteristic curve was measured for matric suctions ranging from 0 kPa (i.e., saturation) to greater than 1,500 kPa using the hanging column, pressure plate cell and thermocouple psychrometer methods.

Initially, the samples were placed in the hanging column and tested in general accordance with MOSA CHP 25 for matric suctions ranging from 0 to 10 kPa. The samples were then placed in a pressure plate cell and the soil-water characteristics measured for matric suctions ranging from 10 to 1,500 kPa using procedures outlined in ASTM D 2325-68. Lastly, a thermocouple psychrometer was used to determine the soil-water characteristics for matric suctions greater than 1,500 kPa, using the method described by Fredlund and Rahardjo (1993). Both drying and wetting curves were measured for all samples to evaluate hysteresis.

3.5 Hydraulic conductivity

Saturated hydraulic conductivity was measured in-situ at seven locations and in the laboratory on four samples. In-situ measurement of saturated hydraulic conductivity was accomplished using a tension infiltrometer designed and constructed by Soil Measurement Systems (1997). A photograph of the tension infiltrometer is presented on Figure 4. Standard procedures outlined in the tension infiltrometer User's Manual (Soil Measurement Systems 1997) were followed during in-situ measurement. Use of the tension infiltrometer for hydrologic studies has been described in detail by Hussan and Warrick (1995) and Maidment (1993). Additionally, saturated hydraulic conductivity was measured in the laboratory for each of the four samples using a falling head test in general accordance with MOSA CHP 28.

Figure 4 Photograph of the tension infiltrometer

3.6 Water content and matric suction

Water content and matric suction were measured for the development of vertical depth profiles and the estimation of the evaporative zone depth used in the HELP modeling. Wall samples were collected from the shallow (5-foot) trenches for the laboratory measurement of water content and matric suction. Samples were placed in glass jars and sealed to prevent water loss. Water content was measured on a mass basis in general accordance with ASTM D2216. Mass-based water contents (gravimetric) were converted to volumetric water contents using the volume-mass relations outlined in Fredlund and Rahardjo (1993) combined with the specific gravity and density data described previously. Matric suction was measured using the filter-paper method and testing procedure ASTM 5298-94.

4   RESULTS

The results of the field and laboratory testing, including texture and specific gravity, density and porosity, runoff, soil-water characteristics, hydraulic conductivity, and moisture content and matric suction are presented below.

4.1 Texture

The particle size distribution for tailings is shown on Figure 5. The tailings range in texture from sandy loam (greater than 75 percent sand and 5 percent clay) to loam (greater than 20 percent clay). Specific gravity ranged from 2.71 to 2.94.

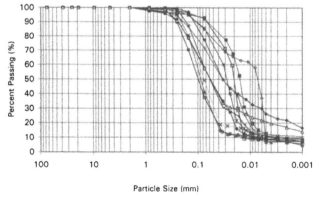

Figure 5 Particle size distribution curves for the tailings samples

247

4.2 Density and porosity

Dry density ranged from 1,300 kg/m^3 to 1,700 kg/m^3. Porosity, estimated based on volume-mass calculations and dry density data, ranged from 44 to 55 percent and averaged 47 percent.

4.3 Runoff

Three rainfall events were recorded using the runoff troughs. The events included both light rainfall (0.02 and 0.18 inches) and heavy rainfall (2.30 inches in 2 hours). As shown on Figure 6, runoff as determined from the runoff troughs varied from 47 percent of rainfall for the heavy rainfall event to 0 to 3 percent for the light rainfall events.

Figure 6 also presents runoff estimated as a percentage of rainfall using the pond level data. As shown on Figure 6, runoff ranges from 0 to approximately 80 percent. This wide range in values was due to the intensity and duration over which the rainfall occurred (i.e., more runoff occurred for a short duration event compared to a long duration event) and surface moisture conditions at the time the rainfall event occurred (i.e., runoff potential increases with the degree of ground surface saturation due, in part, to previous rainfall events).

4.4 Soil-water characteristic curves

The soil-water characteristics for two tailing samples are shown on Figure 7. The measured volumetric water content corresponding to a matric suction of 33 kPa was approximately 35 and 50 percent for samples S6-T13 and S2-T5, respectively. At this matric suction, sample S2-T5 retained more water than sample S6-T13 due to the presence of a higher percentage of silt and clay size particles and a greater density.

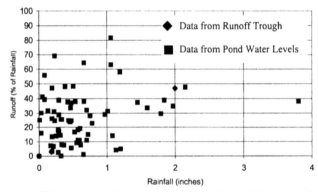

Figure 6 Runoff data obtained from the runoff troughs and the seasonal pond water level data

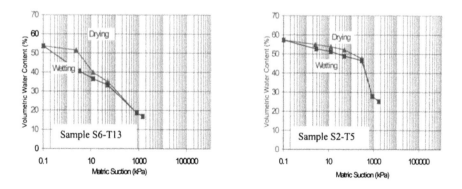

Figure 7 Soil-water characteristic curve data for the tailings samples

248

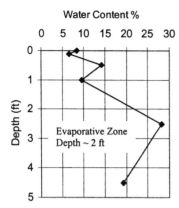
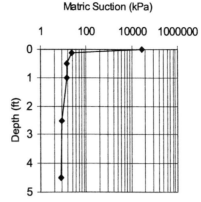

Figure 8 Water content and matric suction data obtained from the shallow trench sample

4.5 Hydraulic conductivity

Saturated hydraulic conductivity measured at the tailing surface using the tension infiltrometer ranged from $2x10^{-3}$ cm/s to $8x10^{-6}$ cm/s. Saturated hydraulic conductivity measured in the laboratory on undisturbed core samples collected at various depths ranged from $2x10^{-4}$ cm/s to $5x10^{-8}$ cm/s. Unsaturated hydraulic conductivity can be accurately predicted based on the saturated hydraulic conductivity and the soil-water characteristic curve (i.e., Fredlund and Rahardjo 1993).

4.6 Water content and matric suction

The measured water content and matric suction profiles for one shallow trench sample location are presented on Figure 8. The evaporative zone depth was inferred from the depth of drying (i.e., depth of lower water content relative to the water content measured at the deepest part of the profile) presented on the water content profile. The matric suction profile provided further resolution to the evaporative zone depth. High values of matric suction are characteristic of dry soil and low values are characteristic of wet soil. As shown on Figure 8, the evaporative zone depth for this tailings profile, ranges from approximately 6 to 24 inches. By comparison, the evaporative zone depths for silt and clay material can range from 8 to 18 inches and 12 to 60 inches, respectively (Schroeder et al. 1994).

In addition to providing further resolution to the evaporative zone depth, matric suction measurements can be used to estimate the hydraulic gradient (direction of flow). Such an estimate cannot be made with water content measurements alone.

5 DISCUSSION

Analysis of the in-situ and laboratory test results revealed that two main types of tailings profiles randomly interspersed throughout the impoundment characterized the surficial unsaturated tailings outside the footprint of the seasonal pond. The first tailings profile is characterized by a thin surface crust underlain by a silt-loam tailings material. The second tailings profile is characterized as a sandy-loam tailing with a lens of lower permeability fine tailing present at a depth of approximately 2 inches. Average saturated hydraulic conductivity values ranged from $2x10^{-6}$ cm/s to $1x10^{-4}$ cm/s and predicted net infiltration rates ranged from $2x10^{-10}$ cm/s to $3x10^{-8}$ cm/s.

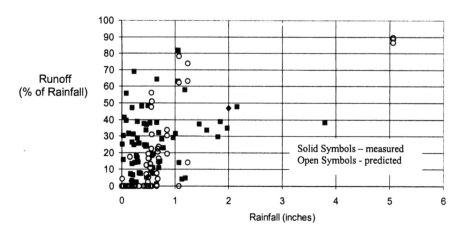

Figure 9 Comparison of computed and measured runoff from the tailings surface

The sections to follow discuss some of the key aspects of the in-situ and laboratory testing data. including extrapolating soil-water characteristic curve data to other site materials based on texture and using runoff data to evaluate model predictions.

5.1 Extrapolating soil-water characteristic curve data to other site materials

Conducting laboratory soil-water characteristic curve testing for each distinct tailings sample based on texture may not be practical due to the extensive testing times (~6 weeks) and associated costs. Therefore, a methodology must be established to estimate the soil-water characteristic curve for these additional materials. To limit lengthy testing times, computer programs based on theoretical models can be used to predict the soil-water characteristic curve using the readily measured particle size curve.

The most recent development for predicting the soil-water characteristic curve combines a theoretical model with a knowledge-based system (Fredlund 1996). This combined approach allows the user to calibrate theoretical predictions to available laboratory soil-water characteristic curve data as a means of improving the accuracy of the theoretical prediction (Swanson et al. 1999). Also, this approach allows the user to build a database of soil-water characteristic curves that are related to particle size. The database knowledge-based computer program SOILVISION (SoilVision Systems Ltd. 1997) was used as the basis for extrapolating laboratory measured soil-water characteristic curves to other locations on the tailings impoundment.

5.2 Using runoff data to evaluate model predictions

Runoff data can be used to evaluate model predictions of net infiltration. For example, runoff predicted from the soil-atmosphere modeling of net infiltration was compared to the runoff data collected from the runoff troughs and the pond level data. As shown on Figure 9, the predicted and the measured runoff show a similar range in values. Such comparisons are an important step in the modeling process to ensure accurate and defensible results.

6   CONCLUSIONS AND RECOMMENDATIONS

Predictions of net infiltration through tailings made using computer modeling techniques should be supported by site-specific climatic and hydrologic data. Hydrologic data can be collected using a combination of in-situ and laboratory testing methods and include particle size, density, soil-water characteristics, hydraulic conductivity, water content and matric suction. Direct measurement of water balance components such as runoff and/or infiltration and soil conditions such as water content and matric suction should also be made to evaluate computer model predictions. Finally, knowledge-based computer programs provide an effective means of extrapolating hydrologic data to other areas/materials within a site.

7 REFERENCES

Fredlund, D. G. and Rahardjo, H., 1993. Soil mechanics for unsaturated soils. John Wiley and Son's, Inc., New York.

Fredlund, M.D. 1996. Design of a knowledge-based system for unsaturated soil properties. M.Sc. Thesis. Department of Civil Engineering, University of Saskatchewan, Canada.

Hussen, A.A., and A.W. Warrick 1995. Tension infiltrometers for the measurement of vadose zone hydraulic properties. Chapter 13, Handbook of Vadose Zone Characterization and Monitoring. Lewis Publishers, Ann Arbor.

Hutchison, I.P.G, and R.D. Ellison, 1992. Mine Waste Management, Chapter 6: Climatic Considerations. Lewis Publishers Inc. Chelsea, Michigan.

Maidment, D.R., 1993. Handbook of Hydrology, Ch. 5: Infiltration and soil water movement. McGraw-Hill Inc.

MOSA CHP 24, CHP25, CHP28, 1986. Methods of soil analyses, part 1. A. Klute ed. American Society of Agronomy, Madison, WI.

Pederson, T.A., A.S. Rogowski R. Pennock, Jr., 1980. Physical characteristics of some mine soils. Soil Sc. Soc. Am. J. 44:321-328.

Schroeder, P.R., T.S. Dozier, P.A. Zappi, B.M. McEnroe, J.W. Sjostrom, and R.L. Peyton. 1994. The Hydrologic Evaluation of Landfill Performance (HELP) model: Engineering documentation for Version 3.0. EPA/600/9-94/xxx, U.S. Environmental Protection Agency Risk Reduction Laboratory, Cincinnati, Ohio.

Soil Measurement Systems 1997. Tension infiltrometer user's manual. Soil Measurement Systems, Tucson, AZ.

SoilVision Systems Ltd. 1997. User's guide for a knowledge-based database program for estimating soil properties of unsaturated soils for use in geotechnical engineering. SoilVision Systems Ltd., Saskatoon, Saskatchewan, Canada.

Swanson, D.A., G. Savci, G. Danziger, R. Mohr, and T. Weiskopf. 1999. Predicting the soil-water characteristics of mine soils. Tailing and Mine Waste 99-Proceedings, Fort Collins, Colorado.

Troxler, 1970. Series COMPAC, surface moisture-density gage manual. Troxler Electronic Laboratories Inc., Raleigh, NC.

USG 1997. User's manual for SoilCover: a 1-D, finite element, soil-atmosphere model. Unsaturated Soils Group, Department of Civil Engineering, University of Saskatchewan, Saskatoon, Saskatchewan, Canada.

Tailings and Mine Waste'00 © 2000 Balkema, Rotterdam, ISBN 90 5809 126 0

Use of a partitioning interwell tracer test (PITT) to measure water saturation in the Vadose Zone

P.E. Mariner & D.R. Donohue
Duke Engineering and Services, Grand Junction, Colo., USA

ABSTRACT: The partitioning interwell tracer test (PITT) is a new technology that offers a mechanism for quantitatively measuring the saturation, volume, and distribution of water within a large volume of unsaturated porous media. This technology may be particularly useful for water saturation measurements at mine, tailings, and other impoundments where pond leakage, cell liner integrity, leachate migration, and infiltration need to be characterized. Results are presented from a vadose zone PITT that was performed within and below a chemical waste landfill to determine the volume, distribution, and saturation of water and nonaqueous phase liquids. Test results revealed water saturations ranging from 13% to 34%, with an estimated standard deviation on the order of 10% of the measured saturation. Use of packers in the injection and extraction wells and multi-port samplers in monitor wells allowed the profiling of moisture distribution with depth. Such large-scale, spatially-integrated measurements cannot be made using conventional methodologies.

1 INTRODUCTION

The partitioning interwell tracer test (PITT) has the potential to revolutionize the measurement of *in situ* water saturation just as the interwell pumping test revolutionized the way *in situ* permeability is measured. The reason is that a PITT, like a pumping test, is not a point measurement but rather a spatially-integrated measurement of a large interwell volume of porous media. Conventional water saturation measurements have relied upon point-source data from boreholes (e.g. neutron probe), *in situ* monitors that provide indirect measurements (e.g. tensiometers), and core samples collected for laboratory analysis. These methods are useful for distinguishing and understanding some of the heterogeneity within a target zone, but even when a high number of point measurements are taken, they often provide a poor basis for calibration of numerical models of ground water flow and transport (Bowman and Rice, 1986; Bronswijk et al., 1995; Roepke et al., 1996). The PITT can be used to measure the water saturation in unsaturated tailings and is especially suited for measurements beneath tanks and liners where conventional measurements are difficult to obtain.

PITT technology, initially developed to measure residual oil saturation for enhanced oil recovery operations, was adapted for the detection and measurement of subsurface nonaqueous phase liquids (NAPL) in the early 1990s (Jin et al., 1994). By 1998, more than 30 such tests had been performed across the United States. While most of these tests were conducted in the saturated zone using aqueous tracers, several used gaseous tracers to measure both NAPL and water saturations in the vadose zone (Studer et al., 1996; Deeds et al., 1999; Simon et al., 1998). This paper describes the vadose zone PITT for water saturation measurement and presents results from two early field demonstrations in New Mexico.

2 THE PARTITIONING INTERWELL TRACER TEST

A PITT is an application of chromatography. Chromatography is the separation of compounds within a gas or liquid carrier stream that flows past a stationary phase. Compounds that have a higher affinity for the stationary phase will spend more time within the stationary phase and will therefore migrate more slowly. In a PITT designed to measure water saturation and volume in the vadose zone, the mobile carrier is air and the stationary phase is water.

Figure 1 illustrates the mechanism of chromatographic separation. The tracer mixture is injected between times t_0 and t_1. As time increases from t_1 to t_4, the non-partitioning tracer remains in the air phase and travels as fast as the air. The water-partitioning tracer also travels as fast as the air when it is in the air phase. However, when it encounters water, it rapidly partitions (distributes) between the mobile air and immobile water, causing its band width (distance from leading edge to trailing edge) to narrow and its overall migration to slow down.

Column experiments confirm that there is an easily quantifiable relationship between the average water saturation and the relative migration velocity of the water-partitioning tracer. A compound's affinity for water relative to air is represented by an equilibrium partition coefficient, K_i. This coefficient is the equilibrium ratio of compound i's concentration in water divided by its concentration in air and is equivalent to the inverse of the non-dimensional Henry's law constant. A high value for K_i implies a high affinity for water.

The mean residence time of the partitioning tracer relative to the non-partitioning tracer provides a measure of the relative chromatographic separation of the tracers. The retardation factor, R_i, is computed directly from the set of tracer breakthrough curves measured at the effluent sampling location and is equal to the mean residence time of tracer i divided by the mean residence time of a non-partitioning, conservative tracer (Jin, 1995). For a water-partitioning tracer, R_i can be defined as:

$$R_i = 1 + \frac{K_i S_w}{1 - S_w} \qquad (1)$$

where S_w is the average water saturation in the swept volume (Jin, 1995).

A vadose zone PITT begins with the injection and extraction of air at opposite ends of a subsurface test zone. Once steady flow is established between the wells, a mixture of the partitioning and non-partitioning tracers is added to the injection stream for a short period of time, followed by continued air injection and extraction. During the PITT, tracer concentrations

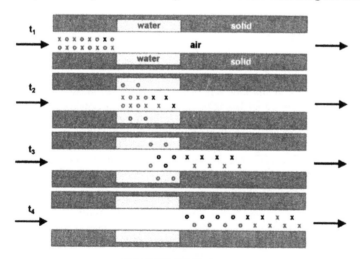

x non-partitioning tracer
o water-partitioning tracer

Figure 1. Schematic illustration of the chromatographic separation of non-partitioning and water-partitioning gaseous tracers.

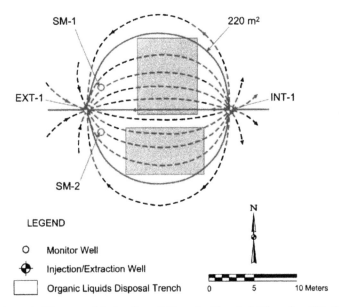

Figure 2. Diagram of design flow field at the Chemical Waste Landfill, Sandia National Laboratories, Albuquerque, New Mexico.

are monitored at the extraction wells and monitor locations. Plots of the tracer concentrations at these locations versus time (or versus cumulative production volume) produce tracer concentration breakthrough curves that are used to quantify average water saturations within the swept regions.

The retardation factor for each breakthrough curve is determined from a first temporal moment analysis. This analysis provides the missing information in equation 1 required to solve for the water saturation. For water volume estimates, the volume of vadose zone swept by the tracers must be determined. The swept volume is determined directly from injection and extraction rates and the recoveries and first temporal moments of the breakthrough curves of the non-partitioning tracer. Detailed discussions of the analysis of breakthrough curves, measurement and estimation of partition coefficients, and protocol for conducting a PITT can be found elsewhere (van Genuchten, 1981; Jin, 1995; Jin, et al., 1995; Jin, et al., 1997; Whitley, et al., 1999; Deeds, et al., 1999).

3 CASE STUDY

The first PITT that measured average water saturation in the vadose zone was conducted in December of 1995 at the Chemical Waste Landfill at Sandia National Laboratories in Albuquerque, New Mexico (Studer, et al., 1996; Mariner, et al., 1999). Because the primary purpose of this PITT was to measure the volume of NAPL (largely trichloroethylene) in the targeted zone, NAPL-partitioning tracers were included with the non-partitioning and water-partitioning tracers. The non-partitioning tracer was sulfur hexafluoride (SF_6), and the water-partitioning tracer was difluoromethane (CF_2H_2). CF_2H_2 has a water-air partition coefficient (K_i) of approximately 1.7 at 20°C.

The PITT design, depicted in Figure 2, involved one injection well (INJ-1) and one extraction well (EXT-1) installed 16.8 m apart on opposite sides of the target zone. Each well was screened and packed off at depths of 3.0 to 10.7, 12.2 to 18.3, and 19.8 to 24.4 m to provide shallow, intermediate, and deep zone measurements. In addition, two monitor wells (SM-1 and SM-2) were installed 3 m to the northeast and southeast of the extraction well to allow separate monitoring of zones to the north and south. Each monitor well accommodated eight sample

255

points evenly spaced at depths from 3.05 to 24.4 m to increase the vertical resolution of NAPL and water saturation measurements.

After steady-state air injection and extraction rates were established, the tracer mixture was injected into the injection well. Concentrations of the tracers were measured at 19 sample locations using two automated stream selectors connected to two online gas chromatographs. Each gas chromatograph (GC) could analyze approximately 30 samples per day. The PITT required a total of 15 days. The large number of sampling locations, limited number of GCs, and time-consuming GC analysis of the NAPL-partitioning tracers limited injection and extraction flow rates. Below-freezing weather on several nights resulted in water condensation inside sample lines, causing the online GCs to malfunction temporarily. Nevertheless, sufficient data were collected over the course of the test to provide useful tracer breakthrough curves.

The breakthrough curves for the shallow zone non-partitioning and water-partitioning tracers are displayed in Figure 3. The average water saturations were measured by moment analysis to be 23 +/- 2.0, 13 +/- 1.2, and 10 +/- 1.9 percent for the shallow, intermediate, and deep zones, respectively. Because the primary purpose of the PITT was to measure NAPL volume, no funds were available to collect water saturation data by conventional means for comparison purposes.

Since this initial PITT, online sampling and analysis of tracers during PITTs has dramatically improved. A vadose zone PITT conducted in 1996 at Kirtland Air Force Base in Albuquerque, New Mexico produced nearly continuous breakthrough curves with very little noise, as shown in Figure 4 (Deeds, et al., 1999). This test used three pairs of injection and extraction wells at three depths for a total of nine pairs. Methane was used as the non-partitioning tracer. According to first moment analysis, the data in Figure 4 correspond to a swept interwell volume of 8.0 m³ and an average water saturation of 30%. The water saturation measurements varied from 8% to 30% for the nine well pairs.

4 DISCUSSION

In many ways, a vadose zone PITT is like a large GC analysis. A mixture of volatile compounds is injected into a gas carrier stream that flows through a medium that acts to separate the mixture prior to analysis at the effluent end. Thus, the interwell unsaturated zone acts as a large, three-dimensional GC column. The only major operational difference is that instead of injecting an unknown mixture, the inverse problem is solved: a known mixture of volatiles (tracers) is injected so that the amount of liquid phase in the medium can be quantified.

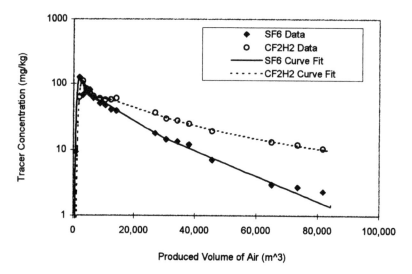

Figure 3. Breakthrough curves for the shallow zone. First temporal moment analysis indicates an average water saturation of 23 +/- 2.0%.

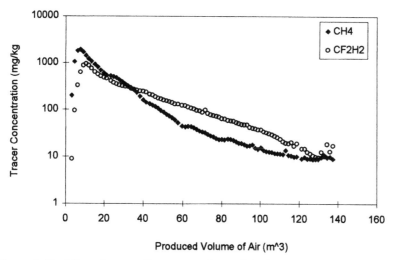

Figure 4. Breakthrough curves for an extraction well in the Kirtland Air Force Base PITT. Moment analysis of these curves indicates an average water saturation of 30%.

Water in liquid form is an excellent phase for quantification because it is essentially stationary and its properties are much different than other predominant phases in the vadose zone.

Although the volume of vadose zone swept by tracers can be computed, the actual pathways the tracers take before they are captured at an extraction or monitor well cannot be determined. Preferred pathways can cause a considerable portion of the tracers to bypass less permeable zones resulting in more than one peak in a breakthrough curve. Nevertheless, the high molecular diffusion of gases in air and the governing transport equations still allow large-scale water saturation measurements in such situations and additionally provide quantitative information on air permeability and its heterogeneity (Jin, 1995). For these reasons, the PITT is not restricted to homogeneous media.

The PITT is also not restricted to equilibrium conditions. High flow rates increase the possibility of nonequilibrium partitioning, but they do not affect the calculated first moment (retardation factor) of a tracer breakthrough curve and, therefore, do not affect the water saturation calculation (van Genuchten, 1981). Instead, nonequilibrium conditions affect the spreading and skew of a breakthrough curve, which are defined by the second and third moments (related to the Peclet and Damkohler numbers, respectively). In practice, however, severe non-equilibrium conditions should be avoided because they may result in long breakthrough curve tails at low concentrations that are below detection limits or are difficult to measure precisely.

5 CONCLUSIONS

The results and concepts discussed in this paper demonstrate the utility of PITTs for the direct measurement of water saturation and volume over a three-dimensional test zone. Unlike point techniques, such as gravimetric core analysis, tensiometers, and neutron probes, the vadose zone PITT offers an *in situ*, spatially-integrated measurement of water saturation over a large zone of interest. With proper design of injection, extraction, and sampling ports, the PITT can also provide information regarding spatial variability in the distribution of water. Execution of sequential PITTs offers a means of determining changes in saturation and moisture content over time and thereby allows monitoring of recharge, leakage from ponds and impoundments, and cover integrity at capped disposal cells. For project applications involving the use of analytical or numerical models which include vadose zone processes, the PITT, coupled with tensiometer data, may be used to determine capillary-pressure curves suitable for use as model input. Such an approach offers the only means of determining a capillary-pressure relationship in an *in situ* environment. This approach also may offer the only means of calibrating a set of *in situ* tensiometers for a field-scale application. Because the PITT is the only known method available for measuring water saturation in large volumes of unsaturated porous media, it has the potential to become a standard approach to field-scale water saturation measurements.

ACKNOWLEDGMENTS

The authors gratefully acknowledge the following individuals for their continuing contributions to the development of vadose zone PITT technology: Gary Pope of the University of Texas at Austin and Duke Engineering & Services colleagues Richard Jackson, Minquan Jin, Neil Deeds, Varadarajan Dwarakanath, and Jeff Silva. PITT technology is patented by the University of Texas and Duke Engineering & Services.

REFERENCES

Bowman, R.S., and Rice, R.C., 1986. Transport of conservative tracers in the field under intermittent flood irrigation. *Water Resources Research*, 22(11): 1531-1536.

Bronswijk, J.J.B., Hamminga, W., and Oostindie, K., 1995. Field-scale solute transport in a heavy clay soil. *Water Resources Research*, 31(3): 517-526.

Deeds, N.E., Pope, G.A., and McKinney, D.C., 1999. Vadose zone characterization at a contaminated field site using partitioning interwell tracer technology. *Environmental Science & Technology*, 33(16): 2745-2751.

Jin, M., 1995. *A Study of Nonaqueous Phase Liquid Characterization and Surfactant Remediation*. Ph.D. Dissertation, University of Texas, Austin, TX.

Jin, M., Delshad, M., McKinney, D.C., Pope, G.A., Sepehrnoori, K., Tilburg, C., and Jackson, R.E., 1994. Subsurface NAPL contamination: partitioning tracer test for detection, estimation and remediation performance assessment. In *Toxic Substances and the Hydrologic Sciences*. American Institute of Hydrology, Minneapolis, MN.

Jin, M., Delshad, M., Dwarakanath, V., McKinney, D.C., Pope, G.A., Sepehrnoori, K., Tilburg C., and Jackson, R.E., 1995. Partitioning tracer test for detection, estimation, and remediation performance assessment of subsurface nonaqueous phase liquids. *Water Resourses Research*. 31(5): 1201-1211.

Jin, M., Butler, G.W., Jackson, R.E., Mariner, P.E., Pickens, J.F., Pope, G.A., Brown, C.L., and McKinney, D.C., 1997. Sensitivity models and design protocol for partitioning tracer tests in alluvial aquifers. *Ground Water*, 35(6): 964-972.

Mariner, P.E., Jin, M., Studer, J.E., and Pope, G.A., 1999. The first vadose zone partitioning interwell tracer test for nonaqueous phase liquid and water residual. *Environmental Science & Technology*, 33(16): 2825-2828.

Roepke, C., Strong, W.R., Nguyen, H.A., McVey, M.D., and Goering, T. J., 1996. Unsaturated hydrologic flow parameters based on laboratory and field data for soils near the mixed waste landfill, Technical Area III, Sandia National Laboratories, New Mexico. SAND96-2090, Sandia National Laboratories, Albuquerque, NM.

Simon, M.A., Brusseau, M.L., Golding, R., and Cagnetta, P.J., 1998. Organic and aqueous phase partitioning gas tracer field experiment. In *Platform Abstracts of the First International Conference on Remediation of Chlorinated and Recalcitrant Compounds*, Monterey, CA.

Studer, J.E., Mariner, P.E., Jin, M., Pope, G.A., McKinney, D.C., and Fate, R. 1996. Application of a NAPL partitioning interwell tracer test (PITT) to support DNAPL remediation at the Sandia National Laboratories/New Mexico Chemical Waste Landfill. In *Proceedings of Superfund/Hazwaste West Conference 1996*, Las Vegas, NV.

van Genuchten, M.T., 1981. Non-equilibrium transport parameters from miscible displacement experiments. Research Report No. 119, U.S. Salinity Laboratory, U.S. Department of Agriculture Science and Education Administration, Riverside, CA.

Whitley, G.A., McKinney, D.C., Pope, G.A., Rouse, B.A., and Deeds, N.E., 1999. Contaminated vadose zone characterization using partitioning gas tracers. *Journal of Environmental Engineering*, 125(6): 574-582.

Tailings and Mine Waste'00 © 2000 Balkema, Rotterdam, ISBN 90 5809 126 0

Metal solubility and solid phase stability in arid, alkaline environments

J.VanMiddlesworth, J.Anderson & A.Davis
Geomega, Boulder, Colo., USA

ABSTRACT: Pit lakes form when open pit mines become inundated by groundwater upon closing. Future pit lake water quality in arid, alkaline environments has been modeled using both observational and theoretical methods. Understanding the geochemical parameters controlling metal solubility and solid phase stability is essential for the realistic prediction of water quality in the mine pit. Testing methods were designed to both elucidate the effects of proportional mixing of influent waters on the pit lake chemistry and also to simulate evapoconcentration, a major influence on pit lake water quality in arid climates. In order to simulate evapoconcentration, pit water was allowed to evaporate to dryness and both solid and aqueous samples were analyzed. Solid equilibrium phases were demonstrated to incorporate elements of concern including Pb, U and Sr, thereby reducing the aqueous concentrations of the elements by 100%, 25% and 10%, respectively, as compared to an ideal solution. The geochemical equilibrium code PHREEQC was then employed to predict future water quality for a solution in equilibrium with the solid phases observed. PHREEQC predicted analogous results to the measured evapoconcentration results.

1 INTRODUCTION

Closure of open pit mines results in the creation of a "pit lake" if the groundwater table is higher than the bottom of the pit after it is decommissioned. It is necessary to predict the resulting pit lake chemistry to (1) assess potential future impacts on water resources adjacent to the pit and (2) identify the potential to affect the health of human, terrestrial, or avian life.

The objective of this study was to use site-specific chemical and hydrologic data in conjunction with laboratory tests and predictive computer modeling to understand pit and aquifer geochemical interactions, and of the evolving pit lake water quality (chemogenesis), as the open pit becomes inundated at the end of mining. To accomplish these objectives, the geochemical attributes of the wall rock in the ultimate pit surface(s) were superimposed on the flow domain to predict the final water quality in the future pit lake

As water enters the pit from various wall rock source regions, *in situ* mixing, precipitation, sorption, and evaporation occur, determining pit lake chemogenesis. The bulk wall rock leachates for this pit are alkaline. Therefore, geochemical mixing results in precipitation of calcite, barite, celestite and gypsum. Chemical mixing, precipitation, sorption, and evapoconcentration were assessed using both observational and theoretical methods.

2 OBSERVATIONAL AND THEORETICAL METHODS

2.1 Column Tests

The primary factor influencing wall rock leachate chemistry is the release of constituents from weathered wall rock that will be flushed by influent groundwater or precipitation. Solutes are released until the groundwater has leached the soluble mass from the wall rock surface. Column tests are frequently used to assess variations in leachate chemistry from wall rock representative of the ultimate pit surface.

Samples representing each of the rock types (carbonatite, gneiss, and alluvium) in the ultimate pit surface (UPS) were collected from the Site. These samples were crushed to a particle size of 1-2 mm diameter for use in column tests to test 5 water:rock interaction scenarios, i.e.:

- alluvial rock flushed with alluvial background groundwater;
- gneissic rock flushed with gneissic background groundwater;
- gneissic rock flushed with impacted groundwater;
- carbonatite rock flushed with possibly impacted carbonatite background groundwater from the Pit well; and
- carbonatite rock flushed with impacted groundwater.

Two kilograms of each sample were placed in 5-centimeter diameter PVC columns approximately 76 centimeters tall. An equal mass (2 liters) of unfiltered site groundwater was passed through each column every 24 hours. Water drained from a reservoir above the column through its top and emerged from the column bottom into HDPE collection bottles. The tests resembled the Meteoric Water Mobility Test (MWMT) procedure (e.g., standard rock particle size, leachate volume based on the mass of the sample, etc.).

Column effluent was sampled daily. Seven indicator analytes (specific conductivity, pH, Eh, total Fe, Fe^{+2}, sulfate, and sulfide) were measured immediately following the collection of the sample. Effluent was then filtered through Gelman Supor polyethersulfone 0.45 μm membrane filters in Nalgene filter holders and analyzed for major and trace elements. Indicator parameters were monitored to determine when the effluent chemistry has returned to the influent chemistry (i.e., leaching is not significantly contributing to solute concentrations). When this condition had been achieved, the columns were stopped.

The major constituents of interest in pit lake chemistry include pH (which controls anion and cation solubility), TDS (a measure of suspended solids in the water), lanthanide series elements, chloride and strontium (which are found in high concentrations in the area), and lead and uranium (from an ecological risk perspective).

The column tests demonstrated little leaching of constituents from site alluvium and bedrock beyond dissolution of small particulates from freshly crushed and sieved samples. On the other hand, the site bedrock (gneiss and carbonatite) demonstrated a limited capacity to mitigate transport of some constituents (e.g., aluminum, boron, fluoride, manganese, molybdenum, nickel, and strontium).

Using the column data, Chemical Release Funtions (CRFs) were computed to mathematically represent the temporal release of solute into solution by coupling data generated by column tests with the background groundwater chemistry for each rock type. Empirical curves fit to the temporal chemical data were then used as the bulk leachate chemistry input to PITQUAL to determine pit lake chemogenesis. The asymptotic limit of the CRFs was determined from the background groundwater chemistry associated with each lithology in the UPS (Table 2-2). This ensures that in computing the leachate chemistry the water quality emanating from thoroughly leached wall rock would be consistent with the background groundwater chemistry, and would not underestimate constituent contributions to the pit lake. CRFs for all relevant chemical constituents were generated using column test data for each rock type.

2.2 Pit Lake Analog Test

The evolving pit lake water quality will be a function of several interrelated processes including (1) the chemistry of recharging groundwater, (2) the chemistry of seepage from tailings impoundments, (3) the chemical reactivity of site bedrock and alluvium, and (4) the mass and reactivity of solid-phase precipitates that form as groundwater, seepage, and wall rock leachate mix and react. To quantify the nature of the chemical reactions that occur when groundwater, seepage, and the local lithology are present, site-specific, bench-scale laboratory pit lake analog tests were performed. The tests involved mixing (and subsequent evaporation) of appropriate proportions of background groundwater, tailings impoundment seepage, and effluent from the column leaching tests.

The primary focus of the analog tests was to identify the chemical reactions germane to the system, e.g., the precipitation of potential solid phases, and the magnitude of sorption processes involving these precipitates. Sediment generated during the analog tests was examined using electron microprobe analysis (EMPA) to identify sediment mineralogy, grain size, chemical compositions, and phase associations (Davis et al. 1992, Link et al. 1994).

The pit lake analog tests were designed to simulate both the juvenile pit lake water quality (at five years after the pit begins refilling) and the mature pit water quality (50 years after the pit begins refilling). Water fluxes to the pit from various sources were allocated on a volume contribution basis with background water quality, impoundment water and leached water combined in appropriate volumes. Pit filling estimates from aquifer recharge were calculated using the Jacob-Lohman equation.

A total volume of 5,000 mL of groundwater, tailings impoundment seepage, and column test effluent were used in both juvenile and mature pit lake simulations. Following mixing of the waters, the predicted evaporative volumes for the juvenile stage (4% of the total volume) and mature stage (42%) were allowed to evaporate. Following evaporation, pH, Eh, and specific conductivity measurements of the solutions were taken approximately every two hours until the readings stabilized, at which time samples were collected, filtered, and submitted for major ion and trace element analysis. The precipitates were collected on the filters and analyzed by (EMPA) to determine mineral phases formed during evapoconcentration.

Both a "dirty" calcite (the primary mineral phase) that contains lead, strontium, manganese, and other impurities (Figure 1) and nickel chloride precipitates control solute concentrations in the mixed waters. For example, coprecipitation of lead with calcite reduced lead concentrations from 0.03 mg/L in the juvenile pit lake to below detection limits (<0.005 mg/L) in the mature pit lake.

2.3 Evapoconcentration Testing

The evolving groundwater quality is a function of several interrelated processes including, (1) the chemistry of background groundwater, (2) the chemistry of tailings impoundment seepage,

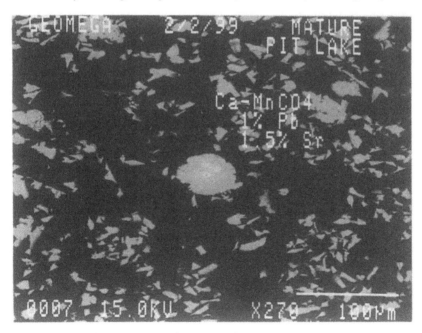

Figure 1. EMPA photomicrograph of precipitates found in the mature pit lake analog test.

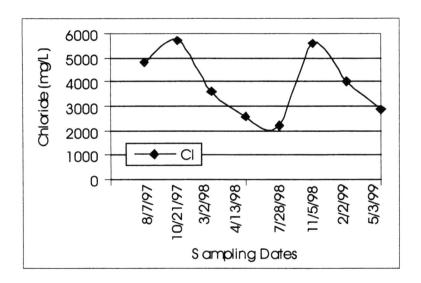

Figure 2. Fluctuations of chloride data over the last two years of sampling.

and (3) evapoconcentration of solutes in surface water and shallow groundwater. The arid climate at the site makes evapoconcentration a driving factor in surface water chemistry and a significant influence in shallow groundwater. For example, solute concentrations in some shallow monitoring wells such as SRK-20U have historically varied by up to an order of magnitude based on the time of year samples are collected (Figure 2)

Evapoconcentration results in increasing concentrations of some constituents (e.g., TDS and salinity) while the pH may become more alkaline (Eugster and Hardie 1970). Other solute concentrations decrease, e.g., calcium due to precipitation of calcite. These evolving solute concentrations may be complex. For example, Eugster and Hardie (1970) introduced the concept of the geochemical divide, which states that precipitation of a binary salt (i.e., $CaSO_4$) will remove the least abundant component and increase the more abundant component. The subsequent water quality is controlled by the Ca:alkalinity; Ca:SO_4; and Mg:alkalinity ratios. For example, if $CaSO_4$ precipitates from solution, and the concentration of Ca is greater than SO_4 at the start of the process, then all SO_4 would be removed from solution before Ca is depleted. The end result is a solution with no SO_4, but a higher concentration of Ca (Drever 1988). Using this concept, Eugster and Hardie (1970) were able to predict the chemistry of numerous saline water bodies in the western United States.

More refined predictions are possible with detailed chemical data which computes chemogenesis as the solution progressively becomes more evapoconcentrated. With detailed analytical data it is possible to predict the concentrations of analytes at any point in the evaporation sequence as well as define the geochemical mechanisms driving the chemical evolution of the water. The chemical divide concept was used to confirm the analytical results.

Because of the potential importance of evapoconcentration on the evolving surface water and groundwater conditions, site groundwater was collected from the pit well and evaporated to dryness in the laboratory. Two evapoconcentration tests were performed simultaneously; one tank contained only the Pit Well (ET) water while another contained the Pit Well water and carbonatite rocks (ETRX). These two tests allow an assessment of the reactivity of the rocks in the mine pit. The evaporating water was monitored for changes in pH, Eh, and specific conductivity (SC) every 7-10 days. Five water samples were collected and submitted to a certified analytical laboratory (NEL Laboratories, Las Vegas, NV). These samples consisted of the unevaporated water, and water after 50%, 68%, 75%, and 97% evaporation. These values cover the potential range of long term evaporation at the site which is calculated to be 58% over 70 years.

The concentration of most solutes rose as the water evaporated, as expected. However,

Table 1 Evapoconcentration Test Results

| Constituents | Percent Evaporation | | | | | Solid |
| | 0% | 50% | 67% | 75% | 97% | |
	Mg/L					mg/Kg
Aluminum	0.043	<0.05	<0.05	<0.05	0.13	4.9
Antimony	<0.005	<0.005	<0.005	<0.005	0.0056	<0.5
Arsenic	<0.005	<0.005	<0.005	<0.005	<0.005	<0.5
Barium	0.084	0.17	0.25	0.31	<0.005	6.7
Boron	0.41	0.87	1.3	1.7	4.2	52
Calcium	510	890	1400	1700	3800	67000
Cerium	0.0059	<0.005	<0.005	<0.005	<0.005	9.3
Cesium	0.014	0.027	0.039	0.059	0.13	2.6
Chloride	1900	3200	5700	6900	18000	Not Analyzed
Iron	<0.1	<0.1	<0.1	<0.1	<0.1	24
Lanthanum	0.0091	<0.005	<0.005	<0.005	<0.005	4.5
Lead	0.0085	<0.005	<0.005	<0.005	<0.005	1.5
Magnesium	300	630	930	1200	2900	40000
Manganese	0.055	0.03	0.029	0.048	0.059	4.2
Mercury	0.0025	0.00044	0.00046	0.00032	0.00078	<0.25
Molybdenum	0.0083	<0.01	<0.01	0.013	0.032	<0.5
Neodymium	0.0051	<0.005	<0.005	<0.005	<0.005	1.7
Potassium	17	37	59	76	190	3500
Praseodymium	0.0055	<0.005	<0.005	<0.005	<0.005	1.2
Selenium	<0.005	<0.005	0.011	0.017	0.036	<0.5
Sodium	260	570	880	1100	3000	71000
Strontium	88	160	200	310	690	5000
Thorium	<0.005	<0.005	<0.005	<0.005	<0.005	<0.5
Uranium	0.093	0.14	0.2	0.27	0.41	0.99
Vanadium	0.0027	ND	0.0094	0.016	0.036	0.4
Zinc	0.17	<0.1	<0.1	<0.1	<0.1	<5.0

elements behaved differently based on their mobilities and the stability of their respective solids under these conditions. Table 1 shows the concentrations of detectable constituents in solution at each stage of evaporation and also in the final solid. Solute concentrations, in general, fell below the ideal concentrations (concentrations expected if none of the solutes formed solid phases and dropped out of solution). Specific conductivity, a measure of the total amount of ions in solution, was lower than for the ideal solution (Figure 3) because of solid phase precipitation. However, more mobile and more concentrated elements do not diverge from ideality until significant evaporation has taken place. For example, calcium and chloride which are both very concentrated in the Pit Well water do not diverge from ideality until approximately 67% of the water has evaporated, due in part to the chemical divide theory of Eugster and Hardie (1970). The ability of the ion to stay in solution is also a factor as precipitating elements will tend to diverge from ideality more rapidly. For example, lanthanum readily coprecipitates with solid phases in these waters and therefore diverges from ideality quickly.

During evaporation, many solutes precipitated from solution. These solids were collected and analyzed by EMPA to establish specific geochemical reactions. Several equilibrium phases were found including calcite ($Ca(CO_3)$), gypsum ($Ca(SO_4)$), barite ($Ba(SO_4)$), celestite ($Sr(SO_4)$) and NaCl. Calcite, consistently containing 0.8% Sr and 0.6% Mg as impurities (Figure 4) was most common, and the first phase to precipitate. Celestite contained 0.12% each of La, Ce and Nd (Figure 5).

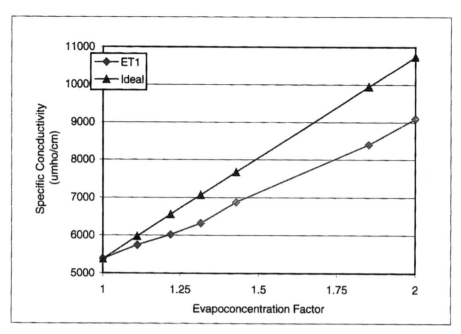

Figure 3. Specific conductivity in versus evapoconcentration factor in experimental and ideal solution.

3 THEORETICAL METHODS

Groundwater flow volumes provided by the numeric model developed for the site were used to calculate mixing ratios of different groundwater chemistry for each rock type (calculated as CRFs). Relevant pit lake water column and groundwater chemistry were used as input to the geochemical model PHREEQC (Parkhurst 1995) to determine the CO_2 saturation level of these waters, and to determine the metal solubility controls in this system. These mixtures of different groundwaters describe the solutions that will fill the future pit lake. Geochemical simulations in PITQUAL used the code PHREEQC (Parkhurst 1995) to model the fluids over time. One-year time steps were employed with the results of each year's simulation used as input into the next year's simulation, over a total simulation period of 70 years. Adsorption processes were not invoked in the model due to the low iron concentrations in the system.

Comparison of the observational (pit lake analog testing and evapoconcentration tests) and theoretical methods provided model validation. The pit lake analog testing quantified the geochemical reactions and *in situ* mixing taking place when seepage from impoundments encounters background groundwater. Evapoconcentration testing quantified the chemical reactions taking place during evaporation from groundwater or surface water.

The model accurately predicted constituent concentrations versus the evapoconcentration factor (the total volume of solution/ the remaining volume of solution) for the evapoconcentration tests (Figure 6). The excellent agreement between the model and observational testing demonstrate that invoking the solid equilibrium phases calcite, barite, gypsum and celestite provides the best explanation for the changing solute concentrations in the evapoconcentration experiment. Furthermore, trace element concentrations, such as uranium, could be accurately represented by invoking an impure form of calcite in the equilibrium model.

4 RESULTS

Both observational and theoretical methods determined that barite, calcite, celestite, and gypsum are the solid equilibrium phases for this system. Gypsum appears to be stable only as the system approaches dryness and therefore is less important to pit lake chemogenesis. Calcite was the most voluminous solid phase in all cases.

264

Figure 4. EMPA photomicrograph of evapoconcentration precipitates.

Figure 5. EMPA photomicrograph of evapoconcentration precipitates.

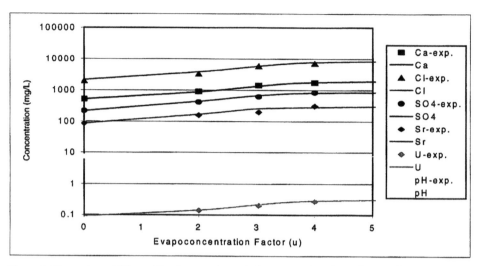

Figure 6. Comparison of Observational versus theoretical concentrations for major constituents.

Impure mineral phases (mainly calcite and celestite) are the main control of trace metal solubility in this pit lake. Analyses of total solids and calcite mineral separates (Table 2) shows that several trace elements (including uranium) were incorporated into the calcite phase (Figures 1, 4 and 5). These data are consistent with the observations of Russell et al. 1994 and Bard et al. 1991 who found that calcite and aragonite shells of marine organisms contain uranium in the

Table 2 Trace element analyses of calcite mineral separates

Constituents	TotalSolid mg/Kg	Calcite Mineral Separate mg/Kg
Barium	6.7	0.75
Cerium	9.3	<0.0005
Dysprosium	<0.5	<0.0005
Cesium	2.6	0.43
Erbium	<0.5	<0.0005
Europium	<0.5	<0.0005
Gadolinium	<0.5	<0.0005
Holmium	<0.5	<0.0005
Lanthanum	4.5	<0.0005
Lead	1.5	0.261
Lutetium	<0.5	<0.0005
Manganese	4.2	<0.0005
Neodymium	1.7	<0.0005
Praseodymium	1.2	<0.0005
Samarium	<0.5	<0.0005
Strontium	5000	786
Terbium	<0.5	<0.0005
Thorium	<0.005	0.008
Thulium	<0.5	<0.0005
Uranium	0.99	1.97
Ytterbium	<0.5	<0.0005
Zinc	<5.0	0.61

shell structure. The quantity of U in the shell material has been linked to the concentration of aqueous U in seawater. Coprecipitation of U and other radionuclides has also been documented (Kitano and Oomori, 1971, Meece and Benninger, 1993). Calcite in this system also incorporates lead, barium and strontium as impurities at 17%, 11% and 16% respectively.

No rare earth elements(REE) were found in the calcite phase. Rather the light REE in the total solids are most likely forming solid solutions with celestite ($SrSO_4$) because lanthanum and other REE were found in celestite (Figure 5).

5 CONCLUSIONS

This investigation has demonstrated that coprecipitation of trace elements with major equilibrium phases is an important factor in the evolving chemistry of the pit lake. Unlike most other pit lakes this is essentially iron-free and the chemistry is largely controlled by the carbonate system. Solid equilibrium phases were identified including barite, calcite, celestite, and gypsum with trace metals incorporated into the calcite and structure. Formation of impure calcite controls the solubility of many trace metals including U, Pb and to a lesser extent Sr. Since many of the chemicals of concern at this site will be sequestered by the carbonates in the system, the future pit lake water quality is not likely to adversely impact water resources.

REFERENCES

Kitano Y. and Oomori, T. (1971) The coprecipitation of uranium with calcium carbonate. *J. Oceanog. Soc. Japan* 27, 34-42.

Meece D. E., and Benninger L. K. (1993) The coprecipitation of Pu and other radionuclides with $CaCO_3$. *Geochim. Cosmochim. Acta* 57, 1447-1458.

Miller, G. C., Lyons, W. B., Davis, A. (1996) Understanding the Water Quality of Pit Lakes. *Environmental Science and Technology*. Vol. 30, no. 3, 118A-123A.

Russell, A. D., Emerson, S., Nelson, B. K., Erez, J., and Lea, D. W. (1994) Uranium in foraminiferal calcite as a recorder of seawater uranium concentrations. *Geochim. Cosmochim. Acta* 58, 671-681.

Tailings and Mine Waste'00 © 2000 Balkema, Rotterdam, ISBN 90 5809 126 0

Mineralogy of precipitates formed from mine effluents in Finland

Liisa Carlson
Geological Survey of Finland, Espoo, Finland

Sirpa Kumpulainen
Department of Geology, University of Helsinki, Finland

ABSTRACT: Four iron oxides and related minerals were identified in ochreous precipitates collected from the surroundings of 12 sulfide mines, ore deposits and talc quarries in Finland. Jarosite is characteristic of low-pH (< 3) environments high in dissolved sulfate. Schwertmannite is typically formed where pH ranges from 3 to 4. Ferrihydrite is found only where acidity is neutralized (pH > 5). Goethite, which is the only stable oxyhydroxide of iron, can be formed either through transformation via solution of metastable minerals such as schwertmannite and ferrihydrite or by precipitation from solution at pH 3 or higher if sulfate or bicarbonate is present. Surface sorption of sulfate is typical of low pH goethite. The minerals often occur as mixtures, e.g., jarosite plus goethite, jarosite plus schwertmannite, goethite plus schwertmannite. Most of the minerals found in mine drainage ochres are poorly ordered, have high specific surface area and high capacity to scavenge heavy metals and other noxious compounds from solution.

1 INTRODUCTION

Sulfide minerals, such as pyrite and pyrrhotite, are weathered in tailings impoundments and waste rock piles exposed to air and rain, and acid mine effluents gather in pools and ditches. Depending on local conditions, such as the type of bedrock and soils, the acid waters can be neutralized. In Finland, however, the bedrock and therefore also soils have low acid neutralization capacity because carbonate rocks are rare and the proportion of mafic rocks is low. Oxidation and hydrolysis of iron, which is the most common metallic cation in waters, leads to the formation of ochreous precipitate. At low pH bacteria such as *Thiobacillus ferrooxidans* calalyze the oxidation which abiotically would be much too slow. The mineral species formed depend on the conditions, such as pH and dissolved compounds other than iron. Jarosite ($KFe_3(SO_4)_2(OH)_6$) and goethite (α-FeOOH) have been known for a long time to occur in mine drainage precipitates. Ferrihydrite ($Fe_5HO_8 \cdot 4H_2O$) was first described from natural milieu by Chukhrov et al. in 1972, but was formerly known, e.g., as amorphous ferric hydroxide. This mineral is typical of environments with neutral pH and with dissolved silica in solution. Therefore, it is not a mineral to be expected in acid mine effluents, unless they are neutralized. Schwertmannite ($Fe_8O_8(OH)_6SO_4$) which was accepted in 1992 (Bigham et al., 1994), is a typical mine drainage mineral formed from low-pH waters high in dissolved iron and sulfate.

2 MATERIALS AND METHODS

2.1 Sampling

Ochreous precipitates were sampled around 9 sulfide mines or ore deposits, where they are mainly formed in ditches and ponds surrounding tailings impoundments and waste rock piles. A few samples are from 3 talc quarries where acid effluents are due to the weathering of sulfides in gangue rock black shale. On the other hand, talc deposits often contain carbonate minerals which have the capacity to neutralize acidity.

Precipitates were collected in plastic bags by hand using gloves to prevent contamination. The samples were kept cool during transportation and sifted in the laboratory to remove organic debris and coarse-grained detrital minerals. They were further gravity sedimented in distilled water and the fine fraction was used for mineralogical determinations. 500 ml of untreated and 100 ml of filtered (< 0.45 μm) and acidified (suprapur HNO_3, pH < 2) water was sampled. Temperature, pH, conductivity, and O_2 were measured in the field.

2.2 Laboratory methods

Water analyses were performed at the Geological Survey of Finland using ICP-AES, ICP-MS and anion cromatographic methods. Precipitates were dissolved in 4 M HCl and analysed using ICP-AES at the Department of Geology, University of Helsinki. Unfortunately, only preliminary chemical data is available at the time of writing this report. Minerals were identified by X-ray powder diffraction (XRD) using CuKα radiation and a Philips PW3050 vertical goniometer equipped with a curved Cu diffracted beam monochromator. The patterns were scanned either from 10 to 70°2Θ with 0.1° steps and counting time 20 sec/step (Fig. 1) or from 2 to 70°2Θ with 0.02° steps and counting time 1 sec/step. Infrared spectra (IR) were recorded using a Perkin Elmer 983 IR spectrometer. 3 mg of sample was mixed with 300 mg of KBr and discs were pressed of 150 mg of the mixture. Selected samples were examined with a JEOL JSM 5900LV equipped with EDS and with a Cameca Camebax SX 50 electron microbrobe equipped with four wavelength dispersive spectrometers (WDS) and a PGT energy dispersive spectrometer.

3 RESULTS AND DISCUSSION

3.1 Description of selected study sites

3.1.1 Luikonlahti

Cu-Zn-Co ore was mined from Luikonlahti in eastern Finland from 1968 to 1983. Processing of talc from a nearby quarry started in 1979 and from 1984 to the present only talc is processed. The tailings pond, ca 0.4 km^2 in extent and 2.2 million m^3 in volume is filled with wastes half from sulfide and half from talc processing. Unfortunately, the dam which surrounds the pond, was built of sulfide-rich tailings sand. The water, which seeps through the dam in one site and down a slope along a small ditch, has pH 2.8. No ocreous precipitate was found on the bottom of the ditch. The water runs below a road and further down a steep slope where it is mixed with water seeping from the ground. Bright yellow crust (dry Munsell color 10YR 6/8) is formed where the water (pH 3.2) runs to a wetland. Trees in the wetland area close to this site are dead. Schwertmannite was identified in the crust. This is a typical environment for ongoing schwertmannite formation, low-pH water with high concentrations of dissolved iron and sulfate is aerated, bringing about oxidation, hydrolysis and precipitation.

3.1.2 Otravaara

Pyrite ore was mined at Otravaara from 1919 to 1922. The small open pit is now filled with water the pH of which is 2.7. Waste rock pile covers an area of ca. 10.000 m^2. Rainwater leaches the pile and flows downhill to a wetland. A ca 10-15 m broad area down to the wetland is covered with ochreous precipitate and is unvegetated. During the rainy summer 1998 dead fish were observed in a lake ca 1 km downstream from the mine site. The environmental authorities allowed two ditches to be dug perpendicular to the flow direction below the waste rock pile. In July 1999 water was standing in the ditch close to the pile, it was red in color and had pH 2.1. In a ditch, which flows along the margin of the unvegetated area, water pH 2.4 was measured ca 40 m from the pile. From there on, a slow rise in pH to 2.6 was observed, and a jump to 4.7 where the ditch coalesces with another ditch with brown humic water. Pyrite is known to decompose much slower than, e.g., pyrrhotite. Therefore, almost 70 years after closing the small mine at Otravaara, effluents are still highly acidic. Concluded from the amount of pyrite still left in the waste this situation is going to remain prevalent.

Ochreous precipitate (dry Munsell color 10YR 5/6) was sampled in two sites on the ground. Both samples consist of jarosite and goethite (Fig 1). Using EDS, potassium was observed in jarosite. Closer to the pile the proportion of jarosite was higher than lower down, probably due to the effect of pH. It is possible that originally jarosite and schwertmannite were formed and schwertmannite was later transformed to goethite whereas jarosite remained unchanged. In natural conditions this transformation can be caused by changes in water composition, e.g., decreasing sulfate content, or in case of dry precipitate on the ground, by leaching with rainwater. In a laboratory experiment the transformation of synthetic schwertmannite stored in distilled water in refrigerator to well crystalline goethite took ca 18 months (Bigham et al. 1996).

3.1.3 Hammaslahti

The Hammaslahti Cu-Zn-Au ore, situated in the Outokumpu black shale belt, was mined from 1972 to 1986. Three open pits are now filled with water and have underground connection with each other through shafts. Water flows from the lowest lying open pit a distance of 1.2 km along a ditch to a river. In the open pit water is turbid and a reddish precipitate (dry Munsell color 5YR 5/8) has settled on the bottom close to the shore and on the steep walls. The ditch bottom is covered with precipitate down to the river. pH of water in the pit and in the ditch ranged from 3.4 to 3.6, concentration of total iron a was 59 mg/L and of dissolved SO_4 was 1410 mg/L. Goethite is the only mineral identified in the precipitate, even if the conditions are typical of schwertmannite formation. According to IR spectra much sulfate is adsorbed on the goethite surface (Fig. 2).

The tailings area and the clarification pool are now levelled and dry, covering an area of 0.4 km^2. Water seeps from the slopes to ditches and small pools and ochreous precipitate is formed on the bottom. In a wetland area below the former clarification pool water pH was 4.4 and schwertmannite was identified in the precipitate. In a ditch pH ranged from 3.8 (vegetated area) to 3.2 (area without vegetation) and schwertmannite with some goethite was found in the precipitate mainly formed on pebbles on the ditch bottom.

Black shale in the Hammaslahti area contains sulfide minerals, iron concentration of groundwater is high and ochreous precipitates are formed in natural cold springs and ditches unaffected by mining. Sampling was conducted on the shore of two small marshy lakes ca 1.5 km from the mine. Water pH ranges from 6.5 (lake) to 7.0 and 7.1 (two small springs). Several species of diatoms covered with iron oxide were observed in one of the spring samples. Poorly crystalline ferrihydrite was the only mineral identified in this sample. Considerably high manganese contents were observed using EDS in couple of ferrihydrite aggregates. Framboidal pyrite aggregates, 10 to 20 μm in diameter, consisting of small roundish crystals of ca 1 μm in diameter were observed using microprobe. They are probably formed in the bog sediments

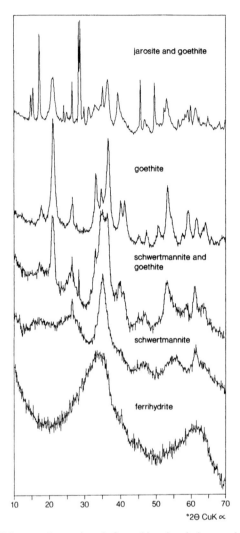

Figure 1. Representative XDR traces from minerals formed in mine-drainage ochres. The samples are from the Otravaara pyrite mine (jarosite & goethite), the Talvivaara Ni-Cu-Zn occurrence (goethite; schwertmannite; ferrihydrite), and the Vuonos Cu-Zn-Co mine (schwertmannite & goethite). A small amount of quartz and feldspar is present in both the schwertmannite samples.

through reduction of sulfate by bacteria active at neutral pH. Reducing conditions often prevail under a thin oxidizing layer in wet sediments rich in organic matter.

3.2 Mineralogy of mine-drainage ochres

Several iron oxides and related minerals have been observed in mine drainage environments. Bigham (1994, Fig. 4.12) summarises what is known about the conditions of formation of these minerals. Jarosite is formed through bacterial oxidation of iron optimally at pH 1.5 to 3.0 and sulfate concentration above 3000 mg/L. If sulfate is leached and pH rises, jarosite is transformed via solution to goethite. Schwertmannite is transformed as well which is optimally formed at pH 3-4, SO_4 concentration 1000-3000 mg/L and oxidation of iron accelerated by

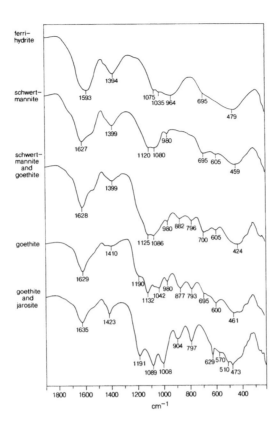

Fig. 2. Representative IR patterns of mine drainage ochres. The samples are from the Hammaslahti Cu-Zn-Au mine (four uppermost patterns) and from the Otravaara pyrite mine.

Thiobacillus ferrooxidans. Goethite can be formed directly or through transformation of other compounds via solution. Two dissolved species contribute to the formation of goethite from solution. SO_4 at pH < 6 and concentrations < 1000 mg/L, and HCO_3 at pH > 6 are known from laboratory experiments to favor the crystallization of goethite (Carlson and Schwertmann, 1990). Ferrihydrite can only be formed at pH > 5 in the presence of dissolved silica. Organic compounds may contribute to its formation, too (Schwertmann et al., 1984). Lepidocrocite has surprisingly been observed in mine drainage environments (Blowes et al., 1994, Milnes et al., 1992). In laboratory experiments many common impurities, such as dissolved silica, sulfate, or bicarbonate prevent its formation. According to Bigham (personal communication) a small amount of lepidocrocite was observed in mixture with other iron oxides in an abandoned coal mine in USA. As soon as no more dissolved oxygen could be measured, lepidocrocite disappeared.

Mineralogical data from mine-drainage environments in Finland as a function of effluent pH is summarized in Fig. 3. Altogether 33 precipitates from 8 mine sites are included. As expected, jarosite is present only at pH < 3 and ferrihydrite at pH 5 and over. Schwertmannite occurrences are mainly grouped between pH 3 and 4, but it can be found at pH up to 5. This can be due to fluctuations in water pH or slowness of dissolution leading to transformation. In Finland, goethite seems to be a mineral formed in mine environments at pH values and even at SO_4 concentrations typical of schwertmannite formation. As a primary component goethite occurred at pH 2.7-4.3. Bigham et al. (1992) only found two mine-drainage ochres with goethite as primary component, both formed around pH 7. Most of samples included in his data are from coalmines in USA.

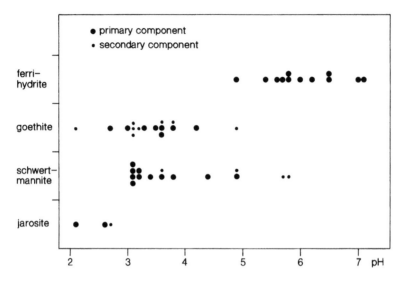

Fig. 3. Distribution of mine-drainage minerals in Finland as a function of effluent pH.

IR spectra (Fig. 2) provides more detailed information about the precipitates. Organic compounds are present in all samples, as shown by the COO-band at 1400 cm^{-1} which is intensive in most of the spectra. The Si-O stretching band at 964 cm^{-1} is characteristic of ferrihydrite (Carlson & Schwertmann, 1981). In addition, a double band at 1075-1035 cm^{-1} is due to silica of diatoms in the ferrihydrite sample. Sulfate is present in all samples containing jarosite or schwertmannite but also in most goethite samples. The presence of four bands in the S-O stretching region (v_3 at 1170-1190, 1110-1140, and 1040-1070 cm^{-1}, and v_1 at 970-980 cm^{-1}) indicates that sulfate forms a binuclear bridging complex on goethite (Harrison & Berkheiser, 1982). All the four bands are clearly present in the IR spectrum of goethite & jarosite from Otravaara, but the shoulder at 1190 cm^{-1} is hardly visible in the two schwertmannite spectra. Further sulfate bands are those at 610-620 cm^{-1} (v SO$_4$) and at 420-430 cm^{-1} (v_2 SO$_4$) (Bigham et al., 1990). OH-bending bands at 880 and 790 cm^{-1} are characteristic of goethite. The H$_2$O deformation band at 1630 cm^{-1} seems to be a broad doublet with maximum at 1590 cm^{-1} in the ferrihydrite sample. In both schwertmannite samples there is a shoulder in the 1630 cm^{-1} band at 1550 cm^{-1}.

The results of this study show that, as expected, jarosite and schwertmannite were formed at low pH values and ferrihydrite where neutralization of mine effluents had occurred. Goethite, which often is a product of the transformation of metastable minerals via solution, was also formed directly from low-pH solutions at sulfate levels characteristic of schwertmannite formation.

REFERENCES

Bigham, J.M. 1994. Mineralogy of ochre deposits formed by sulfide oxidation. In J.L. Jambor & D.W. Blowes (eds) *Environmental geochemistry of sulfide mine-wastes*: 103-132. Waterloo, Ontario: Mineralogical Society of Canada.

Bigham, J.M., Carlson, L., & Murad, E. 1994. Schwertmannite, a new iron oxyhydroxysulphate from Pyhäsalmi, Finland, and other localities. *Mineralogical Magazine* 58: 641-648.

Bigham, J.M., Schwertmann, U., Carlson, L., & Murad, E. 1990. A poorly crystallized oxyhydroxysulfate of iron formed by bacterial oxidation of Fe(II) in acid mine waters. *Geochimica et Cosmochimica Acta* 54: 2743-2758.

Bigham, J.M., Schwertmann, U., Traina, S.J., Winland, R.L., & Wolf, M. 1996. Schwertmannite and the chemical modelling of iron in acid sulfate waters. *Geochimica et Cosmochimica Acta* 60: 2111-2121.

Blowes, D.W., Reardon, E.J., Jambor, J.L, & Cherry, J.A. 1991. The formation and potential importance of cemented layers in inactive sulfide mine tailings. *Geochimica et Cosmochimica Acta* 55: 965-978.

Carlson, L. & Schwertmann, U. 1981. Natural ferrihydrites in surface deposits from Finland and their association with silica. *Geochimica et Cosmochimica Acta* 45: 421-429.

Carlson, L. & Schwertmann, U. 1990. The effect of CO_2 and oxidation rate on the formation of goethite versus lepidocrocite from an Fe(II) system at pH 6 and 7. *Clay Minerals* 25: 65-71.

Chukhrov, F.V., Zvyagin, B.B., Ermilova, L.P., & Gorshkov, A.I. 1972. New data on iron oxide in the weathering zone. *Proceedings of the International Clay Conference, Madrid*, 1972, Vol. 1, pp. 333-341.

Harrison, J.B. & Berkheiser, V.E. 1982. Anion interaction with freshly prepared hydrous iron oxides. *Clays and Clay Minerals* 30: 97-102.

Milnes, A.R., Fitzpatrick, R.W., Self, P.G., Fordham, A.W. & McClure, S.G. 1992. Natural iron precipitate in a mine retention pond near Jaribu, Northern Territory, Australia. In H.C.W. Skinner & R.W. Fitzpatrick (eds) *Biomineralization processes of iron and manganese - Modern and ancient environments*: 233-261. Cremlingen-Destedt, Germany: Catena Verlag.

Schwertmann, U., Carlson, L., & Fechter, H. 1984. Iron oxide formation in artificial ground waters. *Schweizerische Zeitschrift fuer Hydrologie* 46: 185-191.

Tailings and Mine Waste'00 © 2000 Balkema, Rotterdam, ISBN 90 5809 126 0

Arsenic release from oxide tailings containing scorodite, Fe-Ca arsenates, and As-containing geothites

E.A. Soprovich
Environmental Protection Branch, Yukon Division, Environment Canada, Whitehorse, Yukon Terr., Canada

A multi-year geochemical testing program was undertaken to examine arsenic release from oxide mine tailings produced at the Ketza River Mine, a gold mine which operated in Central Yukon. Materials for this program included milled tailings and drillcore from the oxide ore-body. Tailings samples from the exposed beach and from the ponded area were subjected to column testing which emulated exposed (wet/dry) and totally submerged disposal conditions. Crushed ore was also subjected to (wet/dry) column testing. Geologic materials were analyzed for mineralogical and other characteristics as a complement to the column program. Initiating testwork provided for licencing the project had suggested the tailings – then thought to contain exclusively scorodite as the arsenical species – would be stable under neutral pH conditions and given the high Fe:As ratios for these materials. In contrast however; review of water chemistry from the new column program, along with information from related investigations, demonstrate potential for on-going release of soluble arsenic from the oxide tailings. It is clear from the present work that careful thought to long-term disposal conditions for this type of tailings material is required in order to limit arsenic release to the environment.

1 INTRODUCTION

Metal mines are the principle contributors of arsenic to waterbodies (Moore & Ramamoorthy, 1984); mainly through processing ore containing arsenic with discharge of waste tailings to a body of water or engineered impoundment. Once released to a natural waterbody or impoundment, arsenic speciation and solubility is controlled by a combination of redox conditions, pH, adsorption reactions, and biological activity (Matisoff et.al., 1982).

The availability of metal oxides and hydroxides, clay sediments, and organic substances serve to decrease the level of dissolved arsenic present in water. Arsenic is coprecipitated with metallic ions such as aluminum and iron, and is absorbed to clay particles and humic constituents (Ferguson & Gavis, 1972; Gupta & Ghosh, 1953). Dissolved arsenic species are more strongly adsorbed to ferric hydroxides than to the more soluble ferrous hydroxides (Matisoff et.at., 1982), and arsenates (+5) have a higher affinity for adsorption than do arsenites (+3). Regarding the stability of arsenic, therefore, ferric arsenates under aquatic conditions have traditionally been considered to confer a high degree of environmental stability.

While there is general support in the literature for the relative stability of ferric arsenates in aquatic systems – particularly at high Fe:As molar ratios – more recent investigations have countered this paradigm where inclusion of calcium or other available cations into the Fe-As mineral may occur (Swash & Monhemius, 1995), or where the conditions of formation may lead to creation of an amorphous or cryptocrystalline solid (Dutrizac & Jambor, 1988).

2 GEOLOGIC SETTING, RESOURCE USE, AND PROCESS

Geologic materials (core samples and tailings) used for this study were taken from the Ketza River Mine, located in central Yukon approximately 200 km northeast of Whitehorse. The site is situated in alpine terrain of the Pelly Mountain Range at an elevation of 1500 meters. The headwaters of Cache Creek drain the site area: Cache Creek is a high energy, high gradient tributary of the Ketza River which in turn flows to the Pelly River in the Yukon River system.

The Ketza site lies within an uplifted block of Paleozoic-age sediments which have been structurally deformed by fold and fault sequences. These sediments have been subsequently intruded by Eocene to Cretaceous stockworks, resulting with hydrothermal alteration and mineralization of the host rock. The stratigraphic record consists of a basal quatzite of Pre-Cambrian age, overlain by a Lower Cambrian limestone containing mudstone markers and an upper Cambrian carbonaceous shale. The limestone unit is 200 meters thick at this location and is the main host to gold mineralization which comprises the orebody.

The gold deposit exists as an extensive pipe and manto system terminating at the top of the limestone unit. The oxide deposits – the ore delineated for extraction at the Ketza River Mine – formed as a result of oxidation of previously emplaced sulphide mineralogy. According to permitting documents (Canamax, 1987), the oxide ore was thought to consist of fine particulate gold disseminated throughout a variety of iron oxides: mainly a soft mixture of goethite ($HFeO_2$) and hematite (Fe_2O_3) with a host of accessory minerals. Early analysis suggested that arsenic occurred in ore almost entirely in the form of scorodite ($FeAsO_4 \cdot 2H_2O$).

The Ketza River gold mine commenced production in March 1988, extracting gold-bearing oxide ore until mining ceased in November 1990. An average of 364 tonnes/day of ore was mined from mainly underground and some surface open pit workings prior to abandonment.

Ore was processed on-site in a conventional mill utilizing the carbon-in-pulp cyanide extraction (CIP) process. Tailings produced by the mill were discharged as a slurry to a constructed tailings pond located in the Cache Creek drainage basin. Tailings were treated with an SO_2-Air cyanide destruction circuit in order to reduce cyanide concentrations. Alkaline conditions required for cyanidation and cyanide destruction contributed to mobilization of soluble arsenic in the tailings. An additional wastewater conditioning circuit, utilizing ferrifloc for reduction of dissolved arsenic, was implemented for treating excess reclaim water prior to release to the receiving environment.

Concentrations of 2 to 10 ppm dissolved As reported to the CIP and SO2-Air circuits. Alkaline conditions in the tailings pond additionally resulted with arsenic mobilization or remobilization from tailings.

Slightly more than 324 thousand tonnes of tailings were discharged to the Ketza River Mine tailings impoundment over the more than two year operating period. Several attempts by previous mine owners were made to finalize a closure plan for the Ketza site which was acceptable by Regulators. These efforts have been partially confounded by an incomplete understanding of the Ketza River Mine tailings: tailings mineralogy and the conditions under which they are least stable with respect to release of arsenic.

3 TESTING PROGRAMS FOR THE KETZA OXIDE ORE AND TAILINGS TYPES

Initial investigations centered around program components which could be conducted at the modestly equipped Environment Canada laboratory collocated in Whitehorse. Studies entailed initial extraction testing of tailings and followed with a multi-year, multi-component column lysimeter testing program of tailings and also a smaller-scale column test program for ore. These programs provided a number of insights about arsenic release from oxide tailings and ore, particularly with respect to magnitude of release under various disposal scenarios.

Column testing provided insight regarding arsenic release from Ketza oxide, but also raised additional questions regarding the cause of release at the magnitudes noticed, and especially regarding what minerals present in Ketza oxides were possibly responsible for release under the applied test conditions. Because of these new questions coming to light, a decision was made to undertake mineralogical testing of oxide ore and both column-run and raw or untested oxide tailings. Results from the mineralogical testwork are presented first.

278

4 MINERALOGY OF ORE AND TAILINGS

4.1 *Mineralogy program description*

The Canada Centre for Mineral and Energy Technology (CANMET) conducted mineralogical testing and other characterization analyses including:
- scanning electron microscopy (SEM-EDXA), electron microprobe and image analysis;
- semi-quantitative x-ray powder diffraction analysis;
- whole-rock bulk geochemical analysis; and
- other physical testing unavailable at the Whitehorse laboratory.

Procedures for each of these analyses followed ASTM or other accepted scientific methodologies (as detailed in Paktunc et.al, 1996).

4.2 *Results of mineralogy studies*

Prior to this work, there had been only marginal information about ore and host rock mineralogy available in written documentation. Environmental Protection Branch determined that a better understanding of arsenic release mechanisms would be gained via mineralogical characterization testwork of the geologic materials from the Ketza River Mine (ore and also column-tested and non-tested tailings). The mineralogical characterization report identified several arsenic-bearing minerals present in ore and waste tailings: some minerals identified in the report had not been delineated in earlier documentation because a complete mineralogical investigation had not been previously done with Ketza ore, and none had been undertaken before with the mill tailings.

Since earlier work had only identified two arsenic-bearing minerals (scorodite and arsenopyrite) as potentially being present, this mineralogical program proved beneficial to a better understanding of minerals which participate in release of arsenic from Ketza oxides. The new work confirmed scorodite and arsenopyrite present at Ketza, as previously understood, but also identified the presence of other important arsenicals in ore and tailings including: an iron-calcium arsenate hydrate, arenic-bearing (III) hydroxides, and a Fe-Bi-As phase mineral (the latter noticed as a rare occurrence in exposed tailings only).

Of the arsenic-bearing minerals present in these materials; scorodite, the iron-calcium arsenates, and the arsenic-bearing iron (III) hydroxides (certain of the Ca-As goethites) were considered important because of total contained arsenic. Arsenopyrite occurred in minor or trace quantities only, and was highest in ore (at only seven percent of total arsenic versus less than two percent in tailings). Scorodite made up only two to five percent (by weight) of total mineralogy yet contained between 17 to 45 percent of total arsenic occurring in these samples. Iron-calcium arsenate hydrates (likely calcium-pharmacosiderite and yukonite) were also listed as relatively abundant arsenical species: averaging 26.4% of the total arsenic. The amorphous iron (III) oxyhydroxides were the most abundant mineral species identified in samples and (phase-depending) contained significant percentages of arsenic as well. The mineral-type associated with the amorphous iron (III) oxyhydroxides was Ca-As goethite, and the report noted that although arsenic was always present, calcium content ranged from as much as 5 wt% of the mineral to non-present.

Based upon the information contained in the new report on mineralogical characterization and quantification: the average content of arsenic measured in Ketza ore and tailings was 4.0 wt% of the total rock mass. Furthermore, Fe:As ratios were lowest in exposed tailings (at 5.9:1 molar) and highest in submerged tailings (7.9:1 molar) and in ore (8.9:1 molar).

5 COLUMN LYSIMETER TESTING PROGRAM

5.1 *Program description*

5.1.1 *Sampling program*

Exposed tailings were sampled in 1993 from dry to moist exposed beach areas of the Ketza River Mine tailings impoundment. A transect was walked along the entire length of the exposed tailings beach and 60 - 100 gram samples were collected at random from surface exposed dry to slightly moist tailings at three to five meter intervals. Tailings samples were retrieved from the entire interval, at random, between near pond edge to the highest limit of tailings dispersal. An approximate 20 kg composite sample of exposed tailings was then transported to Environment Canada's Whitehorse laboratory for testing.

A decision was made during the 1994 field program to retrieve samples of submerged tailings from the Ketza River Mine tailings pond for column testing purposes. A candidate sample site was chosen in reference to a then concurrent water sampling / pond vertical column profiling program. Sampling occurred in April and was performed through a hole augured in the tailings pond ice. Cores of samples were retrieved using a 50 mm I.D. x 3050 mm long PVC pipe constructed to accept a 50 mm O.D. x 510 mm long thin wall acrylic sample carrier tube. The sample tube was additionally cradled behind the nose cone of a Wildco Model 2404 KB sediment corer, providing clean penetration of the tailings surface. The sampler was directed vertically into the two meters of pond water and pushed into the soft underlying tailings. The sampler was capped to close the system, and sample (including overlying water) was withdrawn. Precautions were undertaken at each step of sample retrieval to prevent exposure of submerged tailings samples to the atmosphere. Three samples averaging 350 mm length were obtained in this manner within very close proximity and then transported to the Whitehorse laboratory for testing.

5.1.2 Column leach program

Columns for exposed tailings testing were constructed from 102 mm O.D. (95 mm I.D.) x 1830 mm long clear acrylic cylinders to which were fitted sampling ports throughout the column and base filter / tailings support (bottom drain). All components of the two constructed columns were acid washed and then thoroughly rinsed with deionized/distilled water prior to being loaded with tailings. Exposed tailings were thoroughly mixed prior to filling the columns to the tailings design height of 400 mm (approx. 4 Kg). One of the two columns was constructed and maintained to provide for one meter of water cover over the tailings sample (submerged) while the second synthesized a flow-through (wet/dry) scenario. This arrangement, therefore, provided for a water covered column to simulate a pond with a diffusion controlled upward flux of arsenic only; whereas the flow-through column simulated a more typical exposed tailings scenario where there is significant downward flux of water and contaminants. Additional sample was available from the mixed tailings for performing shake flask extraction tests and for other investigative purposes.

The sample carrier tubes originally used for retrieving submerged tailings from the Ketza Mine tailings pond provided the basal units (along with specially made tailings support / filtration bases) for two columns constructed to test the flooded tailings. An additional section of prepared sample tube was attached to one of the carrier tubes to become a flow-through column (similar to that for the exposed tailings but of a smaller diameter and tailings mass). A single section of 51 mm O.D. (45 mm I.D.) x 1830 mm long clear acrylic cylinder was fitted with sampling ports at 500 mm intervals, then acid washed and rinsed. This column section was then attached (under nitrogen atmosphere) to the second submerged tailings sample tube and subsequently filled with deionized/distilled water to a level one meter above the tailings core surface. Pore water extractions were performed upon the flooded tailings samples prior to column construction, however, which entailed removal of pond water supernatant in a nitrogen atmosphere, pressurizing the tailings cores with nitrogen gas (to approximately 350 kPa / 50 psi), and collecting pore water from the same filter base ports which would be utilized for the column program component. Table 1 summarizes the test column configurations.

Maintenance and sampling followed regular seven day cycles consisting of: low-rate cycling of column (bottom to top) to turn over one-half the water volume (day three) followed by sample removal (350 ml) from the top and replacement of abstracted volume with deionized/distilled water (day seven) for the water cover columns; and addition of 150 mm equivalent DI/Distilled water (550 ml each on days three and six for the large diameter column – 250 ml on day three only for the smaller one) followed by sampling (day seven) for the wet/dry columns. The larger columns continued the seven day cycle for 22 weeks after which a fourteen day maintenance cycle was adopted which resulted with changing the sampling period to every two weeks (from one) for the water cover columns and with the requirement to composite

Table 1 - Column Lysimeter Test Configurations

Code	- Column Setup or Construction	- Simulation
ET/WC/of	Exposed Tailings / 1 m Water Cover / overflow sample	Diffusion control to water cover
ET/WC/bf	Exposed Tailings / 1 m Water Cover / baseflow sample	Downward flux through tailings under a water cover
ET/FT/ft	Exposed Tailings / Flow-Through / flow-through	Downward flux through tailings in a subaerial wet/dry state
ST/WC/of	Submerged Tails / 1 m Water Cover / overflow sample	Diffusion control to water cover
ST/WC/bf	Submerged Tails / 1 m Water Cover / baseflow sample	Downward flux through tailings under a water cover
ST/FT/ft	Submerged Tails / Flow-Through / flow-through	Downward flux through tailings in a subaerial wet/dry state

larger volumes of sample between analyses for the wet/dry or flow through columns. The smaller diameter columns were placed in a fourteen day cycle almost immediately.

The regular testing program ran between 56 to 78 weeks duration, depending upon the column test scenario and tailings type. Extensions to the regular program, testing other aspects of the tailings stability stretched the programs to 94 weeks. Up to 62 L of deionized/distilled water was input during the regular testing program (for the large column flow-through scenario). Total water throughput in terms of pore volume displacement ranged from 40 to 73 pore volumes during the regular program, while up to fourteen pore volumes were displaced during base flow abstractions from water cover columns near the end of the testing program.

Additional sampling conducted prior to the termination of the column testing program included the extraction and analysis of baseflow (pore water) from the water cover columns (Table 1), water quality testing of the various horizons within the large water column after a period of rest, acidification / caustic leaching of exposed tailings, and other investigations. A short term column program testing unprocessed ore was also undertaken to complement the tailings program. Only some of the results from these other test components will be dealt with here.

Measurements and analyses performed at Environment Canada's Whitehorse laboratory included: sample volume abstracted, laboratory reagent water volume added, conductivity, pH, Eh, dissolved oxygen, sulphate, and initially alkalinity/acidity. Filtered (0.45 μm) and preserved samples were analyzed for dissolved metals content at the Environment Canada laboratory in Vancouver. Other analyses were occasionally requested of the Vancouver laboratory.

5.2 Results of column testing program

5.2.1 Shake flask extractions – observations / results
Analysis of the lixiviant from shake flask extraction testing provided early indication that arsenic has a high affinity for leaching from Ketza oxide tailings. Arsenic release from tailings ranged from 8.9 mg/L to 9.6 mg/L. Liquid to solids ratio for the tests was 20:1. Release occurred when contacted with both neutral (DI/Distilled) and more aggressive (acidified) leachate solutions. There was a large difference in calcium release from the different tests, with calcium release to lixiviant from the acidified shake flask being highest at 1150 mg/L while other shake flask testing with neutral water returned calcium in solution at 29 mg/L. Iron release from each of the shake flask tests was similar with dissolved iron reporting to lixiviant for each in the 0.28 mg/L to 0.47 mg/L range.

5.2.2 Column lysimeter program – observations / results
It is beyond the scope of this paper to include a full presentation of data from the long-running column lysimeter tailings testing program. Key observations will be presented here with emphasis upon arsenic, calcium, and iron release under the different scenario tests.

Leachates monitored during the course of the regular program were characterized by relatively neutral pH: individual pH's falling generally in the range between 6.5 to 7.5 pH units. Some outliers were evident, particularly with respect to the flow-through column containing exposed tailings and to some degree to the water cover column containing exposed tailings, whereby pH's measured during the early period of the testing ranged between 5.6 and 8.9 pH units. Mean pH's were slightly higher by about one-half pH unit in the two columns containing exposed tailings than in their counterpart scenario columns which contained submerged tailings samples.

Both flow-through columns exhibited an initial rapid release of sulphate followed by a tapering-off of concentrations reporting to leachate: this was especially noted for the flow-through column which contained exposed tailings (maximum concentration of 1450 mg/L with long term release on the order of 20 mg/L). In contrast to the flow-through scenario, the water cover columns did not produce significant sulphate release. There was a lesser cyclical release of sulphate from the water cover column containing exposed tailings (maximum 65 mg/L). Very minor release of sulphate to the water column occurred in the case of submerged tailings where a water cover was maintained.

Alkalinity buffering also was evident with both flow-through columns whereby there was a high initial release of alkalinity followed by a long-term elevated reporting of alkalinity to flow-through leachate. Other than an initial concentration spike noticed for the water cover column containing exposed tailings, alkalinity reporting to overflow for both water cover columns was muted in relation to the flow-through counterparts.

Conductivity trends for all the columns during the regular program followed a similar pattern as for release of sulphate. Measured conductivities were steady and very similar for three of the four columns after about 200 days. The water column containing submerged tailings maintaining very low conductivities in overflow leachate throughout the program.

Arsenic was readily leached under the various test conditions (Fig. 1). Concentrations of dissolved As reporting to leachate varied from 0.33 mg/L for the submerged tailings / water cover test scenario (during the non-baseflow period) to a high of 34.7 mg/L observed during the regular period for the flow-through column containing exposed tailings. Long term steady state release of arsenic from the flow-through column containing exposed tailings was in the 30 mg/L range. Both flow-through columns produced considerably more arsenic in flow-through leachate than did the water cover columns from overflow sample (wherein the contaminant release to the water cover is largely diffusion controlled). Worth noting, when base-flow leachate was extracted from the two water cover columns near program end, is that while arsenic concentrations were elevated relative to the overflow values they did not-quite approach concentration levels of the flow-through columns containing corresponding tailings types. Further analysis with respect to arsenic release (as well as release of Ca and Fe) will be provided.

Iron concentrations detected in leachate were greater for the two flow-through columns versus Fe detected in water cover overflow. Similar to the arsenic concentration signature, iron released from exposed tailings was higher in magnitude than that from tailings retrieved from the submerged part of Ketza tailings pond. Iron concentrations detected in all the column leachates were lower than those of arsenic: the highest concentration of Fe reporting to the flow-through column containing exposed tailings, with an initial flush-like release of close to 2.3 mg/L followed by Fe concentrations falling with a number of perturbations to range generally between 0.590 mg/L to 0.035 mg/L. The flow-through column containing submerged tailings sample released relatively little Fe during the testing period, with a maximum measured concentration of 0.082 mg/L measured near the beginning and most analyses falling at the detection limit. In relation to the flow-through counterparts; very little iron was detected in overflow leachate samples from each of the water cover columns. Only slightly more iron was released to base flow leachates from the two water cover columns.

Calcium released by all the columns was significant and, following the trend with As and Fe, was greatest for the flow-through columns with maximum analyzed calcium concentrations of 450 mg/L (exposed tailings) and 192 mg/L (submerged tailings). Long term Ca concentrations for both of the flow-through columns was similar, as was minimum calcium concentrations, at approximately 30 mg/L. The initial flush of calcium was similar to that of sulphate for the two flow-through columns, particularly for that column containing exposed tailings. Dissolved Ca reporting to overflow leachate from the column containing exposed tailings reached a long-term release concentration of approximately 20 mg/L while the column containing submerged tailings was approximately half that level. Baseflow leachate from the two water-cover columns contained dissolved Ca in similar magnitude and concentration to the counterpart tailings-type flow-through tests.

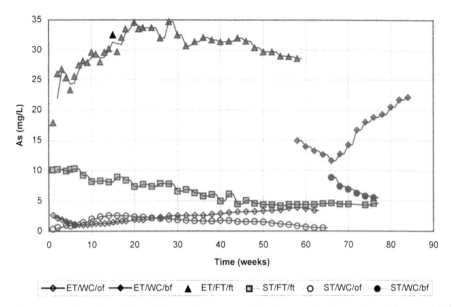

Figure 1 - Arsenic (mg/L) vs. Time. Abbreviations mean: ET = Exposed Tailings; ST = Submerged Tailings; FT = Flow-Through; WC = Water Cover; of = overflow sample; and bf = baseflow sample

6 ANALYSIS AND DISCUSSION

Very little arsenic, iron, and calcium was released to water cover columns during the observed period of the program. This is not surprising given that the transport is largely diffusion controlled for these columns. Based upon the observations, it seems most worthwhile to highlight results and analysis of data from both flow-through columns as well as the baseflow leachate components from the water-cover columns.

Average arsenic concentrations reporting to flow-through or to baseflow leachate for each of the columns (Table 2) trending from highest to lowest were: exposed tailings / flow-through > exposed tailings / water-cover (baseflow) > submerged tailings / water-cover (baseflow) ≈ submerged tailings / flow-through. A comparison of this data based upon flow-weighted concentrations presents one way to compare results where columns differ in tailings mass and volume or where water throughputs are not equal for the entire test period. Flow-weighted average concentrations were calculated as:

Net Release (cumulative mg of analyte / kg of tailings)
Net Water Throughflow (cumulative L / kg of tailings).

The net arsenic release for the column containing exposed tailings and subjected to flow-through testing was 475.48 mg/L, while net water throughflow was 15.68 L/kg. The flow-weighted concentration of arsenic from the exposed tailings under flow-through test conditions was thus 30.39 mg/L (Table 2) and represented again the highest concentrations for arsenic for each of the tailings types and column test scenarios. The flow-weighted arsenic concentration from the water-covered exposed tailings subjected to baseflow extraction was nearly half that of the flow-through tested exposed tailings at a calculated value of 16.69 mg/L. Following in line for flow-weighted concentration were the submerged tailings at 6.36 mg/L (flow-through) and 6.27 mg/L (water-cover (baseflow)): both submerged tailings contributed approximately one-fifth the arsenic to leachate on a flow-weighted concentration basis as did the exposed tailings in flow-through state.

A graphical representation and alternate, perhaps more reasonable, means of comparison is presented in Figures 2 - 5, whereby cumulative release or loading of As, Fe and Ca (mg/kg) is

283

Table 2 - Dissolved Arsenic Concentrations and Calculated Loadings or Releases

Sample Type:	Exposed Tailings Sample			Submerged Tailings Sample		
* Column Code:	ET/WC/of	ET/WC/bf	ET/FT/ft	ST/WC/of	ST/WC/bf	ST/FT/ft
Maximum As (mg/L)	3.86	22.10	34.70	2.59	8.86	10.30
Minimum As (mg/L)	1.05	11.60	17.90	0.33	5.58	4.20
Average As (mg/L)	2.27	16.44	29.91	1.56	6.81	6.63
Median As (mg/L)	2.15	15.80	30.20	1.63	6.81	6.32
Flow-weighted Conc. (mg/L)		16.69	30.39		6.36	6.27
Average Release (mg/kg) / PV		6.62	12.22		1.83	1.90

* where: ET = Exposed Tailings ; ST = Submerged Tailings ; WC = Water Cover ; FT = Flow-Through ;
of = overflow sample ; bf = baseflow sample

graphed against cumulative pore volume water throughput (one pore volume being the displacement within each tailings sample occupied by pore space) for each of the columns.

In taking this approach it is immediately noticeable that the slope of the line denoting arsenic release is relatively straight in each case, suggesting a relatively steady release of arsenic to flow-through and baseflow leachate waters. The slope of the line defines arsenic release per pore volume displacement or water throughput (cumulative load/release in mg/kg divided by cumulative pore volume) and reflects relative consistency with rankings from flow-weighted average concentration calculations whereby exposed tailings subjected to flow-through testing promoted the highest release of As (12.22 mg/kg/PV) followed by exposed tailings /water-cover (baseflow) at about one-half the flow-through level with a calculated value of 6.62 mg/kg/PV (Table 2). Both columns containing submerged tailings gave calculated mg/kg/PV values approximately one-sixth that of the top ranked column: 1.90 mg/kg/PV for the flow-through scenario and 1.83 mg/kg/Pv for the water-cover (baseflow) counterpart (Table 2).

Iron does not report significantly to leachate water quality, as evidenced by the graphs presented in Figures 2 - 5, whereas calcium reports in large quantities to leachates. The steep initial slopes represented in each figure for Ca fits well with sulphate release by each column and is suggestive of an early flush-type release of soluble gypsum: this was particularly evidenced with respect to Ca release from exposed tailings. Formation of gypsum via the SO_2/Air process has been reported by Robbins (1996) and others, so the presence of gypsum in Ketza tailings is not surprising. Concurrent with and following the flush of gypsum, it is proposed that short-term release of calcite and dolomite contributed to reporting of Ca to leachates until at least such time as calcium release stabilized. The alkalinity signature maintained elevated concentrations or ramped up during the latter part of the initial calcium release period beyond the point in which sulphate reporting to leachates began decreasing.

Results from the column testing program are highly suggestive that Fe-Ca arsenates present in Ketza tailings contribute significantly to the release of arsenic to leachates, and given the correct conditions arsenic can be expected to report to leachate at concentrations of up to 30 mg/L (flow-weighted basis). Arsenic release is significant as is release of calcium: although the much higher release of calcium points to other contributing sources, as discussed. The fact that iron reports to leachate water quality at low levels suggests that Fe is attenuated along the flowpath, and this may in turn serve to attenuate and reduce arsenic release from tailings to pore water leachate via adsorption or coprecipitation reactions.

Goethite and similar Fe (III) oxyhydroxides comprised approximately two-thirds of Ketza mineralogy and, along with arsenic-bearing ferric hydroxides, these minerals may act to adsorb some of the arsenic released from the Fe-Ca arsenates. The Ca-As goethites and arsenic-bearing goethites may also contribute, to a much lesser extent, to arsenic release to pore water. Scorodite should contribute to release of arsenic, but at a very reduced rate under the neutral pH test conditions because the Fe released by scorodite dissolution could precipitate as a ferric hydroxide coating on the scorodite particle, further reducing dissolution of the original particle (Nicholson, 1996).

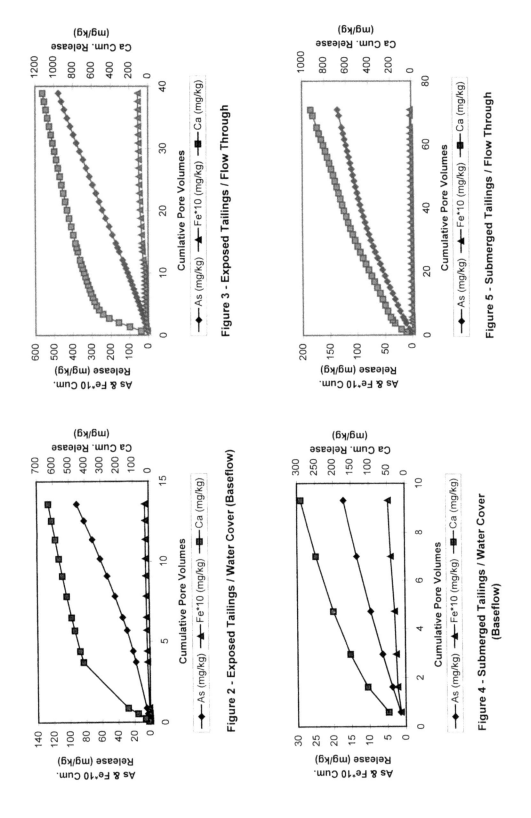

Figure 2 - Exposed Tailings / Water Cover (Baseflow)

Figure 3 - Exposed Tailings / Flow Through

Figure 4 - Submerged Tailings / Water Cover (Baseflow)

Figure 5 - Submerged Tailings / Flow Through

285

The presence of Fe-Ca arsenates in Ketza tailings was one of the most significant findings to come out of the mineralogical research: particularly as it relates to the destabilizing effect calcium imparts to the Fe-Ca arsenate – as suggested by Swash & Monhemius (1995). Because of their relatively high solubilities, the Fe-Ca arsenates are expected to predominate arsenic release at Ketza. An additional consideration, with respect to tailings stability, was the presence of calcium in intimate association with arsenic in some of the iron (III) oxyhydroxides: despite exhibiting high Fe:As ratios (up to 1096:1), the inclusion of Ca in the Ca-As goethites would tend to make these minerals less stable as well. These new findings contrast significantly from early work regarding the important arsenic minerals and release mechanisms at the site, as the presence of these geologic materials at Ketza was previously unknown or at least unreported. Considered along with knowledge regarding relative abundance, this new information has potential to alter decommissioning plans for the tailings in both the short- and long-term.

Research by Robins & Jayaweera (1992), Harris & Monette (1988) and others underline the long-term instability of Ca arsenates and Fe arsenates with conversion of these materials to secondary minerals when exposed to atmosphere in an open aqueous system. By extension, it could be suggested that a similar fate may befall the Fe-Ca arsenates: the increased release of As (particularly noted during flow-through testing) from exposed Ketza tailings would add weight to this proposition.

7 CONCLUSIONS

Column testing demonstrates the capacity for these types of tailings to release arsenic at concentration levels which are considerably higher than what is deemed to be environmentally acceptable: arsenic detected in column leachates were several orders of magnitude higher than that considered protective of aquatic organisms. Flow-through testing produced the highest release of arsenic followed by baseflow extraction beneath a water cover. Release of arsenic to a water cover occurred at the lowest rates and levels for the conditions tested – this is not surprising given the mode of contaminant release. Additionally, results from the column program indicate that atmospheric exposure appears to play a role in further destabilizing these tailings-types: the destabilizing mechanism is particularly suggested for the Fe-Ca arsenates which are expected to be the least stable of the arsenicals found in Ketza River Mine tailings. Each of these findings are intrinsic to long-term decision-making for closure of the Ketza Mine.

The studies presented underline the importance of undertaking mineralogical investigations in concert with column testing or similar programs. Mineralogic testwork indicated which minerals present in tailings could be expected to take part in release of arsenic and serves to explain some of the findings from the column testing program. Results from the column test program and the mineralogical testwork, along with empirical knowledge regarding stability and solubilities of the various arsenic-bearing minerals, serve to place boundaries on what levels of arsenic can be expected to be released from the Ketza tailings under the various disposal conditions which may be contemplated. This information can be used to better understand the dynamics of arsenic release for future modeling and site closure.

The outlined projects also underscore the importance of representative samples for testing purposes. Initial testwork conducted by the licensee used tailings from bench process testing which may not have fully synthesized the quality of tailings produced by the much larger mill. The present column testwork was able to use actual milled tailings which essentially were representative of a much larger 'scaled-up' process sample (reflecting changes due to the larger scale process and including gangue or dilution rock which would report to the mill). Also, this testwork was able to investigate potential effects the post-depositional environment may have upon stability of the Ketza oxide tailings.

Finally, 'theoretical' does not always mesh with 'actual'. Early project proponents had proposed that high Fe:As ratios in feed combined with near neutral pond water would control arsenic release to acceptably low levels. Flawed initial investigations which relied upon incomplete mineralogical knowledge of Ketza ore and tailings have proven, in hindsight, that the theoretical approach does not necessarily mirror reality: good initial investigations are needed.

ACKNOWLEDGEMENTS

The author wishes to thank John Kwong of CANMET for his candid suggestions of what may be the important focus of project data to present in a paper of this nature. Support by Environment Canada and review by Vic Enns of Environmental Protection Branch is gratefully acknowledged. Craig Butt of Environment Canada assisted with the graphs.

REFERENCES

Canamax Resources Inc. 1987. Water licence application supporting information report submitted to Yukon Territory Water Board. Vancouver.

Dutrizak, J.E. & Jambor, J.L. 1988. The synthesis of crystalline Scorodite, $FeAsO_4 2H_2O$. *Hydrometallurgy* 19 (1988): 377-384.

Ferguson, J.F. & Gavis, J. 1972. A review of the arsenic cycle in natural waters. *Water Research* 6: 1259-1274.

Gupta, S.R. & Ghosh, S. 1953. Precipitation of brown and yellow hydrous iron oxide III adsorption of arsenious acids. *Kolloid-Z* 132: 141-143.

Harris, G.B. & Monette, S. 1988. The stability of arsenic-bearing residues. In Reddy, R.G., Hendrix, J.L. & Queneau, P.B. (eds.), *Arsenic metallurgy fundamentals and applications; Proc. TMS ann. mtg., Phoenix, 25-28 January 1988.*

Moore, J.W. & Ramamoorthy, S. 1984. *Heavy metals in natural waters, applied monitoring and impact assessment*. New York: Springer-Verlag.

Nicholson, R.V. 1996. *Personal communication*. 9 October 1996.

Paktunc, A.D., Szymanski, J., Lastra, R., & Laflamme, J.H.G. 1996. *Mineralogical characterization of the Ketza River Mine tailings, MMSL96-016(CR)*. Ottawa: CANMET.

Robbins, G.H. 1996. Historical Development of the INCO SO2/AIR cyanide destruction process. *CIM Bulletin* 89(1003): 63-69.

Robins, R.G. & Jayaweera, L.D. 1992. Arsenic in gold processing. *Mineral Processing and Extractive Review* 9: 255-271.

Swash, P.M. & Monhemius, A.J. 1995. Synthesis, characterization and solubility testing of solids in the Ca-Fe-AsO₄ system. In Hynes, T.P. & Blanchette, M.C. (eds.), *Proc. Sudbury '95, Mining and the environment, Sudbury, 28 May - 1 June 1995*. Ottawa: CANMET.

Surface water quality

Tailings and Mine Waste'00 © 2000 Balkema, Rotterdam, ISBN 90 5809 126 0

Environmental monitoring of the Mantaro River, Central Peru

N. Lozano
New Mexico State University, Las Cruces, N.Mex., USA

ABSTRACT: The Mantaro river drains the mining, grazing and farming regions of Central Peru. The mining industry in this region grew rapidly from 1900 to the late 1970s as a result of the completion of the railroad to La Oroya and the influx of foreign capital. This growth contributed to an ecological imbalance in the region. In this study, water and sediment samples were collected from a 50 mile stretch from Huancayo to La Oroya. The Mantaro river, Yauli river, irrigation canals and smaller tributaries were sampled in the summer of 1994 and 1998. The water samples were analyzed for 32 elements by ICP. The sediments were analyzed by Aqua Regia digestion followed by ICP. The results indicate a decrease in arsenic and lead concentration and well below the U.S.EPA's primary drinking water standards. On the other hand, the sediment samples show large concentrations of toxic metals.

INTRODUCTION

The mining industry has been one of the important sectors of the Peruvian economy since colonial times until the present. Peru's major mineral products are copper, lead, zinc, iron, silver and gold. Copper, zinc and iron are predominantly used in the production of machinery, consumer and construction materials. Copper has been Peru's principal mineral product throughout most of the twentieth century. Mineral export has been Peru's principal source of foreign exchange, but for the world mining industry Peruvian metal production has been small. Peru's share of the world market for copper, lead have oscillated below 7%. Peru is not a major world producer of any metal, but one of the strengths of the Peruvian mining industry is the extreme diversity of its products that includes industrial and precious minerals. Cerro de Pasco had been a mining center since 1630 with insignificant production, but in 1760 the first drainage tunnel was built to mine deeper mineral veins. In the 1780s fifty mine owners at Cerro de Pasco collaborated to build a major drainage tunnel.

In the 1840s, the guano or bird droppings became a major Peruvian export as a fertilizer which generated a substantial revenue for the state. Part of this revenue was spent to initiate the construction of a railway and roads from Lima into the foothills of the central highlands. In 1893, the Peruvian Corporation extended the railroad to La Oroya, which is located in the heart of the mineral region of the central highlands. This railhead at La Oroya sparked a flurry of mining activity in Central Peru. By the early 1900s there were 17 smelters operating in Cerro de Pasco region alone. In 1898 two North American engineers, Jacob Backus and Howard Johnston submitted a proposal to the Peruvian government to develop the mining region around Cerro de Pasco. In 1901 a New York mining syndicate headed by J.P. Morgan, Henry Clay Frick and Cornelius Vanderbilt purchased approximately 80 percent of the mines in the Cerro de Pasco region within a three months period. The syndicate assumed the name of Cerro de Pasco Mining Company. The main advantages of Cerro de Pasco Mining Company compared to the national companies were access to capital and mining technology in the United States. By 1910, the company had invested 25 million dollars in the purchase of mines and the construction of roads, railroads and a smelter in Central Peru.

The mining industry played an important role in providing jobs and choices in the labor market to peasants in the region. These peasants were able to better themselves and many of them were able to educate their children.. The mine workers developed strong labor unions and were able to bargain for better wages, benefits and job stability. The result was improved standard for some peasants and technical training for others that could not have been possible without this industry. The region was brought into the capitalist system and the regional transportation infrastructure was greatly improved.

BACKGROUND HISTORY

The Mantaro river drains Central Peru and later joins the Amazon River. The river's origin is referred to as the Lago Junin. The river runs generally southeast from Lago Junin to La Oroya and later to Huancayo. The major land use along its headwaters consists of mining and grazing. Within the upper drainage basin there are many existing mine tailings, slag tailings, zinc ponds that are exposed to the atmosphere and these are eroded, leached and washed away by runoff and carried into the river. One of the major slag tailing and zinc pond are located in La Oroya near to the shores of the Mantaro river. La Oroya is a smelting and refining center for the mines in central Peru. In contrast the main land use downstream from La Oroya is mainly agriculture. This area is considered the breadbasket of Central Peru. This valley supplies grains, vegetables and dairy products to the region and Lima. The Mantaro river water is used for drinking, watering animals and irrigation by villages and towns along the river. The use of the river water might be contributing to health hazards in the population along the Mantaro river. At the present time there are no measures to decrease discharge of pollutants into the river from mining, smelting and sewage disposals, but companies are required to monitor their discharge.

In the early 1900s, Cerro de Pasco Corporation invested in the construction of the infrastructure, purchased the concession to continue the rail from La Oroya to Cerro de Pasco and to extend the railway to Huancayo into the heart of the Mantaro Valley. Also purchased nearby coal mines to provide fuel for its smelter and began the construction of a railroad to the coal fields. They constructed two electric power plants and a large smelter. This massive investment revolutionized the mining industry by modernizing the shafts and tunnels by introducing ventilation systems and water pumps to drain the mines, construction of a larger smelter and expansion and innovation of the short and long haul transportation systems which made the production process more rational and efficient. From 1906 to 1921 the volume of output per worker in the mining sector increased dramatically. From 1922 to 1928 Cerro's output averaged 80 percent of the region's mining production. In 1923 the Cerro de Pasco productivity increased by 50%.

The smelter was completed in 1922. Opening the new processing plant contributed to two fold increase in metal production compared to the previous year. But within a year after the inauguration of the processing complex at La Oroya, it became evident that the pollution from the plant was lethal. The smelter and refinery dumped waste into the air and waterways. Crops withered or yields decreased, animals died and people became ill in the valley and highlands within a 30 kilometer radius from the smelter city of La Oroya. The river became devoid of aquatic life due to high contamination. Peasants and hacendados were forced to abandon the land from cultivating or grazing. In the mid 1920s, 30 peasant communities and 28 hacendados filed legal claims against the Cerro de Pasco Corporation for damages. This appears to be the first major environmental lawsuit in the country.

The company settled the lawsuit by acquiring the contaminated land from the peasant communities and hacendados. Afterwards the company installed the first pollution control device Cottrell flue in 1927 to capture lead and zinc particles. In the 1930s and 1940s, the corporation installed additional pollution control devices to capture bismuth, sulfuric acid, finer lead and zinc particles for economic reasons. The installation of pollution control devices contributed to the reduction of pollution in the region. The surrounding lands of approximately half a million acres gradually regained their fertility and became productive grazing zones again. The mining company moved into the meat and wool production on the land they had acquired as a result of the environmental degradation lawsuit. The company's holdings were nationalized by a military government in the early 1970s. But the river continues devoid of healthy aquatic life until now.

FIELD SAMPLING AND LABORATORY METHODS

Sampling of water and sediment from the Mantaro river, Yauli , irrigation canal, zinc pond, slag tailing and a drainage ditch that drains the zinc pond and slag tailing were done on January of 1994. The first sampling program was done upstream from Pilcomayo to La Oroya. At this time only the main river and the left margin irrigation canal at Sanos Chico were sampled. The sampling began north of the city of Huancayo. The water and sediment samples were collected from the shoreline of the river.

The second sampling of water and sediments were done on December 26-28 of 1998. The water and sediment sampling began in Chupuro, which is is located south of Huancayo. The main reason to extend the second sampling program to the south edge of the Mantaro valley was to determine if the municipal and waste water dumped into the river contributed to the increase of toxic metals in the river water and sediment. The Mantaro river, major and minor tributaries, irrigation canals, slag tailing , zinc pond and the ditch that drains the slag tailing and the zinc pond were sampled. A total of 28 sediment samples and 28 water samples were collected from Chupuro to La Oroya. The sampling sites were located using geographic positioning system (GPS).

The collected sediment samples were sieved on site to remove particles larger than 4.75 mm. The sampled particle sizes ranged from clay to coarse sand size particles. In the laboratory the samples were air dried and seived again using U.S. Standard sieves # 10, 20, 40, 100 and 200. The portion passing the sieve # 200 in the clay-silt range and the particle sizes between 0.075 to 0.425 mm (fine sand) were used for analysis. The chemical analysis method was done byAqua regia digestion followed by induced coupled plasma (ICP) determination for 32 elements. The water samples were analyzed by optima induced coupled plasma (ICP) for 32 elements. The focus of this study is the to determine the concentration of lead and arsenic in the river water and sediment.

RESULTS

1 Water

The results of the water analysis are as follows: 1) The most abundant element carried by surface water in Central Peru was calcium followed by zinc and sulfur. Zinc and sulfur are found in higher concentration in the Mantaro. The waters of the tributaries are characterized by their low sulfur and zinc concentration. 2) The concentration of lead and arsenic in water have been drastically reduced compared to the samples taken in 1994 (Figure 1). The 1998 samples show that lead and arsenic concentration in the Mantaro river meet the U.S. EPA drinking water standard. (3) The amount of concentration of sulfur in water appears to be a good indicator to distinguish the source of the sampled water (Figure 2). The drain ditch contains the highest concentration of sulfur, followed by the Yauli river, Mantaro river, canals and the tributaries. On the other hand arsenic is a good indicator for water from the Mantaro river, canal and secondary tributaries. The concentration of lead in the main river, tributaries and canals were below detection levels, except for the drain ditch which contained the highest amount of lead, arsenic and sulfur.

2 River Sediment, slag and zinc deposits

The most abundant element found in the 1994 and 1998 sediment samples was manganese followed by zinc, lead, copper, arsenic and strontium. The concentration of lead and arsenic in the sediment along the river from 1994 to 1998 appears to have remained constant at some sites, accompanied by small increases or decreases at others. The zinc pond shows the highest

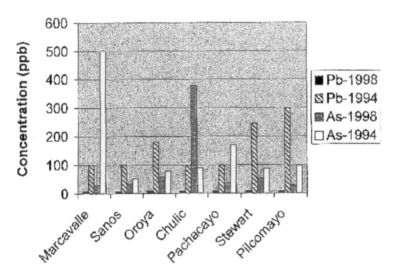

Figure 1. Changes in lead and arsenic concentration in the Mantaro river from 1994 to 1999.

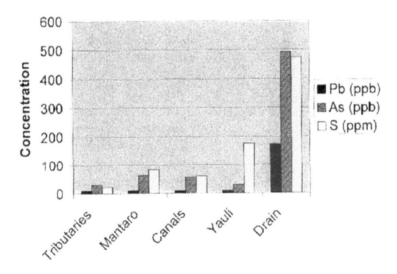

Figure 2. Mean concentration of lead and arsenic (ppb) and sulfur (ppm) in waters of
          Central Peru.

concentration of lead and arsenic followed by the slag tailing. The zinc pond and the slag
tailing and the smelter discharge appears to be the major sources of lead, arsenic, sulfur and
trace elements. The concentration of lead and arsenic in the slag pile is very much the same
compared to the samples were taken from a different location and five years ago.
  Regarding the concentration of lead in sediments of silt and clay size particles appears to be
inversely proportional to the percentage of particle size. In contrast, the concentration of
arsenic in the sediment is not related to the percentage of particle size. In general the average
concentration value of lead and arsenic can be diagnostic of the source of the sediment. The
zinc pond contains the highest amount of lead and arsenic followed by the slag tailing, the
Mantaro river, canals and finally the tributaries. The water in the irrigation canals varied

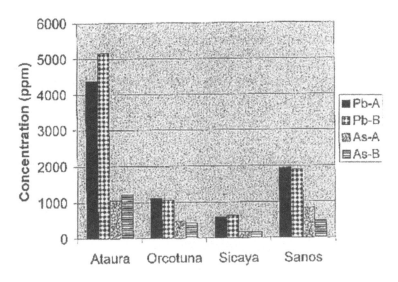

Figure 3. Lead and arsenic concentration in sediment samples from irrigation canals

depending on the location and distance from the point of diversion from the Mantaro river. The canals on the left margin carried water with higher lead and arsenic content compared to the canals on the right margin. Higher concentration of lead and arsenic appears to be related to the distance of point of diversion of the canal from the river (Figure 3).

Lead and arsenic in the tributary streams are below 400 ppm, while the slag tailing contains 6,500 ppm of lead and 3,000 ppm of arsenic. The zinc pond contained 45,000 ppm of lead and about 5,000 ppm of arsenic. In the Mantaro and Yauli rivers lead and arsenic have higher concentration in sand size particles than in silt-clay sized particles. The sediment had a broad range of particle size distribution. Most of the particle sizes ranged from clay-silt sized particles to 8 mm. They were generally fair to poorly graded sediments. The concentration of lead was higher in sediments with low percentage of silt-clay size particles. On the other hand, lead concentration in fine sand particle sizes was higher at higher percentage of fine sand. Arsenic does not follow any trend or relationship with regard to the grain size percentage of silt and clay component.

DISCUSSION

The drainage basin is covered mainly by limestone rock units. According to Turekian (1971) the concentration of lead, arsenic and zinc are 9, 1 and 20 ppm in limestone. The Central Peruvian stream sediments outside of urban areas and undisturbed by mining activities contained lead below 200 ppm. The concentration of arsenic was below 100 ppm. The tributaries that drain urban areas contain higher concentrations of lead, but much less than the Mantaro river or the irrigation canals. Active and inactive mining zones such as the one drained by the Yauli river shows lead and arsenic concentration in thousands of ppm. The limestone was not analyzed for lead and arsenic content. It appears to indicate that the smelter and mining activity contributed to increased presence of lead and arsenic in the river sediments. The rapid weathering of sulfides contributes to high concentrations of dissolved trace elements (Drever, 1988) and the later deposition of the same due to changes in solubility conditions. Additional evidence was offered by the sample above Chulic. The concentration of lead and arsenic in the Mantaro river, upstream from La Oroya drops drastically and compares very closely to the tributaries that are undisturbed by mining activity.

The irrigation canals have lower concentration of lead and arsenic in comparison to the

Mantaro river, but it has a much higher concentration of lead, arsenic and sulfur compared to tributaries undisturbed by mining activity. The lead, arsenic and sulfur are either diluted due to the influx of fresh water and chemical processes along the path of the irrigation canals. This evidence is supported by the canal on the left margin of the Mantaro river. The concentration of lead in Ataura is more than twice than in Sanos Chico which is located farther downstream. Similar pattern is observed for arsenic.

CONCLUSIONS

The contribution of Huancayo and other cities in the region of toxic elements into the river is very small. The smelter, mining activity, slag piles and zinc ponds appear to be the major sources of lead, arsenic and other elements in the water and sediments of the Mantaro river.

The concentration of trace elements and toxic elements decrease downstream along the Mantaro river and the canals.

The tributaries which are undisturbed by mining activities have the lowest average concentration of zinc, lead, arsenic, sulfur and other trace elements.

REFERENCES

Dore, E. 1988. The Peruvian mining industry. Boulder: Westview Press.

Drever, I. J. 1988.The geochemistry of natural waters, 2nd edition. New Jersey, Prentice Hall.

Turekian, K.K. 1971. Elements, geochemical distribution of. Encyclopedia of science and technology, Vol. 4. Mc Graw-Hill, New York pp. 627-630.

Tailings and Mine Waste'00 © *2000 Balkema, Rotterdam, ISBN 90 5809 126 0*

Metals-fixation demonstrations on the Coeur d'Alene River, Idaho – Final results

B.C.Williams
Idaho Water Resources Institute, University of Idaho, Moscow, Idaho, USA

S.L.McGeehan
Analytical Sciences Laboratory, Holm Research Center, University of Idaho, Moscow, Idaho, USA

N.Ceto
EPA Region 10, Seattle, Wash., USA

ABSTRACT: A field demonstration program was conducted to compare *in situ* metals-fixation technologies in fluvially-deposited mine wastes on the Coeur d'Alene River. Groundwater discharge was the targeted metal-loading process to the river. Contaminants of concern were Pb, Zn and Cd. The technologies demonstrated in the field included chemical encapsulation using silica, or fixation by apatite mineralization, Apatite II, or phosphate amendment. Each technology was evaluated according to reduction in metals leachability and changes in soil fertility. The best performer among these technologies depended upon the target metal and the method of assessment (TCLP, SPLP, or field lysimeter).

1 INTRODUCTION

Historic mine waste management practices have resulted in the discharge, or release, of large quantities of tailings and waste rock to surface waters in mining districts across the west. When these materials were transported downstream they mixed with uncontaminated stream sediments. The result is a volume of contaminated sediment far greater than the waste material originally released from the mines. Extensive areas of the riverbed, riverbank and floodplain can be affected. Contamination is often heterogeneous; zones of high metals content are often interbedded with zones of relatively uncontaminated material.

Research over the past decade has lead to a better understanding of the impacts of fluvially-deposited mine waste on groundwater and surface water quality, riparian habitat, ecological receptors, and human health. As environmental impacts are more fully understood there is a new focus on developing strategies for mitigating the affects of mine waste on the environment. Critical to development of cost effective strategies is recognition of the need to manage large volumes of material. This project grew from the idea that *in situ* treatment technologies may offer an opportunity to achieve environmental benefits at a reasonable cost, while also minimizing disruption of the riparian environment.

In April 1998, the U.S. Environmental Protection Agency (EPA) and the Idaho Water Resources Research Institute (IWRRI) invited vendors and investigators to participate in a field demonstration of metals-fixation and metals-removal technologies in sediments mixed with tailings along the Lower Coeur d'Alene River, Idaho.

1.1 *General Discussion of Technologies*

For the purposes of this discussion, *ex situ* methods for remediating metals-contaminated soils will be defined as those involving excavation, treatment and disposal at another location. *In situ* methods will be defined as those in which the contaminants are cleaned or stabilized at their present location. Methods including reworking of the soil for the purpose of combining reagents/amendments are considered *in situ* for the purpose of this study, because the soil material is not transported to another location.

Several general categories of technologies exist (or are emerging) for cleaning or permanently stabilizing soils contaminated with metals, including: chemical stabilization with amendments/reagents, soil washing, electrokinetics, and phytoremediation. An expanded discussion of these technologies can be found in McGeehan and Williams (1999).

Chemical stablization with amendments/reagents has been explored to immobilize or attenuate metals in the soil matrix by mechanisms such as binding (as with Portland cement), precipitation, sorption, and complexation. The stabilizing reagent can be incorporated by excavation and reworking, tilling, augering, or injection grouting depending upon the depth of the target layer. The U.S. EPA (1993) reports binder application costs to be $38/yd^3 for in-place mixing, and $100-200/ton for augering. Specific chemical stabilization techniques include: neutralization (with lime, dolomite, or fly ash), silicate stabilization, phosphate stabilization, and iron stabilization (SME, 1998). All of the demonstrators described in this paper used chemical stabilization with amendments/reagents.

Sodium silicate stabilization provides similar cementing properties as lime or cement, with the dominant mechanisms being cementing and porosity reduction. Associated increased pH may also precipitate many metals as hydroxides or carbonate salts, which are relatively insoluble (SME, 1998). Phosphate stabilization using agricultural phosphate fertilizers, phosphoric acid, polyphosphates, and apatite (phosphate ore) yields relatively stable phosphate salts of such metals as lead, aluminum, iron and manganese (SME, 1998). The SME (1998) review team and other waste management experts have advised that (a) bioavailability of lead phosphates needs to be a near-term research focus, and (b) it must be demonstrated that phosphates do not mobilize other anionic metals, such as arsenic or selenium. Chen et al. (1997) demonstrate the attenuation of leachable Pb, Zn, and Cd from Coeur d'Alene soils using mineral apatite. The apatite was able to reduce the metal concentrations in the TCLP-extractable soil leachates to below US EPA maximum allowable levels.

Soil washing consists of a procedure whereby a washing solution is applied (usually to excavated soils) to extract contaminants. The wash solution may be water, organic solvents, chelating agents (Abumaizar and Khan, 1996; Peters and Miller, 1993), surfactants, acids or bases. Hydrometallurgical or biohydrometallurgical processes may be employed. Soil washing appears to be most successful in sandy soils, and less so in soils with clay.

Electrokinetic remediation is performed by applying a low-level direct current to the soil with electrodes, causing electromigration of ionic species and electroosmosis. Electrokinetics must be performed in the saturated zone.

Phytoremediation refers to the use of plants for the remediation of soils and water contaminated with salts, metals, and organic compounds. Potential limitations of phtoremediation include the need for many growing seasons, the annual costs for harvesting and disposing of plants, and potential requirements to control wildlife foraging (SME, 1998). One of the field demonstrators in this project utilized apatite and phosphate amendment combined with revegetation using, redtop grass (*Agrostis alba*), but the efficacy of the redtop grass in the treatment zone could not be evaluated for this study because the roots had not yet reached the treatment zone.

1.2 *Physical Setting*

The Coeur d'Alene River flows from the confluence of the North and South Forks of the Coeur d'Alene River, westward to its discharge to Coeur d'Alene Lake. Below the confluence of the North and South Forks, the gradient flattens significantly, which has caused deposition of contaminated sediments. The Silver Valley, located on the South Fork, is one of the largest producers of silver, lead, and zinc in the world. Prior to the 1920's, mine and ore processing wastes were discharged directly into the South Fork of the Coeur d'Alene River and its tributaries. Consequently, the Lower Coeur d'Alene River received tens of millions of metric tons of mine wastes that are now intermixed with sediments and deposited in the riverbed, riverbank, and floodplain.

The Coeur d'Alene River has designated uses (IDAPA 16.01.02110,01.ee) of agricultural water supply, cold water biota, primary and secondary contact recreation with salmonid spawning protected for the future (IDEQ, 1998). The constituents lead, zinc and cadmium are target metals in the Idaho Division of Environmental Quality assessment of Coeur d'Alene River for

Total Maximum Daily Loads (TMDL's). As described by IDEQ (1998), metals loading to the Coeur d'Alene River can be attributed to three main mechanisms:
- Accelerated erosion due to long-term changes in river energy.
- Accelerated erosion due to water level management and increased boat traffic.
- Discharge of metals-contaminated groundwater out of the riverbank.

The demonstration projects solicited for this program were focused on the last of these 3 mechanisms – metal loading associated with seeps and discharge of contaminated groundwater.

2 MATERIALS AND METHODS

Invitations to participate in this study were mailed and hand-disseminated to over 200 individuals identified in the literature, the Internet and at Coeur D'Alene Basin-related meetings. A workplan was prepared in the form of a Request for Proposal (RFP) to be completed by potential demonstrators. Respondents to the RFP were asked to provide information including description of process, the metals fixed or stabilized, the long-term stability of the treated phase, whether the process requires excavation, a letter of reference if the process was proprietary, cost per area or volume, and the largest scale at which the technology had previously been applied. Respondents were also required to inform IWRRI of all chemicals used so that they could be checked for risk to the environment, and in the case of pytoremediation technologies, plant lists so that county agents could review them in terms of weed propagation.

2.1 *Measures of Performance*

The performance standards against which the field demonstrations were evaluated fell into two categories, those to be assessed by IWRRI and those to be reported by the demonstrators.

2.1.1 *IWWRI Review*

The quantitative measures of performance for all phases of the comparative studies (bench test, pre-treatment baseline characterization, and post-treatment performance assessment) were directed toward assessing the leachability of metals into solution. Depending upon the phase of the study, one or more of the following procedures were used: 1) the Toxicity Characteristic Leaching Procedure (TCLP, EPA Method 1311), 2) the Synthetic Precipitation Leaching Procedure (SPLP, EPA Method 1312), and 3) *in-situ* testing of pore waters using suction lysimeters. Comparison of the pre- and post-treatment TCLP- and SPLP-leachable metals concentration constituted one measure of performance. In order to evaluate the number of samples required to sufficiently address spatial variability, a modified TCLP[1] protocol was used in conjunction with the standard TCLP test, and a modified-SPLP protocol was developed according to the same criteria. Results were reported with respect to the pre-treatment range of metals found in each plot. Fertility and pH were evaluated as a separate matter, by measuring pH, total nitrogen, phosphorus, potassium and organic matter.

The TCLP test was selected for the bench test phase and follow-up work as it provides a standardized assessment of metals leachability, and was familiar to those vendors who have applied their technologies in landfill (RCRA) settings. The SPLP test was used because, as a pH-neutral leaching solution, it is more suitable to the issue of leaching at Mission Flats. Also, the SPLP test is regularly used by regulators in the mining industry to assess the predicted leachability of mine waste materials during permitting. When it was determined that project funding would permit, *in situ* lysimeter testing was added to the post-treatment suite of analyses.

---

[1] The modified TCLP extracts were obtained using EPA procdure 1311 with the following exceptions: a 5 g/100mL soil:solution ratio was used and the extracts were analyzed for As, Cd, Pb, and Zn. The modified TCLP tests were performed by the UI Soil Science Division Laboratory.

2.1.2 Demonstrator Report

Demonstrators were required to provide a final report that summarized the following information (a) discussion of any potential long-term degradation products from chemicals or biological components applied during treatment (beneficial, neutral, or potentially harmful), (b) cost per volume or per area on a test plot basis as well as projections for cost scaling for larger projects, and (c) operation and maintenance requirements (see McGeehan and Williams, 1999).

2.2 Bench Tests

IWRRI prepared composite samples for demonstrators so that each participant could develop and refine site-specific metals-fixation strategies prior to the field phase of the project. Subsamples, collected at 9 random locations across the demonstration site, were thoroughly mixed to form the composite. Samples were collected from the 0-12" (Composite #1) and 12-24" depth (Composite #2). Composite samples were assessed for Cd, Pb, and Zn using the standard TCLP procedure prior to and after bench treatment by each investigator.

2.3 Field Plot Design and Characterization

The demonstration project was located at Mission Flats on land owned by the Mine Owners' Association between Cataldo and Rose Lake, Idaho. In addition to hosting fluvially-deposited materials, this land also received dredge spoils removed from the river by the Mine Owners' Association from 1929 through the 1960's. The study area is described in detail in McGeehan and Williams (1999).

Six 15 X 15 m plots were installed at the demonstration site during March 1999. Because spatial variability of metals in the fluvially-deposited sediments was expected at many scales, IWRRI document pre-treatment spatial variability for each field plot. Documentation of the variability allowed demonstrators to plan for the full range of metals concentrations and provided plot-specific baselines for post-treatment performance assessment of each technology. Samples were collected at 9 points within each 225 m^2 plot. Each point was sampled to a depth of 24 inches in 6 inch increments (providing 24 to 36 samples per plot). Each sample was analyzed separately for modified TCLP-extractable Cd, Pb, and Zn.

2.4 Technologies

Five demonstrators (Table 1) participated in the bench test phase of the project while only three of the original five completed the field tests. Technologies employed during the field tests fell into two general categories (a) excavation, mixing of reagents for chemical stabilization followed by re-emplacement; and (b) chemical stabilization coupled with phytoremediation. The former approach was employed by the KEECO/Metcalf & Eddy team which utilized a silica microencapsulation reagent and the ICF Kaiser/Forrester Environmental team which employed apatite mineralization. Both of the above technologies were directed at the treatment zone located at a depth of 12 to 24 in. (as discussed in the Spatial Variability section). In each case, the surface 12 in. of the plots was excavated and stockpiled, while the 12-24 in. depth was treated *in-situ*. KEECO/Metcalf & Eddy added their stabilization reagent as a dry powder followed by mechanical mixing with a rototiller. ICF Kaiser/Forrester Environmental applied their reagent as a liquid, accompanied by mixing with a backhoe, to initiate apatite mineralization. Following reagent addition, the stockpiled topsoil was replaced and the treatment area flagged to aid post-treatment sampling.

The Frutchey/UFA team split their plot to compare phosphate fertilizer, applied at agronomic rates, to an Apatite II treatment. Both the phosphate fertilizer and the Apatite II were applied to the surface 12 in. and incorporated using conventional tillage equipment. In both cases, the Frutchey/UFA plots were revegetated with redtop grass (*Agrostis alba*) which has proven resistant to metal toxicity in the Coeur d'Alene River valley.

300

Table 1. List of demonstrators participating in the bench and/or field tests.

BENCH AND FIELD TESTS

Demonstrator(s)	Description of Technology
KEECO, Inc., Lynwood, WA with, Metcalf & Eddy, Santa Barbara, CA	Silica Encapsulation (using silica)
ICF Kaiser, Seattle, WA with, Forrester Environmental Services, Inc. Hampton, NH	Apatite Mineralization (metals fixation by nucleation and coprecipitation during *in-situ* apatite crystallization)
Frutchey, Cataldo, ID with, UFA Ventures, Inc., Richland, WA	Phosphate or Apatite II Amendment followed by revegetation with redtop grass

BENCH TEST ONLY

Demonstrator(s)	Description of Technology
Abdul Majid Inst. for Chemical Processes and Environmental Technology, Ottawa, CAN	Fly Ash amendment
Loren Paripovich ECORx, McCall, ID	metals stabilization using unspecified reagents

3 RESULTS AND DISCUSSION

3.1 *Bench Tests*

Composite Sample #1 exhibited low metals concentrations (TCLP-leachable) in the untreated sample reflecting low contamination in the surface 12 inches (Table 2). Each demonstrator achieved a reduction in TCLP-extractable Cd and Pb relative to the untreated sample. Variable results were obtained for Zn. Arsenic data are not reported because all concentrations were below instrument detection limits.

Initial metal concentrations in Composite Sample #2, collected from the 12 to 24 in. depth, were much higher with Pb exceeding the TCLP regulatory level in the untreated sample. Each participant was again able to decrease the extractable metal concentrations with the most significant changes observed for extractable Pb and Zn. In particular, final Pb values were below the TCLP regulatory level in three of the five remediation processes.

Table 2. TCLP-extractable metals and pH values for bench-tested composite samples.

	Composite Sample #1 (0-12 in.)				Composite Sample #2 (12-24")			
	TCLP-Extract. (mg/L)				TCLP-Extract. (mg/L)			
	Cd	Pb	Zn	pH	Cd	Pb	Zn	pH
Untreated	0.14	0.74	6.9	6.8	0.33	43.8	51.8	6.8
TCLP levels	1.0	5.0	NA	NA	1.0	5.0	NA	NA
Silica Encapsulation	0.13	0.53	3.7	9.8	0.27	13.6	22.0	11.8
Apatite Mineralization	NA	NA	NA	NA	0.17	0.3	24.8	7.5
Apatite II Amendment	0.11	0.49	7.4	7.0	0.19	1.6	21.9	7.0
Fly Ash Amendment	0.19	<0.50	9.4	11.6	0.30	15.0	25.0	11.0
Unspecified Reagents	<0.02	<0.50	<0.02	12.6	0.09	3.6	7.3	9.0

Soil pH did not change significantly for either of the apatite-based technologies (Table 2). However, each of the remaining methods exhibited substantial pH increases with the chemical

301

encapsulation method changing pH from 6.8 to 11.8. None of the technologies adversely influenced other soil fertility parameters such as macronutrient availability and organic matter content (not shown here, see McGeehan and Williams (1999).

3.2 Spatial Variability of Metals in Field Plots

Significant variability in TCLP-extractable Pb and Zn was observed both within and among the plots, and is presented in Williams et al. (1999). These data clearly show the need to collect detailed spatial variability information on a small scale prior to developing metals remediation strategies for sites subject to fluvial deposition.

3.3 Performance Assessment of Field Plots

Each field plot was sampled in April 1998, prior to field treatment, to establish pre-treatment SPLP and TCLP concentrations on a plot by plot basis. Field treatments were applied during July and August, 1998. Post-treatment sampling took place approximately 1 month (9/98) and 8 months (4/99) following field treatment. Soil samples were collected from the 0-12 in. depth for the phosphate and Apatite II treatments (Plots 1 and 2, respectively) and from the 12-24 in. depth for the silica encapsulation and apatite mineralization treatments (Plot 6).

The silica encapsulation field treatment, resulted in lower SPLP-Pb and TCLP-Pb concentrations as compared with the pre-treatment (control) samples (Fig. 1). SPLP-Pb concentrations decreased from 0.32 mg/L in the pre-treatment samples to 0.05 mg/L in the 9/98 post-treatment samples. TCLP-Pb values decreased from 31 mg/L to 17 mg/L in the pre- and post-treatment samples, respectively. SPLP-extractable Zn concentrations did not change significantly following the silica amendment while TCLP-Zn values decreased from 34

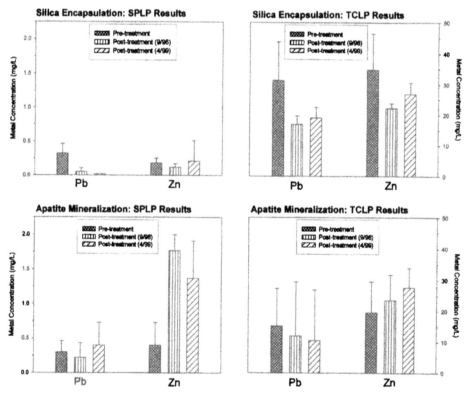

Figure 1. Leachable Pb and Zn concentrations in soils treated by the Silica Encapsulation and Apatite Mineralization technologies, as assessed by SPLP and TCLP procedures.

to 22 mg/L. Comparison of samples collected one month after (9/98) field treatment to those collected eight months after (4/99) the silica encapsulation treatment showed a slight tendency toward increased solubility of both Pb and Zn over time (Fig. 1).

The apatite mineralization treatment, did not significantly influence SPLP-Pb values (Fig. 1). In contrast, TCLP-Pb decreased from 15 to 12 mg/L one month after field treatment and declined further to 10 mg/L eight months after treatment. The apatite mineralization treatment exhibited a marked tendency to increase extractable Zn concentrations. SPLP-Zn increased from 0.40 to 1.77 mg/L while TCLP-Zn increased from 20 to 23 mg/L in the pre- vs. post-treated samples, respectively. Comparison of samples collected one month vs. eight months after field treatment showed a tendency toward increased TCLP-Zn concentrations but a decrease in SPLP-Zn concentrations over time.

The trend in SPLP-extractable metal concentrations were similar for the phosphate- and Apatite II-amended plots (Fig. 2). SPLP-Pb values were not significantly affected by phosphate or Apatite II treatment while SPLP-Zn concentrations were markedly increased. In each case, SPLP-Zn concentrations increased from 0.15 to 1.0 mg/L in the pre- vs. post-treatment samples, respectively. Additional increases in SPLP-Zn were observed between 9/98 and 4/99 (Fig. 2). TCLP-Pb concentrations were relatively low (<5 mg/L) in the pre-treatment samples and were decreased to 3.8 and 1.1 mg/L by the phosphate and Apatite II treatments, respectively. The phosphate treatment was successful in decreasing TCLP-Zn concentrations from 21 mg/L in the pre-treatment samples to 15 mg/L in the 9/98 post-treatment samples. The Apatite II treatment decreased TCLP-Zn from a pre-treatment value of 16 mg/L to a post-treatment value of 9 mg/L. The Apatite II treated plots showed a slight tendency toward increased Zn solubility between the 9/98 and 4/99 samples but no change was observed over the same time period for the phosphate treatment.

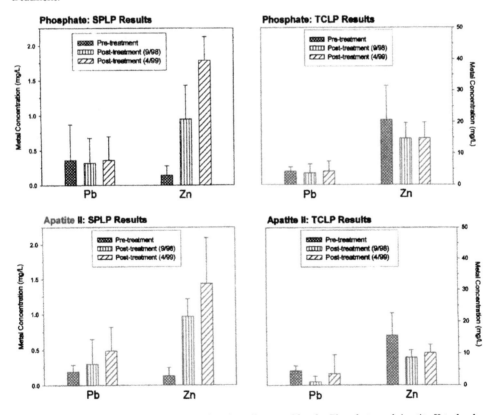

Figure 2. Leachable Pb and Zn concentrations in soils treated by the Phosphate and Apatite II technologies, as assessed by SPLP and TCLP procedures.

Suction lysimeters were used to monitor pore water metal concentrations in each demonstration plot from 4/2 through 5/21, 1999. The ceramic sampling cup of each lysimeter was installed to a depth corresponding to the middle of the treatment zone for each plot - 18" for the silica encapsulation and apatite mineralization treatments and 6" for the phosphate and Apatite II treatments. For this reason, different groups of lysimeters were used as controls for the respective treatment zones. It should also be noted that the control lysimeters were installed immediately adjacent to the plots (within 3 m) in soil that was not disturbed during the excavation portion of the field treatment.

Pore water Pb values were generally low (<0.06 mg/L) in the control lysimeters throughout the sampling period (Fig. 3). Pb concentrations in the silica encapsulation lysimeters were consistently above the controls for all but the final (5/21) sampling while Pb concentrations were below control values in the apatite mineralation treatments for all sampling dates. Zinc concentrations were one to two orders of magnitude higher than Pb values in 18" depth lysimeters. Zinc concentrations were lower than the control for all sampling dates for both the silica encapsulation and apatite mineralization plots. The silica encapsulation treatments exhibited a greater ability to maintain low pore water Zn concentrations when compared to the apatite treatment. Overall, Zn concentrations in the silica encapsulation plots were at least one order of magnitude lower than the apatite mineralization plots.

Control lysimeters for the phosphate and Apatite II plots exhibited much lower Pb and Zn concentrations than the control lysimeters of the silica encapsulation and apatite mineralization plots. This is consistent with the metals trends found in the spatial variability study. Both Pb and Zn concentrations were less than 0.05 mg/L throughout the sampling period for control lysimeters on Plot 6. Note that, due to the shallow location (6") of these lysimeters, some were dry on various sampling dates ('NS' in the figures). Pore water Pb values were less than or equal to the controls in both the phosphate and Apatite II amended plots (Fig. 3). In contrast, Zn values were 2 orders of magnitude higher than control values in the phosphate treated plots, whereas the Apatite II treatments exhibited Zn values at or below the control values.

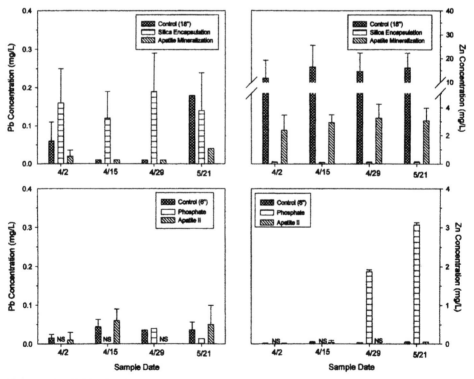

Figure 3. Leachable Pb and Zn concentrations in samples taken from *in situ* suction lysimeters for all four technologies.

4 SUMMARY AND CONCLUSIONS

4.1 *Bench Test Results*

- Pb and Zn were the primary metals of concern at the demonstration site; As and Cd were analyzed but were usually present at concentrations below instrument detection limits.
- Each of the five technologies decreased the TCLP-extractable metal concentrations, relative to the control. The most significant decreases were achieved for Pb with the apatite mineralization and Apatite II technologies.
- None of the technologies examine during the bench test phase appeared to adversely affect soil fertility (N, P, K, organic matter). However, the silica encapsulation, fly ash amendment, and ECORx technologies (using unspecified reagents) produced final soil pH values ranging from 9.0 to 11.8.

4.2 *Field Test Results*

- Significant variability in Pb and Zn was observed within and among the untreated demonstration plots. Both Pb and Zn increased with depth in Plots 1 and 2 reaching maximum concentrations at the 18-24 in. depth. Plot 6 exhibits lower metals concentrations overall and Zn concentrations were greatest near the surface (0-12 in. depth).
- The silica encapsulation technology decreased both SPLP- and TCLP-extractable Pb and Zn concentrations relative to the control. The most signficant decrease occured in Pb concentrations where an 84% and and 45% decrease was observed in the SPLP and TCLP samples, respectively.
- The apatite mineralization technology did not significantly influence SPLP-Pb concentrations but decreased TCLP-Pb by 66%. A significant increase in both SPLP- and TCLP-Zn concentrations was observed in the apatite mineralization plots relative to the control.
- The phosphate/revegetation technology did not change SPLP-Pb values but decreased TCLP-Pb by 24%. Variable results were obtained for Zn with SPLP values increasing and TCLP values decreasing, relative to the control.
- The Apatite II/revegetation technology also did not influence SPLP-Pb concentrations and lowered TCLP-Pb, in this case by 78%. The Apatite II amendment increased SPLP-Zn concentrations and decreased TCLP-Zn.
- Pb and Zn concentrations obtained from the lysimeter study were of the same magnitude as the SPLP concentrations. However, the trends observed with SPLP-Pb and SPLP-Zn in the pre- vs. post-treatment samples were not consistent with comparison of control vs. post-treatment lysimeters.
- None of the field technologies had an adverse effect of soil fertility parameters (N, P, K, organic matter). However, pH values in the silica encapsulation plot were significantly higher (average pH = 11.7) than the untreated soil (average pH = 6.8). In contrast, pH values from the apatite mineralization plot were significantly lower (average pH = 6.1) than the untreated average.

A critical precaution when considering the final results is that these demonstrations were performed in the vadose zone only, and may not be reproducible within the saturated zone or the zone of fluctuation. Also, the results may have been influenced by the mixing procedure, as all three vendors used different techniques. Finally, despite our efforts to assess site variability, and plot-specific baselines, heterogeneity remains as a confounding variable when comparing techniques. Given these precautions, the relative performance of the vendors for each performance measurement (TCLP, SPLP, or lysimeter) are presented in Table 3.

It is recommended that when considering techniques to be employed at a specific location, land managers should select the target metals and the appropriate method of assessment, and then work with the vendor to accurately assess the site hydrologic setting, grain size distribution, and target horizon for optimization of the application procedure.

Table 3. Relative best performers according to each method of assessment.

Metal	TCLP	SPLP	Lysimeter
Lead	Silica encap. / Apatite Min. / Apatite II*	Silica encap.	Apatite Min.
Zinc	Apatite II / Phosph. / Silica encap.*	Silica encap.	Silica encap. / ApatiteMin.

* Silica encapsulation was superior in terms of percentage reduction, and the others in terms of absolute reduction.

5 DISCLAIMER

The information in this document has been funded in part by the U.S. Environmental Protection Agency under cooperative agreement number X-990964. Mention of certain technologies, trade names, or commercial products does not constitute endorsement or recommendation for use.

6 ACKNOWLEDGEMENTS

Financial support for this project was provided by the Environmental Protection Agency. Input received from Earl Liverman of EPA, Region 10, was greatly appreciated. Special appreciation is extended to the demonstrators, who provided their time and technology, and helped us to refine the program design. The work of Robin Nimmer, Karen Zelch, Roy Mink, Dan Kealy, Kathy Canfield-Davis and Peggy Hammel of IWRRI was invaluable. In addition, we thank the Mine Owners' Association for providing the land required for the demonstration projects.

REFERENCES

Abumaizar, R. and Khan, L.I. 1993. *Laboratory Investigation of Heavy Metal Removal by Soil Washing.* J. Air and Waste Manage. Assoc. 46:765-768.

Chen, X., Wright, J.V., Conca, J.L., and Peurrung, L.M. 1997. *Evaluation of Heavy Metal Remediation Using Mineral Apatite.* Water, Air & Soil Poll. 98:57-78.

IDEQ, 1998. Coeur d'Alene River water quality assessment and total maximum daily toad to address trace (heavy) metals criteria exceedences. Idaho Division of Environmental Quality, Northern Idaho Regional Office, Coeur d'Alene, ID.

McGeehan, S.L., and Williams, B.C. 1999. *Demonstration of metals-fixation technologies on the Coeur d'Alene River, Idaho – Final Report.* Idaho Water Resources Research Institute, University of Idaho, Moscow, ID.

Peters, R.W. and Miller, G. 1993. "Remediation of Heavy Metal Contaminated Soil Using Chelant Extraction: Feasibility Studies." In *48th Purdue Industrial Waste Conference Proceedings*, Lewis Publishers, 141-167.

SCS, 1994. *Coeur d'Alene River cooperative river basin study.* U.S. Department of Agriculture – Soil Conservation Service, Boise, ID.

SME (Society of Mining Engineers). 1998. Remediation of Historical Mine Sites – Technical Summaries and Bibliography. SME, Littleton, CO.

U.S. EPA. 1993. *Considerations in Deciding to Treat Contaminated Unsaturated Soils In Situ. Issue Paper.* EPA/540/S-94/500. Office of Solid Waste and Emergency Response, Office of Research and Development. Washington, DC.

Tailings and Mine Waste'00 © *2000 Balkema, Rotterdam, ISBN 90 5809 126 0*

Laboratory simulation of HCN emissions from tailings ponds

Andreas Rubo, Annette Dickmann & Stephen Gos
Degussa-Huels AG, Hanau, Germany

ABSTRACT: The volatilization of hydrocyanic acid from cyanide-containing mine wastes was investigated on a laboratory scale using tailings obtained from leaching an oxide, a transition and a sulfide ore. The main intention of this project was to quantify and assess the potential cyanide releases to the gas phase from tailings pond operations. It was found that, the major amount of total cyanide in solution slowly volatilizes as hydrocyanic acid. HCN releases were the highest with the oxide ore due to the presence of high amounts of free cyanide. However, the HCN concentration above the pulp typically never exceeded 1 ppm. The formation of metal cyanide complexes during leaching of the transition and sulfide ores resulted in reduced volatilization. In the case of the sulfide ore, significant amounts of the complexed cyanide formed OCN^- and SCN^-. HCN volatilization is discussed in more detail regarding aspects of risk, safety and regulations.

1 INTRODUCTION

The fate of cyanide in tailings ponds has recently become an issue of increasing importance, because, although there is a lot of information, very little is of a sound quantitative nature to enable a solid basis for meaningful legislation. Up until recently, reactions in liquid phase effluents were the focus of interest. However, discussions about the introduction of emission regulations, due to uncertainty in this area, has led to gold mines and other cyanide users investigating the losses of cyanide as hydrocyanic acid to the air. Proposed legislation (e.g. TRI) envisages limiting the quantity and concentration of HCN released from gold mining operations.

Some literature has touched on this subject of hydrocyanic acid volatilization from tailings ponds, in particular Smith and Mudder (1991), Griffiths et al (1985), Simovic et al (1987), Rubo et al (1999) and Logsdon et al (1999). However, there was only little experimental foundation to most of these references. These references recognize the fact that HCN volatilizes from tailings ponds, but no sufficient experimental data is presented to be able to quantify the HCN releases.

In large-scale operations, steady state conditions prevail, because mine tailings are continuously discharged into the tailings pond. Although these experiments were performed on a batch scale, they can be considered to give an indication of how a tailings pond behaves.

In addition, it is very difficult to reliably measure and quantify the amount of hydrocyanic acid volatilization from a large-scale tailings pond operation. The HCN concentrations above the surface of the tailings ponds typically do not exceed the detection limit of currently practically applicable measurement devices. This makes the determination of a mass balance very difficult.

The volatilization of hydrocyanic acid from tailings ponds is often explained and described as being natural degradation in the gold mining community. However, there is enough scientific evidence in the literature mentioned above as well as based on these experiments to demonstrate that cyanide is released as HCN.

The formation of hydrocyanic acid from the cyanide ion is a chemical equilibrium reaction and basically dependent on the pH of the solution. It can be seen in Figure 1, that, at a pH where a tailings pond is normally run (pH 8 - 9), approx. 60 – 90 % of the free cyanide is transferred to the hydrocyanic acid side of the equilibrium. Hydrocyanic acid is soluble in water in all proportions, but can easily volatilize as HCN gas because of its high vapor pressure (0.83 bar at 20 °C).

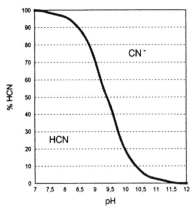

Figure 1: Hydrocyanic Acid / Cyanide Ion (HCN/CN⁻) Equilibrium Curve

This equilibrium is according to the reaction:

$$HCN + OH^- \rightarrow CN^- + H_2O$$
$$pK_a = 9.36$$

Metal cyanide complexes are also in equilibrium with copper like copper cyanides:

$$Cu(CN)_2^- \rightarrow Cu^+ + 2CN^-$$
$$pK_2 = 21$$

$$Cu(CN)_3^{2-} \rightarrow Cu(CN)_2^- + CN^-$$
$$pK_3 = 4.5$$

$$Cu(CN)_4^{3-} \rightarrow Cu(CN)_3^{2-} + CN^-$$
$$pK_4 = 2.15$$

For our discussions, it is important to mention at this stage, that based on this curve, at a pH of 9.5, nearly the same percentage of hydrocyanic acid is formed as free cyanide. Hydrocyanic acid can easily volatilize to the air and, therefore, can contribute to risks in terms of safety and the environment. This pH is not far away from the conditions under which a tailings pond is run in practice because of its high acute toxicity. The lower the pH the higher the formation of hydrocyanic acid.

This testwork endeavored to quantify just how much hydrocyanic acid is actually volatilized from the three ore types with significantly different mineralogies.

2. MATERIALS AND METHODS

2.1 Apparatus

The tests were performed in airtight glass vessels (2 l, 5 l and 6 l in volume). The covers of the vessels were fitted with a stirrer, pH electrode, aeration inlet and outlet and an acid/alkali dosing pipe.

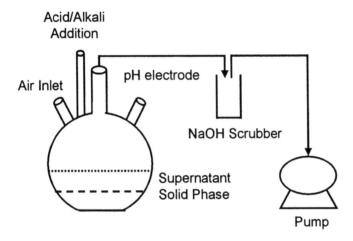

2.2 Conditions of Testwork

To evaluate the influence of the ore type on the cyanide cycle in a tailings pond, all tests were performed under the same conditions on the same oxide, transition and sulfide ore.

40 % ore, by wt.
60 % water, by wt.
pH: 10.5, kept constant by NaOH addition.
After the 24 hour leach, the pH was set to pond conditions (pH 8 - 9).

2.3 Performance of Testwork

The leach testwork was performed to generate and simulate the pulp tailings.

24 h leach:

Aeration: Air (20 l/h) was bubbled through pulp and then through sodium hydroxide solution (c = 1 mole/l). The sodium hydroxide solution was analyzed for cyanide.

After the 24 hour leach, a pulp sample was taken, filtered and the following analytical methods were employed:

Table 1: Overview of Analytical Methods Employed

Analysis	Method
HCN in NaOH	$AgNO_3$-Titration
CN_{TOTAL}	ISO Method
CN_{WAD}	Picric acid
CN_{FREE}	$AgNO_3$-Titration
SCN⁻	HPLC
OCN⁻	HPLC
HCN	Compur HCN Detector

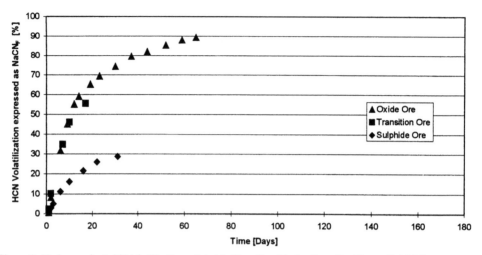

Figure 2: Hydrocyanic Acid Volatilization related to Free Cyanide for three Ore Types (Initial free cyanide concentration in tailings in ppm as NaCN : oxide ore 163 ppm, transition ore 96 ppm, sulfide ore 109 ppm)

After completion of the leach, the pH was decreased by sulfuric acid addition. After a pH of 9 was achieved, the stirring was stopped. During the pond simulation, the air was only drawn over the surface and through a sodium hydroxide solution. The sodium hydroxide solution was frequently analyzed for cyanide.

3. RESULTS AND DISCUSSION

3.1 Hydrocyanic Acid Volatilization from Free Cyanide

It can be observed in Figure 2, that approximately 90 % of the free cyanide volatilizes as hydrocyanic acid in the case of the oxide ore. In the case of the sulfide ore, this is only approx. 30 %. The transition ore is in between.

3.2 Hydrocyanic Acid Volatilization from Complex Cyanides

Even if free cyanide can no longer be measured, hydrocyanic acid volatilization can still be observed. This observation underlines the fact, that heavy metal cyanide complexes also form hydrocyanic acid from free cyanide in equilibrium. This is, however, at a much lower rate.

Therefore, for a successful waste management, the following aspects need to be kept very clear. Firstly, free cyanide is normally present in the pulp tailings after leaching. This can be measured and it will largely volatilize. After a while (in the case of batch state conditions) there is another source of hydrocyanic acid. This is the mixture of complex heavy metal cyanides. Due to the stability of these complexes, the concentration of free cyanide is below the analytical detection limit of approximately 0.2 ppm. However, even at these low free cyanide concentrations, hydrocyanic acid can still be formed and volatilize. This is illustrated in Figure 3.

The brackets in Figure 3 indicate HCN formation only from complex cyanides, because no free cyanide could be detected any more in the solution for these experimental points.

In the case of the oxide ore, there is still a discrepancy to make up 100 % of cyanide in the mass balance. It is assumed that this is due to cyanate (OCN^-) formation. Looking at the sulfide ore, the same phenomenon should be expected, but to a different extent. Indeed, cyanate could be analyzed over the whole period of investigation in the case of the sulfide ore. Since

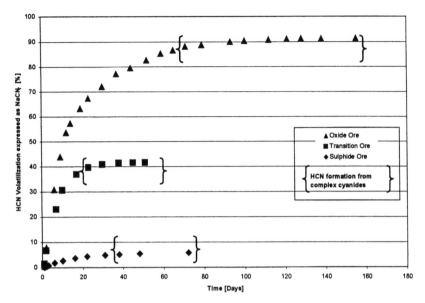

Figure 3: Hydrocyanic Acid Volatilization related to Total Cyanide for Different Ore Types (Initial total cyanide concentration in tailings in ppm as NaCN; oxide ore 200 ppm, transition ore 200 ppm, sulfide ore 650 ppm)

the pH was low enough to cause a rapid hydrolysis of the cyanate, it can be assumed that this is also a very important mechanism for the reduction of the cyanide concentration observed.

3.3 Discussion of the Results

Hydrocyanic acid volatilization was observed in the case of all ores, even after it was no longer possible to measure any free cyanide in solution. Therefore, it can be assumed that hydrocyanic acid is also formed from the heavy metal complexes.

There are numerous cyanide reactions possible, some can run simultaneously. Firstly, there is the formation of different cyanide complexes with the ions of heavy metals. In these complexes, the cyanide is in the form of a ligand and is not chemically changed. These reactions, like the formation of hydrocyanic acid, are not defined as degradation. A degradation of cyanide is obtained by oxidation, where the cyanate then forms ammonium compounds and carbonate.

The mineralogy of the ore mainly influences the reactions of the cyanide. If the ore only contains low metal and sulfide concentrations, the cyanide exists mainly as free cyanide and, therefore, can largely volatilize as hydrocyanic acid, depending on the pH.

3.3.1 Volatilization rate

In the simulated tailings ponds, the cyanide in solution is eventually completely removed without the use of any chemical treatment.

Based on these experiments, the time frame for the removal is as follows:
1. Free cyanide   up to 9 weeks
2. Weak acid dissociable cyanide   up to 6 months
3. Total cyanide   approx. 1 year

The free cyanide volatilizes as hydrocyanic acid. This is especially true in the case of the oxide ore. If heavy metals or sulfides are present, the rate of hydrocyanic acid volatilization is reduced.

The products of the cyanide reactions depend on the chemical composition of the ore. Oxide and sulfide ores generate different products of the cyanide reactions due to the different mineralogies.

3.3.2 Behavior of the Oxide Ore

The oxide ore did not generate cyanate or thiocyanate to the same extent as the transition or the sulfide ore. There was also no formation of weak-acid dissociable cyanide that could not be accounted for as free cyanide. The main reaction product was hydrocyanic acid. More than 90 % of the initial total cyanide content volatilized as hydrocyanic acid.

3.3.3 Behavior of the Transition and Sulfide Ores

The formation of cyanate and thiocyanate was observed. After the 24 hour leach, the cyanide existed as free cyanide and metal cyanide complexes. The percentage of cyanide reduction obtained by volatilization of hydrocyanic acid in the simulated ponds was 40 % in the case of the transition ore. In the case of the sulfide ore, the percentage of hydrocyanic acid volatilization was less than 10 % of the initial total cyanide content (see Figure 3).

In addition to the cyanide concentration reduction in solution by hydrocyanic acid volatilization and the formation of cyanate and thiocyanate, there is also the mechanism of the precipitation of the copper ferrocyanide complex. This mechanism directly contributes to a long-term deposition of cyanide species, which do not form free cyanide and, therefore, do not appreciably contribute to the overall hydrocyanic acid volatilization.

4. SIGNIFICANCE OF THESE CONCLUSIONS WITH REGARD TO POTENTIAL RISKS

From these results, it can be concluded that cyanide releases from the tailings pond surface do not pose an acute hazard for humans and the environment. The results of this testwork also suggest that minimizing HCN releases and thus reducing the potential of long-term risks associated with cyanides in tailings ponds could be achieved by controlling the free cyanide level in tailings ponds.

The concentration of free cyanide or other cyanide species in the pond is lower (typically in an average range between 20 and 500 ppm total cyanide) and the pH is usually between pH 8 and 9. The surface of the pond is, however, large. Due to the exposure to air, carbon dioxide is continuously absorbed by the alkaline pond and the equilibrium is shifted to the hydrocyanic acid side. This can be volatilized to the air. Depending on the pH, the cyanide concentration, the surface of the pond and various other parameters, hydrocyanic acid is volatilized from the surface continuously, but at a low concentration level of typically less than 1 ppm (Detection limit of standard HCN detectors). As a consequence, the hydrocyanic acid concentration above the surface of the pond is usually so low that it is not detectable with today's standard detectors to measure the hydrocyanic acid concentration. Consequently, the potential risk from HCN even for unprotected people in the vicinity of the pond should be low. As opposed to the low personal risk potential, the amount of hydrocyanic acid volatilized from the pond is considerably higher compared to all the other mining process steps, where cyanide is involved. Based on the laboratory experiments, tailings pond from an oxide ore operation in particular are likely to lead to a significant amount of hydrocyanic acid volatilization to the air.

Therefore, a true destruction of all of the cyanide being discharged into a tailings pond is only possible by keeping the cyanide in solution, which is the only phase that it is possible to destroy the carbon-nitrogen bond in the CN^- ion and form less toxic products, by oxidation, for example.

This is important, since any cyanide in solution can easily be volatilized from the solution into the air.

5. RESUME

It can be concluded that significant amounts of cyanide escape as HCN from the simulated tailings ponds of the three different ore types. The emission comes from the free cyanide as well as from the metal cyanide complexes. However, there are generally no acute risks to be expected because of the low concentrations that result from the volatilization.

There is no information or data to indicate that there are negative long-term risks associated in the operation of a tailings pond with regard to HCN releases. However, field tests are recommended under large scale conditions to further enable more precise quantification and less known reactions like biodegradation, for example.

It can be estimated that the basic behavior in terms of the volatilization of hydrocyanic acid should be similar and that, therefore, these tests can be seen as a basic simulation of tailings pond behavior with regard to the fate of cyanide and the hydrocyanic acid volatilization. During the life time of a mine, where oxide, transition and sulfide ores are leached with cyanide, the conclusion for the mines management is, that, especially in the beginning of the operation in the case of the oxide ore, mostly hydrocyanic acid is volatilized.

These findings and the basic conclusions need to be taken into consideration in the overall water management of the mine. This is of significant importance if any kind of regulations or limitations concerning the emission of hydrocyanic acid have to be met.

It would be difficult to exactly quantify what amount of this chemical is emitted during a mine life operation, since this mass balance would require a much more highly sophisticated model to analyze and calculate the respective releases.

This of course is also a challenging approach for the future for the mining industry and also for the responsible authorities, with the aim of finding a practice based regulatory basis for any limits.

For further studies, it is recommended that the long-term fate of cyanide and resulting compounds be investigated in more detail. This would not only enable a more complete balance of cyanide, but also enable a better understanding of the fate of these compounds themselves, which may also be of environmental significance in the future.

6. REFERENCES

Griffiths et al 1987. *The detoxification of gold-mill tailings with hydrogen peroxide, J. S. Afr. Inst. Min. Metall., vol. 87, no. 9. Sep.* Pp. 279-283

Rubo et al 1999. *Effects of Responsible Care on the Risks, Legal and Costs Aspects of Cyanides, Randol Gold Forum*

Simovic et al 1985. *Natural Removal of Cyanides in Gold Milling Effluents – Evaluation of Removal Kinetics, Water Poll. Res. J. Canada, vol. 20, no. 2, Pp. 120-135*

Smith and Mudder 1991. *Chemistry and Treatment of Cyanidation Waste, Mining Journal Books Ltd., ISBN 0 900117 51 6*

Logsdon et.al 1999. *The Management of Cyanide in Gold Extraction, International Council on Metals and the Environment, ISBN 1-895720-27-3*

Remediation and reclamation

Tailings and Mine Waste'00 © 2000 Balkema, Rotterdam, ISBN 90 5809 126 0

Improved metal phytoremediation through plant biotechnology

E.A.H.Pilon-Smits[1]
Department of Biology, Colorado State University, Fort Collins, Colo., USA

Y.-L.Zhu[1] & N.Terry
Department of Plant and Microbial Biology, University of California, Berkeley, Calif., USA

ABSTRACT: Phytoremediation, i.e. the use of plants for environmental cleanup, offers an attractive approach to remediate metals from mine waste. The goal of this study is to use genetic engineering to increase heavy metal tolerance and accumulation in plants, so as to create better plants for metal phytoremediation. The chosen strategy is to overproduce the heavy metal binding peptides glutathione and phytochelatins. Glutathione (γ-Glu-Cys-Gly, GSH) plays several important roles in the defense of plants against environmental stresses, and is the precursor for phytochelatins (PCs): heavy metal-binding peptides involved in heavy metal tolerance and sequestration. Glutathione is synthesized in two enzymatic reactions, catalyzed by glutamylcysteine synthetase (ECS) and glutathione synthetase (GS), respectively. To obtain plants with superior metal accumulation and tolerance, we overexpressed the *E. coli* ECS and GS enzymes in *Brassica juncea* (Indian mustard), a particularly suitable plant species for heavy metal remediation. The transgenic ECS and GS plants contained higher levels of glutathione and phytochelatins than the wildtype plants. In metal tolerance and accumulation studies using cadmium, the ECS and GS plants accumulated 1.5 to 2-fold more cadmium in their shoots than wildtype plants, and also showed enhanced cadmium tolerance. As a result, the total cadmium accumulation per shoot was ~3-fold higher. We conclude that overexpression of the ECS and GS enzymes is a promising strategy for the production of plants with superior heavy metal phytoremediation capacity. Present and future studies include analysis of the ECS/GS plants with other heavy metals, and phytoremediation studies using metal-polluted mine waste.

1. INTRODUCTION

Mining-related heavy metal pollution of soils and waters is a major environmental problem worldwide. An attractive new approach to deal with this problem is phytoremediation, i.e. the use of plants to clean up polluted soils and waters. Plants can be used to remove or detoxify heavy metals in various ways. In the technology of phytoextraction, plants are used to accumulate metals in their harvestable tissues, followed by removal of the plant tissue from the site, ashing of the plants and, if economically feasible, recycling of the metals. Plants can also be used to biotransform toxic forms of metals into non-toxic forms, e.g. forms that are biounavailable. In this case the metals are left on the site (phytostabilization).

The ideal plant to use for phytoremediation has a large biomass, grows well in polluted environments, and accumulates pollutants to high concentrations. Naturally occurring metal hyperaccumulators - heavy metal accumulating flora collected from metal-rich sites - offer one option for the phytoremediation of metal-contaminated sites. However, hyperaccumulators

---

[1] both authors contributed equally to this work

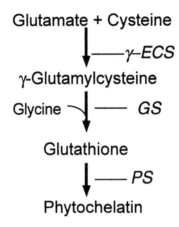

Figure 1. The biosynthetic pathway of glutathione and phytochelatins.

tend to grow slowly and produce little biomass. An alternative approach is to genetically engineer fast-growing species to improve their metal tolerance and metal accumulating capacity. A suitable target species for this strategy is *Brassica juncea* (Indian mustard). Indian mustard has a large biomass production, a relatively high trace element accumulation capacity, and can be genetically engineered.

The goal of this study is to enhance heavy metal tolerance and accumulation in Indian mustard by means of genetic engineering. To this aim, Indian mustard plants were engineered to overproduce the heavy metal binding peptides glutathione and phytochelatins. Glutathione (γ-Glu-Cys-Gly, GSH) plays several important roles in the defense of plants against environmental stresses, and is the precursor for phytochelatins (PCs): heavy metal-binding peptides involved in heavy metal tolerance and sequestration. Glutathione is synthesized from its constituent amino acids in two enzymatic reactions, catalyzed by γ-glutamyl-cysteine synthetase (γ-ECS) and glutathione synthetase (GS), respectively (Fig. 1). Upon exposure to heavy metals, glutathione is aggregated to phytochelatins (structure: $(\gamma\text{-glu-cys})_n\text{gly}$), by phytochelatin synthase (PS).

To enhance the plant's capacity to produce glutathione and phytochelatins, we overexpressed the *E. coli* ECS and GS enzymes in *Brassica juncea* (Indian mustard). The transgenic ECS and GS plants were compared with untransformed Indian mustard plants with respect to their Cd accumulation and tolerance, and their levels of heavy metal binding peptides.

2. MATERIALS AND METHODS (Zhu et al., 1999)

Hypocotyls from 3d old Indian mustard seedlings were transformed using *Agrobacterium tumefaciens*, with DNA constructs containing the *Escherichia coli gshI (ECS)* or *gshII* (GS) gene, under the control of the CaMV 35S promoter (Noctor et al. 1996, Strohm et al., 1995). For Cd tolerance and accumulation experiments, hydroponically grown Indian mustard plants were exposed to 100µM CdSO4 for 14 days, after which growth and Cd accumulation were measured. Cadmium was analyzed by Inductively-Coupled Plasma Atomic Emission Spectroscopy (ICP-AES) after acid digestion. Glutathione and phytochelatins were measured by HPLC, on a C18 reverse phase column. Total non-protein thiols were measured according to Ellman (1959).

3. RESULTS AND DISCUSSION

3.1 Overexpression of γ-glutamylcysteine synthetase

ECS transgenic seedlings showed better growth on Cd-containing agar medium (0.2 mM, Table 1), and had higher concentrations of phytochelatins, glutathione and total non-protein thiols (Table 2), compared to wildtype seedlings. When mature plants were tested in a hydroponic (nutrient film technique) system, ECS plants accumulated more cadmium than wildtype plants: shoot cadmium concentrations were 40-90% higher (Table 1). There were no significant differences in root cadmium levels. The ECS mature plants also showed increased Cd tolerance, as judged from less growth inhibition (Table 1). Thus, γ-ECS appears to be a rate-limiting enzyme for the biosynthesis of glutathione (both in the absence and presence of cadmium), as well as for phytochelatins biosynthesis in the presence of heavy metals. The higher levels of phytochelatins and/or glutathione, in turn, probably enhance cadmium tolerance and accumulation.

Table 1: Cadmium tolerance and accumulation in wildtype, or ECS Indian mustard plants (average of 10 plants $\pm$ s.e.). All values shown were significantly different ($P<0.05$) between ECS and wildtype plants. *Note:* there were no significant differences in growth under unstressed conditions. Seedling growth was measured as root length; mature plant growth as fresh weight production. Seedlings were grown at 200μM Cd; mature plants at 50μM Cd.

	Seedling growth (% of unstressed)	Mature plant growth (% of unstressed)	Shoot [Cd] ($\mu g.g^{-1}$ DW)
Wildtype	39 ± 7	30 ± 3	350 ± 48
ECS	94 ± 9	40 ± 5	653 ± 97

Table 2: Concentrations of glutathione (GSH) and total non-protein thiols (NPT) in cadmium-treated (+ Cd, 200μM) or unstressed (- Cd) wildtype or ECS Indian mustard seedlings (average of 12 plants $\pm$ s.e.). All values shown were significantly different ($P<0.05$) between ECS and wildtype plants. All concentrations are expressed as $nmol.g^{-1}$ FW.

	[NPT] -Cd	[NPT] +Cd	[GSH] -Cd	[GSH] +Cd	[PC] +Cd
Wildtype	1500 ± 100	5500 ± 200	140 ± 10	75 ± 10	871 ± 8
ECS	2000 ± 150	10000 ± 500	380 ± 20	130 ± 30	1128 ± 61

3.2 Overexpression of glutatione synthetase

The transgenic GS seedlings showed better growth on Cd-containing agar medium, compared to wildtype seedlings. Mature GS plants also showed enhanced tolerance to cadmium, and accumulated significantly more cadmium than wildtype plants (Table 3). Shoot cadmium concentrations in GS plants were 44% higher, and total harvestable cadmium per shoot was 3-fold higher. There were no significant differences in root cadmium levels. Cadmium-treated GS plants had higher concentrations of glutathione, phytochelatin, and total non-protein thiol than wildtype plants (Table 4); unstressed GS plants, however, did not have enhanced levels of these thiols. Therefore, in the absence of heavy metals the GS enzyme does not appear to be rate-limiting for GSH synthesis, but in the presence of Cd the GS enzyme appears to become rate-limiting for the biosynthesis of glutathione and PCs.

319

Table 3: Cadmium tolerance and accumulation in wildtype, or GS Indian mustard plan (average of 10 plants ± s.e.) grown at 50µM Cd. All values shown were significantly differen (P<0.05) between GS and wildtype plants. *Note:* there were no significant differences i growth under unstressed conditions. Mature plant growth was measured as fresh weight pro duction.

	Mature plant growth (% of unstressed)	Shoot [Cd] ($\mu g.g^{-1}$ DW)	Total Cd/shoot (μg)
Wildtype	28 ± 2	485 ± 48	2540 ± 480
GS	44 ± 4	698 ± 86	7484 ± 650

Table 4: Concentrations of glutathione (GSH), total non-protein thiols (NPT), and phyto chelatins (PC) in cadmium-treated (+ Cd, 100µM) or unstressed (- Cd) wildtype or GS ma ture Indian mustard plants (average of 10 plants ± s.e.). All concentrations are expressed a nmol.g^{-1} FW.

	[GSH] -Cd	[GSH] +Cd	[NPT] +Cd	[PC] +Cd
Wildtype	290 ± 50	93 ± 40	5000 ± 300	780 ± 70
GS	360 + 40	456 + 80*	9100 + 300*	1260 ± 210*

* Values were significantly different (P<0.05) between GS and wildtype plants.

4. CONCLUSIONS

Since the Cd tolerance and accumulation capacity of Indian mustard was correlated wi GSH and PC levels, these results confirm the importance of these compounds in Cd toleranc and accumulation. The key enzyme controlling the production of these metal-binding pep tides appears to be ECS, although in the presence of metals GS also becomes rate-limiting.

We conclude that overexpression of ECS and GS offers a promising strategy for the pr duction of plants with superior heavy metal phytoremediation capacity. Present and futu studies are focusing on simultaneous overexpression of ECS and GS, on testing these tran genics for their tolerance and accumulation of other heavy metals besides Cd, and on the ef ciency of these transgenics to clean up contaminated environmental soil.

5. REFERENCES

Ellman, G.L. 1959. Tissue sulfhydryl groups. Arch. Biochem. Biophys. 82:70-77.

Noctor, G., Strohm, M., Jouanin, L., Kunert, K.-J., Foyer, C.H., & Rennenberg, H. 199 Synthesis of glutathione in leaves of transgenic poplar overexpressing glutamylcysteine sy thetase. Plant Physiol. 112:1071-1078.

Strohm, M., Jouanin, L., Kunert, K.-J., Pruvost C., Polle, A., Foyer, C.H., & Rennenber H. 1995. Regulation of glutathione synthesis in leaves of transformed poplar overexpressi glutathione synthetase. Plant J. 7:141-145.

Zhu,Y./Pilon-Smits, E.A.H., Jouanin, L., & Terry, N. 1999. Overexpression of glutathio synthetase in *Brassica juncea* enhances cadmium tolerance and accumulation. Plant Physic 119:73-79.

Tailings and Mine Waste'00 © 2000 Balkema, Rotterdam, ISBN 90 5809 126 0

Successful revegetation of neutralized heap leach material

R. Cellan
Reclamation, Homestake Mining Company, San Francisco, Calif., USA

A. Cox
Environmental Affairs, Homestake Mining Company, San Francisco, Calif., USA

L. Burnside & R. Gantt
Harding Lawson Associates, Carson City, Nev., USA

ABSTRACT: Revegetation in the arid environment of the Great Basin represents a challenge . of available technology. The resultant chemical and physical characteristics of neutralized heap leach material compounds this natural challenge. In late 1995, Homestake Mining Company embarked on what would turn out to be a five-year program to implement the most prudent means to attain the highest potential for regulatory reclamation success at their Santa Fe/Calvada Reclamation Site near Hawthorne, Nevada. A combination of field and laboratory tested techniques and materials provided positive guidance for Homestake's successful progress. In addition, Homestake's multidisciplinary project planning and implementation has yielded valuable reclamation planning insight for all phases of mining operations, and is demonstrating marked progress toward attaining successful regulatory site stability through revegetation.

1 INTRODUCTION

Corona Gold, Inc., a subsidiary of Homestake Mining Company, commenced mining operations in 1988 and the Calvada expansion was completed in 1992. The project was a conventional open pit leach pad gold operation consisting of four main pits and four leach pads constructed on approximately 135 acres. Precious metals were recovered from pregnant solutions using a conventional carbon adsorption process plant. Ore was generally crushed to nominal 3/4 inch and most was agglomerated before being placed on the leach pads. Cyanide leaching was terminated in 1994 with 17.6 million tons of ore mined and leached over the life of the project.

Homestake Mining Company has established policies that dictate very high internal reclamation standards for all company mining projects worldwide. Regulatory reclamation success standards for the Santa Fe/Calvada Mine in Nevada consist of the Standards for Successful Revegetation of Exploration and Mining jointly approved by the Nevada Division of Environmental Protection (NDEP), the Bureau of Land Management (BLM) and the USDA Forest Service (FS) in September 1998. An evaluation of reclamation success according to the Standards requires that mines take every prudent step possible, based upon current technology, to re-establish 100 percent of the pre-mine vegetation cover. If such steps are taken and well documented, NDEP, the BLM and the FS will consider the mine to be in compliance with regulatory requirements regardless of the degree of success of the revegetation effort. However, Homestake Mining Company would not be satisfied with the reclamation effort unless revegetation was successful in establishing an ecologically functioning plant community capable of supporting the designated post-mining land uses. Closure and reclamation objectives for the Santa Fe project were designed to restore the site to pre-operation land uses. Objectives were to re-establish vegetation and wildlife/grazing habitat, and insure protection of surface and groundwater resources.

From 1989 through 1994 the mine waste rock dumps and haul roads were reclaimed. The heap leach operation at the Santa Fe/Calvada Mine was decommissioned in 1995 and reclamation activities to revegetate and stabilize the heap leach pads began. Leach pads have been regraded to slopes of 3:1 (horizontal to vertical) steepness, which range from approximately 240-430 feet in length. The Santa Fe/Calvada Mine represents one of the first attempts to revegetate leach pads in this extreme arid region of Nevada. In 1997/1998, remediation of previously revegetated areas, ancillary areas and the heap leach pads were revegetated. Homestake prides itself on active pursuance of successful concurrent reclamation. To date, 717 of the 797 acres disturbed have been revegetated and are involved in a monitoring program. The remaining 80 acres were reclaimed and revegetated in November 1999 and monitoring will begin during the 2000 growing season.

2 ENVIRONMENTAL HISTORY

The Calvada Reclamation Plan filed in 1991, and the Santa Fe Reclamation Plan filed in 1993 prescribed full application of topsoil for the total surface area of the heap leach pads. In September of 1997, HMC filed a minor modification to the Santa Fe/Calvada Reclamation Plans that focused on a shortfall of stockpiled growth medium required to be in compliance with the reclamation plans. The minor modification outlined chemical and physical characterization of the heap material conducted in 1995. In addition, the process employed to determine and implement appropriate revegetation was revised due to insufficient topsoil resource. Based on the characterization report, in 1996 and 1997 HMC implemented a program to test revegetation without the application of growth medium and evaluated alternatives for growth medium application.

3 SITE SPECIFIC CHALLENGES

The Santa Fe/Calvada Mine is located 25 miles east of Hawthorne, Nevada, in the Gabbs Valley Range at an elevation of 6,400 feet. Average annual precipitation is 5 inches with effective precipitation generally falling in the winter and early spring. Average maximum and minimum temperatures are 70 and 53 degrees Fahrenheit with daily highs of 100 degrees being normal during the summer growing season. Summer convection storms are common, however, during periods of high daily temperatures, high evapo-transpiration results. HMC, as with most Nevada mines, did not have the luxury of plentiful topsoil for revegetation. Therefore, it became integral to success to determine the rooting depth and above and below ground growth form of the native plants for growth medium requirements and seed mix design.

4 ADDITIONAL CHALLENGES

During operation of the Santa Fe/Calvada Mine, Corona Gold, Inc., had conducted concurrent reclamation of waste rock dumps which provided good evidence of regional reclamation potential. However, revegetation potential of the dumps varied significantly from the potential of the heap leach pads. Homestake recognized that reclamation of the leach pads would take significantly different measures and extended effort to accomplish successful revegetation.

Precious metal recovery leaching with a basic (pH ≥ 9.0) sodium cyanide solution results in residual electrical conductivity levels toxic to plants, sodium levels that result in deflocculated or soils impervious to infiltration, material totally void of organic material, and a nutrient rich environment (water soluble nitrate). Physically, the heap material at this site ranged from medium brown to black and milled to run-of-mine ranging in size from nominal ¾ inch for the crushed ore up to small cobbles (8 inches dia.). By design heap leach pads are efficient regarding infiltration and percolation. This efficiency equates to a very low water holding capacity and available water for plant establishment and survival. The last challenge was designing a seed mix that emphasized native plants that were commercially available and adapted to both drought and saline/sodic conditions.

5 HOMESTAKE'S 5 YEAR PROGRAM

5.1 Test Plot Program

In 1995, after a field survey of growth medium stockpiles was conducted HMC determined that it would be prudent to attempt to revegetate the neutralized heaps in the absence of a growth medium application. Field and laboratory characterization of the material was completed in early 1996 and specifications were developed to implement a test plot program. The test plot program was designed to evaluate revegetation techniques and plant species adapted to site specific conditions. Techniques emphasized heap material amendment to develop a suitable growth medium, and a seed mix was designed based on the characteristics of the native plant communities, climatic conditions, and heap material characteristics. Approximately 10 acres representing the variety of material and aspect was dedicated to this program. In addition, the heap pad areas outside the test plots were also treated to prepare the materials for revegetation following evaluation of the test plots.

The heap material across all 4 pads was amended with a gypsum application at 2 tons/acre. Green small grain straw was chopped and blown on and incorporated with the gypsum to 18 inches. The green manure application ranged from 2400 to 24,000 lbs./acre, the maximum simulating the organic matter content of an acre furrow slice of arid environment soils. Within the test plots, the material was broadcast seeded with a site specific mix, and a slow release fertilizer was applied. Straw mulch was blown on and crimped. On the remainder of the pad surfaces, barley seed was broadcast to germinate a green crop for incorporation the following growing season. The pads were drip irrigated over the first growing season.

Limitations of the heap material and low precipitation, and ineffective irrigation resulted in minimal response in the test plots. The most sever limitations were the black color, gradation and chemical conditions resulting from the cyanide leach process. Initial germination was successful, however, the significantly higher surface temperatures (20 degrees) of the black pad material and the ineffective irrigation resulted in unsustainable establishment.

Based on the minimal response of the test plots, and the remaining short fall in stockpiled growth medium for revegetation, HMC introduced the concept of bringing the field to the lab where bench scale soil amendment techniques could be tested at a significant cost and time savings.

5.2 Bench Test Program

Samples were collected from each of the four heaps that included the surface salt affected layer and subsurface material. Native growth medium used during testing was collected in a composite sample from three on-site native growth medium stockpiles. Sufficient on-site well water was collected and transported to the laboratory with heap and growth medium samples for the testing program.

Baseline solids analysis focused on constituents that limited the response of the 1996 revegetation test plots. All solids were for major and micronutrients and elements toxic to plants by an agricultural laboratory using standard procedures.

Objectives for the testing program were to determine if:
1. application of well water and gypsum to neutralized heap surfaces was effective in leaching salts to below the rooting zone, and thus support vegetation growth; and,
2. normal meteoric and climatic cycles would cause salts contained in the heap surface layer to capillate into surface applied native growth medium thus impeding vegetation growth, and what depth of growth medium is required to prevent salt capillation, such that vegetation growth is not impeded.

Seven column leaching tests were conducted on various neutralized heap material composites and surface salt layer samples from the site to accomplish objective 1. Four meteoric/climatic cycle simulation column tests were conducted on select neutralized heap material composites, surface salt layer samples, and varied depths of native growth medium to accomplish objective 2.

All tests were conducted in 12-inch diameter (30 cm) clear acrylic or PVC columns of varied height. Columns were loaded by placing heap material (26 or 30-inch beds) onto a perforated punch plate and compacting to an equivalent on-site bulk density (110 lb/ft^3). Surface salt layer samples and/or native growth medium was placed on top of the heap material and was similarly compacted to an equivalent on-site bulk density. A gypsum application equal to 2 tons/acre was applied to the surface of the salt load leach test column charges.

For the column leach tests three quantities of minesite well water (5, 10, and 15 inches) were used in the leach tests. Well water used for the salt load leach tests was slightly alkaline at pH 7.24, and contained alkalinity (344 mg/l), Calcium (290 mg/l) and several other elevated constituents. However, general chemistry of the well water was acceptable for supporting vegetation establishment.

For meteoric/climatic cycle simulation tests column charges were loaded as described earlier, and bedded in increments during column loading procedures to allow temperature and moisture probes to be installed. Two annual climatic cycles were simulated, using deionized water, for each column test. About 70 days were required to complete two simulated annual climatic cycles.

Solids sampling of the column charges for both the leach and meteoric tests required column sacrifice. Sampling increments were based on the layers of material as loaded in the column and the test objectives. Solids for each leach column were sampled in the salt layer (0 to 4 inches), and in 5-inch increments through the heap material. Meteoric column sampling focused on the growth medium material to determine if salts migrated upward. Growth medium was sampled in 3-inch increments. The salt layer and heap material were sampled as separate composites to determine change in salt content.

Overall effluent analysis results from the seven column leach tests indicated that at least 5 inches, but not more than 10 inches, of well water must be applied to the various heaps to leach the salt load to an acceptable depth to support vegetation growth. Solids analysis indicated that leaching was a viable alternative for this site if sufficient well water was available. The 5-inch application of well water provided a wetted front and initiated the leaching process. The 10-inch application accomplished adequate leaching within the rooting zone, and the 15-inch application moved the area of accumulation deeper in the heap material. Figure 1 illustrates the layers of sample materials as they were loaded in the columns for testing, baseline conditions and results of leaching for Pad 1 South materials.

Initial indications of worst case conditions justified leaching and solids analysis for the cumulative leaching of a total of 15-inches of well water. Concentrations of the three constituents in the surface salts decreased by 89 to 97 percent. Within the rooting zone (0 to 14-inches), sodium decreased an average of 87 percent. Nitrate and EC decreased an

Solid Material	Depth, Inch	Before Leach			After Leach		
		Na	NO$_3$	EC	Na	NO$_3$	EC
Surf. Salt	0-4	2760	240	62	313	6.1	2.9
Heap	0-5	1702	65	10	230	3.0	2.6
Heap	5-10	1702	65	10	248	2.3	2.7
Heap	10-15	1702	65	10	373	205	2.8
Heap	15-20	1702	65	10	506	307	3.0
Heap	20-Bot.	1702	65	10	506	209	3.1

Note: Na and NO$_3$ conc. in mg/kg, EC in mmhos/cm^2.

Figure 1. - Summary Results, Salt Load Column Leach Test, Pad 1 S Surface Salts, Pad 1 S Heap Material, 15" Well Water Applied

average of 96 and 81 percent, respectively. Below the rooting zone (14 to 26 inches) sodium decreased an average of 74 percent, NO_3 accumulated in this zone by an average of 269 percent, and EC decreased an average of 70 percent. Salt load leach test results verify that inhospitable constituents can be effectively leached with varying applications of low quality well water.

In comparison to agricultural standards, final concentrations were still high for EC, yet in a manageable range for revegetation. For sodium, all quantities of well water applied brought the concentration in the heap material to equivalent or lower levels than that of the native growth medium. Resulting sodium and EC levels in the solids indicate that the 5-inch application initiated the leaching process, the 10-inch application decreased critical constituents to manageable levels, and the 15-inch application is optimal to insure deep leaching if well water is readily available. These results do provide one technique to progress toward creating a suitable growth medium from neutralized heap material. Additional amendments may be required.

The task of moving earthen material is an expensive cost for even the smaller mining sites. Therefore, when heap material is not wholly conducive as suitable growth medium, and must be covered with suitable material, there must be reasonable assurance for revegetation success to justify the cost. Assurance should focus on protecting the surface applied growth medium from the plant growth inhibiting constituents of the heap material.

In the case of the Santa Fe/Calvada heap material, the concern was twofold. If growth medium was applied directly on top of the heap material, would the sodium and salts migrate up into the growth medium? And, if upward migration did occur, what depth of growth medium would be required to protect the rooting zone? Under meteoric conditions in low precipitation zones (4 to 6 inches), it was expected that a wetted front sufficient to induce capillary rise from the heap material would not form, and the growth medium would be protected.

For both the 6 and 12-inch application depths for growth medium on Pad 1 South with in-situ surface salts and EC did not increase or migrate up into the growth medium. To the contrary, overall Na, NO_3 and EC became vulnerable to the leaching process. With a 6 inch application of native growth medium (Figure 2), Na decreased an average of 57 percent, EC 50 percent, and NO_3 83 percent throughout the growth medium layer. In the surface salt layer the three constituents decreased between 92 and 99 percent. Leaching also occurred in the heap material, although to a lesser degree. The decrease in the three constituents varied from 14 to 62 percent.

Although leaching was not the objective for meteoric testing it generally occurred throughout the rooting zone of the tested material. More importantly, meteoric tests verified that a wetting front sufficient to induce upward migration of Na and salts would not develop in the rooting zone under these particular meteoric conditions.

5.3 *Multidisciplinary Alternatives Analysis*

The minimal response of the field test plots and the very promising results of the bench scale testing program provided HMC a firm technical basis. The multidisciplinary team of experts

Solid	Depth,	Na, mg/kg		NO_3 mg/kg		EC mmhos/cm^2	
Material	Inch	Before	After	Before	After	Before	After
Growth	0-3	460	202	5.9	7.7	1.4	0.7
Growth	3-6	460	193	5.9	1.0	1.4	0.6
Surf. Salt	0-4	2,760	230	240	1.4	62	2.7
Heap	26	1,702	1,472	65	25	10	4.6

Figure 2. - Summary Results, Meteoric/Climatic Cycle Simulation Column Test, 6" Native Growth Medium, Pad 1 S Surface Salt, Pad 1 S Heap Material

then proceeded to define the optimum reclamation inputs to be evaluated to accomplish successful long-term reclamation of a mine site in closure.

From the outset the following were known:

1. No further action must be considered as an alternative in case site constraints could not be overcome.
2. Heap material must be amended if not fully covered.
3. Temporary irrigation was necessary given the extreme arid conditions.
4. Growth medium application would ameliorate extreme surface temperature of the black material.
5. Combinations of these inputs must be considered.

At the other end of the spectrum, the ultimate represented doing everything possible to approximate pre-disturbance conditions. Our goal was to determine the most ecologically effective, cost effective, and regulatory effective measures to accomplish successful long-term reclamation of a mine site in the closure phase. This goal had 2 objectives:

1. Regulatory compliance: State and Federal Regulations require documentation and demonstration that state-of-the-art techniques and products have been applied. No specific numerical success standards are required.
2. Meet Homestake Mining Company's standard to leave a site in better ecological condition than when they entered the site. HMC intends to accomplish this goal by re-establishing a functioning plant community.

The alternatives evaluated by the team include the following:

1. No Action
2. Partial growth medium application on the dark pad surfaces, and Gypsum application and leach light colored surfaces
3. Same as Alternative 2 with the addition of temporary irrigation
4. Growth medium application across all pad surfaces
5. Same as Alternative 4 with the addition of temporary irrigation
6. Targeted restoration which included gypsum and leach all pad surfaces, OM amendment, 12 inches of growth medium, seed and temporary irrigation

Evaluation Criteria Included:

1. Ease in justification of Regulatory Compliance for Bond Release
2. Revegetation Success
3. Ecological Functions
4. Site Stability
5. Estimated Costs, and Time Frame

The comparison of alternatives was based upon their potential for success. "Success" was defined within the bounds of Homestake's reclamation standards and current requirements for regulatory compliance. Thus, success was defined as the process of returning a drastically disturbed site to a self-sustaining condition equal to or better than occurred prior to disturbance in terms of biological organisms, ecological functions, and physical processes.

Reclamation conditions, predicted for a six-year period were analyzed for each alternative. How did the alternatives compare? Although all of the alternatives could meet our definition of success. Only two of the Alternatives (4 and 5) would meet our objectives of Regulatory Compliance and HMC's Standards for Success. Alternative 1 was not considered any further given HMC's objective. Alternative 6 – the Ultimate, we believed that maximizing reclamation inputs and costs would tilt the successional pathway toward some resemblance of pre-disturbance condition. However, restoration was not HMC's goal, nor is restoration required or realistic. The result of this multidisciplinary analysis was that HMC determined that Alternative 5 would satisfy regulatory compliance, and would meet their corporate standards for success. HMC's decision was based on cost weighed against potential success and risk of interim remediation during the monitoring period.

6 IMPLEMENTATION

In late 1997 Nielsons, Inc. of Cortez, Colorado began implementation of Alternative 5 plus remediation of sulfide areas on the overburden piles. The majority of site preparation occurred on Pads 1 and 4. Pad 4 was regraded to facilitate safe conveyance of stormwater runoff. In addition, the north slope was recontoured from 2.5:1 to a 3:1 slope and benched for additional slope stability. The north slope of Pad 1 was subject to final recontouring to establish accessible and stable slopes. The East and West Overburden Pile tops were recontoured to establish conveyance swales to facilitate drainage off the piles into the natural drainage system. Areas on the overburden piles and the low grade sulfide ore stockpile were treated with magnesium hydroxide and magnesium carbonate and capped with a 6-inch layer of compacted clay.

The heap leach pads and the remediation areas of the overburden piles and sulfide ore stockpile were then covered with a minimum of 8 inches of growth medium. Growth medium was obtained from three on-site stockpiles and the Welsh Canyon borrow site. Approximately 167,000 cubic yards (cyds) of growth medium was applied with scrapers and/or dump trucks and graders. Green waste consisting of composted grass clippings and chipped trees was obtained from Sacramento, California. The green waste was applied on the heap leach pads using a hydraulic manure spreader at a rate of 10 cyds per acre. The green waste was ripped in on the contour to a minimum depth of 8 inches to incorporate the material as well as providing a roughened seedbed.

All revegetation areas were broadcast seeded at a rate of 46 PLS (pure live seed) pounds per acre followed by a spring toothed harrow to ensure soil contact with the seed. Small grain straw mulch was then applied at 2 tons per acre with a tub de-baler of flat surfaces and with a standard straw blower on sloped surfaces. The straw was secured in place with a standard crimper perpendicular to the contour. Revegetation activities across 297 acres were completed on February 15, 1998.

A temporary irrigation system was constructed in the spring of 1998 to provide supplemental soil moisture on the heap leach pads. The micro-sprinkler system had heads spaced every 20 feet, effectively providing full coverage. Approximately 145 acres were irrigated using 16,000 micro-sprinklers. The micro-sprinklers delivered water at an extremely low application rate of less than 0.07 inches per hour. This low rate minimized the potential for erosion on side slopes. Water was distributed to the micro-sprinklers using a network of polyethylene tubing laterals connected to polyvinyl chloride submains and mainlines. Water was pumped to the lower three pads from existing ponds using the same pump system used previously in the leaching activities. Water was pumped to Pad 4 using a submersible pump located in the collection pond at Pad 4. Water was supplied to the ponds from a well approximately 6 miles down slope of the reclamation site. This system was designed to deliver water in several short sets early in the growing season with a transition to longer sets less often later in the season. This allowed for a deeper wetting of the soil and the intermittent drying of the surface training young plants to grow deeper root systems. To mimic the heat dormancy of this area, irrigation was suspended during the hot period (approximately July 1) to allow dormancy to occur prior to the onset of winter.

7 SUCCESS MONITORING & RESULTS

Baseline or reference area monitoring was conducted in 1996 to document undisturbed native vegetation cover. The Natural Resource Conservation Service (NRCS) Soil Survey of Mineral County generally divides the Santa Fe/Calvada site into two vegetation communities, black sagebrush and salt desert shrub communities. The dividing line between these communities runs generally north to south across the site in the area of the main access road. Areas west of the main access are the salt desert shrub community, and areas to the east are black sagebrush community. Representative baseline transects within these communities measured cover in the black sagebrush at 23 percent and in the salt desert shrub community at 17 percent. Average vegetation cover for undisturbed native communities is the target cover value for revegetation success. Although undisturbed cover values are used to quantify revegetation success, there are inherent differences between the growth medium found on undisturbed areas and revegetated areas.

From 1993 through 1995 concurrent revegetation was implemented at the Santa Fe/Calvada site. These activities addressed the East and West Overburden Piles, the Slab and the Calvada Pit areas. In 1997/1998, areas that required re-treatment and final

Table 1.   Pre-1995 Revegetation Success

Vegetation Community	Control %	Reclaimed %	Percent Success %
Black sagebrush	23	7	33
Salt desert shrub	17	4	25

reclamation were conducted on approximately 296 acres. Revegetation was segregated into irrigated areas (heap leach pads), and non-irrigated areas (overburden piles, borrow sites and other areas). In addition, a supercell shrub-planting program was completed at the site. In total over three phases, 3,930 shrub seedlings have been planted. Seed was collected locally by Comstock Seed of Reno, Nevada and the Nevada Division of Forestry Nursery propagated the plants at their nursery in Washoe Valley, Nevada.

In 1996, cover data was collected in all areas revegetated between 1993 and 1995. Table 1. presents the results of the average relative cover data for the baseline (control) and pre-1995 revegetated transects and the derived percentage success.

El Nino affected the timing and length of the 1998 growing season across western Nevada. Although local weather reports stated that it had been the wettest spring on record, it had also been cold relative to normal temperatures for that time of year. Although soil moisture had been optimal for seed germination and establishment, establishment growth started slowly. The irrigation schedule for the heap leach pads called for three two-hour sets for one month (April 15 – May 15). Irrigation would then be changed to four-hour sets occurring twice a week for one month (May 16 to June 15). Irrigation would then be suspended for the season on July 15. This irrigation schedule was designed to apply approximately 5.5 inches of water – the average annual rainfall total through irrigation. HMC personnel on-site collected daily precipitation data. As of June 30, 1998, the long-term average annual precipitation of 5 inches had been exceeded by 3.7 inches. In addition, to carry revegetation establishment through the hot dry periods, 1.5 inches was delivered via the temporary irrigation system for a total of 10.1 inches of precipitation realized for the heap leach pads.

Areas revegetated in 1997/1998 were monitored by ocular inspection to determine germination success by plant class in June 1998. Ocular observations of "green-up" were documented and germination overall in the non-irrigated as well as the irrigated areas was determined successful. Representatives of all three plant classes (grass, forb and shrub) seeded were observed. As of the end of the first growing season, areas that received native growth medium were supporting upwards of one foot of vegetative growth of perennial grasses and forbs. The second growing season quantitative monitoring was completed in May 1999. Monitoring consisted of establishing ninety-nine line intercept cover transects within the 1997/1998 revegetation areas. Percent revegetation success was calculated by comparing desirable vegetation cover for revegetated areas with the undisturbed baseline cover for the representative vegetation community.

The Line Intercept Cover Transect Method is described in the BLM Technical Reference (TR 4400-4) Sampling Vegetation Attributes (BLM 1996). The quantity and location of transects was verified with the Nevada Department of Environmental Protection Bureau of Mining Regulation and Reclamation and the BLM. In the field, transect locations were "field fit" considering all environmental variations, including slope and aspect, as well as cultural variations, including irrigation, seed mix, fertilizer applications and representation of the general condition of revegetation for the component being sampled.

At the present time, monitoring for revegetation success in Nevada does not require a test of statistical adequacy. However, as a general rule of thumb, 20 percent of each area sampled is the target to demonstrate representative sampling was achieved. The line intercept method documents relative cover by species. This method employs the temporary installation of a 100-foot tape with rebar stakes to hold the ends in place above the highest projection of the vegetation canopy. Cover sampling is planned to occur when plant phenology is in the reproductive stage to facilitate accurate species identification. The 1999 revegetation monitoring occurred from May 17th to May 20th. Perennial grasses across the site ranged from seed-heads developed but not headed out, to fully developed and headed out. This is also the stage at which vegetation provides the maximum soil surface protection via canopy and basal cover.

Vegetative cover that intercepts the vertical plane of the 100-foot transects was recorded in increments to the nearest $1/100^{th}$ of a foot. The sum of the measured intercept value, by species, is calculated when the data is input into an EXCEL spreadsheet. Photographic documentation is also completed at each transect including a plan view of the sample area and a vertical view of the sample intercepting the tape.

Irrigated revegetation on heap leach Pads 1 through 4 received 3.57 inches of water in addition to 1.80 inches of precipitation during the 1999 growing season (March 19th through June 28th). Bottlebrush squirreltail *(Elymus elymoides)*, crested wheatgrass *(Agropyron cristatum)*, tall wheatgrass *(Elytrigia elongatum)*, and thickspike wheatgrass *(Elymus lanceolatus ssp. dasystachyum)* were the most frequently encountered grass species. Prostrate kochia *(Kochia prostrata)* was the most frequently encountered forb species and four-wing saltbush *(Atriplex canescens)* and Nevada ephedra *(Ephedra nevadensis)* were the most frequently encountered shrub species. Average relative vegetative cover for each of the four irrigated components ranged from 2 to 5 percent. Average relative cover across the irrigated sites was 4 percent. The irrigated areas are performing very well and as expected for the second growing season. Although currently dominated by herbaceous species, diversity of established seeded species is good. Woody species were documented, however, their normally slower establishment and growth renders them less conspicuous. In addition, supplemental irrigation appears to favor quicker establishment of the herbaceous species. New seedlings are also becoming established in between the seed rows, which indicates that ecological functions are becoming established. The top and sideslopes of the Pads are stable with no evidence of soil movement. Overall, the irrigated areas are progressing toward meeting HMC's objective for reclamation success specifically with regard to biological organisms, ecological functions, and physical processes.

On the non-irrigated components (overburden pile re-treatment areas, growth medium stockpiles, borrow area and roads) relative cover averages ranged from 1 to 4 percent across these components. Average relative cover for the non-irrigated sites was 3 percent. Vegetation cover on these components is dominated by crested wheatgrass, thickspike wheatgrass, bottlebrush squirreltail, Indian ricegrass *(Oryzopsis hymenoides)*, and prostrate kochia in the herbaceous plant class. Shrub cover is dominated by Nevada ephedra, fourwing saltbush, Gardner saltbush *(Atriplex gardnerii)*, rubber rabbitbrush *(Chrysothamnus nauseosus)* and black greasewood *(Sarcobatus vermiculatus)*. Non-irrigated areas are performing as would be expected in comparison with the irrigated areas. Visual observation indicates that non-irrigation appears to favor the woody species. Although both herbaceous and woody species are becoming established, the canopy of the woody species dominates cover. As well, herbaceous species are germinating between the seed rows indicating that soil is becoming ecologically functional and no soil movement was documented. Table 2. presents the results of the average relative cover data for the baseline (reference areas) compared to the areas revegetated to date.

The supercell shrub plantings have been ocularly monitored for survival since the first installation in the spring of 1997. Of the six species propagated and planted Wyoming big sagebrush *(Artemesia tridentata wyomingensis)*, black sagebrush *(Artemesia arbuscula nova)*, black greasewood *(Sarcobatus vermiculatus)*, green rabbitbrush *(Chrysothamnus vicidiflorus)*, fourwing saltbush *(Atriplex canescens)* and shadscale *(Atriplex confertifolia)*, the shadscale has performed the poorest. Shadscale has done very well on the on-site seeded areas, however propagation from seed at the nursery was less than expected. Plants that did germinate were very small and consequently survival on-site was minimal. The remaining five species have put on an average eight inches of vertical growth over two growing

Table 2.   1996-1998 Revegetated Area Success

Vegetation Community	Control %	Reclaimed %	Percent Success %
Black sagebrush	23	6	28
Salt desert shrub	17	4	22

329

seasons. Overall survival is estimated at 80 percent. During the 1999 growing season evidence of reproductive growth was documented. Thus, HMC's goal to provide a supplemental seed source for shrubs has been met. DRiWATER was field tested during the 1997/1998 growing season. Given the abnormally high growing season precipitation, test results were inconclusive.

Homestake Mining Company completed final site reclamation in October/November 1999. Final reclamation included 65 acres of revegetation and 16 acres of remediation that included second growing season fertilization. In addition, the final 2,400 containerized supercell shrubs were planted in September 1999.

General observations of revegetation success in this arid region include: seed germination and establishment has typically been greater in harrow rows and the inner portion of the contour ripping benches. These techniques increase the potential to capture and retain moisture. Supplemental irrigation appears to favor the earlier establishment of herbaceous species of woody species. Revegetation of the flatter surfaces of the overburden pile tops subject to direct exposure to sun and wind is slower than the roughened and sloped sides. Based on the observations and data collected to date HMC, NDEP and the BLM are pleased with the site's progress toward meeting the goal of successful reclamation.

In September 1999, Homestake Mining Company of California received the Nevada Division of Minerals and Nevada Mining Association's Overall Excellence in Mine Reclamation Award for their success at the Santa Fe/Calvada Reclamation Project.

Tailings and Mine Waste'00 © 2000 Balkema, Rotterdam, ISBN 90 5809 126 0

Alternatives analysis for mine waste relocation repositories

B. R. Grant, M. Cormier & W. H. Bucher
Maxim Technologies Incorporated, Helena, Mont., USA

ABSTRACT: The Streamside Tailings Operable Unit, Silver Bow Creek/Butte Area, Montana NPL site requires disposal of mine and milling waste materials in mine waste relocation repositories. The Record of Decision (ROD) for the site requires that the mine wastes be amended with lime and revegetated, and that the repositories be sited in a protective location, potentially near Silver Bow Creek. If the near-stream setting for repositories are not sufficiently protective of human health and the environment, then the ROD provides for locating the repositories off-site where conditions would be more protective. An analysis was completed for the purpose of comparing the near-stream and off-stream settings, as well as to assess what types of design elements could be incorporated into the repository to raise the level of protectiveness to the environment. The completed analysis explored liner system alternatives, leachate generation rates, synthetic leachate characteristics, soil attenuation characteristics, estimated contaminant concentrations (particularly arsenic) in groundwater, and the cost associated with each of the alternatives.

1 INTRODUCTION

The Streamside Tailings Operable Unit (SSTOU) is one of several Operable Units that make up the Silver Bow Creek/Butte Area NPL Site. This NPL Site encompasses the historic mining and milling operations within the vicinity of the town of Butte, Montana as well as the mining and milling wastes present in Silver Bow Creek from its headwaters to the terminus of the site where Silver Bow Creek leaves the Warm Springs Ponds system. The SSTOU is defined as Silver Bow Creek, its present stream channel, current and historic floodplain, and railroad beds and embankments adjacent to the floodplain, and extends from the edge of Butte to the Warm Springs Ponds, a distance of about 40 km. Numerous studies have been completed to date as part of remedial investigation activities. Among others, soil sampling has documented the extent and magnitude of waste materials and levels of metals and arsenic in the waste and revegetation studies have been conducted to evaluate lime amendment methods and plant establishment techniques.

The remedial action for the SSTOU consists of removing mine and milling waste from the floodplain and areas adjacent to the stream. Mine waste relocation repositories (MWRRs) are a central component of the remedial action. The Record of Decision (MDEQ, 1995) for the site requires that the mine wastes be amended with lime and revegetated and that the MWRRs be sited in a protective location, potentially near Silver Bow Creek.

The evaluation which this paper presents was completed to study whether placing the MWRRs near Silver Bow Creek was a viable remedial action, or if locating the MWRRs further away from the stream in a more upland setting would provide a higher level of protectiveness that the near-stream setting could not afford. Determining the exact sites and design requirements for MWRRs was not a specific goal for the study. The evaluation was completed for Subarea 1 of the SSTOU which includes the upper reaches of the stream (Figure 1) in which six proposed, near-stream MWRRs were located. Total waste to be removed within the subarea is estimated at 380,000 m³. The waste is characterized as a silty sand, with elevated levels of copper, lead, zinc, cadmium and

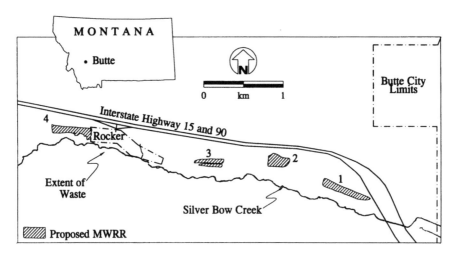

Figure 1. SSTOU, Subarea 1, upper reach.

Table 1. Groundwater Standards

Constituent	Concentration ($\mu g/l$)
Copper	1000
Lead	15
Zinc	5000
Cadmium	5
Arsenic	18
Mercury	0.14

arsenic, which has been fluvially deposited within the approximate 100-year floodplain of the creek.

Design criteria used to evaluate near-stream and off-site settings included geology, surface water hydrology, groundwater hydrology, community impacts and land use. The principal criterion was groundwater quality. Discharges to groundwater may not exceed Montana DEQ Circular WQB-7 standards for groundwater (Table 1). The arsenic concentration goal is quite low in comparison to the U.S. EPA maximum contaminant level for drinking water supplies of 50 $\mu g/l$.

2 METHODS

Essential components of the evaluation were determining leachate quality from the tailings waste and adsorption characteristics for soil which will underlie the repositories. Methods used for assisting in determining these characteristics are described in this section.

To approximate leachate quality, a synthetic leachate was derived. Waste material used to derive the leachate was acquired by obtaining a composite sample prepared from selected soil depth intervals from 9 test pits. Test pit locations and sample depths were determined using an existing database for 435 test pits. Preparation of the composite was patterned toward the average soil grain size (silty sand) and metal and arsenic concentrations that would approximate the third quartile levels of all samples collected from the 435 test pits. The composite sample was analyzed for saturated paste pH, arsenic, metals and several other analyses for determining the lime amendment rate.

Synthetic leachate solutions were prepared for amended waste and for unamended waste. Preparation of the synthetic leachate for unamended tailings was performed following EPA Method 1312 using a 60/40 mixture of nitric and sulfuric acid at an unbuffered pH level of 4.2. Preparation of the synthetic leachate for the lime amended waste was performed using reagent grade deionized water since the lime amended tailings are expected to produce a buffered leachate. Each synthetic leachate was prepared in triplicate batches with the extract fluid from each filtered and analyzed for pH, arsenic and metals.

Sorption testing was completed on subsurface soil materials from Settings 1 and 2 (discussed below). The Setting 1 sample was a composite sample collected from a boring located at proposed MWRR-2. The Setting 2 sample was collected from beneath the surface of an excavation cut immediately north of Rocker. Developed soil horizons were excluded from each soil sample. The sorption testing consisted of introducing 800 milliliters of synthetic amended tailings leachate at dilutions of 0, 50, 75 and 90 percent to 40 grams of soil. The tests were conducted in a manner similar to the leaching test by placing the solution and soil in a rotating drum for 18 hours. The extract fluid from each test was filtered (0.45 micron filter media) and the filtrate analyzed for total arsenic and copper.

3 DEVELOPMENT OF ALTERNATIVES

The process used to develop alternative repository designs is similar to the feasibility study process outlined in EPA guidance documents for evaluating National Priority List sites (USEPA, 1988). The first step of the process was to complete an initial screening of potential design components. These design components provided the building blocks for several different design options which were then applied to a selected set of environmental conditions. The initial screening of design components brought forward only the most likely components that could potentially meet or exceed repository design and siting criteria, either separately or in combination with other components.

3.1 Repository Design Components

The general elements of a waste repository include a cover system, the waste materials, and a leachate collection system. Design components were developed for each of these three elements during the initial screening process, with several components being selected for use above and below the mine wastes. The design components selected for incorporation into design options are described below.

3.1.1 Cover System

The function of a cover system is to protect the waste material enclosed within the repository from climatic effects of wind and precipitation. The major concerns associated with these effects are infiltration and erosion. The design components for a cover system that passed the initial screening are the following:

• Directly Amended Waste Materials - This design component would be used where no cover system is proposed. Mine waste would be amended with lime and revegetated. The revegetated surface would provide protection from erosion. Completed studies have shown that the SSTOU amended tailings can be revegetated (RRU and Schaefer & Assoc., 1993).

• Cover Soil - Cover soil applied above the amended waste can improve the revegetated cover such that moisture can be stored and evapotranspired more readily than with directly amended waste. In addition, cover soil further minimizes potential wind and water erosion of the waste and provides a barrier for direct contact with animals.

• Drainage Layer - A drainage layer is often included immediately above a low permeable liner. Commonly used material for drainage layers include moderate to highly permeable soil (e.g. sand and gravel) and synthetic materials such as a drainage net.

• Low Permeable Cover Liner - Liner materials include recompacted clay, geomembranes (such as HDPE), and geosynthetic clay liners (GCL). A GCL was selected for this project as a design component for options that contain a low permeable liner.

3.1.2 Waste Material

For design components, only treatment or non-treatment of the waste materials were considered for inclusion in the design options.

• Treated Waste - mixing of a lime amendment with waste materials to reduce the mobility of metals.
• Untreated Waste - waste materials placed without amendment.

3.1.3   Leachate Collection System

Low permeability liner systems at the base of a repository are often used to intercept and collect leachate percolating down through the waste material. Leachate collection systems can consist of the following elements.

• Drainage Layer - A drainage layer immediately above a low permeable base liner, constructed similarly to that described above for cover systems, provides a pathway for the leachate to be removed from beneath the repository. This leachate is collected and disposed by evaporation, incineration, or transport to an approved waste management facility.

• Low Permeable Liner - A low permeable base liner beneath the drainage layer reduces the amount of leachate entering the subsurface beneath the repository. Low permeable liners can consist of recompacted clay and synthetic products such as geomembranes or GCL. Only geomembranes were considered further in this evaluation.

3.2   Repository Design Options

Several designs options were constructed using the design components described above (Figure 2). The design options include the repository design as described in the ROD and presented in an Intermediate Design Report (ARCO, 1997) along with four alternate designs. The following discussion explains the components of each design option.

• Design 1 - Revegetated Amended Waste Material (ROD Design) - Waste material would be placed in the repository and amended with lime. Amended waste material at the surface of the repository would be revegetated to reduce infiltration and erosion.

• Design 2 - Improved Cover, Amended Waste Material - A soil cap consisting of at least 18 inches of borrow soil would be placed over the amended tailings. The use of 18 inches of soil cover was selected for this design to provide adequate rooting depth and to improve soil moisture storage capacity in the cap. Plant roots would be able to extend into the underlying amended tailings. The soil would be well suited for establishing a vegetative cover and provide adequate protection from wind and water erosion.

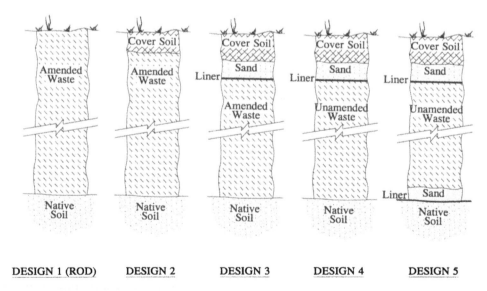

Figure 2. Repository Design Options

• Design 3 - Improved Cover with Low Permeable Liner, Amended Waste Material - This design is similar to Design 2 but incorporates a low permeable liner in the cover system to reduce infiltration into the repository. A one to two foot drainage layer is included immediately above the cover liner to transmit water away from the repository. The cover soil thickness was increased to two feet to provide additional soil moisture storage and rooting material since a geotextile filter fabric would need to be placed between the cover soil and drainage layer.

• Design 4 - Improved Cover with Low Permeable Liner, Unamended Waste Material - This design is similar to Design 3 but evaluates placing unamended tailings in the repository. In this design, higher metal concentrations may be produced in leachate since the waste is unamended. A low permeable liner in the cover will reduce overall leachate volume as compared to Designs 1 and 2.

• Design 5 - Improved Cover with Low Permeable Liner, Unamended Waste Material, Leachate Collection System - This design is similar to Design 4 but includes a leachate collection system at the base of the repository below the unamended waste materials. In this design, leachate would be collected and managed for either further treatment, evaporation, or disposal.

3.3   *Site Environmental Settings*

In order to evaluate the alternative MWRR designs, two potential site settings were assumed to be the most likely for siting the repositories. The differences between the two site settings is primarily subsurface soil texture, depth to ground water, and distance from the 100-year floodplain. The following describes the general environmental conditions expected to be encountered at the two sites.

• Setting 1 - Near-Stream Repository Site. Near-stream repository sites would be located immediately adjacent to the 100-year floodplain as proposed in the ROD. A typical setting for these sites is on moderate slopes adjacent to Silver Bow Creek. Subsurface soil may be composed of sandy loam soil and/or weathered bedrock (grus) with quartz monzonite bedrock at a relatively shallow depth. This site setting has a minimum separation from ground water of 10 feet at the base of the repository. Hydrogeologic characteristics are:

	Soil Type	medium sand
	Saturated Permeability	10^{-2} cm/sec
	Groundwater Gradient	0.01 m/m

• Setting 2 - Off-stream Repository Site. These sites are located in an upland setting underlain by Tertiary sediments and located well away from the 100-year floodplain. Tertiary sediments north of Silver Bow Creek typically exhibit a higher content of silt and clay than near-stream sites. The minimum separation from ground water in Setting 2 is assumed to be greater than 40 feet. Hydrogeologic characteristics for this setting are:

	Soil Type	sandy silt
	Saturated Permeability	10^{-4} cm/sec
	Groundwater Gradient	0.01 m/m

3.4   *Number of Repository Sites*

The ROD identified six near-stream locations for MWRRs within Subarea 1. Figure 1 shows the location of four of the MWRRs. Designing numerous repositories in a near-stream location minimizes haul distances and reduces cost but has disadvantages as well. Using a greater number of repositories will have greater area to volume ratios, requiring more land to be overlaid with tailings and greater disturbance area for construction. The numerous repositories will require an increase in the amount of vegetation needed for erosion protection and, if the design includes cover materials, greater cover soil material volumes and associated costs. In addition, higher monitoring and maintenance costs may result from numerous repositories with lesser volumes of tailings per repository.

In the near-stream setting (Setting 1), construction of three repositories for designs other than Design 1 was selected to provide a balance between cost of construction, hauling tailings a greater distance, and maintenance of the repositories. In the off-stream setting (Setting 2), two repositories were selected by extension of the reasons given for Setting 1.

Table 2. Repository Design Alternatives

Alternative No.	Design/Setting	No. of Sites	Alternative No.	Design/Setting	No. of Sites
1	Design 1/Setting 1	6	5	Design 2/Setting 2	2
2	Design 2/Setting 1	3	6	Design 3/Setting 2	2
3	Design 3/ Setting 1	3	7	Design 4/Setting 2	2
4	Design 1/Setting 2	2	8	Design 5/Setting 1	2

3.5 *Repository Design Alternatives*

Repository design alternatives were developed by combining the various design options with the two environmental settings (Table 2). Design 4 was not considered to comply with the ROD if placed in Setting 1 without a base liner. Design 5 was not considered for Setting 2 because Setting 2 is considered to provide an added depth to groundwater and offered native materials that could provide attenuation of metals and arsenic in leachate.

4 GROUNDWATER PROTECTION ANALYSIS

Physical and chemical characteristics associated with each of the alternatives affect metal and arsenic concentrations in groundwater downgradient of the MWRR. The repository designs are analyzed with respect to their effectiveness in protecting groundwater quality. To perform the analysis, information on leachate characteristics was obtained. The various repository designs are then evaluated with respect to contaminant loading to the subsurface and attenuation of metal and arsenic concentrations by soil sorption and dilution/dispersion in groundwater.

4.1 *Leachate Characteristics*

Amending tailings material placed in the MWRR would result in minimizing the potential for acid conditions to develop. Most mobile metal cations in the tailings (Cu, Pb, Zn, Cd) would form relatively stable precipitates as a result of the lime addition. At high pH levels, similar to which may be found in the amended tailings, arsenic may exhibit an increase in solubility (Dragun, 1988).

Table 3 presents metal and arsenic concentrations in the composite waste sample, which was used for preparing the synthetic leachates, along with statistical data for these elements collected across the site (ARCO, 1995a). The pH of a saturated paste made from the waste was 4.5. Table 4 compares metal concentrations for fluids from amended waste obtained from recent SSTOU studies and from the synthetic leachate study. Arsenic concentrations in leachate are approximately one order of magnitude greater than the WQB-7 standard; cadmium is only slightly elevated above the standard.

Table 3. Metal Concentrations in Composite Waste Sample (mg/kg)

Description	Copper	Lead	Zinc	Cadmium	Arsenic
Composite tailing sample	1717	2243	3337	22	1059
Subarea 1, mean values	739	540	2400	8	278
Subarea 1, third quartile	2458	1048	3560	17	491

Table 4. Metal Concentrations in Leachate from Amended Tailings Material (µg/l)

Sample Location	Copper	Lead	Zinc	Cadmium	Arsenic
Lysimeter 07D14; 40 and 90 cm depth, collected 1991 and 1992; mean concentrations[1]	121	2.5	1530	9.6	208
Lysimeter SPW3-S, depth 45 cm, mean concentrations[2]	182	7	717	9	175
Synthetic Leachate, Amended Tailings	130	<150	<30	<15	210
WQB-7 Standards	1000	15	5000	5	18

Data Source: [1] RRU and Schaefer, 1993; [2] Schaefer, 1997

Table 5 presents pore water metal concentrations from lysimeters in unamended tailings and from synthetic leachate. The unamended tailings exhibit higher concentrations of several elements with exceedances of WQB-7 standards possible for copper, lead, zinc, cadmium and arsenic. With the exception of arsenic, amendment of the tailings with lime results in a noticeable improvement in leachate quality.

4.2 Performance Evaluation of Alternative Designs

An evaluation of the alternative designs was completed using the generalized hydrogeologic characteristics for the two types of settings to gain an understanding of what effects leachate metal and arsenic concentrations may have on groundwater quality. This evaluation was performed for the two types of hydrogeologic settings defined.

4.2.1 Leachate Production Estimates

Each of the MWRR designs was evaluated using the Hydrologic Evaluation of Landfill Performance (HELP) model, version 3.05 (Schroeder, et. al., 1994) to predict the amount of leachate which would percolate through the base of each design. The results are summarized in Table 6. The model was run using 100 years of local precipitation data.

4.2.2 Attenuation of Metals by Sorption

Metal cations and arsenic anions in leachate flowing through the subsurface will be attenuated by sorption which includes the processes of adsorption, fixation and precipitation. Sorption testing was performed using the amended tailings synthetic leachate for soil samples collected from Settings 1 and 2. Fluid from the testing was analyzed for copper and arsenic because only these two elements were present at significant concentrations in the synthetic leachate. Sorption testing was not completed using unamended tailings synthetic leachate.

Figure 3 presents the isotherms generated from the sorption testing using Cs (sorbed mass) versus Ce (equilibrium concentration) graphs. The arsenic isotherms indicates that the soil sorption affinity decreases at lower equilibrium concentrations and that the Setting 1 soil more readily adsorbs aresenic than Setting 2 soil. Arsenic sorption did not appear to occur at concentrations below about 0.024 mg/l for the Setting 1 sample and 0.018 mg/l for the Setting 2 sample. The lower limit of arsenic sorption in the test are similar to arsenic concentrations observed in a nearby background well (well C-15) which exhibits an arsenic concentration near 11 µg/l (ARCO, 1995b). The copper isotherms indicate that the metal is adsorbed be the soil. Several of the copper equilibrium concentrations were below laboratory detection.

Table 5. Metal Concentrations for Leachate Solution from Unamended Tailings (µg/l)

Sample Location	Copper	Lead	Zinc	Cadmium	Arsenic
Highest measured concentrations in lysimeters installed within Clark Fork River untreated test areas, collected 1992 [1]	1050	nm	222	nm	23
Lysimeter SPW3-D, depth 45 cm, mean concentrations [1]	3076	7	15,225	60	135
Synthetic Leachate, Unamended Tailings	35,500	3820	28,300	39	350
WQB-7 Standard	1000	15	5000	5	18

nm = not measured
data source: [1] Schaefer, 1997

Table 6. Leachate Production Estimates

Design No.	Leachate (cm/yr))
1	5.5
2	1.6
3	0.08
4	0.08
5	0.02

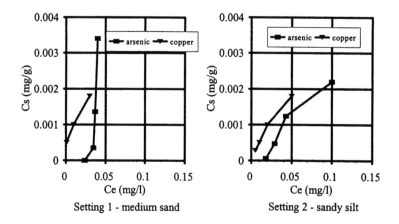

Figure 3. Sorption Isotherms

4.2.3 *Dispersion/Dilution*

Leachate which enters the groundwater system beneath the repository will be dispersed and diluted. Leachate and attenuation test data indicate that repositories constructed with amended tailings will show relatively low concentrations of arsenic in the leachate which reaches groundwater. Copper, lead, zinc, and cadmium will not likely be present due to amendment of tailings with lime and/or sorption of the metals with soil. Table 7 presents estimates for arsenic concentrations in groundwater for Alternatives 1 through 6 (amended waste). To prepare these estimates we assumed that the repository is 4 ha in area with a length to width aspect of 1:1 and that leachate is mixed within the uppermost 3 m of the aquifer. Estimates of arsenic in leachate reaching the groundwater were obtained from the attenuation testing of synthetic leachate (24 and 18 µg/L), and to demonstrate the sensitivity of the mixing calculation, estimates were also prepared under the assumption that attenuation of arsenic is relatively ineffective in the subsurface and the leachate contains 100 µg/L or 200 µg/L of arsenic. This could be the case if very little attenuating material is located under the repository or if the arsenic attenuation capacity of the subsoil is exhausted over time. A background arsenic concentration in groundwater was selected using data for well C-15.

Results shown in Table 7 indicate that arsenic may not exceed WQB-7 groundwater standard provided certain design components are added to the repository. For Setting 1 and using the attenuation test results, incorporation of a low permeable liner into the repository cover system

Table 7. Estimated Arsenic Concentration in Groundwater After Dilution/Dispersion of Leachate (µg/l)

| Alternative No. | Attenuation Test Results | | | | | |
	Setting 1 As=24 µg/l	Setting 2 As=18 µg/l	Setting 1 As=100 µg/l	Setting 2 As=100 µg/l	Setting 1 As=200 µg/l	Setting 2 As=200 µg/l
1	13	--	21	--	31	--
2	12	--	14	--	17	--
3	11	--	12	--	12	--
4	--	17	--	93	--	185
5	--	16	--	79	--	156
6	--	12	--	25	--	40
7	na					
8	na					

WQB-7 Standard = 18µg/l
Background Groundwater = 11.4 µg/l

-- not applicable, the alternative does not apply to this setting
na = not analyzed (unamended tailings)

does not appreciably affect arsenic concentrations as indicated by the 2 µg/l difference between Alternative No. 1 and Alternative No. 3. As compared to Setting 1, the Setting 2 alternatives result in higher arsenic concentrations in large part because of slower groundwater velocities and therefore less dilution of the leachate. For Setting 2 (Alternatives 4, 5, 6) inclusion of a liner in the cover shows a more distinct advantage in lowering arsenic concentrations in groundwater.

It is interesting to note the sensitivity of this calculation to the leachate quality. If we assume that leachate has 100 µg/L or 200 µg/l of arsenic, only Alternatives 2 and 3 provide sufficient protection to meet the Circular WQB-7 standard after mixing. This analysis demonstrates the value of the improved cover designs within Setting 1 if attenuation capacity can not be relied upon to reduce arsenic concentrations.

5 SUMMARY

The most significant points resulting from the evaluation include the following.

• Arsenic appears to be the only constituent in the amended tailings case which has the potential to exceed regulatory levels in groundwater downgradient of the repositories.

• Alternatives 1, 2 and 3 (Setting 1) appear to provide adequate protection to groundwater resources as compared to Alternatives 4, 5, 6 and 7 which are located off-site (Setting 2).

• Alternatives 2 and 3, which provide an improved repository cover over Alternative 1, do not substantially reduce the arsenic concentration in groundwater because of elevated background arsenic concentration in groundwater. However, cover improvements contained in Alternatives 2 and 3 do reduce the contaminant load reaching the groundwater system and provide greater assurance that groundwater quality standards will not be exceeded.

• Alternative 1 includes several repositories some of which are smaller in size than the size evaluated by this study (4 ha). The smaller repositories are less prone to result in arsenic exceedances in groundwater due to a smaller area over which leachate is entering groundwater. One of the main disadvantages of this alternative is the smaller repositories are less efficient in storing tailings so that more land area is disturbed. The overall area for Alternative 1 repositories is approximately 24 to 28 ha and the overall area for Alternatives 2 through 8 repositories is approximately 6 to 8 ha.

The calculations of leachate production, leachate loading, attenuation, and groundwater mixing completed in the study are subject to a number of limitations due to limited data on which the analysis is based and the inherent uncertainty in geochemical and hydrogeological parameters used in the calculations. Results of the attenuation study indicate that arsenic concentrations in leachate that have been subject to attenuation may exceed the Circular WQB-7 standard because arsenic does not appear to completely sorb to soil at relatively low solution concentrations. Because of limitations in data and uncertainties in calculations, the quantitative results of the evaluation were used more to gauge the relative protectiveness of the alternatives rather than their absolute capabilities.

REFERENCES

Atlantic Richfield Company (ARCO) 1997. Silver Bow Creek/Butte Area NPL Site (NPL). Streamside Tailing Operable Unit. Intermediate Design Report. July 1997.

ARCO, 1995a. Silver Bow Creek/Butte Area National Priorities List Site (NPL). Streamside Tailing Operable Unit. Draft Remedial Investigation Report. January 1995.

ARCO, 1995b. Silver Bow Creek/Butte Area NPL Site. Streamside Tailings Operable Unit. Draft Feasibility Study Report, Revision 2.

Dragun, J., 1988. *The Soil Chemistry of Hazardous Materials*. Silver Spring, Maryland: Hazardous Materials Control Resources Institute

Montana Department of Environmental Quality (MDEQ), 1995. Record of Decision, Streamside Tailings Operable Unit of the Silver Bow Creek/Butte Area National Priorities List Site.

Reclamation Research Unit of Montana State University (RRU) and Schaefer & Associates, 1993. Streambank Tailings and Revegetation Studies, STARS Phase III Final Report. Montana Department of Health and Environmental Sciences. Helena Montana.

Schaefer & Associates, 1997. Clark Fork River Governor's Demonstration Monitoring (1993-1996). Final

Report.

Schroeder, P.R., Dozier, T.S., Zappi, P.A., McEnroe, B.M., Sjostrom, J.W., & Peyton, R.L., 1994 *The Hydrologic Evaluation of Landfill Performance (HELP) Model: Engineering Documentation for Version 3*, EPA/600/9-94, U.S. Environmental Protection Agency Risk Reduction Engineering Laboratory, Cincinnati, OH.

United States Environmental Protection Agency, 1988. *Draft Guidance for Conducting Remedial Investigations and Feasibility Studies under CERCLA.* Office of Emergency and Remedial Response and Office of Solid Waste and Emergency Response.

Radioactivity and risk

Tailings and Mine Waste'00 © 2000 Balkema, Rotterdam, ISBN 90 5809 126 0

Speciation of ^{226}Ra, uranium and metals in uranium mill tailings

S. Somot
URSTM-Université du Québec en Abitibi-Témiscamingue, Rouyn-Noranda, Qué, Canada

M. Pagel
Departement des Sciences de la Terre, UMR Orsay-Terre, Université Paris XI, Orsay, France

J. Thiry & F. Ruhlmann
SEPA Bessines et Service Qualité Sécurité Environnement, COGEMA/BU, France

ABSTRACT: Mineralogical speciation of ^{226}Ra, uranium and metals in uranium mill tailings is of prime importance in order to choose best long-term tailing pound remediation options. Methods of sequential and selective leachings are simple methods to quantify ^{226}Ra, uranium and metals fractions that are bound to mineralogical phases. More, results allow a better understanding of radium behaviour in the uranium mill tailings studied. These methods may also allow the prediction of the environmental impact of the tailings. For example, it may predict consequences of an increase in permeability, an acidification or a reduction of uranium mill tailings.

1 INTRODUCTION

The management of great quantities of uranium mill tailings raises the problem of long-term control of ^{226}Ra, U and metals migration from the impoundment area to the environment. Mobility of radio-nuclides and metals depends mainly on their mineralogical speciation coupled to pore water composition, hydrologic conditions and nature of active bacteria populations. Physical, mineralogical and chemical parameters of the geologic environment for radium have changed from the ore deposits to mill tailings ponds and are changing with time within mill tailing storages. In order to evaluate (1) the way mill tailings evolve with time and (2) the mineralogical speciation of radio-nuclides and metals, a characterization of uranium mill tailings and of their pore water was investigated at various depths of sampling within mill tailing impoundments [Somot & al., 1995; Somot & al., 1997]. Modelling of diagenetic evolutions were deduced from the simultaneous study of mill tailings samples and of their pore waters, but the behaviour of ^{226}Ra remained uncertain. It was then decided to study the distribution of ^{226}Ra on various mineralogical phases present in mill tailings. A qualitative mapping of ^{226}Ra, by coupling alpha auto-radiography and fission tracks mapping on thin sections of tailings, suggests that radium was essentially linked to sulfates and iron oxy-hydroxides. A selective and sequential method of speciation [Gupta & Chen, 1975] of radium and accompanying elements (including potentially toxic metals) was developed in order to quantify ^{226}Ra proportions which are respectively bound (1) to easily leached elements, such as exchangeable ions [Coleman & al., 1959], as exchangeable acidity [Espiau & Peyronel, 1976] or as easily dissolved salts, (2) to acidic and complexing leachable phases [Tamm, 1934], (3) to reducing and complexing leachable phases [Mehra & Jackson, 1960] and (4) to residual minerals. Leaching tests with water was also applied to these samples, as recommended by the AFNOR X 31-210 norm [1988], and results were compared.

2 MATERIAL AND METHODS

Uranium mill tailings that were compared in this study were sampled in three mining sites of France : Ecarpière in Loire-Atlantique, Jouac in Haute-Vienne and Lodève in Hérault. Main differences between them are (1) the nature of the host rocks of the ore, (2) the nature of the milling process and (3) the age of mill tailings deposits.

2.1 From the ore to acid and alkaline uranium mill tailings

Processed ores have been described by geologists for exploration purposes. At Ecarpière, ores were extracted from granitic to metamorphic host rocks. That of Jouac were extracted mainly from the episyenitic host rock of Le Bernardan and from Le Piégut mine hosted in gneissic rocks. The ores of Lodève were hosted in Permian volcano-sedimentary rocks. The milling processes of Ecarpière and Jouac ores were an acidic one whereas, that of Lodève was alkaline because of the high content of carbonates in the ore. Main reactants of the whole acidic milling process were sulfuric acid, sodium chlorate, whereas main ones for the alkaline process were sodium bicarbonates, sulfuric acid and soda. At Jouac, sludge from the neutralization of effluents were continuously added to mill tailings and influenced their mineralogical and geochemical composition. It is important to remember that the pH values of all these mill tailings range from 6.2 to 9.9, because of the addition of lime for neutralization at the end of the acidic process.

2.2 Sampling methods of uranium mill tailings and of pore waters

In Ecarpière, a core (C1-samples) has been drilled within mill tailing impoundment by CO-GEMA and analyzed for mineralogy and geochemistry. Another drilling core (C7-samples), at about 32 m far from C1 has been drilled. Drilling cores from Ecarpière (C7), Jouac and Lodève have been sampled using a specific open barreled corer, allowing to collect structurally, texturally undisturbed samples at various depth, in order to study simultaneously pore waters and mill tailings. Immediately after the drilling, transparent tubes that enclosed cores were properly sawed at various depths in sections of about 10 cm length, depending on gamma radioactivity, color and texture of samples. Sampling has been performed all along the profile from the top of the mill tailing impoundment to the underneath substratum. These samples contain informations on diagenetic evolutions of mill tailings from their disposal in impoundment until the date of sampling, that is for about 35 years in Ecarpière, 12 years in Lodève and 11 to 16 years in Jouac. Mineralogical and geochemical characterizations lead to distinguish the ore hosting rocks influence from the mill tailings own characteristics. Immediately after their segregation, mill tailings samples were squeezed in a stainless device [Manheim, 1966 ; Kalil & Goldhaber, 1973] and pore-water was drained to syringes within less than an hour. Dissolved oxygen, Eh and pH values were measured immediately after the collection of pore water with daily calibrated electrodes, taking into account salinity of water, temperature and altitude.

2.3 Chemical analysis of mill tailings and pore waters samples

Pressed solid samples (C7) were isolating from atmospheric oxygen in air evacuated and sealed plastic bags for transport to laboratory, before being dried at 40 °C and were then homogeneized. A small fraction was crushed, fused with $LiBO_2$ and dissolved in HNO_3 before analysis. ^{226}Ra activities were determined by gamma spectrometry (SEPA/COGEMA, Bessines, France). Other analysis were performed at the SARM (CRPG/CNRS, Nancy, France). Major and minor elements were analyzed by ICP-ES. Trace and Rare Earth elements were analyzed by ICP-MS. International geostandards were used for quality control of aqueous and solid analysis. Carbonates were analyzed by a coulometric CO_2 analyzer. Sulfates dissolved in boiling water and total sulfur were determined by a gravimetric method.

344

Petrography and mineralogy of samples were studied by scanning electron microscopy combined with back-scattered electrons detector and an energy dispersive spectrometer (University of Nancy, France). X-ray diffraction and mineralogical studies were performed by COGEMA on samples from the C1 drilling [Reyx et Pacquet, 1992].

The analysis of dissolved cations and metals in pore waters were performed on acidified samples. Samples for the quantification of anions and isotopic determinations were not acidified, but they were stored at 6°C in chemically inert bottles, filled to the brim and covered with paraffin films before analysis. Dissolved ^{226}Ra was analyzed by a scintillometric counting method after radon emanation. Dissolved cations (Ca, Na, K, Mg, Fe, Si, Al, Mn) were analyzed by atomic absorption spectrophotometry (AAS), dissolved anions (Cl, SO_4, HPO_4, ...) were determined by ion chromatography and other dissolved elements were analyzed by inductively coupled plasma mass spectrometry (ICP-MS).

2.4 *Mineralogical speciation of uranium, radium and metals*

Uranium and most of potentially toxic metals contents are generally sufficiently high in mineralized rocks to be observed and quantified by microscopic technics as separated minerals, whereas radium content is so low in nature that radiferous minerals can not be quantified by any usual microscopic technique. Many studies showed that during grounding, ^{226}Ra was concentrated in the fine sized minerals preferentially to coarser ones [Seeley, 1977]. It has been shown by various methods that in uranium mill tailings ^{226}Ra can be bound to sulfate minerals [Shearer & Lee, 1963; Kaiman, 1977; Snodgrass, 1982; Nirdosh & al., 1984; Snodgrass, 1990; Gnanapragasam & Lewis, 1995], to feldspars [Latham & Schwarz,1987; Landa & Bush, 1990], to clays [Ames & al., 1983; Bassot, 1997], to amorphous silica [Benes & al., 1981; Nirdosh & al., 1987; Valentine & al., 1987], to oxi-hydroxides [Ebler, 1915; Nirdosh & al., 1989], to magnetic fractions and sulphides [Haque & al., 1986] and to organic matter [Benes & al., 1981; Landa, 1982; Nirdosh, 1987]. But, no quantitative mineralogical speciation of radium has been made in uranium mill tailings that consist in a mix of several of these phases.

First of all, ^{226}Ra behaviour was approached by comparing its activity profile with depth to other element contents profiles. No general explanation concerning the behaviour of radium in uranium mill tailings was entirely satisfactory.

The major isotopic form of radium is ^{226}Ra, resulting from successively naturally occuring ^{238}U and ^{230}Th disintegration, and which direct daughter is ^{222}Rn. Radium is indirectly quantified by its daughter, which is a rare gas and which half life time is about 3 days only. That is in a non aerated system, ^{222}Rn activity becomes fast equal with ^{226}Ra activity. Following this assumption, frequent alpha particles emissions from radon desintegration are representative of high contents of ^{226}Ra. In uranium mill tailings, the aim of the study was first to speciate radium from uranium containing minerals. Uranium is also an alpha emitter, but only it is able to give fissions tracks. Coupling the mapping of both fission tracks and alpha emitters for a polished sample of mill tailings, it was possible to distinguish radium enriched zones from uranium enriched ones. This oriented the interpretations to the mobilization of radium from inherited minerals to gypsum, iron oxy-hydroxides and perhaps carbonates during the milling process and/or during diagenesis of disposed mill tailings in impoundment. But this method only allowed a qualitative study.

Sequential and selective leachings were performed in order to quantify ^{226}Ra which was bound to various mineralogical phases and especially to sulfates and iron oxy-hydroxides. Various methods of sequential leachings [Gupta & Chen, 1975; Tessier & al., 1979] have been proposed by scientists. The following leaching methods were chosen because of their respective higher selectivity as regarding easily dissolved salts and iron oxy-hydroxides. They have been applied sequentially on a same sample. Experimental procedures were applied aiming the dissolution :

- of easily leached elements, such as exchangeable ions [Coleman & al., 1959] or as exchangeable acidity [Espiau & Peyronel, 1976] or as easily dissolved salts (optimization of the procedure for gypsum dissolution), in a solution of potassium chloride,
- of acidic and complexing leachable phases such as amorphous and bad crystallized iron oxy-hydroxydes [Tamm, 1934] in an oxalic acid buffer at pH 5. A part of sulfides, ferrous chlorite, biotite and illite [Arshad & al., 1972] are also altered.
- of reducing and complexing leachable phases, such as well crystallized oxy-hydroxides in a solution of citrate-bicarbonate-dithionite solution [Mehra & Jackson, 1960]. Barite is perhaps also leached in these highly reducing conditions.
- and finally, of residual minerals in a triacid solution of HF, HNO_3 and $HClO_4$ or by $LiBO_2$ fusion.

Leachates were successively filtrated through a 0.45 µm pore size membrane and analyzed for dissolved ^{226}Ra, major and trace elements after appropriate dilutions or concentrations. Experimental proportions are presented without bringing them to a total of 100 % (Tab. 5). When mass balance error exceeds 25 %, the analysis is excluded. Experimental errors include the possibility of material loss when filtering and manipulating. Moreover, successive errors on each leaching and analysis of concentrated solutions are cumulated, justifying the acceptation of a maximum global error of 25 %. However, most of presented results show a good mass balance near to 100 %. The AFNOR X 31-210 procedure [1988] recommends three (or more for this study) successive leachings of 100 grams of raw material, in an agitated bottle, with one liter of demineralized water, during sixteen hours. Leachate is then filtrated on 0.45 µm pore size membrane before chemical analysis.

3 RESULTS

3.1 *Main characteristics of studied mill tailings*

In mass proportions, main minerals of uranium mill tailings are quartz, feldspars, micas, clays, sulfides and depending on mines, carbonates. Inherited minerals from the ores are also rutile, tourmaline, zircon, apatite, monazite, uraninite, fluorite, barite, coffinite, mineralized organic matter. Studied acid uranium mill tailings are characterized by their volumetric abundance of gypsum and their reddish to orange color (except samples resulting from gneissic hosting rock ore). Barite is abundant in Jouac samples because sludges of the effluent treatment process are added to mill tailings. Studied alkaline ones are characterized by the abundance of carbonates, the diversity of sulfides and their black to greenish color. Authigeneous uraniferous minerals and clays have been observed [Reyx & Pacquet, 1992; Somot & al., 1995]. Mean contents of particular elements of interest with respect to radium behaviour in mill tailings are gathered in Table 1.

Table 1. Uranium, ^{226}Ra and metals contents (mean values on n samples) in studied uranium mill tailing samples

		Ecarpière	Jouac	Lodève
		n = 11	n = 8	n = 9
^{226}Ra	(Bq / g)	25.7	41.4	27.5
U	(g / kg)	129	134	313
CaO	(%)	8.0	4.7	6.6
Sr	(g / kg)	69	69	191
Ba	(g / kg)	218	1020	529
Pb	(g / kg)	142	204	417
As	(g / kg)	122	99	178
Fe_2O_3	(%)	4.1	3.8	6.8
MnO	(%)	0.1	0.1	0.1

3.2 Radium activities profiles with depth

3.2.1 Radium activities profiles in mill tailings

At Jouac, the ^{226}Ra activities profile is similar in shape with many other elemental content profiles because ^{226}Ra activity and many other elemental contents are lower in samples resulting from the process of Le Piégut Mine. At Ecarpière and Lodève, the ^{226}Ra activities profiles with depth are distinct from all other elemental contents in mill tailings. This results mainly from the mineralogical speciation of elements.

3.2.2 Dissolved radium activities in pore waters

In all studied sites, dissolved ^{226}Ra activities profiles with depth are approximately similar in shape with dissolved sulfates concentrations profiles, but better similitudes in profiles shapes are observed, such as :
- In Ecarpière : similarity of profiles between ^{226}Ra activities profile and Sr concentrations one (Fig. 1a), and in less extent with that of sulfates, K, Mg. There are no correlation with As concentrations or with Eh values profiles.
- In Jouac : similarity of the profile of ^{226}Ra activities, with that of As concentrations (Fig. 1b) and with that of dissolved oxygen concentrations one. There is no correlation with Sr concentrations profile.
- In Lodève : similarity of profiles between ^{226}Ra activities profile and Eh values one and in less extent with that of Sr, Ca, K, Mg. There is no correlation with As concentrations profile.

The link between sulfates and radium is suspected, but the lack of systematic correlations did not allow to draw clear conclusions on the behaviour of radium in acid and alkaline uranium mill tailing ponds at this stage.

3.3 Uranium and radium mapping

Results have been obtained on several samples of Ecarpière and only semi-statistical analysis brought to these conclusions. Uranium and radium mapping showed that the radioactivity of finer mill tailings was higher than that of coarser ones as mentioned by Seeley [1977]. Whatever the granulometry, a part of alpha emitters is dispersed in the matrix, such as gypsum is. Moreover, in a convenient and demonstrative fragment of mill tailings, an accumulation of oriented gypsum crystals was bordering an oxidized microfracture. On the other side of the microfracture, oriented clays rich in iron and magnesium were deposited. Alpha tracks map of this piece of mill tailings was easily compared to mineralogical composition. It showed that alpha

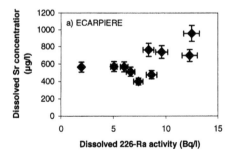 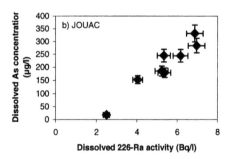

Figure 1 : Correlations between dissolved 226Ra activities in uranium mill tailings pore waters and a) dissolved strontium concentrations in Ecarpière and b) dissolved arsenic concentrations in Jouac.

tracks were concentrated in the oxidized microfracture and that they could not be related to dispersed residual uraniferous minerals only. Radiating groups of alpha tracks were also revealed at the azimut of isolated residual uraniferous minerals, but also where radiferous barite was observed and where pyrite was partially oxidized as amorphous ferric hydroxides.

The limitation of fast and effective mineralogical speciation of radium by this technique is due to fine granulometry of minerals compared to the size of revealed alpha tracks and to the complex mix of several minerals able to trap radium in mill tailings (gypsum, iron oxy-hydroxides, clays,...).

3.4 Sequential and selective leachings

Sequential and selective leachings were applied on four to six of studied samples of stored mill tailings from Ecarpière, Jouac (oxidized facies) and Lodève. Knowing the initial composition of leached mill tailings, mass balance have been verified for each analyzed elements in order to validate results of leachings. Corrections were made in regards with major elements concentrations from residual pore waters in initial samples, which are dissolved in the first leachate.

3.4.1 Leaching with KCl

In these conditions of a high ionic strength solution, all gypsum of acid mill tailings is dissolved. Proportions of ^{226}Ra that are dissolved differ from a mining site to another and from a sample to another (Tab. 2). Results obtained in this study on uranium mill tailings have been compared with experimental results of Gnanapragasam & Lewis (1995) obtained on synthetic radiferous gypsum. Then, in all studied acid mill tailings samples it has been demonstrated that dissolved ^{226}Ra in pore waters was in equilibrium with rediferous gypsum. Proportions of radium controlled by gypsum does not exceed 20% in the studied acid mill tailings.

Gypsum is not abundant in alkaline mill tailings (except in samples from the bottom of the pile) but mirabilite has been observed and carbonates could have been partially dissolved at this step. When gypsum is abundant, dissolved ^{226}Ra control by gypsum is evidenced both by dissolved sulfates profile with depth within the tailings pond and by high proportions of radium leached from deeply buried samples (around 30 %). Comparison of these results on alkaline mill tailings with experimental results of Gnanapragasam & Lewis [1995] show the role of carbonates in ^{226}Ra mobilization when gypsum is not abundant, but chemical data are less accurate for carbonates than for gypsum and conclusions are uneasy to draw with certitude.

3.4.2 Leaching with oxalic buffer

Oxalic buffer is used in order to dissolve amorphous iron or manganese oxy-hydroxides and elements bound to them. The pH solution is buffered to 5. In these acidic conditions, great pro-

Table 2. Maximum and minimum (from n values) proportions (w%) of uranium, ^{226}Ra and metals leached from mill tailings samples in high ionic strength conditions (KCl, 1M, L/S=20, ambiant temperature)

	Ecarpière	Jouac	Lodève
	n = 4; n = 2*	n = 4; n = 2*	n = 5; n < 5 *
^{226}Ra	9 – 18	3 – 5 *	9 – 43
U	0 –2 *	0 – 1	14 – 17 *
Ca	85 – 90	63 – 82	2 – 6
Sr	27 – 37 *	14 – 19 *	--
Ba	1 – 5 *	1 *	1 – 2 *
Pb	0 – 1 *	0 *	0 *
As	0 – 1 *	1 – 2	1 – 2
Fe	0	0	0
Mn	6 – 20 *	4 *	0

Table 3. Maximum and minimum (from n values) proportions (w%) of uranium, ^{226}Ra and metals leached from mill tailings samples in acidic conditions (Oxalic buffer pH 5, L / S = 40, obscurity, ambiant temperature)

	Ecarpière	Jouac	Lodève
	n = 4; n = 2 *	n = 4; n = 2 *	n = 5; n < 5 *
^{226}Ra	39 – 48	64 – 72 *	17 – 34
U	69 – 82 *	48 – 61	5 – 7 *
Ca	0	0	0
Sr	10 – 24 *	0 *	--
Ba	8 – 10 *	11 – 12 *	4 – 7 *
Pb	7 – 9 *	10 – 12 *	4 – 8 *
As	40 – 81 *	27 – 45	72 – 82
Fe	45 – 58	13 – 26	30 – 39
Mn	45 – 54 *	59 – 61 *	26 – 41

Table 4. Maximum and minimum (from n values) proportions (w%) of uranium, ^{226}Ra and metals leached from mill tailings samples in reductive and complexing conditions (CBD, L/S = 40, 80°C)

	Ecarpière	Jouac	Lodève
	n = 4; n = 2 *	n = 4; n = 2 *	n = 5; n < 5 *
^{226}Ra	34 – 45	35 – 45 *	14 – 28
U	7 – 8 *	8 – 31	0 *
Ca	9 – 13	11 – 18	25 – 43
Sr	0 – 21 *	0 *	--
Ba	0 – 14 *	11 – 15 *	14 – 25 *
Pb	6 – 10 *	0 *	25 – 29 *
As	0 – 6 *	15 – 21	0
Fe	9 – 16	34 – 44	5 – 6
Mn	10 – 11 *	28 – 31 *	12 – 19

portions of radium, iron, manganese and arsenic are leached from acid mill tailings samples (Tab. 3). High proportions of uranium are leached from acid mill tailings samples. In alkaline mill tailings samples, less radium is controlled by amorphous oxy-hydroxides.

3.4.3 Leaching with citrate-bicarbonate-dithionite(CBD)

This reactant is used to dissolve crystallized iron and manganese oxy-hydroxides. Proportions of iron and manganese that are leached from crystallized oxy-hydroxides are lower in Ecarpière than in Jouac, however leached radium proportions are of the same order of magnitude in Ecarpière than in Jouac (Tab. 4). High concentrations of carbonates in the reactant could have partially dissolved carbonates (see Ca, Sr, Ba) of the tailings samples. Also barite (see Ba, Sr, Pb) could have been partially dissolved. The CBD reactant is not sufficiently selective in regards with radiferous mineralogical phases of mill tailings to evaluate the partitionning of radium between barite, carbonates and oxides, but it shows that in reductive and complexing conditions, great amounts of ^{226}Ra could be leached. Great proportions of iron, manganese and arsenic are also leached from Jouac samples.

3.4.4 Final residual fraction of mill tailings

Two more leaching methods have been applied to alkaline mill tailings in order to quantify the role of organic matter on ^{226}Ra retention. Proportions of ^{226}Ra leached by sodium bicarbonate and sodium pyrophosphate solutions ranged from 5 to 13%. For comparison purposes between acid and alkaline mill tailings, proportions of leached elements by these two methods were added to the final residual fraction. Major proportions of baryum, strontium, lead from mill tailings samples were not leached by successive leachings (Tab. 5). Major proportions of

Table 5. Maximum and minimum (from n values) proportions (w%) of uranium, ^{226}Ra and metals from mill tailings samples after leaching tests

	Ecarpière		Jouac		Lodève	
	n = 6; n = 2*	Exp. Total %	n = 4; n = 2*	Exp. Total %	n = 5; n < 5*	Exp. Total %
^{226}Ra	3 – 7 *	96 - 121	0 *	111 - 113	30 – 48	80 - 124
U	7 *	85 - 97	14 – 22	83 - 111	76 – 80 *	89 - 96 n = 3
Ca	0 – 2	99 - 103	3 – 10	106 - 108	55 – 72	95 - 98
Sr	46 – 71 *	114 - 122	79 – 94 *	97 - 103	--	--
Ba	74 – 80 *	89 - 103	78 – 101 *	101 - 111	70 – 78 *	94 - 113 n = 2
Pb	77 – 89 *	92 - 107	75 – 84 *	94 - 96	67 *	99 - 107 n = 2
As	18 – 40 *	80 - 106	39 – 58	98 - 105	16 – 25	92 - 104
Fe	30 – 40	97 - 98	44 – 51	102 - 106	55 – 64	92 - 102
Mn	28 – 29 *	93 - 123	13 – 18 *	108 - 115	45 – 96	99 - 106

Table 6. Maximum and minimum (from n values) proportions (w%) of uranium, ^{226}Ra and metals leached from mill tailings samples by demineralized water as recommended by AFNOR X 31-210

	Ecarpière	Jouac	Lodève
	n = 3	n = 3; n = 12 *	n = 3; n = 10 *
^{226}Ra	0.3 – 0.6	0.2 – 0.3 *	0.7 – 1.05 *
U	< 0.5	< 0.4 – 2.5 *	27 – 57
Ca	-	-	-
Sr	-	-	-
Ba	< 0.3	0.3 – 0.5 *	0.3 – 0.65 *
Pb	< 0.3	-	0.7
As	-	-	7.5 – 10.5 *
Fe	-	< 0.002	-
Mn	-	2.9 – 3.5 *	-

uranium from mill tailings samples were found in the residual fraction of Lodève. Other elements have been effectively leached during sequential and selective leaching methods.

3.5 Successive leachings with demineralized water

Demineralized water is used according to the french leaching test AFNOR X 31-210. This test has been applied to 3 samples of Ecarpière (C1-samples), and to more samples from Jouac and Lodève (Tab. 6). Less than half part of gypsum is dissolved in these conditions in Ecarpière samples [Somot & al., 1997]. Despite the fact that experimental conditions are far from natural leaching, a very low proportion of ^{226}Ra is leached in these stringent conditions. This is due to re-adsorption of radium on other minerals as demonstrated by Chuiton [1983] in absence of sufficiently complexing solutions [Somot & al., 1997]. Aged and fresh tailings have been leached in same conditions and results were compared (Fig. 2). Twice more radium is extracted from fresh tailings than from aged ones after nine successive leachings.

4 CONCLUSIONS

In actual acid mill tailings that have been studied, dissolved radium activities are controlled by gypsum. Acid uranium mill tailings pore-water is actually in equilibrium state with respect to gypsum. In alkaline mill tailings of Lodève, where gypsum is not abundant, dissolved radium activity seems to be controlled by carbonates but results are less demonstrative. Whatever the milling process, the major proportion of radium in mill tailings is bound to iron and/or manganese oxy-hydroxides.

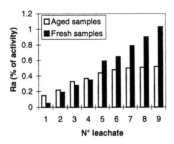

Figure 2 : Cumulative proportion of radium (% of initial activity) leached by demineralized water according to AFNOR X 31-210 test from fresh and aged mill tailings samples.

Oxy-hydroxides are stable in most of surface environments. An important fraction of residual uranium is also linked to these phases in acid mill tailings and authigeneous and uraniferous minerals have been described. Baryum and radium have a similar chemical behaviour, but barite precipitation rate is fast and its solubility is very low. Radium fraction that is mobilized in studied uranium mill tailings has not been co-precipitated with baryum. Lead is also known to have a behaviour similar to that of radium but in studied uranium mill tailings, major fraction of lead is not leached when radium is, except at Lodève when aqueous conditions are reducing and carbonated. Similarity in shapes of dissolved radium activities and (1) of dissolved strontium concentrations at Ecarpière is explained by equilibrium state of strontiferous and radiferous gypsum [Somot at al., 1997], (2) of dissolved arsenic concentrations at Jouac is explained by the lower Eh of green samples resulting from Le Piégut ore and by a different mineralogical speciation of radium in these samples, (3) of Eh values at Lodève is probably resulting from equilibrium states of carbonates in these mill tailings. Mineralogical speciation of radium, uranium and metals is of prime importance for long term uranium mill tailings management because, if the evolution of physical, bacterial and chemical conditions of pore waters with time is evaluated, leaching of radio-nuclides and toxic metals from mill tailings impoundment to environment by rain water can be avoided. For example, selective leachings showed that oxy-hydroxides are of prime importance for the retention of radium and of most metals. A fast establishment of acidic or reducing conditions within mill tailings would induce the dissolution of oxy-hydroxides and the leaching of great proportions of radium and other metals. Iron and sulfate reducing bacteria would be a nuisance for the evolution of these mill tailings, as shown by various authors [Mc Gready & al., 1980; Fedorak &al., 1986; Lovley & al., 1990; Landa & al, 1991]. Actually, the pH of mill tailings is still near the neutrality even after 35 years of evolution, limiting the release of radium and metals. The stability of gypsum is maintained since the liquid to solid ratio, or permeability of tailings, is low. In studied acid mill tailings, permeability is very low and gypsum reached equilibrium state with pore water, limiting radium leaching. Presently, in order to limit environmental impact of uranium mill tailings, remediation for these mill tailings impounds includes the following main options : covering the tailings with compacted layers of natural material (to reduce infiltration or L/S), mine site drainage water flux management and selective treatment of seeping water due to compaction (fixation of radium and metals by adsorption or co-precipitation, decantation), avoiding the growth of long roots vegetation (limiting the adding of bioavailable nutrients).

ACKNOWLEDGMENTS

This study was part of the FISRAMUT Project (CEE Brite-Euram). Financial support was provided by both COGEMA and the CEE. Sampling was performed on COGEMA mining sites and experimentation was performed both at CNRS / CREGU and CNRS / CRPG (Nancy, France). The ``Service Environnement des Sites`` from COGEMA is acknowledged, especially J.L. Daroussin whose interest lead to constructive discussions.

References

AFNOR X 31-210 norm (1988) – Déchets – Essai de lixiviation – section 1 : Recommandations générales pour l'échantillonnage des déchet au niveau de leur production ou de leur site de dépôt (Échantillonnage sur site). *J. Off.*

Ames L.L., Mc Garrah J.E. & Walker B.A. (1983) – Sorption of trace constituents from aqueous solutions onto secondary minerals. II Radium. *Clays clay Minerals*, Vol. 31, n° 5, p. 335-342.

Arshad M.A., Saint-Arnaud R.J. & Huang P.M. .(1972) – Dissolution of trioctaedral layer silicates by ammonium oxalate, sodium dithionite-citrate-bicarbonate, and potassium pyrophosphate. *Can.J. Soil Sci.* 52, 19-26

Bassot S. (1997) – Mobilité du radium et de l'uranium dans un site de stockage de résidus issuds du traitement de minerais d'uranium. *PhD Thesis*, CEA-R-5761, IPSN, 154.

Benes P. Sedlacek J., Sebesta F., Sandrik R. & John J. (1981) – Method of selective dissolution for characterization of particulate forms of radium in natural and waste waters. *Water Res.* 15, 1299 - 1304.

Chuiton G. (1983) – Identification des formes physico-chimiques du radium 226 contenu daans les résidus de traitement de minerais d'uranium. Étude réalisée sur les résidus des mines Dong-Trieu. Rapport interne COGEMA/STP n°234, ARD1/841/MS

Coleman N.T., S.B. Weed & R.J. Mc Cracken (1959) – Cation-exchange capacity and exchangeable cations in Piemont soils of North Carolina. *Soil Sci. Soc. Amer. Proc.*, 23, 146-149.

Ebler E. (1915) – Manufacture, isolation and enrichment of radioactive substances by adsorption from solutions. *U.S. Patent 1*,142,153.

Espiau P. & Peyronel A. (1976) — L'acidité d'échange des sols. Application à une séquence altitudinale des sols du Massif du Mont Aigoual. *Science du Sol*, 3, 161-175.

Fedorak P.M., Westlake D.W.S, Anders C., Kratochvil B., Motkosky N., Anderson W.B. & Huck P.M. (1986) – Microbial release of 226Ra2+ from (Ba, Ra)SO4 sludges from uranium mine wates. *Applied and Environmental Microbiology*, p. 262-268

Gnanapragasam E. K. & Lewis B. A. G. (1995) – Elastic strain energy and the distribution coefficient of radium in solid solutions with calcium salts. *Geochem. et Cosmochim. Acta.* , vol. 59, n° 24, 5103 - 5111.

Gupta S.K. & K.Y. Chen (1975) – Partitioning of trace metals in selective chemical fractions of nearshore sediments. *Environ. Lett. 10 (2)*, 129-158.

Haque K.E., Lucas B.H. & Ritcey G.M. (1986) – Hydrochloric acid leaching of an Elliot Lake uranium ore. *CIM Bull.*, 73, 819, 141- 147.

Kaiman S. (1977) – Mineralogical examination of old tailings from the Nordic Lake Mine, Elliot Lake, Ontario, *Report ERP/MSL 77-1901/IR* CANMET,.

Kalil E. K. & Goldhaber M. (1973) – A sediment squeezer for removal of pore waters without air contact. *Journ. of Sedimentary Petrology*, 43, 2, 553-557.

Landa E.R. & Bush C.A. (1990) – Geochemical hosts of solubilized radionuclides in uranium mill tailings. *Hydrometallurgy, 24*, Elsevier Science Publishers B.V., Amsterdam, p. 361-372.

Landa E.R. (1982) – Leaching of radionuclides from uranium ore and mill tailings. *Uranium, 1*, Elsevier Scientific Publishing Compagny, Amsterdam, p.53-64.

LandaE.R., Phillips E.J.P. & Loveley D.R. (1991) – Release of 226Ra from uranium mill tailings by microbial Fe(III) reduction. *Appl. Geoch.* 6, p. 647-652

Latham A.G. & Schwarz A.C. (1987) – The relative mobility of U, Th and Ra isotopes in the weathered zones of the Eye-Dashwa Lakes Granite Pluton, Northern Ontario, Canada. *Geoch. Cosmoch. Acta*, Vol. 51, p. 2787-2793.

LovleyD.R., Chapelle F.H. & Phillips EJP (1990) – Fe(III)-reducing bacteria in deeply buried seiments of the Atlantic coastal plain. *Geology* n°18, p. 954-957.

Manheim F.T. (1966) – A hydraulic squeezer for obtaining interstitial water from consolidated and un consolidated sediments. *Professional Paper 550-C*, United States, Geological Survey, pp. C-256-C261.

Mc Gready R.G., Bland C.J. & Gonzales D.E. (1980) – Preliminary studies of the chemical, physical and biological stability of Ba/RaSO4 precipitates. *Hydrometallurgy* 5, p. 109-116.

Mehra O.P. & M.L. Jackson (1960) – Iron oxide removal from soils and clays by a dithionite-citrate system buffered with sodium bicarbonate. *Clays Clay Min.* ,11, 189-200.

Nirdosh I. (1987) – A review of recent developments in the removal of 226Ra and 230Th from uranium ores and mill tailings. *Hydrometallurgy 12*, 151 - 176.

Nirdosh I., Muthuswami S.V. & Baird M.H.I. (1984) – Radium in uranium mill tailings – some observations on retention and removal. *Hydrometallurgy 12*, p. 151-176.

Nirdosh I., Trembley W.B., Muthuswami S.V. & Johnson C.R. (1987) – Adsorption-desorption studies on the radium-silica systems. *The Canadian Journal of Chemical Engineering*, Vol. 65, p. 928-934.

Nirdosh I., Trembley W.B & Johnson C.R. (1990) – Adsorption-desorption studies on the 226Ra-hydrated metal oxide systems. *Hydrometallurgy 24*, p. 237-248.

Reyx & Pacquet (1992) – Rejets miniers de l'usine SIMO de l'Écarpière (Vendée), sondage C1, sondage C2, Étude 8508. Rapport interne COGEMA.

Seeley F.G. (1977) – Problems in separation of radium from uranium ore tailings. *Hydrometallurgy 2*, p. 249-263.

Shearer S.D., Lee Jr. & G.F. (1963) – Leachability of radium-226 from uranium mill solids and river sediments. Health Physics Pergamon Press 1964, Vo. 10, p.217-227.

Snodgrass W.J. (1990) – The chemistry of 226Ra in the uranium milling process, In The Environmental Behaviour of Radium, Vienna 1990, 5-25.

Snodgrass W.J., Luch D.L. & Capobianco J (1982) – Implication of alternative geochemical controls on the temporal behaviour of Elliot Lake tailings. Management of Wastes from Uranium Mining and Milling. *Proceeding of an International Symposium, Alburquerque, May 1982*, IAEA, Vienna nov. 1982, STI/PUB/622, 285-308.

Somot S., M. Pagel, A. Pacquet, J. Reyx & J. Thiry (1995) – A mineralogical and geochemical study of the Ecarpière uranium mill tailings : Their diagenesis over the last thirthy years In J. Pasava, B. Kribek & K. Zak.(eds), Mineral Deposits : From Their Origin to Their Environmental Impacts, *Proceeding of the third biennial SGA meeting, Prague*, 705-708.

Somot S., Pagel M. & J. Thiry (1997) – Spéciation du radium dans les résidus de traitement acide du minerai d'uranium de l'Écarpière (Vendée, France). *C.R. Acad. Sci. Paris, Sciences de la terre et des planètes*, 325, 111-118.

Tamm O. (1922) – Um best äming ow de oorganiska komponenterna i markens gel complex. *Medd. Statens Skogsförsökanst.*, 19, 385-404.

Tessier A., Campbell P.G.C. & Bisson M. (1979) – Sequential extraction procedure for the speciation of particulate trace metals. *Anal. Chem.*, 51 (7), 844-851.

Valentine R.L., Mulholland T. & Splinter R.C. (1987) – Radium removal using sorption to filter sand. *Jour. AWWA*, 79 : 4 : 170.

352

Tailings and Mine Waste'00 © 2000 Balkema, Rotterdam, ISBN 90 5809 126 0

Distribution of radionuclides in the tailings of Schneckenstein, Germany

T. Naamoun, D. Degering & D. Hebert
Institute of Applied Physics, T.U. Bergakademie, Freiberg, Germany
B. Merkel
Institute of Geology, T.U. Bergakademie, Freiberg, Germany

ABSTRACT: The uranium tailings Schneckenstein I and II (Germany) cover an area of 4.5, 1.5 ha and 600, 105 Tm^3 of volume, respectively. Uranium ores were treated with different methods (flotation, acid and alkaline leaching). These techniques changed over time. Radionuclides from the Uranium decay series were analysed in tailing material from four bore holes by different methods. ^{234}U and ^{235}U was measured by a semiconductor alpha spectrometer. For the measurement of ^{238}U, ^{230}Th, ^{226}Ra and ^{210}Pb from ^{238}U chain and ^{227}Ac from ^{235}U chain a semiconductor gamma spectrometer was used. ^{226}Ra is approximately in equilibrium with ^{230}Th and ^{227}Ac which is a proof of its immobility in the tailing environment. On the other hand, the diffusion of ^{222}Rn in the tailing material is obviously and therefore its participation in the water cycle has to taken into account.

1 INTRODUCTION

Since the middle ages the western Erzgebirge/ Vogtland has been famous for its mining industry especially for the mining of iron ore. Mainly the promising mineralised veins were exploited here. In the late fourties of this century, mineralisations containing uranium were discovered and the exploitation was being carried out by the Soviet joint-stock company Wismut (SDAG Wismut) for ten years between 1947 and 1957. Besides ore from the Schneckenstein deposit [(700g U/t)], Uranium ore from Zobes, Johanngeorgenstadt, Schlema-Alberoda, Culmitzsch and Schmirchau were treated in the processing plant Schneckenstein (object 32)/ (Wismut 1994). According to their uranium content the ores were classified into four main categories:
1) Ore with uranium content more than 3 %
2) Ore with uranium content between 1 and 3 %
3) Ore with uranium content between 0,3 and 1 %
4) Ore with uranium content between 0,015 and 0,30 %
The uranium ore belonging to the first two categories was only gravimetrically separated and directly carried to the former Soviet Union without any further processing. The other two ore categories were processed chemically (by acid and mainly alkaline leaching). The rocks containing less than 0,005 % of uranium were considered as trivial and without economic importance (Wismut 1946-1955).
The extracted uranium was transported altogether to the former Soviet Union as a compensation after the second world war. At the Schneckenstein site, during the whole period of production 660 tons uranium were extracted from 1.200.000 tons of ore at a capacity of 150.000 tons of ore per year. (It corresponds approximately to 10% of uranium output of the Vogtland and it is less then that of the 1% of the production in the whole area of the Erzgebirge/Vogtland (Bartel 1993).
The aim of this paper is the investigation of equilibria or disequilibria between ^{238}U and its daughter products ^{234}U, ^{230}Th, ^{226}Ra, ^{222}Rn, ^{210}Pb in the tailing material in order to estimate their mobility and the risk of their participation in the water cycle in the tailing environment .

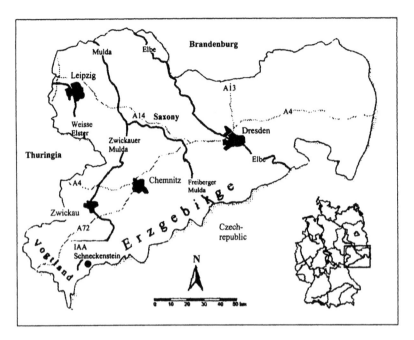

Fig. 1: The area of investigation- Uranium Tailings Schneckenstein

2 LOCATION AND MORPHOLOGY

The area of investigation, i.e. the uranium tailing Schneckenstein, is located in the southwest of Saxony: in the Boda valley north of the district of Schneckenstein in the village of Tannenbergsthal / Vogtland (Fig. 1)

In the northwest the Boda valley is surrounded by the Runder Hübel mountain (837 m above sea level) and in the southeast by the Kiel mountain (943 m asl). The valley exposition is specified by 35. 4° north. The area of investigation is located at an altitude of 740 to 815 m above sea level.

The topographic relief is clearly characterized by tectonic and lithofacies features. Besides the mineralogical composition and grain size as well as orientation of joints also the type and dip and strike of faults determine the relief forming erosion channels of the rainwater running off at the surface. Deep valley cuts alternate with steep heights and flat domeshaped formations.

Among others also anthropogenic accretions in form of mining dumps and industrial settling basins are relevant to the surface configuration.

3 EXPERIMENTAL

3.1 Sampling

Four sediment cores were taken at the tailing site by drilling four bore holes with different depths; two bore holes in each tailing (Fig. 2). The first and the second bore hole (GWM 1/ 96; GWM 2 / 96) were drilled down to the granite foundation in Tailing 2 (IAA I). The third bore hole (RKS 1/ 96, Tailing 1(IAA II) was sunk for about eight meter depth but the granite foundation was not attained in this case due to some technical problems. The fourth bore hole (RKS 2/ 98, Tailing1) is 12 m deep. The cores (diameter 50 mm) were cut into slices of 1 m length and transported in argon filled plastic cylinders to avoid contact with air.

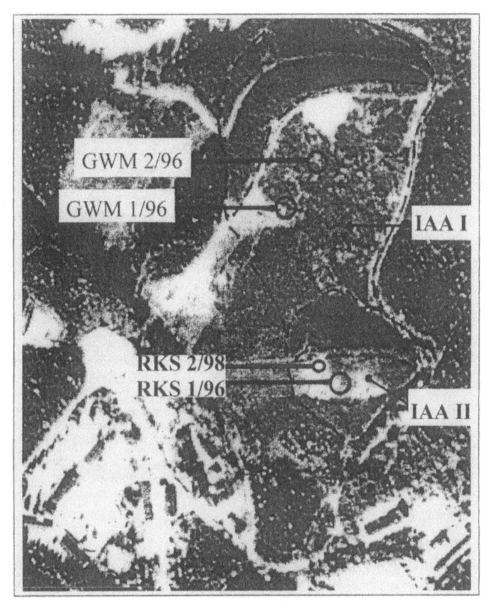

Fig. 2: Location of the bore holes in the tailing sites IAA I and IAA II.

3.2 Sample praparation and analysis

About 200 g of the lower part of each core section was used for the analysis by alpha and gamma spectrometry. The samples were dried at about 90 °C to prevent the loss of Pb and after this filled into gasproof measuring container of cylindrical shape. A minimum of two weeks of rest were necessary to restore the radioactive equilibrium between ^{226}Ra and ^{222}Rn. In general, the optimum measuring time of the γ-spectra was about 24 hours. The utilised gamma spectrometer contains a p-type high purity Ge-detector with 36% relative efficiency and 1.8 keV line width (FWHM) at 1.3 MeV. It is surrounded by a passive shielding of 9 cm lead, 2 cm

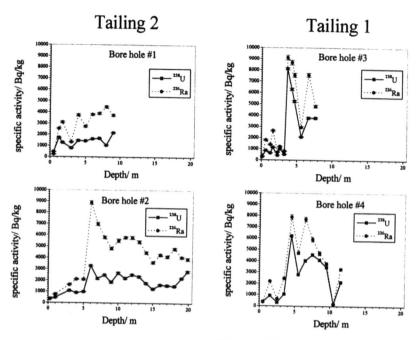

Tailing 2   Tailing 1

Fig.3 Distributions of the specific activity of ^{238}U and ^{226}Ra in the bore holes #1 - #4 of the tailings from Schneckenstein. Profile #1 and #2 reach the granitic foundation; #3 and #4 were not drilled down to the foundation.

mercury and 4 cm electrolytic copper. Each gamma spectrum was analysed by comparing the peak count rates of a calibration source with that of the sample at the same measuring geometry. In the calculation of the specific activity, the effect of self absorption in the samples was taken into account. For calibration the IAEA reference materials RGU was used. The following nuclides were determined: from the ^{238}U-serie- ^{234}Th and ^{234m}Pa (for ^{238}U), ^{230}Th, ^{226}Ra, ^{214}Pb and ^{214}Bi (for ^{226}Ra), ^{210}Pb; and from the ^{235}U-serie- ^{227}Th, ^{223}Ra and ^{219}Rn (for ^{227}Ac).

4 RESULTS

4.1 Gamma spectrometric measurements

^{238}U (Fig. 3): In the bore holes of tailing 2 (#1 and #2) the specific activity of ^{238}U did not exceed 3500 Bq/kg (280 ppm U). The contents in tailing 1 (hole #3 and #4) were higher with a maximum of 8100 Bq/kg (660 ppm U). In the covering layer the concentrations are generally below 450 Bq/kg (36 ppm U).

^{226}Ra (Fig.3): In general, the specific activity of ^{226}Ra is higher than that of ^{238}U. Maximum values of about 9000 Bq/kg were reached in both tailings. The profiles of the radium content can be divided into several clusters, separated by distinct minima. This is clearly seen in bore hole #2 were one can distinguish between an interval of low ^{226}Ra activity (0-5 m) and three of higher concentrations (6-9 m; 10-15 m; 16-20 m). This sequence is not observed in hole #1 since it was drilled down near the border of the tailing 2 where a lower thickness of the sediment occurs. The activity plots of bore hole #3 and #4 are nearly parallel. The ratio ^{238}U/^{226}Ra (Fig.4) in the uppermost layer of both tailings is about 1, except in hole #1 where this ratio was found to be about 2. In tailing 2, this value decreases down to 0.5 and stays constant in the deeper region. For Tailing 1 this ratio varies between 0.5 and 0.9. In the cases, where the bore holes were drilled down to the foundation (#1 and #2), the ratio increases up to 1 at the bottom of the profile.

356

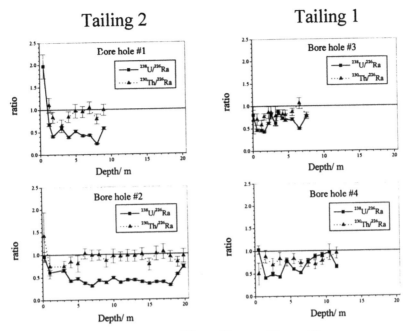

Tailing 2

Tailing 1

Fig. 4. Ratios of the specific activities of ^{238}U to ^{226}Ra and ^{230}Th to ^{226}Ra for the bore holes #1 - #4 of the tailings from Schneckenstein

^{230}**Th:** The profiles of this nuclide are in all sediment cores similar to that of ^{226}Ra. The ratio of the specific activities ^{230}Th/^{226}Ra (Fig.4) in Tailing 2 is < 1 in the depth range with ^{238}U/^{226}Ra > 0.5. An equilibrium between ^{230}Th and ^{226}Ra was observed were ^{238}U/^{226}Ra is about 0.5. In tailing 1 the ^{230}Th content is lower than that of ^{226}Ra. Because a γ-line of weak intensity was used for the analysis, the error of the ^{230}Th activities is larger than for the other nuclides.

^{210}**Pb:** The activity concentrations of the Radon-successor ^{210}Pb are in most layers lower than that of ^{226}Ra (Fig.4). The ^{210}Pb/^{226}Ra ratio varies between 0.7 and 0.9. Only in some layers this value reaches 1 or is even >1 (cf. bore hole #4). In general, the ratio is higher for deeper parts of the profiles.

^{227}**Ac:** This nuclide is the first precursor of the analysed nuclides from the ^{235}U-serie with a half life > 1 y ($T_{1/2}$ = 21.8 y). Therefore we assign the determined activities to this nuclide. In natural samples it can be assumed that the activity ratio ^{238}U/^{235}U (= 21.7) is not disturbed. So we multiplied all activity ratios ^{227}Ac/^{226}Ra by this value to get the plots more clearly. ^{227}Ac is for most of the samples within the error bars in equilibrium with ^{226}Ra. Only in the covering ^{227}Ac seems to be a little enriched compared to ^{226}Ra.

5 INTERPRETATION

As mentioned above, the material of the two tailings originates from different deposits and was treated by different methods. This heterogenity of the sediment leads to a variety of Uranium contents and of leaching efficiency. Since radium was fixed during the processing by the precipitation of $BaSO_4$ or $BaCl_2$ it was recovered to about 99 % in the material which was deposited in the tailings (Snodgrass 1990). Indeed, high barium concentrations were measured in the sediment samples (Gmelin 1977). From this and from the fact that in ores the uranium series should be in equilibrium we can conclude that the specific activity of ^{226}Ra (half life 1600 y) reflects the original uranium content. The ^{238}U/^{226}Ra ratio is then a measure of the efficiency of the leaching. No more than about 50 % of the total uranium content was therefore removed during the processing of the ore. Thereby the processed ore in hole #1 below 2 m and in #2 below

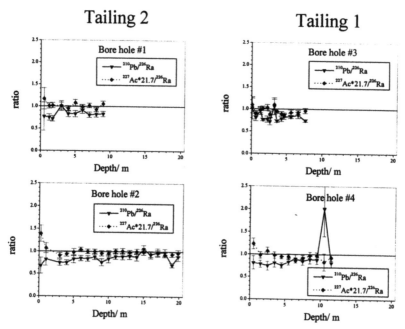

Fig. 5. Ratios of the specific activities of ^{210}Pb to ^{226}Ra and ^{227}Ac to ^{226}Ra for the bore holes #1 - #4 of the tailings from Schneckenstein. The latter was multiplied by the natural activity ratio of $^{235}U/^{238}U$ (=21.7).

5m depth is relatively homogeneous as seen by the nearby constant activity ratio compared to the profiles of tailing 2 where a broad variation of this ratio occurs. The determined ^{226}Ra activities correspond to uranium concentrations in the range 0.0052 % to 0.073 % except in the first layer of all profiles and in the depth of 10 –11 m of hole #4 where lower values occured. These layers showed an equilibrium between ^{238}U and ^{2226}Ra indicating that non-leached material was utilized for the covering and also deposited in hole #4 (the high ratio in the uppermost material of profile #1 is not clear; probably it contains contaminated soil). At the bottom of holes #1 and #2 the trend of $^{238}U/^{226}Ra$ towards 1 shows the influence of the foundation.

According to the literature (IAEA 1993) thorium is soluble during acid leaching but not in alcaline environment. ^{230}Th (half life 7.54 10^4 y) must also show this behaviour. Thus, activity ratios $^{230}Th/^{226}Ra$ below 1 will occur in acidic leached material whereas the equilibrium is maintained in alcaline leaching. This can be especially seen in hole #2. The material in the range 1 to 5 m depth shows a coarse grain size contrary to the deeper layers and was therefore not milled prior to the processing which is possible only at acidic treatment. Indeed the $^{230}Th/^{226}Ra$ ratios are < 1 in this interval. About 20 to 30 % of the Th content are lost during this process in agreement with literature data (IAEA 1993).

^{227}Ac (half life 21.8 y) is a nuclide of the ^{235}U serie. Its parent nuclide is ^{231}Pa with a half life of 3.3 10^4. For that reason the $^{227}Ac/^{226}Ra$ ratio reflects in deeper (older) layers rather the $^{231}Pa/^{226}Ra$ ratio. But in the whole tailing material the equilibrium to radium is not disturbed. This shows that ^{227}Ac as well as ^{231}Pa are not influenced by the treatment of the ore.

On the other hand, the equilibrium of ^{226}Ra with the nuclides of the different elements Th (at least in the acaline leached material), Ac and Pa is a hint that this nuclide as well as the others three does not migrate through the tailing. Migration of Ra or other nuclides would have some influence on the distribution of their corresponding activities. For example, if radium is soluble and moves through the tailing, this leads to an increasing of Ra concentration in material with low content and vice versa to a decrease in radium rich layers. The ratios of the investigated nuclides to ^{226}Ra must therefore show significantly lower values in Ra poor material and higher values for higher Ra contents. For none of the above mentioned nuclides this or the inverse dependence was observed. Their mobility in the tailings must therefore be low at present.

The last analysed nuclide ^{210}Pb (half life 22.3 y) is that successor of the radioactive noble gas nuclide ^{222}Rn (daughter of ^{226}Ra) with the longest half life. Its distribution in the profiles is an indicator of diffusion and transport processes in the tailings. In general, a radon loss of the sediment can be deduced from the ^{210}Pb/^{226}Ra ratios < 1 whereby this effect is lower for deeper layers. The large ^{210}Pb excess in interval 10–11 m in hole #4 is an example of the slow diffusion of ^{222}Rn (half life 3.8 d) from layers of high Ra content into a region with low Ra concentration. The radon decay results in an additional deposition of ^{210}Pb in this interval. The observed mobility of ^{222}Rn leads to its outgassing from the tailing and it may also be distributed in the aquifer of the tailings. This results in a possible radiation risk caused by this nuclide.

6. CONCLUSIONS

Gamma spectrometric investigations of members of the ^{238}U and ^{235}U decay series in material of the tailings of Schneckenstein reflect the different types of ore processing and of the variation in the leaching efficiency caused by the material of different origin. Nuclide ratios show that U, Ra, Th, Ac and Pa is at this time not mobile in the sediment. On the other hand, the migration of Rn through the material could be proved and must be taken into account in the radiation risk assessment of the former tailings.

REFERENCES

Barthel, F.H. 1993. Die Urangewinnung auf dem Gebiet der ehemaligen DDR von 1945 bis 1990. Geologisches Jahrbuch A142. Hannover.

Gmelin-Handbuch"Radium", (ed. 8)1977.Element Verbindungen.Berlin- Heidel berg:Springer- verlag: pp. 147.

IAEA. 1993. Uranium extraction technology. Technical reports series N°359: 110- 111.

Snodgrass, W.J 1990. "The chemistry of 226Ra in the uranium milling process" The environmental behaviour of radium. Technical reports series No. 310, Vol. 2, IAEA, Vienna: 5-26.

Wismut archiv. Razrabotka mestarazdenia uranovi rud sovietska germanskim akcion ernim obchestvom Vismut za 1946-1955 godi.

Wismut. 1994. Teilbericht: Strahlenexposition in den Aufbereitungsbetrieben und Beprobung szechen der SAG/SDAG Wismut.

Tailings and Mine Waste'00 © 2000 Balkema, Rotterdam, ISBN 90 5809 126 0

Identification of long-term environmental monitoring needs at a former uranium mine

R.C. Lee, R. Robinson, M. Kennedy & S. Swanson
Golder Associates Limited, Calgary, Alb., Canada

M. Nahir
Public Works and Government Services Canada, Edmonton, Alb., Canada

ABSTRACT: The former Rayrock uranium mine in the Canadian Northwest Territories (NWT) was abandoned in 1959 and remediated in 1996. Canadian government agencies were interested in developing short- and long-term monitoring plans for the site to satisfy Atomic Energy Control Board (AECB) requirements and to ensure that potential human health risks from ionizing radiation remain minimal. This study determined whether potential risks are likely to remain minimal, identified data gaps, and recommended a long-term plan. A high-level analysis was conducted in conjunction with risk estimations to determine sensitivity of risks to uncertainties associated with radionuclide concentrations or radiation levels in exposure media including gamma radiation, radon, soil, water, and consumption of wild game/fish. Potential human health risks under current conditions are minimal under any conceivable exposure scenario. Recommendations for long-term monitoring focus on the exposure pathways of greatest concern, combined with the feasibility of further data collection.

1 INTRODUCTION

The former Rayrock Mine is located approximately 145 km northwest of Yellowknife, NWT. Underground mine exploration and development activity began in 1955. The deposit was mined by Rayrock Mines Ltd. between 1957 and 1959, and abandoned in operable condition in 1959. During the operation of the mine and associated mill, 70,903 tonnes of un-neutralized tailings were discharged. The tailings piles are weakly acid generating, and are potential sources of radionuclides of the uranium-thorium decay series and heavy metals (including copper) to the surrounding area.

The Rayrock site underwent a remediation program in 1996. The goals of the remediation program were to stop further mobilization of contaminants from the site into the surrounding environment, and to reduce hazards to humans potentially visiting the area. The remediation was not intended to remove contaminants and restore pristine conditions in the immediate area of the mine. Remediation included sealing the mine adit and ventilation shafts, removal of radioactive material from the dump and disposal of the dump material across the tailings ponds, and thick-capping of tailings ponds with silty-clay material borrowed from an old airstrip near the mine site.

A short-term environmental monitoring program was designed (Golder Associates Ltd. 1996) and implemented between 1996 and 1998 by the Department of Indian Affairs and Northern Development (DIAND), Public Works and Government Services Canada (PWGSC) and the Low Level Radioactive Waste Management Office (LLWRMO) of the Atomic Energy Control Board (AECB). A report by LLWRMO (1999) describes the results of this program.

The results of the short-term monitoring program indicated that the remedial activities were successful in reducing gamma radiation and radon emissions (the main radiation sources) to acceptable regulatory limits. The results of multiple-pathway radiation dose estimations in several different exposure scenarios indicated that even in extreme exposure scenarios doses were minimal and well below regulatory dose limits.

The present analysis includes an evaluation of the relative impact of exposure routes on total doses, identification of areas of uncertainty that are likely to be important to the proposed long-range monitoring plan at the site, recommendations regarding long-range monitoring, and identification of possible risk management strategies at the site.

2 EXPOSURE AND UNCERTAINTY ANALYSIS

The LLWRMO (1999) found that external gamma radiation is the most potentially important source of ionizing radiation at the Rayrock site, although the total dose is minimal. However, uncertainty associated with the doses was not assessed. In the following analysis, the dependency of dose on time is examined, as well as the uncertainty associated with contaminant concentrations/activity. Doses resulting from multiple simultaneous exposure routes (e.g. gamma plus water plus food) are not assessed here because the LLWRMO determined that there is no unacceptable residual risk remaining at the site even in an extreme worst-case scenario. The purpose of the present analysis is to guide further data-gathering efforts and possible further mitigation strategies.

2.1 *Exposure Models*

The general form for the exposure model is:

$$D_i = (C_{i:s} - C_{i:b}) \times IR_i \times ED \times CF \qquad (1)$$

where D_i = site-related incremental radiation dose from exposure route i; $C_{i:s}$ = site media concentration/activity from exposure route i; $C_{i:b}$ = background or control concentration/activity from exposure route i; IR_i = intake or exposure rate associated with exposure route i; ED = exposure duration; and CF = unit and/or dose conversion factor(s).

Specific models that represent the exposure routes of interest are listed in Table 1.

Table 1. Radionuclide dose equations. All equations are based on LLWRMO (1999). Variable definitions are listed in Table 2.

Exposure Pathway	Dose Equation (Incremental)
Ingestion of meat or fish	$D_{m/f} = (C_{m/f:s} - C_{m/f:b}) \times IR_{m/f} \times DCF_1 \times ED$
Ingestion of water	$D_w = (C_{w:s} - C_{w:b}) \times IR_w \times DCF_2 \times ED$
Inhalation of dust	$D_d = (C_{d:s} - C_{d:b}) \times InhR \times TSP \times DCF_3 \times UCF_1 \times ED$
Inhalation of radon	$D_r = (C_{r:s} - C_{r:b}) \times DCF_4 \times ER \times UCF_2 \times ED/WM$
External exposure to gamma radiation	$D_g = (DR_{g:s} - DR_{g:b}) \times UCF_2 \times UCF_3 \times ED$

Mitigation decisions are normally based on the incremental dose because it is not possible to mitigate background ionizing radiation exposures. AECB is in the process of adopting the International Commission on Radiation Protection (ICRP) annual dose limit for the public of 1 mSv, which is incremental to background (AECB 1998). This dose limit is used here for comparative purposes.

The variables in the exposure models are heterogeneous across space or a population ("variable"), and the parameters (e.g. mean, variance) of these distributions are uncertain (inaccurate and/or imprecise). Variability is an inherent feature of space or populations and is irreducible; uncertainty stems from inadequate knowledge of a variable and is reducible given more data and/or more representative information. It is not possible, given the current data, to adequately characterize the true variability associated with potential doses at the Rayrock site. The LLWRMO analysis covered a wide range of potential exposure scenarios, and thus addressed at least some of the variability in dose that is likely to exist. The current assessment examines uncertainty because reduction of uncertainty is one goal of long-term monitoring.

Each of the variables in the models above can be represented as a distribution of values (other than the unit conversion factors, which are deterministic by definition). These variables are expected to be independent, and thus there is no need for a defined correlation structure. Uncertainty associated with physiological variables is minimal, and is not quantitatively evaluated here (mean point estimates are used instead). Estimates of these variables are listed in Table 2. Development of contaminant concentration/activity uncertainty distributions is described below.

Table 2. Radionuclide dose variables and factors. Distributions are represented as U(x,y), where U = uniform distribution, x = lower bound, and y = upper bound.

Variable or Factor	Definition	Values	Units	Reference	Notes
$C_{m/f:s}$	Current site-related activity of uranium in meat and/or fish	U(0,0.0007)	Bq/g	LLWRMO 1999, Table 5.2	1
$C_{m/f:b}$	Current background activity of uranium in meat and/or fish	U(0,0.058)	Bq/g	Swanson 1996, in Golder 1996, Table 6	2
$C_{w:s}$	Current site-related concentration of uranium in surface water	U(0,263)	µg/L	LLWRMO 1999, Table 4.3	3
$C_{w:b}$	Current background concentration of uranium in surface water	U(0,0.6)	µg/L	LLWRMO 1999, Table 4.3	4
$C_{d:s}$	Current site-related concentration of uranium in dust	U(0,65)	µg/g	LLWRMO 1999, Table 4.12	5
$C_{d:b}$	Current background concentration of uranium in dust	U(0,16)	µg/g	LLWRMO 1999, Table 4.12	6
$C_{r:s}$	Current site-related activity of radon	U(0,794)	Bq/m^3	LLWRMO 1999, Table 4.13	7
$C_{r:b}$	Current background activity of radon	U(0,135)	Bq/m^3	LLWRMO 1999, Table 4.13	8
$DR_{g:s}$	Current site-related dose rate of gamma radiation	U(0,1.6)	µSv/hr	LLWRMO 1999, Table 4.14	9
$DR_{g:b}$	Current background dose rate of gamma radiation	U(0, 0.1)	µSv/hr	LLWRMO 1999, Table 4.14	10
$IR_{m/f}$	Ingestion rate of meat and/or fish	Mean = 270, variable 0 to 1500	g/d	O'Connor 1997	11
IR_w	Ingestion rate of water	Mean = 1.5, variable 0 to 4	L/d	O'Connor 1997	12
InhR	Inhalation rate	Mean = 16	m^3/d	O'Connor 1997	13
TSP	Concentration of total suspended particulates	50	µg/m^3	LLRWMO 1999	
ED	Exposure duration	Variable (0 to 20 days)	d	Professional judgement	14
DCF_1	Dose conversion factor 1 (ingestion of meat/fish)	0.0025	mSv/Bq	LLRWMO 1999	15
DCF_2	Dose conversion factor 2 (ingestion of water)	3.2×10^{-5}	mSv/µg	LLRWMO 1999	15
DCF_3	Dose conversion factor 3 (inhalation)	0.0017	mSv/µg	LLRWMO 1999	15
DCF_4	Dose conversion factor 4 (radon)	0.0011	m^3×mSv/Bq	LLRWMO 1999	15

Variable or Factor	Definition	Values	Units	Reference	Notes
ER	Equilibrium ratio (radon daughters)	0.1	--	LLWRMO 1999	16
WM	Working month	170	hr	LLRWMO 1999	17
UCF_1	Unit conversion factor	0.000001	g/µg	LLWRMO 1999	--
UCF_2	Unit conversion factor	24	hr/d	LLWRMO 1999	--
UCF_3	Unit conversion factor	0.001	mSv/µSv	LLRWMO 1999	--

Table 2 General Note: Maximum value represents maximum at any location on the site or background locality according to referenced data unless otherwise noted.

Table 2 Specific Notes:

1. Maximum value represents organ levels in a pike caught from Sherman Lake.
2. Background concentrations not available from LLRWMO (1999). Historical maximum value represents average tissue activity in white suckers taken from lakes near Uranium City (^{210}Pb activity).
3. Sample taken from Lake Alpha. A conversion was applied to measured uranium levels by the LLWRMO (1999) to account for greater activity of some of the uranium series radionuclides.
4. Sample taken from Maryleer Lake.
5. See text regarding "dust" assumptions. Surface sample taken from downwind of north tailings pile. A conversion was applied to measured uranium levels by the LLWRMO (1999) to account for greater activity of some of the uranium series radionuclides.
6. See text regarding "dust" assumptions. Surface sample taken from near Maryleer Lake.
7. See text regarding breathing zone assumptions. Average concentration west of the south tailings area.
8. See text regarding breathing zone assumptions. Average concentration at Maryleer Lake. Emil River samples not assessed by LLWRMO due to uncertainties in data (see LLWRMO 1999).
9. Measurement at north tailings area.
10. Measurement at Maryleer Lake.
11. Mean represents average wild game ingestion rate for Native Canadian adults of both sexes (includes Inuit populations). Maximum value represents the 99th percentile of a lognormal distribution of wild game ingestion rates for Native Canadian adults of both sexes (includes Inuit populations). Note that approximately 70% of respondents in the original survey stated that they ate no wild game. Ingestion rates are lower for fish consumption.
12. Mean represents average water ingestion rate for Canadian adults of both sexes. Maximum value represents the 99th percentile of a lognormal distribution of water consumption rates for Canadian adults of both sexes.
13. Represents the mean of a lognormal distribution of 24-hour inhalation rates for adults of both sexes.
14. Maximum represents the longest reasonable period that a person would spend continuously at the site, based on professional judgement.
15. See LLWRMO 1999 for detailed explanation of dose conversion factor.
16. ER represents the ratio of radon daughter products to radon gas.
17. Standard radiological protection assumption.

2.2 Contaminant Concentration/Activity Uncertainty Distributions

The uncertain variables that are of interest in terms of long-term monitoring are exposure media concentrations/activity of radionuclides. Media concentration/activity distributions used here represent the uncertainty associated with doses from defined exposure media to an individual who wanders at random through the site.

Table 2 contains the distributions used for media concentration/activity variables along with the sources of data. In all cases, uniform distributions are defined with a lower bound of zero (a physical limitation) and an upper bound of the maximum measured concentration/activity as determined in the LLWRMO (1999) short-term monitoring study. Uniform distributions are used, in accordance with the principle of maximum entropy, when knowledge is limited to the bounds of a distribution (Lee & Wright 1994). Use of uniform distributions implies that a true average and/or statistical spread is not known for site contamination, and this distribution maximizes uncertainty within reasonable bounds.

Direct measures relating to potential human exposure such as consumption of wild meat and/or fish and surface water are used in this assessment rather than attempting to model food chain or indirect uptake pathways. Groundwater is not a viable direct exposure pathway at this site. Any exposures to contaminants in groundwater would be mediated through surface water pathways. It is possible that direct exposures to sediments could occur, perhaps in fishing activities, but it is unlikely that receptors would receive appreciable doses from this route in terms of dermal absorption or gamma radiation exposure.

Use of the distributions defined here may introduce *a priori* conservative biases (toward safety) in the following ways:

- Because statistically designed sampling has not been performed at this site, these distributions do not represent spatial variability in contaminant concentrations/activity. Rather, they represent uncertainty associated with exposure media concentrations/activity to which a random individual may be exposed. Due to site characteristics and topography, it is less likely that the individual will be exposed to relatively high contamination areas than to lower contamination areas. The assumption of uniform probability over the range of values specified is a conservative assumption.
- Exposure concentrations/activity of dust and ambient radon are highly dependent on season and weather conditions. A simple soil-to-dust conversion is used here (the same as used in LLRWMO 1999), and measured concentrations of radon are assumed to be the same as in the breathing zone of the receptor (the same assumption as used in LLWMO 1999). Both of these assumptions are highly conservative, particularly in the winter months when snow covers the ground.
- The maximum surface water values used reflect sampling results from Lake Alpha, a shallow lake near the mine. This lake is not considered to be a desirable potable water source in comparison to the much larger, more distant, and less contaminated Lake Sherman, which is connected to Lake Alpha.

2.3 *Methodology*

The point estimates and distributions defined in Table 2 are combined in the appropriate equations listed in Table 1. The dose conversion factors used here were calculated by the LLWRMO by application of standard health physics assumptions (LLWRMO 1999). Monte Carlo simulation (using the Excel® add-in Crystal Ball®) is used to subtract background media concentration/ activity distributions from site-related media concentration/activity distributions. Ten-thousand random draws are used for each simulation; this number of iterations allows a reasonable degree of confidence to be placed in upper percentiles of the output distributions (a greater than 99% confidence level in the 99th percentiles) without requiring an excessive amount of computer time (IAEA 1989). A Boolean statement is used to convert negative values resulting from the subtraction operation to zero. Thus, if a large number of subtractions result in negative values, the distribution will incorporate a "spike" of zero values. Percentiles of the "net" site-related concentration/activity distributions are incorporated into exposure algorithms; the results are distributions of possible incremental doses that can result from particular exposure routes.

Depending on the exposure route, different variables can be used as independent variables to provide an indication of the point when the AECB/ICRP annual dose limit of 1 mSv is attained. Exposure duration (ED) is used as an independent variable here. Calculations are performed for a series of time steps up to 20 days (a reasonable maximum ED; see Table 2). The median, 5th percentile, and 95th percentile of the uncertain cumulative dose for each exposure route are then calculated. Intake rate (IR) is also used as an independent variable in the cases of water and meat and/or fish ingestion. Dose calculations are performed for daily intake rates up to a reasonable maximum for each route (4 L per day for water, and 1500 g per day for wild meat and/or fish, see Table 2).

2.4 *Simulation Results*

Table 3 presents the results of the simulated site-related concentration/ activity distributions from exposure media, and Table 4 and 5 present the results of the simulated doses associated with those media.

Table 3. Results of contaminant concentration/activity modelling.

	Gamma (μSv/hr)	Radon (Bq/m^3)	Dust (μg uranium/g)	Water (μg uranium/L)	Meat/Fish (Bq/g)
Statistic					
Mean	0.72	382	25	132	0
Standard Deviation	0.44	230	18	76	N/A
Coefficient of Variation	0.61	0.60	0.72	0.58	N/A
Percentile					
5%	0.03	24	0	14	0
25%	0.33	182	8.2	66	0
50%	0.72	385	24	131	0
75%	1.10	580	40	197	0
95%	1.40	740	54	250	0

Table 4. Estimated exposure-route-specific incremental doses using average values for exposure assumptions (see Table 2). All doses given in mSv.

Exposure Route	Percentile	Exposure Duration (days)				
		1	*5*	*10*	*15*	*20*
External exposure to gamma radiation	50th	1.7×10^{-2}	8.6×10^{-2}	1.7×10^{-1}	2.6×10^{-1}	3.5×10^{-1}
	95th	3.4×10^{-2}	1.7×10^{-1}	3.4×10^{-1}	5.1×10^{-1}	6.7×10^{-1}
Inhalation of radon	50th	6.0×10^{-3}	3.0×10^{-2}	6.0×10^{-2}	9.0×10^{-2}	1.2×10^{-1}
	95th	1.2×10^{-2}	5.7×10^{-2}	1.2×10^{-1}	1.7×10^{-1}	2.3×10^{-1}
Inhalation of dust	50th	3.3×10^{-5}	1.7×10^{-4}	3.3×10^{-4}	5.0×10^{-4}	6.6×10^{-4}
	95th	7.4×10^{-5}	3.7×10^{-4}	7.4×10^{-4}	1.1×10^{-3}	1.5×10^{-3}
Ingestion of surface water	50th	6.3×10^{-3}	3.1×10^{-2}	6.3×10^{-2}	9.4×10^{-2}	1.3×10^{-1}
	95th	1.2×10^{-2}	6.0×10^{-2}	1.2×10^{-1}	1.8×10^{-1}	2.4×10^{-1}
Ingestion of meat and/or fish	50th	0	0	0	0	0
	95th	0	0	0	0	0

Table 5. Estimated exposure-route-specific incremental doses using extreme values for ingestion exposure assumptions (4 L/d for water and 1500 g/d for meat/fish), and highest measured values for exposure media concentrations/activity (without subtracting background concentrations) (see Table 3). Note that this is a *highly unrealistic* scenario. All doses given in mSv.

Exposure Route	Exposure Concentration	Daily Dose	Days to Reach 1 mSv Annual Incremental Dose Limit
External exposure to gamma radiation	1.6 μSv/hr	3.8×10^{-2}	26
Inhalation of radon	794 Bq/m^3	1.2×10^{-2}	83
Inhalation of dust	65 μg uranium/g	8.8×10^{-5}	11,400
Ingestion of surface water	263 μg uranium/L	3.4×10^{-2}	29
Ingestion of meat and/or fish	0.0007 Bq/g	2.6×10^{-3}	384
Simultaneous exposure to all routes and associated concentrations/activity as defined above	N/A	8.7×10^{-2}	11

There are differences between the doses as calculated by the LLWRMO (1999) and the doses calculated here due to subtraction of background concentrations/activity in this analysis, differences in intake values, and differences in exposure durations. However, the general conclusions are approximately the same as those made by the LLWRMO. The following ranking of pathways applies to daily doses across percentiles and exposure durations:

1. External exposure to gamma radiation (highest);
2. Ingestion of surface water from water bodies near the mine site;
3. Inhalation of radon;
4. Inhalation of dust; and
5. Ingestion of wild meat or fish (lowest).

366

It was not possible to perform traditional statistical analyses (due to the lack of data from a statistically-designed sampling plan) to address the question of whether site contamination in particular media exceeded that of background concentrations/activity to be expected in a mineral-rich area. However, it is informative to examine the percentiles of the simulated site-related distributions. For example, approximately 10% of the dust concentration distribution is below zero, and 95% of the wild meat/fish distribution is below zero. This provides an indication that the site-contamination-related wild meat/fish distribution is no different than background. This conclusion is uncertain due to the limited number of samples, but nonetheless may indicate that it is likely that little if any appreciable off-site contamination is presently occurring that is reflected in native foods.

The coefficients of variation (which provide a means of comparing relative uncertainty) of site-related radionuclide concentration/activity distributions are similar (Table 3). The coefficient of variation for the meat/fish concentration distribution is noninformative due to the large number of zero values.

2.5 Discussion

Gamma radiation accounts for approximately 3 times the dose from the next most important pathway (water ingestion). It would take almost a month of continuous exposure to reach the proposed incremental annual dose limit of 1 mSv for a 95th percentile exposure to gamma radiation (Table 4). Using average exposure assumptions, it would take 17 days of continuous simultaneous exposure to the 95th percentile of all exposure routes evaluated to reach the dose limit.

As an extremely conservative worst-case scenario, total doses (as opposed to incremental doses) were calculated using maximum values for all exposure media (without subtracting background levels), as well as maximum values for water (4 L/d) and wild meat/fish ingestion (1500 g/d). It would take 11 days to reach the exposure limit under this scenario (Table 5), which is virtually impossible. These results indicate that there is minimal remaining risk to humans from radionuclide exposure at the Rayrock site under any conceivable circumstance under present conditions.

The majority of the minimal potential risk from radionuclide exposure at the Rayrock site is associated with pathways that are directly associated with on-site contamination (e.g., gamma radiation from capped tailings). Potential off-site exposure pathways such as consumption of migratory wild game or consumption of water from off-site sources will result in much lower doses than on-site pathways. Therefore, there is a high degree of certainty that exposure from off-site sources would pose even lower potential risk than the minimal risk from on-site exposure.

There are a number of considerations that should be applied to interpretation of these results:

- The results are indicative of current conditions only;
- The calculations of site-related doses are dependent on limited data, especially for indicators of potential off-site exposure such as fish;
- Maximum values used here may not be representative of true maximums at the site;
- Appropriate background values are uncertain due to the natural heterogeneity of radionuclides in a mineral rich area;
- Realistic exposure patterns and behaviour of potential receptors are uncertain; and
- Uncertainty is associated with a number of standard health physics assumptions that were used by the LLWRMO in their dose conversions and repeated here.

As indicated above, not all sources of uncertainty were quantitatively addressed in this analysis. However, quantitative characterization of these sources of uncertainty would not be expected to change the conclusions of this report appreciably. Further characterization of background concentrations is not informative under current conditions because both incremental and total potential doses at the site are minimal. The long-term monitoring recommendations are targeted toward assuring that mitigation efforts performed to date continue to keep the potential for human radionuclide exposure at a minimal level.

3 LONG-TERM MONITORING RECOMMENDATIONS

Based on the LLWRMO assessment and the current analysis, a long-term monitoring plan is recommended to ensure that the mitigation efforts performed to date continue to keep the potential for human radionuclide exposure to a minimal level. The long-range plan is designed to focus on the media that offer the most appreciable potential for unacceptable exposures and that have appreciable associated uncertainty, balanced with the feasibility of sample collection based on the experience gained in the short-term sampling.

As indicated earlier, the ranking of media according to potential for radiation exposures is:

1. External exposure to gamma radiation;
2. Ingestion of surface water from water bodies near the mine site (e.g. Lake Alpha);
3. Inhalation of radon;
4. Inhalation of dust; and
5. Ingestion of wild meat or fish.

The uncertainty associated with contaminant concentrations/activity is similar across the exposure media as indicated by the respective coefficients of variability (Table 3). The exceptions are levels in meat and fish, which are highly uncertain due to lack of data. It is not possible to estimate a coefficient of variability for this pathway due to the apparent lack of difference between site-related concentrations and background. Due to the nature of caribou movements, it is unlikely that these animals would accumulate sufficient incremental doses from environmental media at the Rayrock site so that the meat would pose a potential hazard. However, fish have more direct and food-chain contact with potentially contaminated media, and it is possible that the true variability in radionuclide levels was not captured in the short-term program.

The feasibility and cost of collecting additional information vary dramatically across the exposure media. The following criteria can be used to score the different media as to feasibility of additional data collection in a remote Northern area:

- Necessity for use of helicopters as opposed to float planes for access;
- Necessity for use of other transport equipment such as all-terrain vehicles or snowmobiles as opposed to access on foot;
- Necessity for use of large sampling equipment such as drilling rigs;
- Time of year that samples should be collected;
- Objective hazards associated with sampling (e.g., necessity of traversing hazardous terrain, potential for adverse weather, etc.); and
- Number of person/hours necessary for sample collection.

Based on discussions with the site manager and with environmental professionals who have experience in different types of sample collection schemes, qualitative "feasibility scores" were assigned to the different exposure media according to the criteria above.

"Dose scores" that reflect the 50th percentile of daily risks (Table 4) were assigned to appropriate media. Table 6 is a matrix of estimated dose vs. feasibility of sampling scores.

The long-range sampling plan should optimally focus on those media that pose the greatest potential for doses along with ease/low cost of sample collection. The media that satisfy these criteria include gamma radiation, surface water, radon and fish.

Gamma radiation is associated with the highest potential doses, and it is relatively straightforward to collect these data. Also, gamma radiation measurements provide a measure of the

Table 6. Matrix of estimated doses and feasibility of sample collection.

Dose Exponent	Feasibility		
	Low	*Medium*	*High*
>10^{-2}			Gamma
$10^{-2} - 10^{-3}$		Radon	Surface water
$10^{-3} - 10^{-4}$			
$10^{-4} - 10^{-5}$	Dust, wild meat	Fish	

integrity of the existing tailings caps. Therefore, collection of gamma measurements is important to the long-range sampling plan. Personal dosimetry of sampling personnel is an important occupational safety measure.

Surface water ranks relatively high in terms of potential doses. The hydrology of the site is currently uncertain due to limited information regarding transport pathways (i.e., surface flow vs. groundwater input). Therefore, it is useful to perform surface water sampling, in both summer and winter, as well as sampling of groundwater from an important groundwater well downgradient from the mine to determine time trends and relative contributions from transport media to surface water bodies. Surface water sampling in winter will eliminate the contribution of surface water flow to total contaminant load, and will allow the possible contribution of groundwater flow to surface water bodies to be assessed.

Radon ranks high in terms of potential doses in the current assessment. However, the highly conservative assumption was made that radon measurements reflect breathing zone concentrations. Although radon is unlikely to be associated with appreciable exposures under current conditions, radon measurements provide a means of monitoring integrity of caps on shafts and adits.

Although fish appear to pose minimal risks, the large uncertainty associated with true site-related concentrations and the relative feasibility of collecting additional data contribute to this media's inclusion in the list. The fish sampling program outlined here includes provision for sampling fish in two lakes: a nearby reference lake that is unaffected by potential contamination from the Rayrock mine site; and Sherman Lake, which is a likely source of fish used for consumption.

Additional sampling of the remaining media will not be informative nor feasible. Further characterization of background concentrations will not be informative under current conditions because both incremental and total potential doses at the site are minimal, and extensive characterization of background heterogeneity is not feasible. Because wild meat appears to be associated with minimal exposures, further sampling of intermediate media such as lichens and macrophytes will also not be informative. Surface water and fish (direct exposure pathways) sampling should be sufficient to characterize the water environment. Sediment sampling will not be required unless significant increases in radionuclide levels in surface water and/or fish over time are evident.

The recommended plan includes periodic monitoring of the following components for a time scale of years:

1. Visual inspection of tailings and adit/shaft cap integrity;
2. Gamma radiation surveys and dosimetry in tailing areas;
3. Radon measurements in the vicinity of adits/shafts;
4. Surface water sampling in Lake Alpha and in the southwest arm of Sherman Lake; and
5. Groundwater sampling in a monitoring well downgradient from the mine.

Additionally, the following one-time data collection and interpretation exercise is recommended to improve the data upon which further recommendations may be based:

6. Fish collection in a reference lake and Sherman Lake.

There are several approaches for assessment of whether contaminant levels are increasing appreciably or significantly in a statistical sense over time. A dramatic increase (e.g., 100%) in a measured value in a particular medium over the previous measured value would be an obvious trigger for further investigation. Assuming that trends in contaminant levels are likely to be more gradual over a long period of time, a nonparametric statistical trend test such as the Mann-Kendall test (Gilbert 1987) can be employed. Use of such a test is recommended after the first 5 and 10 years, and every 10 years thereafter, to determine whether significant trends in contaminant concentrations are occurring. Statistically significant upward trends in contaminant levels over time will trigger re-evaluation of the monitoring and risk management plan.

Potential exposure media that are not listed above do not appear to pose appreciable risks, and reduction of uncertainty associated with exposures to these media will not be informative. However, further sampling of these media may be necessary if the recommended inspections and sampling indicate a change in current physical conditions. Changes in conditions associated with the monitored media, such as significant upward trends in surface water radionuclide concentrations, may be difficult to assess without an extensive statistically designed monitoring plan. However, a consistent increasing trend in surface water concentrations may indicate that a change in the physical condition of the site is occurring, and that the monitoring plan should be revised. The monitoring plan should be revised in any event in 100 years due to long-term uncertainties and possible changes in land use.

4 RISK MANAGEMENT

Both the LLWRMO (1999) and current assessments indicate that the potential for unacceptable radiation exposures at the Rayrock site is very low as long as the existing physical barriers remain intact. The long-range monitoring plan will provide information regarding the continued integrity of these low risks, and risk management will consist of ensuring that existing remedial approaches stay intact. Any compromise of existing physical barriers should be assessed and mitigated in an expedient fashion.

Additional recommendations can be made for management of non-radiation-related risks that may exist at the site. Long-term liability management is important at a site such as Rayrock where wastes have been left *in situ*. Implementation of legal land use restrictions, after consultation with any affected stakeholders, is recommended. It appears unlikely that the site can be released for unlimited public use in the near future, due to the potential for disturbance of remedial measures and for exposure to non-radiation physical hazards. Erection of "No Trespassing" signs and maintenance of caps on adits and shafts is important to avoid the potential for accidents.

5 REFERENCES

AECB (Atomic Energy Control Board), 1998, *Assessment and Management of Cancer Risks from Radiological and Chemical Hazards*, Minister of Public Works and Government Services Canada, Ottawa.

Gilbert, R.O., 1987, *Statistical Methods for Environmental Pollution Monitoring*, Van Nostrand Reinhold, NY.

Golder (Golder Associates Ltd.), 1996, *Environmental Monitoring Program for Assessing Remediation Efforts at the Rayrock Uranium Mine, Northwest Territories*, prepared for Public Works and Government Services Canada, October.

IAEA (International Atomic Energy Agency), 1989, *Evaluating the Reliability of Predictions Made Using Environmental Transfer Models*, Safety Series No. 100, Vienna.

Lee, R.C. & Wright, W.E., 1994, Development of Human Exposure-Factor Distributions Using Maximum-Entropy Inference, *J. Exp. Anal. & Env. Epid.* 4:329-341.

LLWRMO (Low-Level Radioactive Waste Management Office), 1999, *Short-Term Environmental Monitoring Program, Rayrock Uranium Mine: Revision 1*, prepared for Indian and Northern Affairs Canada, March.

O'Connor (O'Connor Associates Environmental Inc.), 1997, *Compendium of Canadian Human Exposure Factors for Risk Assessment*, Ottawa.

Tailings and Mine Waste'00 © 2000 Balkema, Rotterdam, ISBN 90 5809 126 0

Waste waters remediation from ^{226}Ra removal

E. Panţuru, D. Filip, D. P.Georgescu, N.Udrea, F.Aurelian & R.Rădulescu
*National Company of Uranium Research and Design Institute for Rare and Radioactive Metals,
Bucharest, Romania*

ABSTRACT: Radioactive on extraction and processing represented the most dougerous works involving an important risk and a high pollution potential on the environment. The principale aim of protection measurement is to provent uranium and its product diffusion in the environment. The results of this fact is to stop supllimentary iradiation of people.The most important ways to radioactive elements transportation in the environment ore air and water.For these resous, the most important factor for reduction the pollution at the uranium mine are waste waters contaminated with ^{226}Ra. Radium contaminated waste waters represent the major biological risks, too. The present paper study the sorbtion on activated carbons mecanisme and the influence of the physical and chemical characteristics of the support and the fluid. It has been studied the ^{226}Ra removal from pond water at the uranium on processing plent using eight types of indigenous activated carbons The paper presents tables and graphics with experimental results for each types of activated carbon and their removal from waste waters.

KEYWORDS: activated carbon, adsorption, ^{226}Ra separation, contaminated waste waters, pond waters, wet uranium tailings;

1 INTRODUCTION

The waste pond waters resulted within the uranium ores alkaline processing industry must be decontaminated prior to flowing into a large inflow river ^{226}Ra separation and recovery as a stable product is required by the present domestic and international regulations. Generally some procedure of adsorption separation on activated carbon for different heavy metals species are known to data. For the specific case of metal ion adsorption the nature of metal aqua species is of great importance. The separation process involving adsorption on activated carbons is due mainly to physical interaction between metals ions and the different size pores surface. Some specific processes for activated carbons obtaining lead to surface structures that enhance the heavy metal adsorption. We have not found detailed references for ^{226}Ra separation and recovery using activated carbons although this problem is of great importance for the environmental protection within the uranium processing industry. The present paper deals with the pond water decontamination by ^{226}Ra separation on different activated carbon types and recovery of the contaminant under the form of a stable product. Minimization of the water treatment plant size and surface necessary in the same time with ensuring a high decontamination rate are the main goals of this study.

2. EXPERIMENTAL

^{226}Ra adsorption on activated carbons was studied as a variant to the direct precipitation – setting of the Ba(Ra)SO$_4$ system.

The early experiments were undertaken using a sample of pond water collected after the uranium ion exchange decontamination module, with an 5,90 pCi/l (0.218 Bq/l) activity, and a sample of clarified waste water solution having an 7 pCi/l (0.259 Bq/l) activity, from the uranium RIP sorption process.

All the activated carbon types used were washed two times by distilled water and then samples of 0.50 dm^3 were boiled during 1 hour, at a solid / liquid ratio of 3/1, under a continuos slow mixing. Dried samples (160 °C during 4 hours) of granular adsorbents were used for ^{226}Ra separation from both aqueous solutions. The activated carbons selected are known as MN, MN+P,MDC, AG, CGN, CGN+COCS, WITCARB.

The experiments made under dynamic conditions had a water flow rate of 12 BEV/h for all the activated carbons. The adsorption columns were filled with the same volume of 0.50 dm^3 of adsorbent.

The analytical results obtained after sampling the effluents during the experiments are presented in tables 1 and 2.

Table 1.

Activated carbon type	Number of BEV processed	^{226}Ra in effluent, pCi/l*	^{226}Ra loaded on activated carbon ,pCi/l	F,adsorbent /water concentration ratio
AM	100	1.0	1.58	5.9
	200	1.1	3.13	5.4
	300	2.1	4.35	2.8
	400	2.4	5.48	2.4
	500	3.1	6.39	1.9
MN+P	100	1.1	1.6	5.4
	200	1.2	3.17	4.9
	300	1.7	4.57	3.5
	400	1.7	5.97	3.5
	500	2.0	7.26	2.9
MDC	100	1.3	1.27	4.5
	200	1.9	2.39	3.1
	300	2.2	3.42	2.7
	400	2.2	4.44	2.7
	500	2.6	5.36	2.3
AG	100	0.6	1.29	9.8
	200	0.7	2.56	8.4
	300	0.7	3.83	8.4
	400	0.9	5.05	6.5
	500	0.9	6.27	6.5
CGN	100	0.5	1.20	11.8
	200	0.5	2.40	11.8
	300	1.3	3.42	4.5
	400	2.1	4.27	2.8
	500	3.0	4.91	1.9
CGN+COCS	100	1.8	1.05	3.2
	200	2.2	2.0	2.7
	300	2.4	2.89	2.5
	400	2.5	3.77	2.4
	500	2.8	4.56	2.1
WITCARB	100	1.3	1.12	4.5
	200	1.3	2.24	4.5
	300	2.1	3.17	2.8
	400	2.4	4.02	2.5
	500	4.5	4.37	1.3

*1pCi = 0.037 Bq

Table 2

Activated carbon type	Number of BEV processed	^{226}Ra in effluent, pCi/l [*]	^{226}Ra loaded on activated carbon, pCi/l	F,adsorbent /water concentration ratio
MN	100	3.1	1.26	2.3
	200	3.1	2.52	2.3
	300	3.9	3.52	1.8
	400	5.0	4.16	1.4
	500	7.0	4.16	1.0
MN+P	100	1.1	1.97	6.4
	200	1.1	3.39	6.4
	300	1.5	5.77	4.7
	400	1.9	7.47	3.7
	500	3.9	8.5	1.8
MDC	100	1.1	1.64	6.4
	200	1.7	3.11	4.1
	300	1.7	4.58	4.1
	400	2.4	5.86	2.9
	500	3.4	6.94	2.0
AG	100	1.0	1.46	7.0
	200	1.1	2.9	6.4
	300	1.1	4.34	6.4
	400	1.3	5.73	5.4
	500	1.3	7.12	5.4
CDC	100	3.1	0.9	2.3
	200	3.2	1.9	2.2
CGN	100	2.3	1.12	3.0
	200	2.4	2.21	2.9
	300	2.4	3.3	2.9
	400	3.1	4.24	2.3
	500	4.3	4.88	1.6
CGN+COCS	100	1.8	1.53	3.8
	200	2.2	2.94	3..2
	300	2.4	4.29	3.3
	400	2.5	5.62	2.8
	500	2.8	6.85	2.5
WITCARB	100	1.3	1.58	5.4
	200	1.3	3.16	5.4
	300	2.1	4.52	3.3
	400	2.4	5.80	3.3
	500	4.5	6.50	1.55

Comparing the obtained values for the activated carbons ^{226}Ra loading capacity, for all the adsorbents studied we found that the MN+P and AG types had a $7 - 8pCi/dm^3 (0.259 - 0.296$ Bq/dm^3) capacity, after the decontamination of a 500 BEV waste water volume. All the activated carbons studied had loading capacities greater than a reference adsorbent, the dry peat respectively, used in previous ^{226}Ra separation experiments, made also under dynamic conditions. The obtained ^{226}Ra water decontamination curves for the experimental involved adsorbents are presented in figures 1 and 2.

Concerning the decontamination efficiency one may see that the performances are reached by the AG and MN+P types activated carbons. The effluents resulted after processing 500 BEV wastewater had a residual activity under 1 pCi/l (0.037 Bq/l), for the pond water, as shown in figures 3 and 4.

Considering the results found during the experiments undertaken were selected for supplementary testes the AG and MN+P activated carbons. For this aim, using AG adsorbent in dynamic conditions was processed a 1355 BEV volume of waste pond water for establishing the

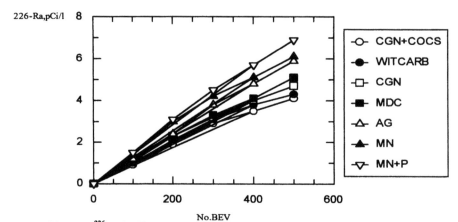

Figure.1. ^{226}Ra loading curves for adsorption from pond waste waters

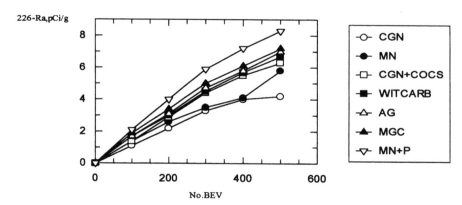

Figure.2. ^{226}Ra loading curves for adsorption from depleted uranium leach solutions

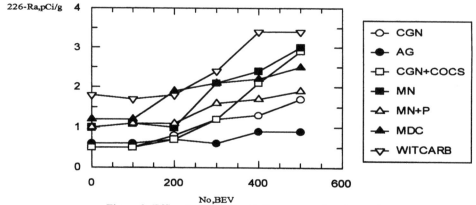

Figure.3. Effluents residual activity curves (pond water)

^{226}Ra-separation yield on this activated carbon. The results obtained are presented in table 3 and in figure 5

We found that the AG type adsorbent loading capacity has a higher value comparing to peat and other activated carbons. The loading capacity of the AG adsorbent is twice the capacity of

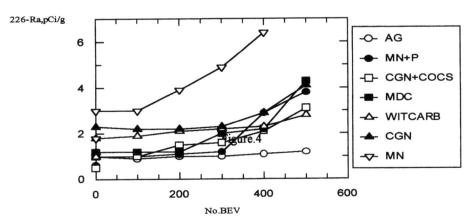

Figure.4. Effluents residual activity curves

Table.3.

Water volume , BEV	^{226}Ra in effluent, pCi/l	^{226}Ra loaded acti- vated carbon, pCi/g	F.adsorbent/water concentration ratio
100	0.6	1.29	9..8
200	0.7	2.56	8.4
300	0.7	3.83	8.4
400	0.9	5.05	6.5
500	0.9	6.27	6.5
870	0.9	10.78	6.5
1125	1.1	13.77	5.3
1355	2.6	15.62	2.3

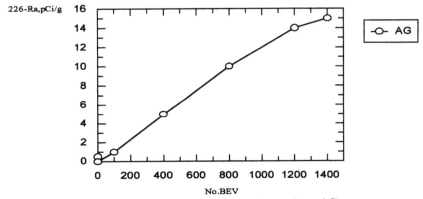

Figure.5. Activated carbon loaded curve (type AG)

peat, the ^{226}Ra activity in effluent being always under 3 pCi/l (0.111 Bq/l). The elution – regen-eration for the AG type activated carbon was made using a 30 BEV volume of an aqueous volu-tion having HCl / H$_2$O = 1/ 1. The ^{226}Ra contaminated eluate resulted had an activity of 201 pCi/l (7.437 Bq/l), the elution yeld being over 99 %.

Within the adsorption – elution cycle was obtained a decrease of the total contaminated water volume by some 45 times, the further processing of such a smaller volume being possible by the classical barium coprecipitation – settling, obtaining a precipitate with a Ba (Ra)SO$_4$ content.

Another experiments consisted in the study of ^{226}Ra adsorption on a MN+P activated carbon bed, during two cycles of adsorption – elution –regeneration to find if the eluant solution does

Table.4

Cycle	Water volume, BEV	^{226}Ra in effluent, pCi/l	^{226}Ra loaded on activated carbon, pCi/g	F, adsorbent/water concentration ratio
I	70	1.2	1.028	4.9
	145	1.7	2.012	3.5
	245	1.9	3.262	3.1
	315	2.3	4.049	3.0
	375	2.6	4.771	2.3
	464	2.7	5.671	2.1
	514	2.8	6.155	2.1
II	62	0.7	1.007	8.4
	152	1.7	2.189	3.5
	228	1.7	3.186	3.5
	308	2.2	4.111	2.7
	388	2.3	5.011	2.5
	500	2.5	5.755	2.4

not decrease the loading capacity of this adsorbent. The preliminary tests cared out on the MN+P type adsorbent lead to good separation yields.

The experimental results obtained are presented in table 4.

The adsorption conditions for ^{226}Ra separation on activated carbon were similar, the pond water having a radium content of 5.9 pCi (0.218 Bq) and the water inflow in the column being 12 BEV/h. After flowing in a 514 BEV pond water volume the activated carbon was eluted by an aqueous solution of HCl/H$_2$O (1/3 in volume), using an inflow of 3 BEV/h, when an elute having an 199.4 pCi/l (7.377 Bq/l) activity was obtained with an 99 % recovery yield.

Then another 500 BEV pond water volume was treated on the regenerated activated carbon, with the same 12 BEV/h inflow. From the obtained data we can state that the activated carbon elution – regeneration using a HCl/H$_2$O (1/3) solution have not affected the adsorbent separation capability for ^{226}Ra.

The activated carbon leading, after treating a 500 BEV water volume, being similar with the leading obtained with in the first adsorption cycle.

3.CONCLUSIONS

The experimental data obtained for activated carbons of AG and MN+P type impose their use in ^{226}Ra separation, having a high pond water processing capability and a good separation yield, obtaining effluents with a residual activity value of less 1 pCi/l (0.037Bq/l). The high content of ash in the activated carbons may enhance the separation yield during the first adsorption cycle.

The researches had for find result a proposal for technical sheet process for ^{226}Ra separation from the pond waters, with the following main operations:

- activated carbon cleaning by boiling with distilled water for one hour, drying at 160 °C for four hours;
- ^{226}Ra adsorption on activated carbon, in column, from the waste pond water, at a inflow volume of 12 BEV/h ;
- water washing of the loaded activated carbon ;
- elution – regeneration with a HCl/H$_2$O (1/3) solution at an inflow of 3 BEV;
- contaminated elute processing by chemical treatment and precipitation with BaCl$_2$, precipitate setting and separation.

4.REFERENCE

[1] Miers,A.C, 1985,*Journal of American Water Wortes Association*,77 (5)
[2] Snoeying,V.L, 1987, *Journal of American Water Wortes Association*,79 (8)
[3] Snoeying,V.L, 1985, *Raport EPA,/600/ 2-85 /006*
[4] Hatch, M. J, 1983, *European Patent EP 71810*, U. S. 289615

[5] Wang,S. R, 1985,*Journal of Radioanalytical and Nuclear Chemistry,* 96(6), 593-9

[6] Georgescu,P.D, Stoica, L,Rădulescu, R, Filip,D, 1995, Uranium Mining and Hydrometalurgy *,Proceding of the International Conference and Workshop,* Freiberg, Germany, 197

[7] Stoica,L, Filip,D, Filip,Gh, Rădulescu,R, Răzvan,A, 1998,*Journal of Radioanalytical and Nuclear Chemistry,* 229, 139-142___

Policy and procedures

Tailings and Mine Waste'00 © 2000 Balkema, Rotterdam, ISBN 90 5809 126 0

Communication with the public about tailings projects

Jeremy Boswell & Nanette Hattingh
Bohlweki Environmental (Pty) Limited, Midrand, South Africa

David de Waal
Afrosearch (Pty) Limited, Pretoria, South Africa

ABSTRACT: The increasing focus in democracies around the world on environmentalism has had significant impact on the way in which tailings projects are now performed. The development and management of tailings disposal facilities is required to consider the following interfaces with the public:
a. Public involvement in the siting of new disposal facilities.
    Public involvement in the operation, monitoring, auditing, revegetation and closure of existing facilities, especially where they are located near residential areas and/or have significant impacts on humans.
b. The preparation of flood notification procedures and emergency warning systems where changes in legislation, increased risk or public perception demand.
c. Occupational health and safety, occupational hygiene and health risk aspects relating to public or worker exposure to tailings and impacts from tailings facilities.
A number of high profile failures of tailings dams in recent times have only served to heighten public concern about tailings and mobilise public opposition to tailings projects in much the same fashion as hazardous waste landfills or other so called "lulu's" (locally unwanted land uses) such as airports, highways and the like.
A central facet of tailings communication involves the communication of risk. After consideration of a brief literature scan and citing a number of case histories, the authors show the importance of the involvement of the recipient of the message in the design of the medium and the content of the message that is used in the communication about risk. Unless the receiver is involved in the process that is designed to communicate risk the message will fall on deaf ears. The authors outline in some detail, the factors that should be considered in the communication of risk, and so provide the practising tailings engineer with useful tools in straddling the divide between engineers and the public in communication regarding tailings projects. It is suggested that very few new tailings projects will be allowed without significant involvement of interested and affected parties, especially the public.

1   INTRODUCTION

The South African economy depends to a large extent on mining and minerals processing. The negative environmental impacts associated with these activities, are countermanded by the positive effects these activities have on the national economy (Chamber of Mines of South Africa,1996). It is acceptable practice for residues of mining to be disposed of in the form of tailings dumps and tailings dams. These are generally associated with negative impacts such as visual impact, dust, soil infertility and health impacts.

For the layman the word 'tailings' immediately conjures up images of failures of tailings dams. High profile media event include such failures as the Los Frailes tailings dam failure

(Website, 1999) on 25 April 1998 in Spain in which some 5 million m^3 of lead-zinc tailings entered the Rio Agrio and further threatens the Doñana National Park, a UN World Heritage Site. Another failure of a 31 m high tailings dam near Merriespruit, South Africa, (Wagener *et al,*1997*)* left hundreds homeless and 17 people were killed. Failures such as this, have caused the general public to associate tailings, tailings dumps, tailings dams or slimes dams with high risk.

These high profile failures of tailings dams in recent times have only served to heighten public concern about tailings and mobilise public opposition to tailings projects in much the same fashion as hazardous waste landfills or other so called "lulu's" (locally unwanted land uses) such as airports, highways and the like.

Recent changes in South African legislation, have made provision for the public to be taken into consideration when embarking on new developments. The Bill of Rights, contained in the Constitution of the Republic of South Africa (Act No 108 of 1996), states (loosely paraphrased) that everyone has the right to an environment that is not harmful to their health or wellbeing. Furthermore, everyone has the right to have the environment protected through reasonable legislative and other measures that prevent pollution and ecological degradation.

As stated by Petts and Eduljee, (Petts and Eduljee, 1994): "Opposition to waste activities reflects a complex mix of psychological, cultural, and socio-economic factors which underly responses to issues in general. Superimposed onto these are concerns which relate to any industrial activity". This will clearly affect the response of the public towards proposed activities such as the siting of a new tailings disposal facility, or the closure and rehabilitation of these facilities.

If the perception exists that a tailings project could have health implications, or holds a certain element of risk, the proposed project could be jeopardised by potential social mobilisation resulting from a negative attitude towards the development. It is therefore becoming increasingly important to involve the public in the development and management of tailings disposal facilities, in order for these projects to succeed.

The increasing focus in democracies around the world on environmentalism has had significant impact on the way in which tailings projects are now performed. The development and management of tailings disposal facilities is required to consider the following interfaces with the public:

a. Public involvement in the siting of new disposal facilities.
b. Public involvement in the operation, monitoring, auditing, revegetation and closure of existing facilities, especially where they are located near residential areas and/or have significant impacts on humans.
d. The preparation of flood notification procedures and emergency warning systems where changes in legislation, increased risk or public perception demand.
e. Occupational health and safety, occupational hygiene and health risk aspects relating to public or worker exposure to tailings and impacts from tailings facilities.

2   LITERATURE SURVEY

A brief review of literature undertaken during the writing of this paper, indicated widely differing approaches to public participation. Turning to the local scene in South Africa, definitive references which would guide the prospective environmental practitioner regarding public participation have not, to the authors' knowledge, been developed. The recently published South African code SABS 0286 (Code of Practice : Mine Residue 1998) requires of developers public participation, but gives little detailed guidance on how to approach public participation. Perhaps the lack of detailed guidance on public participation is not surprising, if one considers the dramatic changes that have been taking place in South Africa over the last five years. Nevertheless, the authors have found two documents particularly useful that have been published locally (Department of Environment Affairs, South Africa 1992 and SAICE

Environmental Engineering Division, 1993). The Department of Environment Affairs provides (Department of Environment Affairs, South Africa 1992) some background to public participation and the associated volumes in the series provide some very useful checklists for use in integrated environmental management and environmental impact assessment, but they are unhelpful with regard to how public participation and scoping should take place. Again, this is not surprising, since they are focussing on the methodology of scoping and environmental impact assessment rather than the full process of public participation. The Environmental Engineering Division of the South African Institution of Civil Engineers (SAICE, 1993) provides more detailed and more useful background information for public participation, but this reference considers civil engineering projects in general. There are a number of specifics relating to waste management facilities, which the document does not address.

From a brief review of currently available documentation on public participation in the waste management field and mining industry, clear guidance as to how to avoid the pitfalls of public participation and a comprehensive methodology of what to do, are not available.

Several case studies on tailings projects were reviewed. A summary of these case studies are attached as Appendix 1.

3   DEFINITIONS

3.1   *Public participation defined*

Public participation is the process of engaging and involving the public, in the scoping, planning and management of facilities. The public participation process is aimed at enabling people to participate in the project evaluation process, to share information and to identify concerns. It also helps them to grasp a better understanding of what the project entails. The process of public participation will allow for informed decisions based on a clear understanding of how various constituencies would be likely to respond to the project and any initiatives undertaken.

A number of other activities are often used in the same context as public participation, but they cannot be considered to be public participation in the truest sense of the word. It is important to exclude these from the process that one is contemplating. Among them would be:

a. *Public relations*: Public participation in no way replaces or precludes the use of conventional marketing, selling and PR techniques towards best advancing the cause of a particular company. While public participation may refer to these parallel activities of a company, it is the authors' experience that extensive use of PR is inappropriate in a public participation project.

b. *Lobbying by itself*: is not public participation. It is important for those making decisions and those influencing decision makers to be made aware of the issues at stake, but once again, extensive use of lobbying in a public participation forum is invariably counter productive.

c. *Issue Management and Crisis Communication*: In the authors' experience, the need to use such methodologies is heightened in the absence of professional public participation. However, just because the public have been fully engaged does not mean to say that should the company hit a crisis, a public issue will not result. Public participation tends to work better when followed in a routine and planned manner and is less successful in times of crisis.

d. *Mediation, arbitration and litigation*: May be used instead of public participation, but they cannot replace it, nor can they be fully replaced by public participation. The use of litigation in parallel with public participation will tend to frustrate both processes.

A fundamental aspect of public participation, which is not generally indicated in the literature, is the effect of proper public participation on project planning. A failure to recognise the requirement for public participation at early stages of the project has the potential to torpedo the project altogether. Should the project survive the lack of an early public participation process or programme, it will in any event be much more expensive and take much longer. World history over the last twenty years is littered with examples of such failed environmental projects which failed to take into account the views of the public at the right stage in project planning. If, however, the notion of public participation is fully included at the planning stages of a project, it may make the project more time consuming and more expensive, but it is far more likely to ensure that the project will reach finality. A project in the first case scenario runs a very severe risk of never being completed at all.

The authors have found that the most effective structure relies on the establishment and project management of a team of four functions as shown in Figure 1.(Boswell 1996)

The public participation programme may be conducted by one party who then employs an independent facilitator, especially in the case of controversial projects. The authorities may also want to get involved directly in the facilitation and/or public participation aspect. In the absence of clear guidance and legislation in South Africa the involvement of the authorities in this way has been found to be a significant advantage and has been encouraged.

Public participation needs to identify interested and affected parties very early and use them in designing an agreed planning or process framework. Public participation cannot replace normal democratic processes and the functions of normal civil society such as the courts, the police and the legislature, but hopefully it obviates the need to resort to these avenues.

An extensive list of interested and affected parties should be developed. This should be the result of a desktop process supported by a networking mechanism. Listings of interested and affected parties should be updated and maintained on an ongoing basis. The preparation of documentation, describing the full extent of the project and relevant details, may then be circulated to the interested and affected parties. It is vital that prior to the finalisation of any participative process, the interested and affected parties concerned have been consulted on the nature of the process and ideally have indicated their support.

The public participation programme should determine the structure of the project programme, project activities and milestones to coincide with the public participation mechanisms, such as workshops, public meetings, open days, etc.

It is vital that at the very earliest juncture, the process by which the project will be managed is agreed with the interested and affected parties. No-one will just rubber stamp a process developed by someone else. Empowerment and the granting of legitimate responsibility in terms of influencing the decision is critical to the success of the project.

4   SELECTED CASE STUDIES

In order to better explore the psychology of risk and public communication for tailings projects, perhaps a few brief examples should be considered first.

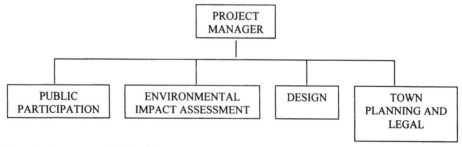

Figure 1: Structure of Project Management

4.1 *Tailings dam failure at Merriespruit, South Africa* (Wagener *et al, 1997*)
Introduction

On the night of 22 February 1994 a 31 m (100 ft) high tailings dam upslope of Merriespruit, a suburb of the town Virginia, in the Free State Goldfields of South Africa, failed with disastrous consequences. The dam failed a few hours after 30 to 50 mm (1-2 inches) of rain fell in approximately 30 minutes during a later afternoon thunderstorm. The failure resulted in some 600 000 m^3 (785 000 cuyd) of liquid tailings flowing through the town, causing the deaths of 17 people, wide-spread devastation of property and environmental damage.

Most likely mode of failure

Rain falling in the impoundment flowed towards the pool which had moved away from the penstock towards the north wall. The pool already contained some 40 000 to 50 000 m^3 of water pumped from the mill. The impoundment had insufficient capacity to contain the additional inflow of about 50 000 m^3 that had resulted from the rainstorm and started overtopping at the lowest point of the north wall. The spilling water, together with what had accumulated on the northern berms of the dam, eroded the loose tailings that had been pushed into earlier sloughs on the lower slope. This led to small slip failures occurring with sloughing tailings being removed by the overtopping water, and preventing the buttressing effect of any materials that would otherwise have accumulated at the toe of the unstable slopes. This resulted in a series of slip failures retrogressively moving up the slope in domino fashion ultimately leading to a massive overall slope failure that released the flow slide.

Preventing similar occurrences in future

The Merriespruit disaster has provided an impetus for both the South African mining industry and the state to take a more responsible and serious approach to disposal of tailings. A fundamental re-assessment of the philosophies for design, management and operation of tailings dams has been initiated in South Africa. The mining industry has carried out independent self-audits, now viewing tailings dams and associated problems far more seriously than in the past.

In 1995, work started to draft an obligatory Code of Practice for the design, operation management, rehabilitation and closure of tailings dams. It was published in 1998 as SABS 0286: Code of practice for mine residue.

4.2 *Premier Mine – No 7 Slimes Dam. Preparation of flood notification procedure*

No. 7 Dam is the main slimes disposal facility for Premier Diamond Mine (Premier 1999). Presently the dam is supported by an 80 m high wall, which impounds 80 million cubic metres of water and suspended slimes. An inspection by expert consulting engineers has found the dam to be structurally sound. Until recently, a large slimes dam such as No 7 Dam was not subject to the same legal regulations imposed on a similar water storage dam. The regulations have however, recently been amended and No 7 Dam has been classified as a Category III water storage dam. In compliance with the recent changes in applicable regulations under Section 9C(6) of the Water Act (Act 54 of 1956), De Beers Consolidated Mines Limited is required to prepare flood notification procedures and a flood warning system for No 7 Dam, as required by the Department of Water Affairs and Forestry.

A process of public participation has been initiated in order to receive comments, issues and concerns that may affect people and the environment. These inputs will be appropriately recorded, discussed, and addressed. Intensive consultation is envisaged with those people residing in the primary affected area, which stretches for 28 km downstream from No 7 Dam to the farm Haakdoornfontein, at an average width of 5 km across the Premiermynloop and the Pienaars River valley. A further phase will cover a secondary area, which extends an additional 52 km downstream from No 7 Dam.

4.3 *Other case histories*

A number of other case histories were to be used to illustrate the sensitivity of tailings projects with regard to the public. Unfortunately permission could not be obtained within the publishing deadline to report in more detail on these additional case histories. This aspect only serves to highlight the extent to which public sensitivity of tailings projects can be critical.

5   WHY PEOPLE REACT THE WAY THEY DO

A clear understanding of the perceptions of risk by the affected community/stakeholders can contribute in managing such issues in a proactive manner. It is necessary to understand why people often react the way they do. It is critical that a clear distinction be drawn between factual risk assessment conducted by engineers and scientists on the one hand and the perception of risk by the community on the other. It is important to realise that a discrepancy exists between scientific "truths" and what ordinary people perceive as "truth". The latter base their beliefs and "truths" on the way things are likely to turn out.(Schoeman *et al* 1994). It is essential that these beliefs and perceived risks should be treated as real, with real consequences as they may occur at any stage of the project.

There are some key variables that determine community perception of risk and community anger at agencies that may treat these factors as irrelevant: (The following section is largely based on Chloorkop Report, Specifically Hance, Chess and Sandman (1988) as discussed in (Schoeman and Brugge, 1994).

5.1 Voluntary risks are accepted more readily than those that are imposed, hence:
- lack of choice leads to anger;
- community perception of coercion evokes fierce reaction;
- as processes are questioned, risks are ultimately perceived to be more "risky".

   This is the "decide and defend" approach: where a locally unwanted land use (such as an airport, tailings dump, landfill or highway) is sited first, before engaging the public. It is always much more difficult to manage than a utility which has been sited using community input. This may also be described as the "not-invented-here" syndrome.

5.2 Risks under individual control are accepted more readily than those under government control.

   Hence most people feel safer with risks under their own control, most of us feel safer driving than riding as a passenger - our feeling has nothing to do with the data, i.e. our driving record vs. the driving record of others.

5.3 Risks that seem fair are more acceptable than those that seem unfair. Hence:
- a coerced risk will always seem unfair;
- a community that feels stuck with the risk and gets little benefit will find the risk unfair - and thus more serious.

5.4 Risk information that comes from trustworthy sources is more readily believed. Hence on going battles with communities erode trust and make the message far less believable.

5.5 Risks that seem ethically objectionable will seem more risky than those that do not. Hence:

- to many people pollution is morally wrong;
- speaking to some people about an optimal level of pollution is like talking about an acceptable number of child molesters.

5.6 Natural risks seem more acceptable than artificial risks. Hence a risk "caused by God" is more acceptable than one caused by people, since natural risks provide no focus for anger.

Here for example, the loss of life as a result of earthquake, flood or fire is seen as an "act of God" and therefore in some strange way, more acceptable. The loss of life associated with man-made structures such as buildings, dams or bridges is somehow seen to be less acceptable since it is supposedly more foreseeable. The fact that more people are killed in South Africa through criminal violence or by road accidents does not mitigate the loss of life associated with a civil engineering disaster.

5.7 Exotic risks seem more risky than familiar risks.

5.8 Risks that are associated with other, memorable events are considered more risky (e.g. Chernobyl or Three Mile Island).

5.9 Risks that are 'dreaded' seem less acceptable than those that carry less dread. Hence an industrial emission that may cause cancer seems much less acceptable to many people than one that may increase the risk of emphysema, even though both diseases can kill.

5.10 Risks that are undetectable create far more fear than detectable risks (an expert war correspondent said at Three Mile Island Nuclear Plant that 'at least in war you know you have not been hit yet'). Hence risks with effects that may take years to show up are more likely to be feared.

The clean-up of asbestos contamination in South Africa is proving to be extremely problematic in that many cases of mesothelioma, lung cancer and related asbestos diseases have only been detected 10 years after exposure (sometimes 10 years after the cessation of exposure). This clearly has a dramatic effect on increasing public concern.

5.11 Risks that are well understood are more acceptable than those that are not. Hence risk that scientists can explain to communities are more acceptable than those about which they have to admit a great deal of uncertainty.

The greater the number and seriousness of these factors, the greater the likelihood of public concern about the risk, regardless of data.

6   LESSONS LEARNED FROM PRACTICAL IMPLEMENTATION

6.1 *The process of public participation that is followed*

Petts and Eduljee 1994, give some valuable insights into classic 'decide - announce – defend approaches and other approaches to siting problems. Consideration should be given to involve the public as early as possible, especially in the design of the process that is to be followed. Transparency is clearly of critical importance.

6.2 *Do your homework*

There can be no excuse for an engineer who fails to establish the existence of fatal flaws an early stage of the project, who fails to understand and acknowledge the role players in a project and the background to their approach. Wholehearted commitment to the project is vital. The emotional response to interested and affected parties which is so often and. for good reason, discouraged, should instead be translated into emotional commitment to making progress in an ethical and environmentally responsible way.

Know exactly what is to be done and what cannot be done. The consequences of ignorance are most damning for the developer.

Apart from knowing what you are doing, it is necessary to know what you may need to do, what your non-negotiables are and how to make compromises with fairness.

6.3 *Use independent experts*

This applies to both facilitation and study of the different impacts.

6.4 *Avoid public meetings as an isolated event*

Public meetings ideally should be an event in a participative process. In practice however, these are often used as a procedural mechanism to comply with consultation requirements. In isolation, this can easily become a rubberstamping and coercion mechanism, rather than transparent consultation.

Only make use of public meetings if you really know what you are doing and you are advised by somebody who has been there before. A carefully prepared and planned workshop or a tour of facilities, followed by a presentation, may often achieve the objectives with far less risk and animosity.

6.5 *Build trust*

Honesty is most certainly the best policy. Petts and Eduljee (Petts and Eduljee 1994) have some useful guidance on how to build trust. Openness, frankness and clear communication channels all assist in building trust, especially when trying to establish the need and desirability of a project. The developer should endeavour, as far as possible, to talk directly to interested and affected parties, not to use consultants or other agents. Should interested and affected parties have to communicate through gate keepers, this can severely disrupt the process.

In the authors' experience, one cannot conduct public participation in isolation from the rest of the company business nor the social and related issues within the interested and affected groupings. Such things as affirmative action policy, social responsibility programme, trade union relations, company environmental policy and training programmes are important.

6.6 *The use of legal means*

The use of legal means and the law courts is usually counter-productive to good public participation. They may even be mutually exclusive.

6.7 *The role of the media*

The role of the media must be carefully considered. They are powerful and their needs must be satisfied. A failure to do so may lead to an outcry. Information needs to be provided honestly, regularly and quickly. Once a public participation programme is in motion, it must be very clearly understood by the members who are participating, who is to speak to the media and what their terms of reference are. This is especially true in the case of disputes and in mediation. The use of the media is an essential adjunct to promote honest two way communication between the parties.

6.8 *The structure, nature and methodology of appeals procedures for sensitive environmental projects*

The structure, nature and methodology of appeals procedures for sensitive environmental projects must be well known. A tribunal panel set up to hear an appeal must be assisted or have direct access to relevant environmental technology though experts. The cross examination of witnesses tends to be counter productive.

6.9 *The role of elected leaders, politicians and decision makers*

The lack of clearly defined decision making responsibilities and structures for implementation can frustrate the development. Politicians should avoid making dramatic statements which are later difficult to defend or reverse. They should need to intervene rarely, but where necessary, decisively.

In the authors' experience, where active involvement of the authorities has taken place in monitoring groups, public participation, mediation and workshopping, progress has usually resulted.

6.10 *Make progress without creating losers*

It is important to give the public victories. The focus on win-win strategies is vital. Where necessary, separate meetings should be convened to workshop issues. An awareness of deadlock breaking mechanisms is always useful. The expertise in this field is established and credible .

6.11 *Educate and inform wherever possible*

The active involvement of an informed public in tailings management in South Africa will be a breakthrough. Build bridges of trust in advance of the floods of concern. Use can be made of facility tours, lectures, information dissemination, lectures at schools and research at places of higher learning. We have a long way to go in this regard. A simple measure of our progress to date is to ask the average person what they understand by the term 'tailings dam'.

7 RECOMMENDATIONS

7.1 For public participation in tailings projects to be effective, the following principles should be followed:
   a. Public involvement in the process must occur early and be ongoing in nature.
      Public participation takes more time and effort up front, but will result in better decisions which often have less negative long-term perceived impacts.
   b. Public trust is earned through openness, outreach, consistency and results.
      Public involvement is integral to sound decision-making.
   c. Public dialogue will increase understanding among all interests affected by environmental decisions. Public dialogue can aid in helping the public understand their individual expectations, resulting in more workable and widely acceptable solutions.
   d. Public input should be solicited from all sectors of society.
      Equal opportunity for comment and equal consideration of comments from the private

and public sectors should be provided.
  e. The public deserves substantive responses to all comments they submit.

7.2 Information to the public should be made available in an informative manner, and in such a way that all levels of society will be able to understand the content of the message. The following list is not meant to be complete, and only serves as an indication of the type of information that should be included:
  a. Tailings supernatant lagoons: location, size, water composition, dissolved loads and concentrations in the sediment, sediment quantities.
  b. Ecosystem vulnerability: type, uniqueness, connectivity of lagoons to wetlands, risk.
  b. Overtopping spill potential: lagoons into nearby rivers, dam type, geographical setting.
  c. Transport and storage capacity of nearby river systems: streamflow regime, sediment type, known deposition areas, river morphology.
  e. Possibility of health risks.
  f. Proposed mitigation measures.
  g. Proposed management structures.
  h. Proposed safety measures.

CONCLUSION

Risk communication carries with it the risk of adverse public reaction. It must be recognised that no matter how reassuring the technical evidence regarding the real risk of failure, the act of communicating unwelcome information to interested and affected parties will cause initial alarm.

On the other hand, effective risk communication also carries with it opportunity. A degree of alarm can assist the process by accelerating it, and ensuring that maximum awareness and consultation is achieved without major expense.

A central facet of tailings communication involves the communication of risk. After consideration of a brief literature scan and citing a number of case histories, the authors show the importance of the involvement of the recipient of the message in the design of the medium that is used in the communication about risk. Unless the receiver is involved in the process that is designed to communicate risk the message will fall on deaf ears. The authors outline in some detail, the factors that should be considered in the communication of risk, and so provide the practising tailings engineer with useful tools in straddling the divide between engineers and the public in communication regarding tailings projects. It is suggested that very few new tailings projects will be allowed without significant involvement of interested and affected parties, especially the public.

ACKNOWLEDGEMENTS

The permission of Premier Diamond Mine and Fraser F. Alexander Tailings for use of material in Paragraph 4, is graciously acknowledged.

REFERENCES

Begassat, P, Valentis, G and Weber, F. 1995. *Requirements for Site Selection and Public Acceptance Criteria for Waste Storage Facilities*. Proc. Sardinia, Cagliari, Italy. pp 77-78 Vol. 2.
Boswell, J E S. *The Identification and Development of Waste Management Facilities While Incorporating Public Participation and Environmental Impact Assessment.*

Conference on Mining and Industrial Waste Management. June 1997, Midrand, South Africa.

Boswell, J E S. *Public Participation in the Waste Management Industry.* Proc. 2nd Int Conference on Environmental Management, Technology and Development. October 1996, Fourways, South Africa.

Boswell, J E S, and De Waal, Dr D. 1999. *Social Impact Assessment in Civil Engineering Projects.*

Chamber of Mines of South Africa. 1996. *Guidelines for Environmental Protection.* Chamber of Mines of South Africa, Johannesburg.

Department Of Environment Affairs, South Africa, 1992. *Guidelines for Scoping.* Department of
Environment Affairs, Pretoria.

Department Of Water Affairs And Forestry. *Waste Management series. Minimum Requirements
for Waste Disposal by Landfill.* Vol. 1, 1994 pp 12, 32.

Fuggle, R F and Rabie, MA. 1992. *Environmental Management in South Africa.* Juta & Co. Cape Town. pp 493-521 and 748-761.

Gertz, C P. *Status of the Yucca Mountain Site Characterisation Programme.* First Int. Congress on Environmental Geotechnics, July 1994, Edmonton, Canada. p 827.

Homer, J D. Attorney, Parkowski, Noble and Guerke, Dover, Delaware, U.S.A. *Personal communication 1994.*

McDonald, T. U.S.E.P.A., Washington. *Personal communication 1994.*

Premier Diamond Mine Background Information Document on Flood Notification Procedure for No. 7 slimes dam, September 1997).

Petts, J, and Eduljee, G. 1994. *Environmental Impact Assessment for Waste Treatment and Disposal Facilities.* John Wylie & Sons, Chichester, England. pp 388 - 434.

Philpott, M J. Then Technical Director, Shanks & McEwan, Woburn Sands, Buckinghamshire, U.K. *Personal communication 1995.*

Premier Diamond Mine Background Information Document on Flood Notification Procedure for No. 7 slimes dam, September 1997

Schoeman, G and Brugge, K. 1994. *Assessment of the potential Psycho-Social impact of the proposed Class 1 Waste Disposal site at Chloorkop.* Evidence to the Commission of Enquiry tasked to provide a ruling on the pre-conditions for approval of the site. Johannesburg, South Africa.

Sol, V M, Peters, S W M and Aiking H. 1999. *Toxic waste storage sites in EU countries. A preliminary risk inventory.* IVM Report number: E-99/02. World Wide Fund for Nature.

South African Bureau of Standards SABS 0286 Code of Practice : Mine Residue 1998 South African Institution Of Civil Engineers, Environmental Engineering Division. 1993. *Guidelines for Public Participation in the Planning of Civil Engineering Projects.* S.A.I.C.E., Yeoville.

Striegel, K H. 1993. *Guidelines for Industrial and Urban Waste Disposal in Latin America.* Proc. Sardinia, Cagliari, Italy. pp 1867-1881.

Tidball, J R, and Lopes, R F. *Framework for Approval of Landfills in Ontario, Canada.* Proc. Sardinia '95, Cagliari, Italy Vol.III pp 79-92.

Viste, D. President, Warzyn, Madison, Wisconsin. *Personal communication 1994.*

Wagener, Dr F von M. 1997. *The tailings dam flow failure at Merriespruit, South Africa: Causes and consequences.* Proceedings of the Fourth International Conference on Tailings and Mine Waste 1997, Fort Collins, Colorado, USA.

Website on Los Frailes Tailings Dam Failure http://www.nl/wise-database/ uranium/ mdaflf. Html 27/9/99.

Tailings and Mine Waste'00 © 2000 Balkema, Rotterdam, ISBN 90 5809 126 0

Trends in the stewardship of tailings dams

T. E. Martin & M. P. Davies
AGRA Earth and Environmental Limited, Burnaby, B.C., Canada

ABSTRACT: Tailings impoundments frequently represent the most significant environmental liability associated with mining projects, both during the operational and decommissioning phases of a project. A spate of recent and well-publicized incidents involving tailings impoundments has placed the mining industry in general, and those responsible for tailings impoundment design and safety in particular, under intense scrutiny. Frequently, the seeds for such incidents are sewn as a result of a combination of design flaws and inadequate stewardship practices. Proper stewardship is critical in that, through implementation of quality assurance practices such as independent peer review of designs or risk assessments, it can protect the owner from faulty designs. For proper designs, good stewardship provides a tailings facility designer with reassurance that the facility will be constructed, operated, and monitored in accordance with that design.

This paper discusses recent trends in the stewardship of tailings facilities, both in Canada and world-wide. These include proactive initiatives by industry associations, and individual mining companies specifically addressing stewardship issues. The paper also discusses the roles that tailings dam design consultants can play in assisting the mining industry in this regard

1 INTRODUCTION

Tailings dams typically represent the most significant environmental liability associated with mining operations. They have been in the news frequently in recent years for unfortunate reasons, as a result of a series of well-publicized failures subjected to largely biased and sensationalized reporting in the media. These recent failures, together with previous ones, have put the mining industry under increasing pressure and scrutiny in regard to the safety of tailings impoundments. This scrutiny, while more intense in recent years, is nothing new, as evidenced by the quotation below:

"The strongest argument of the detractors of mining is that the fields are devastated by mining operations...further, when the ores are washed, the water used poisons the brooks and streams, and either destroys the fish or drives them away...thus it is said, it is clear to all that there is greater detriment from mining than the values of the metals which the mining produces" (Agricola, 1556).

This quotation, over 400 years old, unfortunately is often repeated by many today, in spite of the enormous benefits that mining has and continues to provide society. The mining industry response to this blind criticism can only be that its tailings facilities are well-designed and are well-managed. For the most part this is indeed the case, but it is the failures, rather than the successes, that garner the publicity, all of it negative.

For the purposes of this paper, stewardship is defined as the direction and implementation of all design (conceptual through detailed), construction, operations, inspection, surveillance, review, and managerial aspects (corporate policies, training, roles and responsibilities, documentation and reporting, etc.) involved in seeing a tailings facility through from conceptual design through closure. No one of these issues can be said to be more important than another (Szymanski, 1999). This paper will focus primarily on those aspects of stewardship of facilities that apply once a tailings facility is in operation.

Over the past several years, initiatives have been taken on a variety of fronts to improve the stewardship, and therefore the safety, of tailings impoundments. In addition, design practice for tailings dams continues to evolve and improve. Purely technical design aspects are extensively covered in the geotechnical literature and are not the subject of this paper, which instead focuses on evolving practice and initiatives related to the stewardship of tailings facilities. Stewardship is of equal significance to technical design aspects because under proper stewardship design errors can be detected and, in some instances, prevented from manifesting themselves as failures. However, no stewardship will be sufficiently robust to cover for all potential design flaws, and no design is sufficiently robust for the most negligent stewardship.

The initiatives discussed in this paper have been spear-headed by regulatory agencies, the United Nations Environment Programme (UNEP), the Canadian Dam Association (CDA), the Mining Association of Canada (MAC) and, most importantly, individual mining companies. The importance of the mining industry, and individual mining companies, leading these efforts cannot be overstated, because if the industry cannot properly manage its tailings facilities, then someone else (e.g. regulators) may impose proper management. More failures will serve to strengthen the hand of mining's detractors. Tailings dam design consultants can take a significant role in assisting mining companies in development and implementation of proper stewardship programs.

2 TAILINGS DAM FAILURES – A BRIEF REVIEW AND PERSPECTIVE

Tailings dam failures are, despite the recent spate of publicity, rare events that are unrepresentative of modern mining industry success in safe tailings disposal. The reporting of such events is incomplete and heavily biased, and there is no worldwide database of failures, nor is there one in Canada. This is illustrated by Figure 1, which appears to suggest that the United States has experienced the most tailings dam failures, but what it really indicates is that the United States has had the best means of reporting and documenting failures. For example, the authors have collective documentation on more than a dozen "failures" in Canada alone, three more in Central America, and another five in South America (using the USCOLD/UNEP definition of failure) that do not show up on Figure 1. This is undoubtedly the case for other jurisdictions. The database is further biased in that it does not account for the number of tailings dams in each country.

Figure 2 plots the frequency of significant tailings impoundment failures against time. This plot shows a rapid increase in the number of reported failures through the 1960's and 1970's, probably reflecting increased reporting, increased and larger scale mining developments and larger tailings impoundments, raised at increasing rates. This period also corresponds with the rapid development of tailings dam design as a formal engineering discipline, and the growing realization on the part of the mining industry that, with tailings impoundments becoming ever more larger and having to meet increasingly stringent environmental performance standards, they required a considerable degree of attention to maintain their safe operation. These two factors probably underlie the possible trend of decreasing frequency of failures since 1980. Figure 2 also indicates that the future trend may continue to project downward, but this is dependent on the continuation and implementation of improving stewardship practices.

It is estimated that there are in the order of 3500 active tailings impoundments worldwide. Major failures occur at a frequency of less than 2 to 5 per year (i.e. about 0.1%), and minor failures at a frequency of about 35 per year (i.e. 1%). While low, these figures are still unacceptably high. Peck (1980) quoted reviews of the history of conventional (water storage) earth dam failures that concluded that the probability of catastrophic failure of a conventional earth dam during any one year is about one chance in 10,000 (i.e. 0.01%). The authors contend that, with the increasing emphasis on proper stewardship practices for tailings facilities, and implementation of these practices, the favorable downward trend should continue to the point that the frequency of tailings dam failures (normalized to the number of tailings dams in existence) is no higher

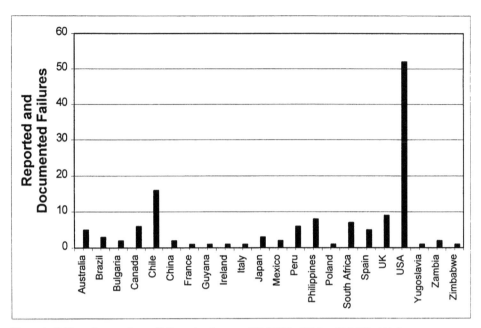

Figure 1. Tailings Impoundment Failures by Country (USCOLD, 1994 and UNEP, 1996)

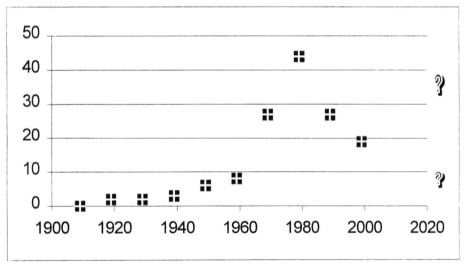

Figure 2. Frequency of Significant Tailings Impoundment Failures (USCOLD, 1994; UNEP, 1996 and authors' files)

than that of conventional dam failures (see Figure 3). This is an ambitious goal given that tailings dams must last for hundreds of years, while conventional dams have a shorter design life. Therefore, it can be argued that the number of conventional dams worldwide could remain the same, while the number of tailings impoundments will increase. This increase, in terms of tailings dam safety, is particularly significant given that the water cover option is so often selected for tailings dam closure. In many cases, a low permeability tailings dam without the water cover option still has roughly the same attributes as a dam with the water cover.

3 UNIQUE FEATURES OF TAILINGS DAMS

Stewardship practices for conventional dams are well-established. Some aspects of these practices are applied to tailings dams, but many are not. This is as it should be, since tailings dams differ significantly in many respects from conventional dams. To properly understand the specific requirements for stewardship of tailings facilities, it is necessary to consider the unique features of tailings dams relative to conventional dams. The key differences between tailings dams and conventional dams, and the significance of these differences, are discussed in detail by Szymanski (1999), and are summarized below.

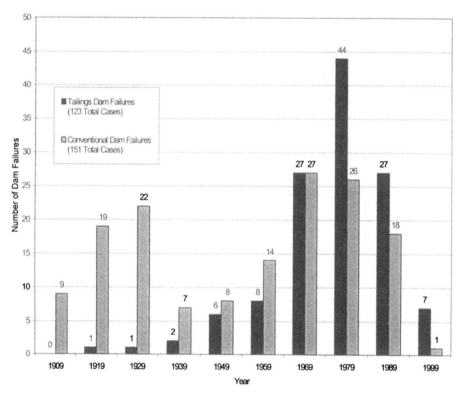

Figure 3. Tailings Dam Failures vs. Conventional Dam failures (ICOLD, 1995)

- Tailings dams are typically constructed in stages or on a continuous basis over many years, while conventional dams are usually constructed in a single stage in a short time period. As a result, the condition of the tailings facility is continually changing, and so its safety must be continually re-evaluated. In some respects, this renders tailings dam stewardship more onerous than is the case for conventional dams.
- Tailings dams are in many instances constructed on a continuous basis by mine operators, who are in the business of extracting wealth from the ground, not in dam-building. Their condition, safety, construction, and operating practices must therefore be monitored and re-evaluated on a continuous basis, whereas these requirements for conventional dams are less demanding.
- Conventional dams are typically owned by a state, province, public utility company or water resource authority. These dam owners typically have substantial resources at their disposal, and have a different relationship with the public in that the public benefits directly from the dams. Contrast this to tailings dams, which are owned and operated by mining companies, with the public perceiving no direct benefit from the tailings dam. As a result, mining companies tend to be "punished" more severely than conventional dam owners when failures do occur.

- Conventional dams are viewed as an asset, so their construction, operation, and maintenance receives a high degree of attention from owners, who often retain in-house dam engineering expertise. Contrast this to tailings dams, which are often viewed by their owners as an un-profitable, money-draining part of the mining operation, and as a "necessary evil". The significance of this aspect is that with such attitudes a mining operation would be naturally less inclined to expend effort in the management of its tailings facility than the owner of a conventional dam. Mining companies typically do not retain in-house dam engineering expertise, relying instead on consultants
- Tailings dams typically retain materials (solids and water) that would be considered "contamination" if released. The need to prevent release of these materials in terms of environmental impact is not a consideration for conventional dams, and represents an aspect in which tailings dam design and performance requirements differ significantly from conventional dams.
- Conventional dams generally do not need to be designed to last forever, as they have a finite life. Tailings dams, in contrast, because they have a closure phase as well as an operational phase, have to be essentially designed and constructed to last "forever", and require some degree of surveillance and maintenance long after the mining operation has shut down, and generation of cash flow and profit has ceased.

The common thread of each of the above differences is that they demand application of unique, and comprehensive, stewardship practices for a tailings facility throughout its long service life. Furthermore, since many tailings dams are unique, there is no "off the shelf" approach suitable for every tailings facility, although certain basic elements should be common to all.

4 RECENT INITIATIVES AND TRENDS

4.1 *Mining Association of Canada*

The Mining Association of Canada (MAC), has recently published a document entitled "A Guide to the Management of Tailings Facilities" (MAC, 1998). This document, reviewed by industry and consultants, is very valuable, providing a framework of management principles, policies, and objectives, and checklists for implementing the framework through the life cycle of a tailings facility. It is general in nature, but this is a strength in that it recognizes the need for individual mining companies and mining operations to establish their own specific programs that meet their own specific needs.

The checklists provided in the MAC guide identify six key elements for ensuring the effective implementation of the management framework:
- Management Action – from the management framework
- Responsibility – person responsible and accountable for delivery of a management action
- Performance Measure – indicator of progress toward a target or objective
- Schedule – time frame for completion of significant milestones of a particular management action
- Technical Considerations – relevant technical aspects requiring consideration
- Other References – additional technical, managerial and regulatory considerations related to the management action.

The MAC guidelines emphasize the need to "close the loop" in the management process, which includes confirming that management actions have been implemented, and which seeks to continually improve the management framework.

4.2 *Canadian Dam Association*

The Canadian Dam Association (CDA) recently updated its dam safety guidelines (CDA, 1999). The update focussed in large part on incorporating elements specific to the safety of tailings dams. There are recommendations regarding the following issues that are very relevant to tailings dam stewardship:

- Responsibility for dam safety
- Scope and frequency of dam safety reviews, which includes a review of operations and maintenance, and dam surveillance program
- Operation, maintenance and surveillance, including need for an Operations Manual
- Emergency preparedness, including elements of an emergency preparedness plan (EPP)

4.3 *International Committee on Large Dams and Related Organizations*

The International Committee on Large Dams (ICOLD), and related organizations, have published numerous materials with regards to tailings dams. Bulletin 74 (ICOLD, 1989) presents guidelines for tailings dam safety. Bulletin 104 (ICOLD, 1996) specifically addresses monitoring of tailings dams. Other ICOLD publications pertaining to tailings dams are as follows:
Bulletin 97 (1994). Tailings Dams – Design of Drainage – Review and Recommendations.
Bulletin 98 (1998) Tailings Dams and Seismicity – Review and Recommendations
Bulletin 101 (1995) Tailings Dams – Transport, Placement and Decantation.
Bulletin 103 (1996). Tailings Dams and Environment. Review and Recommendations
Bulletin 106 (1996) A guide to Tailings Dams and Impoundments – Design, Construction, Use and Rehabilitation
The United States Committee on Large Dams (USCOLD) published a compendium of tailings dam incidents (USCOLD, 1994). This was the probably the first attempt to catalogue and assess <u>published</u> information on tailings dam incidents. The document categorizes the various incidents in terms of technical causation, but there is no discussion of the extent to which inadequate stewardship played a role in the various incidents. The value of this document is that is discusses failure modes for different dam types, which is of benefit in scoping out requirements for a dam safety program, particularly dam surveillance.

4.4 *United Nations Environment Programme/International Council on Metals and the Environment*

The United Nations Environment Programme, Industry and Environment (UNEP), and the International Council on Metals and the Environment (ICME) have been active in recent years in sponsorship of seminars, and publication of case studies (UNEP-ICME, 1997 & 1998), related to tailings management. Many of the topics covered directly address stewardship issues. Mining companies provided most of the contributions to these publications, thereby making these forums an excellent means for dissemination of knowledge and experience to the international mining community. Key topics covered that relate to stewardship of tailings dams include:
- Corporate policies and procedures regarding stewardship of tailings facilities
- Evolving regulatory climates and trends
- Definitions of roles and responsibilities
- Application of risk assessment techniques
- Environmental management systems
- Emergency preparedness and response
- Education and training

Blight (1997), in the 1997 workshop sponsored by UNEP-ICME, and Wagener et al (1998), provide an in-depth discussion of the 1994 failure of the Merriespruit tailings dam in South Africa, in which 17 people lost their lives. Technical issues aside, these discussions are extremely valuable in that they demonstrate how the inadequacies in the management aspects of the stewardship of the tailings dam allowed the technical factors that caused the failure to develop. All too often, published case histories on tailings dam failures focus solely on technical issues, without addressing the contribution of inadequate stewardship of the facilities to the failures. The authors are aware of numerous such examples where inadequate stewardship practices were the principal factor precipitating a chain of events leading to failure of tailings dams. The authors contend that such a link likely exists for most tailings dam incidents, and this is why adequate stewardship of tailings facilities is so critical to their safety. Even the best designed-facility is susceptible to failure if not managed properly. Conversely, even a facility whose design is flawed can be operated successfully with good stewardship practice. Good stewardship, in the form of an ongoing dam safety evaluation program, should in fact allow a mining operation to detect any such design flaws and correct them in advance of any serious incident occurring.

4.5 Initiatives by Mining Companies

4.5.1 General
The most encouraging trend in terms of tailings dams stewardship is that it is the mining industry in general, and individual mining companies in particular, that are leading the way in improving state of practice and, equally as important, in sharing and publishing information. The following sections discuss some representative examples of proactive stewardship policies and practices being followed by mining companies.

4.5.2 Corporate policies and management issues
Several major Canadian-based mining companies have established corporate policies and procedures to ensure that all personnel involved in stewardship of tailings facilities, from the corporate level to the operators, clearly understand their roles and responsibilities (e.g., Siwik, 1997). Such an understanding, and enforcement (performance measurement)of those roles and responsibilities is vitally important. The authors have reviewed many tailings disposal facilities where there was no such understanding, and no "ownership" of key stewardship functions. A number of companies have also established policies with respect to degrees of training and competency required for the various roles involved in tailings facility stewardship (e.g. Siwik, 1997, Brehaut, 1997 and Maltby, 1997). This is also extremely important, especially for tailings dam operators, because they have the most frequent exposure to the facility, and usually are responsible for dam surveillance. It is essential that these personnel understand what to look for and why, what constitutes unfavorable conditions, and what to do about it.

Some companies have also established formalized dam safety programs (e.g., Coffin, 1998). Some of these programs include classification of each dam in terms of the consequences of a potential failure, which facilitates the dam safety review process and corporate prioritization of corrective measures, if required. Some of these programs include a detailed inspection and review of their tailings facilities by specialists (e.g. Coffin, 1998). Still others require that Operations Manuals be maintained for their tailings facilities (e.g. Maltby, 1997). In the province of British Columbia, in fact, Operations Manuals and annual inspections/reviews by specialists are a regulatory requirement.

4.5.3 Auditing of tailings facilities
Brehaut (1997) describes how, in an internal evaluation of its management systems, Placer Dome recognized that its tailings management systems were a priority for enhancement, and that tailings management was an issue of great import at a corporate level. Placer then embarked on development of guidelines to cover the design, construction, operating and closure phases of tailings management systems. Placer also determined that the application of risk assessment techniques was an essential next step in the review and enhancement of its tailings dam stewardship policies and procedures.

However, no sooner had this process been initiated than the Marcopper incident occurred in the Phillipines, involving release of about 2 million tons of tailings into a local river system. As is widely known within the industry, but not appreciated without, Placer's response to the incident was exemplary, and Brehaut (1997) indicated that the total cost to Placer Dome was estimated to be $43 million after insurance and tax recovery. The total cost to present far exceeds that value. Spurred on by the Marcopper incident, Placer Dome quickly initiated formal risk assessments of the tailings facilities at all of its operations. In many instances, these risk assessments were carried out, and/or facilitated by, geotechnical consultants who were not the engineers of record for the various facilities audited. The findings of these risk assessments (Brehaut, 1997) were that any design deficiencies identified were of minor significance, and the greatest weakness was related to management aspects sof the stewardship of the facilities.

Other mining companies have implemented similar risk assessment programs for many of their tailings facilities, including Cominco, Breakwater Resources, and Rio Tinto, to name but a few.

4.5.4 *Geotechnical review boards*

Syncrude, Kennecott Utah Copper, and Inco, and numerous other mining companies, retain a board of eminent geotechnical consultants to provide independent review and advice in terms of the design, operation, and management of their respective tailings facilities. Such review boards are independent of the design engineers, be these consulting firms or geotechnical personnel the mining company has on staff. Review boards are now considered to be the state of dam steward-ship practice for owners of major water dams. Dunne (1997) describes how Kennecott Utah Copper retained a geotechnical review board as a means of providing cost-effective quality as-surance and risk management for the design of a major expansion of a 95 year old tailings im-poundment near Salt Lake City.

Inco has a geotechnical review board for its Copper Cliff tailings facilities in Sudbury, On-tario. This tailings facility has been in use since the 1930's, and will not reach capacity until about 2030 (McCann, 1998). The review board represents a means of Inco applying its "fail-safe" review process to the design, construction, operation, and management of this large, his-toric tailings facility.

McKenna (1998) describes how Syncrude Canada Ltd., a large oilsands company in northern Alberta, has benefited from its geotechnical review board over the last 25 years, summarizing these benefits as follows:

- The board provides expert assistance in terms of assessing and managing risk.
- It ensures that all of the bases are covered (i.e. posing the question "has anything been missed?").
- Review board members bring to Syncrude a vast amount of varied practical experience and expertise.
- The board provides reassurance to senior management that an acceptable balance between risk-taking and conservatism is maintained in an operation where failure consequences are extreme.
- Independent review by pre-eminent specialists gains the trust of regulators and the public, and facilitates the regulatory processes.
- Design engineers benefit through in-depth review of their work by pre-eminent specialists.

Another very important benefit afforded by a geotechnical review board is the continuity it provides. For example, over the 25 year period during which Syncrude has maintained a review board, there has likely been considerable turnover in staff and consultants. However, the current members of Syncrude's review board have been on the board, more or less continuously, since the board was first struck in 1972.

In summary, a review board can provide an objective view as to the potential, consequences, and cost of a potential failure, and help the owner ensure that decisions on design alternatives are not based solely on minimizing capital and operating costs.

4.5.5 *Information database*

McCann (1998) describes systems that Inco has implemented to develop an information matrix to maintain records, in an easily retrievable manner, pertaining to the design, construction, op-eration and monitoring of its Copper Cliff tailings facilities, in use since the 1930's. For such a facility, given the inevitable turnover of operations personnel, management personnel, and de-sign consultants, a good database is essential to maintaining continuity.

The authors cannot overemphasize the importance of this point, because they are aware of at least two recent tailings dam incidents that can be attributed in large part to the lack of such an accessible historical database, and/or inadequate appreciation of that database. Davies et al. (1998), discuss the static liquefaction failure of a portion of the Sullivan Mine active Iron tail-ings dyke, without off-site impact. A similar failure had occurred in 1948 and during that earlier failure, tailings flowed into the nearby town of Marysville. Museum records show the general public support for the mine and sympathy to the clean-up. Had the 1991 incident progressed off-site, it is without doubt that the community response would have been dramatically different.

The authors are aware of another tailings impoundment, in operation for over fifty years, that underwent a partial dam wall failure primarily as a result of key historical information not being readily available, not being adequately documented, and not being taken into account in design. Good record keeping, and maintaining those records in good and easily accessible order, is an important aspect of stewardship of tailings facilities.

From carrying out more than 30 risk assessments of tailings dams worldwide, the authors have reviewed the entire available "tailings library" at many mines. It is almost a certainty that any operation over 10 years of age will demonstrate "tailings database amnesia" (TDA) and will repeat costly studies, ignore essential design criteria or unknowingly re-invent a tailings management plan without appreciation of the "forgotten" earlier information. Maintaining the same consulting organization does not seem to stem the onset of TBA unless the mine itself is an active partner in tailings dam stewardship. The authors' review work also shows that tailings dams that have had "incidents" in their past are often remarkably well-placed to have them occur in the future.

4.5.6 *From audits to operations manuals to implementation*

On August 29, 1996, a portion of an upstream-constructed tailings dam collapsed at the Porco Mine in Bolivia, owned by Compania Miñera del Sur S.A. (COMSUR), with a resultant release of 400,000 tonnes of tailings. Following this incident, COMSUR initiated dam safety and environmental audits of its tailings facilities (active and inactive) in Bolivia and Argentina. This was the first step in a process that saw COMSUR, within a year, implement a formal stewardship program for its tailings facilities, including training, monitoring and surveillance programs, and Operations Manuals for each operation. An environmental management system (EMS) was developed and implemented concurrently. Based on observations made by one of the authors during the site visits, training seminars on environmental monitoring, tailings management and surveillance of tailings facilities were developed, specific to each of the operations. These training seminars, presented on site to mine management and operators, were also used as a forum to begin the development of Operations Manuals, and provided a means to involve both operations and management personnel in the development of these manuals, and, equally as important, to gain their buy-in to the process and the end product. There was also corporate level participation in the development of these manuals, an in the training seminars.

The authors believe that the inter-active development (not just the end product) of a comprehensive Operations Manual for a tailings facility can be the single most important and pro-active measure in formalizing and implementing good stewardship practices of tailings facilities. This certainly was the case for the COMSUR tailings facilities. The Operations Manuals were drafted to capture the following key elements of a comprehensive tailings dam stewardship program:

- Project administration, and responsibilities for facility operation, safety and review (including corporate level roles and responsibilities).
- Design overview and key design criteria.
- Tailings deposition plan and water management plans.
- Planning requirements (reviews, construction, operation, training).
- Training and competency requirements.
- Operating systems and procedures.
- Dam surveillance, including checklists, signs of unfavorable performance, and responses to unusual reading/events/observations.
- Reporting and documentation requirements.
- Emergency action and response plans.
- Construction and QA/QC requirements.
- Standard formats for monthly status reports for tailings facilities, and for performance reviews.
- Reference reports and documents.

The benefits of having an Operations Manual in place are as follows:
1. It provides a concise, practical document that can be used by site operating personnel for operation and surveillance of the tailings facilities.
2. It serves as a useful training document for new personnel involved in tailings management and operations.
3. Its existence provides reassurance to senior level management, and to regulators, that formalized practices are in place for the safe operation of the facility.
4. It demonstrates due diligence on the part of the owner.

Reliance on the preparation of an Operations Manual by a consultant only should be discouraged, as this does not foster an intrinsic understanding of each part of the manual by operations personnel. Further, it is essential that an Operations Manual takes into account the perspective, knowledge and experience of operations personnel.

4.6 The Role of Tailings Dam Design Consultants

4.6.1 General
Tailings dam design consultants have an essential role to play in promoting good stewardship of tailings facilities, besides the obvious technical role of providing safe, cost-effective, practical and enduring designs. Consultants obviously have common cause with the mining industry in this regard, and the following sections discuss a number of ways consultants have, are, and should be contributing to this effort.

4.6.2 Publications and participation in conferences
Tailings dam consultants work on a variety of projects, in many countries and for many clients. By so doing, they amass a wide array of varied experience from which the mining industry and other consultants can and should benefit. Conferences such as the annual Tailings and Mine Waste series provide a forum for interchange of ideas and sharing of experiences. More emphasis, however, needs to be placed on stewardship issues as opposed to the purely technical topics that typically dominate such conferences.

Consulting engineers specializing in tailings dams have also contributed through publication of entire books on the subject. The first such book with widespread distribution was written by Steve Vick, entitled "Planning, Design, and Analysis of Tailings Dams" (Vick, 1983). It is an excellent treatise recommended for all persons responsible for some aspect of tailings management. More recently, Maciej Szymanski published a book entitled "Evaluation of Safety of Tailings Dams" (Szymanski, 1999). This book provides a detailed discussion of elements of a comprehensive dam safety program specifically tailored to tailings dams. Issues related to good stewardship are discussed throughout.

4.6.3 The design product
The design product provided by tailings dam engineers to the owner is, all too often, rich in discussion of the finer points of soil mechanics, but poor in terms of detailed guidance for operation and surveillance. The condition of a tailings facility is governed by how it is operated and constructed, not necessarily by how it was designed. Likewise, its safety is better judged based on surveillance than design analyses in the appendix of some design report. The design report must therefore include operational and surveillance requirements. Ideally, an Operations Manual, or at least most elements of one, should be provided with the design, otherwise the design is not complete.

4.6.4 Risk assessments
Consultants are increasingly applying techniques that can be broadly categorized as "risk assessments" to various facts of tailings management, most notably in reviews of existing facilities. Mining companies are also applying such techniques in the stewardship of their tailings facilities. A risk assessment, by the authors' definition, provides answers to the following questions:
1. What can go wrong?
2. How likely is it that it will happen?
3. If it does happen, what are the consequences?
4. What can/should be done to reduce the likelihood and/or consequences of this potential occurrence?

The authors have made frequent use of failure modes and effects analysis (FMEA) in workshop settings that include mine management and operating personnel. FMEA, and most other qualitative risk assessment methods, are nothing more than organized judgement, common sense with a fancy name. Risk assessment techniques can be used to audit any number of technical and managerial aspects of tailings dam stewardship.

As an example, the FMEA technique, carried out in a workshop setting, is particularly effective in scoping out requirements for dam surveillance, and ties in well to the correct application of the observational approach (see Figure 4) that is so fundamental to tailings dam design and safety. The FMEA process captures the key elements of comprehensive dam surveillance, including:

- Identification of potential failure modes
- Identification of warning signs for failure modes
- Consideration of how quickly failure could occur, and how potential problems can be detected well in advance of their developing into incidents
- Consideration of the significance of temporal trends as opposed to single measurements/observations
- Allows "green light" (safe) versus "yellow light" (caution) versus "red light" (stop) limits/criteria to be established

The workshop format, involving personnel responsible for dam surveillance, as well as management personnel, provides the following:

- Forum for interchange of ideas and concerns
- Technical, operational, environmental, and management input
- Transfer of essential knowledge from the designers to "front-line" personnel, and vice versa
- Development of a team approach to dam surveillance
- Buy-in from responsible parties

The FMEA process is illustrated schematically in Figure 5, and provides the following:

- A structured, repeatable, and documented process
- Assessment of current surveillance practices in terms of scope, frequency, reporting and interpretation, response to unusual conditions, and resources available versus resources required
- Identification of aspects requiring improvement
- Justification for allocation of resources to dam surveillance
- An action plan that evolves directly from the process

Special emphasis is placed on the subject of dam surveillance by the authors because, with the exception failures triggered by earthquakes or major storm events, all types of failure give some warning signs (Smith, 1972), making dam surveillance a critical aspect of proper stewardship. Assisting an owner develop a comprehensive dam surveillance program is a vitally important responsibility borne by the designer.

4.6.5 Training seminars

Consultants are increasingly being called upon to provide on-site training to operators, and this is a very welcome trend. This facilitates transfer of key knowledge from the design engineer to the operators, who represent the designer's "eyes and ears". Such seminars are as beneficial, if not more so, for the designer as the operators, providing the designer with a reality check on the constructability of the designs and the practicality of the operating requirements imposed by the designs. Further, it is the authors' experience that designers can learn more of practical benefit to themselves and their client from a day on site with experienced operators than a week of reading technical papers. Tailings dam operators truly appreciate such seminars, which facilitates their buy-in, understanding, and commitment to good stewardship practices.

4.6.6 Designer Expertise and Perspective

As noted earlier, tailings dams and impoundments are unique structures in the engineering world. Commensurately, design consultants should have appropriate educational and practical experience directly applicable to tailings dam design. Moreover, a proven construction and operations history for the designer(s) projects is of extreme importance. There are a limited number of qualified designers and organizations (more than one designer in house) and owners should share information with one another as much as possible regarding design consultants. It is an unfortunate fact that "all designers are not born equal". Independent peer reviews, or review boards, are means by which mining companies can protect themselves from questionable – designs, and design consultants should proactively encourage their clients to adopt these protective measures, which, as discussed previously, are also very much to the design consultants benefit.

Designers themselves need to maintain an appropriate balance between the real world (i.e. case history experience, both good and bad) and theory (e.g. laboratory testing and visco-

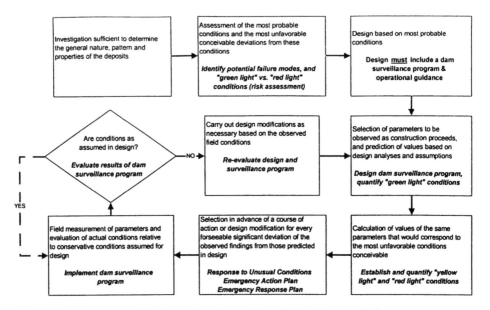

Figure 4. Risk assessments and the observational approach applied for dam surveillance

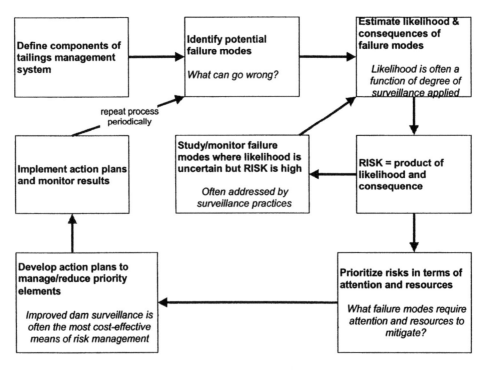

Figure 5. Risk assessment process

plasticity theory), with the real world always being the most important. Prior to the birth of tailings dam engineering as a formal discipline in the late 1960's, tailings dams were designed and constructed based almost entirely on experience (trial and error), without the benefit of soil me-

404

chanics theory. Today, the pendulum may have swung too far in the opposite direction, with many tailings dams being designed "in the laboratory" and in computer simulations. Terzaghi is reputed to have said that nature has no contract with mathematics; she has even less of an obligation to laboratory test procedures and results. Peck (1980) pointed out that theory can inhibit judgement if used without discrimination and without critical evaluation. The following excerpt from a government inquiry into why the United States intelligence services were "caught napping" at Pearl Harbor stands as a poignant warning to design consultants entranced by the finer points of theory at the expense of practicality and case history experience:

"Science was king, and his subjects were suspicious of anything that could not be observed under a microscope or demonstrated in repeated laboratory tests. The imaginative interpretation of intelligence is much more an art than a science. So the American intelligence community as scientists triumphed in technology and cold logic – magic and cryptanalysis – while as artists it failed in interpretation of the facts thus gathered."

4.7 *Regulatory Trends*

4.7.1 *General*
Regulatory agencies typically do not "prescribe" stewardship practices to the industry, apart from some basic requirements (in some jurisdictions) like requiring a dam surveillance program, requiring annual reports, operations manuals, and so on. It is in terms of environmental issues (water quality, for example) that regulations are, necessarily, prescriptive. This is as it should be, because mining companies themselves, supported by their design consultants as appropriate, are best qualified to design stewardship programs appropriate to their particular facilities. An attempt by regulators to impose a uniform "code of stewardship" would be unsuccessful because each mining company, and each tailings facility, have their own unique requirements, resources, and constraints. So much of stewardship relates to the corporate and mine-specific organizations and personnel, aspects that cannot be regulated.

4.7.2 *Regulatory Trends in Developing Countries*
An interesting regulatory trend is that codes and standards with respect to tailings disposal are gradually becoming "harmonized" internationally. Many developing companies, Bolivia being one example, have only recently enacted regulations covering tailings disposal. These regulations are largely modeled after World Bank guidelines and regulations in North American jurisdictions. Unfortunately, there appears to be a trend among developing companies to make their regulations technically prescriptive, with some (e.g. Peru) actually presenting a very elementary course in soils mechanics, describing methods of stability analysis and liquefaction potential screening methods. This is due in part to the comparative lack of skill and expertise in tailings dam engineering on the part of the regulators, and due in part to their understandable desire to have tailings facilities in their respective countries conform to "international" (i.e. ICOLD) standards, which really do not exist in any tangible, readily-referenced form. This is a misguided trend, because it can provide a false sense of security to the regulators and, worse, the mining companies themselves. Many failure case histories likely involved facilities that were in conformance with all regulations, except the most important of all (the dam failed).

Reliance on regulations full of prescriptive design criteria that meet "international standards" and full of elementary soil mechanics do nothing to address actual operating practices and stewardship issues. As discussed previously, even the most robust design can fail if not stewarded properly. As regulations generally do not address stewardship issues, it is incumbent on the mining industry to take the lead in this regard. The mining industry, and its consultants, could be of great assistance to regulators in developing countries by providing training and workshops for regulatory personnel responsible for tailings facilities. It is in the mining industry's best interest to achieve a condition of "co-regulation", whereby mining companies regulate themselves to a greater degree than do the regulatory agencies.

4.7.3 *Regulatory Trends in Developed Jurisdictions*
In more developed jurisdictions (for example, the authors' home jurisdiction of British Columbia), the regulators employ geotechnical engineers experienced in tailings management. Here,

regulations are not prescriptive, neither in the technical nor the stewardship sense, because regulators have the expertise to assess the design of each tailings facility, and the manner in which it is operated, on a case by case basis. The B.C. regulations do require that an Operations Manual exist for each tailings facility, but the contents of those manuals are up to the owner. Similarly, a dam surveillance plan is also required (typically included in the Operations Manual). However, the details of tailings dam stewardship are left to the individual operations. An annual review report, prepared by a qualified geotechnical engineer, is also a requirement.

In B.C., the regulators have actually assisted the industry by publication of a document entitled "Tailings Dam Inspection Manual". This document is not a set of regulations, but rather is intended to assist regulatory and mine personnel in inspection of tailings dams. B.C. regulators are currently working with regulators in Peru to achieve a "knowledge transfer" to Peruvian regulators. This too is a welcome trend. The mining community is now international, and includes mining companies, its consultants, and regulators. The more these three bodies can assist one another in facilitating good stewardship practices, the better off everyone will be.

5 CONCLUSIONS

The safety of tailings impoundments is inextricably linked to the stewardship practices applied to their management and operation, from the concept stage through to closure. Positive and proactive trends in stewardship of tailings dams, initiated primarily by the mining industry as a whole and by individual mining companies, provide the basis for considerable optimism that the trend of decreasing tailings dam failures will continue. Tailings dam design consultants can and should be of considerable assistance to the industry in this regard, placing as much emphasis and effort on stewardship issues as on technical issues in their practice. The goal of the industry, its consultants, and regulators must ultimately be to endure fewer failures than their colleagues dealing with conventional dams, and to demonstrate and publicize mining industry successes to regulators and the public.

REFERENCES

Blight, G. 1997. Insights into tailings dam failure – technical and management factors. *Proceedings of the International Workshop on Managing the Risks of Tailings Disposal.* ICME-UNEP, Stockholm, pp. 17-34.
Brehaut, H. 1997. Tailings management – a Placer Dome perspective. . *Proceedings of the International Workshop on Managing the Risks of Tailings Disposal.* ICME-UNEP, Stockholm, pp. 35-46.
Canadian Dam Association (CDA) 1999. *Dam Safety Guidelines.*
Coffin, V. 1998. Risk assessment of tailings areas at Noranda. *Case Studies on Tailings Management*, UNEP-ICME, pp. 44-45.
Davies, M.P., Chin, B.G. & Dawson, B.G. 1998. Static liquefaction slump of mine tailings – a case history. *Proceedings, 51ˢᵗ Canadian Geotechnical Conference*, Edmonton.
Dunne, B. 1997. Managing design and construction of tailings dams. . *Proceedings of the International Workshop on Managing the Risks of Tailings Disposal.* ICME-UNEP, Stockholm, pp. 77-88.
International Committee on Large Dams (ICOLD) 1989. *Tailings Dam Safety Guidelines.* Bulletin 74.
International Committee on Large Dams (ICOLD) 1994. *Tailings Dams – Design of Drainage – Review and Recommendations*, Bulletin 97.
International Committee on Large Dams (ICOLD) 1995. *Dam Failures Statistical Analysis.* Bulletin 99.
International Committee on Large Dams (ICOLD) 1995. *Tailings Dams – Transport, Placement and Decantation.* Bulletin 101.
International Committee on Large Dams (ICOLD) 1996a. *Tailings Dams and Environment – Review and Recommendations.* Bulletin 103.
International Committee on Large Dams (ICOLD) 1996b. *A Guide to Tailings Dams and Impoundments – Design, Construction, Use and Rehabilitation.* Bulletin 106.

International Committee on Large Dams (ICOLD) 1998. *Tailings Dams and Seismicity– Review and Recommendations*. Bulletin 98.

International Council on Metals and the Environment (ICME) and United Nations Environment Programme (UNEP) 1997. *Proceedings of the International Workshop on Managing the Risks of Tailings Disposal*. Stockholm.

International Council on Metals and the Environment (ICME) and United Nations Environment Programme (UNEP) 1998. *Case Studies on Tailings Management*.

Maltby, J. 1997. Operational control. *Proceedings of the International Workshop on Managing the Risks of Tailings Disposal*. ICME-UNEP, Stockholm, pp. 161-169.

Mining Association of Canada 1998. *A Guide to the Management of Tailings Facilities*.

McCann, M. 1998. Sustaining the corporate memory at Inco's Copper Cliff operations. . *Case Studies on Tailings Management*, UNEP-ICME, pp. 55-56.

McKenna, G. 1998. Celebrating 25 years: Syncrude's geotechnical review board. Geotechnical News, Vol. 16(3), September, pp. 34-41.

Peck, R.B. 1980. Where has all the judgement gone? The 5th Laurits Bjerrum Memorial Lecture, *Canadian Geotechnical Journal17(4)*, pp. 584-590.

Siwik, 1997. Tailings management: roles and responsibilities. . *Proceedings of the International Workshop on Managing the Risks of Tailings Disposal*. ICME-UNEP, Stockholm, pp. 143-158.

Smith, E.S. 1972. Tailings disposal – failures and lessons. *Tailings Disposal Today, Proceedings of 1st International Symposium*, Tucson.

Szymanski, M.B. 1999. *Evaluation of Safety of Tailings Dams*. BiTech.

U.S. Committee on Large Dams (USCOLD) 1994. *Tailings Dam Incidents*.

United Nations Environment Program (UNEP) 1996. Tailings Dam Incidents 1980-1996.

Vick, S.G. 1983. *Planning, Design and Analysis of Tailings Dams*, John Wiley & Sons.

Wagener, F.M., Craig, H.J., Blight, G., McPhail, G., Williams, A.A.B., Strydom, J.H. 1997. The Merriespruit tailings dam failure – a review. Proceedings, *Tailings and Mine Waste '98*, Fort Collins, Colorado, pp. 925-952.

Tailings and Mine Waste'00 © *2000 Balkema, Rotterdam, ISBN 90 5809 126 0*

The Canadian Mine Environment Neutral Drainage 2000 (MEND 2000) program

G.A.Tremblay
MEND 2000, CANMET, Natural Resources Canada, Ottawa, Ont., Canada

ABSTRACT: Acidic drainage has long been recognized as the largest environmental liability facing the Canadian mining industry, and to a lesser extent, the public through abandoned mines. The Mine Environment Neutral Drainage (MEND) Program, formed in 1989, was the first initiative to develop scientifically-based technologies to combat acidic drainage. This 9-year volunteer program established Canada as the recognized leader in research and development on acidic drainage for hardrock metal mines. Through MEND, Canadian mining companies, the federal and provincial governments have reduced the liability due to acidic drainage by an estimated $400 million. This is an impressive return on an investment of $17.5 million over nine years.

MEND 2000 is a three-year program that started in 1998. The key to MEND 2000 is technology transfer – providing state-of-the-art information and technology developments to users. MEND developed technologies are being verified by long-term monitoring of large-scale field tests.

Although not all of the research program objectives have been yet achieved, tremendous progress has been made. MEND, and now MEND 2000 are described as a model way for governments and industry to cooperate in technology development. A summary of the results that have been achieved, the liability reduction, the lessons learned, and the opportunities for future actions are given. Case studies depicting Canadian full-scale applications of water and dry cover technologies will be presented and discussed. Through these efforts a further reduction of the environmental liability associated with acidic drainage will be realized.

1 INTRODUCTION

One of the most significant environmental issues facing the global mining industry today is acidic drainage which affects all sectors of the mining industry including coal, precious metals (gold, silver), base metals (copper, nickel, zinc, lead) and uranium. Acidic drainage is the result of a natural oxidation process whereby sulfur bearing minerals oxidize upon exposure to oxygen and water. The net result is the generation of metal laden effluents of low pH, that can potentially cause damage to ecosystems in the downstream environment. Acidic drainage is not only caused by mining activities but also civil works. Remedial measures are currently in place at an international airport in Canada after construction of a runway exposed sulphide minerals which in turn resulted in acidic drainage. Road and pipeline construction are also periodic contributors.

Although the issue of acidic drainage is not new and has an extensive history spanning decades (and even centuries in Europe), it is not fully understood. In the past 10 years, changes in socio-economic expectations and heightened environmental awareness have made the management of waste an increasingly pressing issue in the mining industry (Price 1995), and have made the mining industry one of the more intensively regulated and scrutinized of our industries. Extensive liabilities have been generated in countries such as Canada, the United States, Australia, Sweden and Germany by the inability to adequately deal with acidic drainage issues. Other countries such as Brazil, are now discovering their own problems with acidic drainage. These liabilities are essentially the costs incurred by the property owner/manager during or after the life of the mine to ensure that the impact to the environment is minimized and consistent with environmental regulations. Costs typically include: the collection and treatment of acidic drainage; construction of engineered structures to contain mine wastes; relocation of mine wastes to containment areas; and rehabilitating the mine, mill and containment areas after operations have ceased. Some operations, at closure, may require treatment in perpetuity.

2 SCOPE OF THE PROBLEM

In the United States approximately 20,000 kilometers of streams and rivers have been impacted by acidic drainage, 85-90% of which receive acidic drainage from abandoned mines (Skousen 1995). Although there are no published estimates of total U.S. liability related to acidic drainage, some global examples may help to quantify the dimensions of the problem:

- The Leadville site, also in Colorado, and also a Superfund site, has an estimated liability of $290 million due to the effects of acidic drainage over the 100-year life of the mine.
- The Summitville Mine in Colorado has been declared a Superfund site by the Environmental Protection Agency (EPA) which estimated total rehabilitation costs at approximately ~$100 million (US).
- At an operating mine in Utah, U.S. regulators estimate liability at $500-$1,200 million (Murray 1995).
- The Mineral Policy Center in the US has estimated that there are 557,000 abandoned mines in 32 states, and that it will cost between $32 - $72 billion to clean them up (Bryan 1998).
- Liability estimates for Australia in 1997 and Sweden in 1994 were $900 million and $300 million respectively (Harries 1997, Gustafsson 1997).
- The total Canadian liability has been estimated to be between $2 and $5 billion (CDN) (Feasby 1994).

Considering the above data, the number of new mining projects currently under development plus existing mining projects in other countries not mentioned above (e.g. South America, South Africa), one might anticipate the total worldwide liability to be in the region of $100 billion (US), or even beyond.

3 RESPONSE TO THE PROBLEM

Twenty years ago, rehabilitation was regarded primarily in terms of physical stabilization and the establishment of a self-sustaining vegetative cover. It was generally thought that the surface addition of alkalinity and the establishment of a vegetative cover would alleviate acidic drainage problems from these sites, and allow mining companies to abandon them without further liability. However, monitoring of the quality of the water draining from revegetated acid-generating waste sites clearly showed in the years following, that acidic drainage remained a concern at many of these sites. In some cases, property owners were faced with the prospect of continuing to operate and maintain lime treatment plants indefinitely. There was need for a better understanding of processes involved, and for new remedial technology to be developed and demonstrated.

As a consequence, in 1988 the Canadian mining industry, 5 provincial governments and the government of Canada came together to form a tripartite consortium called the Mine Environment Neutral Drainage (MEND) program. Over the succeeding ten years, the two level of government together with the Canadian mining industry spent over $17 million within the MEND program to find ways to reduce the estimated liabilities.

4 ORGANIZATION

MEND was an unusual consortium driven primarily by the 130 volunteer representatives of the different participating agencies: regulators, mining company managers and engineers, and government officials and scientists who freely contributed their time and expertise to the program. The program adopted an organizational structure that included a Board of Directors, a Management committee and several technical committees and a coordinating secretariat. Roles were simple. The Board of Directors provided vision and approval of yearly plans and budgets; the Management committee provided "hands-on" management of the program; and the technical committees addressed technological issues and solutions. The Secretariat was essentially the "hub" of the organization and ensured coordination of the elements within, and external to MEND. It should be emphasized that the MEND program was focussed entirely on technology development to reduce the liability associated with acidic drainage.

Planned funding for MEND was divided equally between the three major partners; the mining industry, the federal government and five provincial governments. When MEND ended in December 1997, the federal government had contributed 37% of the funding, the provinces 24%, and the industry 39%.

5 MAJOR ELEMENTS AND RESULTS OF THE CANADIAN RESEARCH

MEND organized its work into 4 technical areas: prediction, prevention and control, treatment and monitoring. Over 200 projects have been conducted or are in the final stages of completion and this from across Canada. Some of the technical results and observations to date include the following:
- Prevention has been determined to be the best strategy. Once sulphide minerals start to react and produce contaminated runoff, the reaction is very difficult to stop. Also, at some mine sites acidic drainage was observed many years after the waste pile had been established. With many old mine sites, there may be no "walkaway" solutions.
- Laboratory and field prediction tests for waste rock and tailings have been investigated and further developed. These tests include static and kinetic tests, mineralogical evaluations and oxygen consumption methods.
- Models that will predict the performance of dry and wet covers on tailings and waste rock piles are being developed and evaluated.
- In Canada, the use of water covers and underwater disposal are being confirmed as the preferred prevention technology for unoxidized sulphide-containing wastes.
- Innovative "dry" cover research is indicating that a range of materials, including low cost waste materials from other industries and non acid-generating tailings, may provide excellent potential for generating moisture retaining, oxygen-reducing surface barriers.
- Several other disposal technologies that will reduce acid generation are being investigated including permafrost in northern environments and waste organics from cities (crude compost) as oxygen-consuming covers for mine tailings

At the conclusion of the MEND program, a "tool box" of technologies had thus been developed to assist the industry in addressing its various concerns related to acidic drainage, and in significantly reducing its estimated liability. A particular important outcome has been the development of a common understanding among participants, inasmuch as it has allowed operators to take actions with greater confidence and to gain multi-stakeholder acceptance more quickly.

Subaqueous tailings disposal (the placement of tailings under a water barrier) has been confirmed as an effective technology for unoxidized sulphide-containing wastes. The work leading to this conclusion consisted of three major phases: preliminary studies to scope the effectiveness of subaqueous disposal at several sites, more detailed geochemical studies at two specific sites, and finally development of design guidelines. Operating factors that make this method an effective long-term solution include:

- the lower oxygen concentration available for reaction;
- the tendency for sediments to create an environment in which sulphides are formed naturally and remain in their most stable state; and
- the burial of tailings by natural sediments, which act as oxygen scavengers and have a retarding effect on the ability of reactive tailings to affect water quality.

A critical concern was the potential for trace metals, contained in the tailings porewater, to leach into the overlying water column. Extensive studies were performed on lakes where tailings had been disposed in the past. These studies established that the tailings are geochemically stable and that any transfer of metals is mostly downward from the water column into the sediments. It is estimated that about 50% of the new mine sites in Canada with sulphide-bearing materials could use subaqueous disposal for the tailings - and where feasible this will likely be the most cost-effective method for environmentally sound disposal.

An information dissemination program was conducted during the subaqueous disposal project, involving the various regulators and other stakeholders. This was regarded as a critical step in obtaining preliminary qualified public and regulatory endorsement. A generic design guide which outlines basic guidelines and direction on where and how to apply this method of tailings disposal has also been developed (MEND 2.11.9). The guide outlines the factors involved in achieving physically stable sediments, and discusses the chemical parameters and constraints that need to be considered in the design of both impoundments, and operating and closure plans.

Underwater disposal of mine wastes (tailings and waste rock) in man-made lakes is presently an option favored by the mining industry to prevent the formation of acidic drainage. Because of the controversy associated with using natural lakes for subaqueous disposal of tailings, there is considerable interest in evaluating whether artificial basins offer similar advantages to natural lakes. At the Louvicourt Mine (Québec) fresh, sulphide-rich tailings have been deposited in a man-made impoundment since 1994. Laboratory and pilot-scale field tests to parallel the full-scale operation and evaluate closeout scenarios are ongoing. The tests are developing economic methodologies for use at future man-made, underwater disposal sites.

The use of water covers to flood existing oxidized tailings can also be a cost effective, long lasting method for prevention of acid generation. Both the Quirke (Ontario) and Solbec (Québec) tailings sites which were subjects of MEND field and laboratory investigations, have been decommissioned with water covers and are presently being monitored. Further, results-to-date on a water cover placed on uranium tailings show that oxidation is effectively arrested and techniques have been developed to minimize water contamination so that treatment plants can be shut down in a few years. Where mining wastes are significantly oxidized, laboratory results have shown that the addition of a thin sand or organic-rich layer over the sulphide-rich materials can prevent or retard diffusion of soluble oxidation products into the water column. These barriers should be properly placed so that they are undisturbed by either wind or ice action.

7 DRY COVERS AS A CLOSURE TECHNIQUE

As previously stated the potential for acid generation from tailings and waste rock may be reduced by placing them under a column of water. Water covers, however, may require construction and maintenance of structures which are costly. In some instances the topographical, and hydrological conditions at the site may not favour this particular solution.

Dry covers are an alternative where flooding is not possible or feasible. MEND has extensively investigated multilayer earth covers for tailings and waste rock (e.g. Waite Amulet and Les Terrains Aurifères (tailings) and Heath Steele (waste rock): 3-layer systems). These type of covers utilize the capillary barrier concept and although they are effective, they are also very costly to install in many areas of Canada. Most covers used to isolate wastes are constructed with natural (clay, till) or industrial materials (geosynthtics).

The objectives of cover systems are to provide a low hydraulic conductivity barrier to minimize the influx of water and provide an oxygen diffusion barrier to minimize the influx of oxygen. To achieve these objectives a capillary barrier effect is created in the three-layer cover system. The concept is to keep the central compacted fine grained soil near saturation and the sand layers on either side of this fine grained layer in an unsaturated condition. A two-layer cover was used on the waste rock pile at Equity Silver (Wilson, 1997). The cover system consisted of a compacted till lower layer and an upper layer made of the same material but placed in a loose state. The resulting loose/compact profile provided a soil cover which allows rapid infiltration and storage of precipitation at the surface while also providing a lower barrier with a low hydraulic conductivity that remains saturated which minimizes oxygen fluxes to the waste rock.

Innovative "dry" cover research is indicating that several materials, including waste materials from other industries provide excellent potential at lower cost for generating moisture retaining, oxygen-consuming surface barriers. Recent studies at l' École Polytechnique in Montréal have shown that clean (non-acid generating) tailings, often available close to problematic sites, can be used as a lower cost alternative for the fine material in layered cover systems (MEND 2.22.2a, MEND 2.22.2b). Laboratory studies have confirmed that sulphide-free fine tailings offers some promising characteristics as cover materials. For example, if high saturation ($\geq 90\%$) can be maintained in the cover through capillary barrier effects, then a layer of fine material (i.e. tailings) sandwiched between two sand layers will effectively reduce the oxygen flux to the reactive tailings materials by a factor of 1000 or more. Theoretically, the efficiency of a dry cover then becomes comparable to that of a water cover of the same thickness (MEND 2.22.2a).

To further demonstrate the efficiency of covers built from non-reactive tailings, six experimental cells were constructed on a site near Val d'Or, Quebec. Data obtained confirm the behaviour of the capillary barrier concept. As expected, the fine grained material has stayed close to saturation serving as an oxygen barrier while the coarse material below and above drained rapidly and remained almost dry. The multilayer cover system, using tailings as the fine grained layer, was selected as the closure option for a 60-hectare tailings site at Barrick's Les Terrains Aurifères (LTA) property, located near Malartic, Quèbec. The cover, constructed during the winter to allow for heavy machinery to work on site, is composed of 0.5 m layer of locally available sand, a 0.8 m layer of non-acid generating tailings from the older tailings impoundment (as the fine material) and a 0.3 m sand and gravel layer which was seeded. The preliminary estimate of closure costs for the site, including vegetation, is $93,500/ha (Cdn $), of which $65,000/ha (Cdn $) is directly related to the construction of the cover (Tremblay 1999).

At the Poirier site in Northern Quèbec, a geomembrane liner (60 mil HDPE), with a one metre protection layer made of locally available soil, was used to reclaim the site. The mine operated between 1965 and 1975 producing about 5 million tonnes of acid-generating tailings, which have contributed acidity and metals to the local watershed. The final closure option selected involved moving a significant amount of contaminated material (e.g. spilled tailings) from the general vicinity of the impoundment to the tailings basin prior to the placement of the liner. The objective of the liner is to substantially reduce infiltration and limit oxidation of the tailings thereby reducing the release of metal loadings to the environment.

The restoration of the Poirier site required industry and the provincial government (Quèbec) to work together to develop a closure plan appropriate for site specific conditions. Through these efforts, the reclamation plan for the Poirier site will provide considerable improvements to the surrounding habitat.

Both LTA and Poirier have the distinction of being first full-scale demonstration projects of closure techniques in Canada (e.g. multilayer covers, geomembrane). The multilayer engineered oxygen barrier used at LTA has the advantage of using mine tailings as part of the reclamation solution.

8 TECHNOLOGY TRANSFER

Technology Transfer activities within MEND have been significantly expanded in recent years and this will continue in MEND 2000. MEND has an Internet site (**http://mend2000.nrcan.gc.ca**) for on-line retrieval of annual reports and summaries of MEND documents and other related information. Workshops based on MEND results have been presented at various locations across Canada and have been well received.

A popular output of MEND has been the production of videos in English, French, Spanish and Portuguese, describing technology advances relating to the prediction, prevention and treatment of acidic drainage from mine sites. These videos are available free of charge and can be ordered directly through the Internet.

A "MEND Manual" is being prepared. This manual, to be available in both English and French, will summarize all of the MEND and MEND associated work on acidic drainage from mine wastes and openings.

9 OTHER ASPECTS OF MEND'S SUCCESS

Aside from its technical successes, MEND represents an innovative method of partnering for technological research and development. Reasons for this include:
- The high return on the investment targeted and achieved, in terms of knowledge gained and environmental and technical awareness of the scope of the problem and credible scientific solutions.
- The partnership and improved mutual understanding developed between the two levels of government and the mining industry in search of solutions to a major environmental problem.
- The small dedicated secretariat group which coordinated activity, managed the accounting, reporting and technology transfer, and was the "glue" which held the program together.
- The extensive peer review process that was both formal and informal, and resulted in enhanced credibility of the information base.
- The aggressive approach taken for transferring the knowledge gained during MEND.

In large part as a result of MEND, it was shown that new mines are able to acquire operating permits faster and more efficiently than before since there are now accepted acidic drainage prevention techniques. As an example, the Louvicourt mine in northern Québec adopted MEND subaqueous disposal technology and has been able to progress from the exploration phase to an operating mine within 5 years, with a reduced liability of approximately $10 million. Similar impacts are reported for existing sites in the process of decommissioning. MEND has also fostered working relationships with environmental groups ensuring they are an integral part of the process

10 MEND 2000

MEND concluded on December 31, 1997. However, the MEND partners agreed that additional cooperative work was needed to further reduce the acidic drainage liability and to confirm field results of MEND-developed technologies. MEND 2000 is a three-year program that officially started in January 1998. The program is funded equally by the Mining Association of Canada (MAC) and Natural Resources Canada, a department of the Canadian government. The objectives of MEND 2000 are:
- to transfer and disseminate the knowledge gained from MEND and other related acidic drainage projects;

- to verify and report the results of MEND developed technologies through long-term monitoring of large scale field tests;
- to maintain links between Canadian industry and government agencies for information exchange and consensus building; and
- to maintain linkages with a number of foreign government and industry driven programs.

An important function for MEND 2000 is technology transfer. A MEND 2000 Internet site has been established and is regularly updated with current information on technology developments. Report summaries, national case studies on acidic drainage technologies, a publication list, and conference and workshop announcements are provided. Further, MEND 2000 hosts several workshops per year on key areas of technology (case studies, risk assessment, etc.).

11 CONCLUSIONS

The successes of the MEND program have come through the sharing of experiences, the thorough evaluation of technologies and their incremental improvement. No dramatic technological breakthrough other than water covers has been achieved. Nonetheless, Canadian industry reports that a significant reduction in liability is confidently predicted. An evaluation of MEND was conducted in 1996 (Young and Wiltshire 1996) and concluded that the estimated liability had been reduced by $340 million for five Canadian mine sites only. It is also acknowledged that the reduction in liability is significantly higher than this quoted value, with a minimum of $1 billion commonly accepted. The same study concluded:
- there is now a much greater common understanding of acidic drainage issues and solutions;
- the research has led to less environmental impact;
- there is increased diligence by regulators, industry and the public;
- MEND has been recognized as a model for industry-government cooperation; and
- the work should continue with strong international connections.

As a result of MEND and associated research, technologies are in place to open, operate and decommission a mine property in an environmentally acceptable manner, both in the short and long term. This can have a major impact on new mine financing and development. Moreover, mining companies and consultants have acquired a great deal more capability to deal with water contamination from mine wastes, including acid generation.

MEND is thus a good example of a successful, multi-stakeholder addressing a technical issue of national importance, and has been a model for cooperation between industry and various levels of government.

12 REFERENCES

1. Price, B. 1995. Defining the AMD Problem II. An Operator's Perspective. *In Proceedings of the Second Australian Acid Mine Drainage Workshop*, N.J. Grundon, L.C. Bell Eds. March 28-31: 17
2. Skousen, Jeff, Ziemkiewicz, Paul. 1995. *Acid Mine Drainage Control and Treatment*, West Virginia University: 13.
3. Murray, G., Ferguson, K., Brehaut, H. 1995. Financial and Long Term Liability Associated with AMD. *In Proceedings of Second Australian Acid Mine Drainage Workshop*, N.J. Grundon, L.C. Bell Eds. March 28-31: 165.
4. Bryan, V. Personal Communication.
5. Harries, J. 1997. Estimating the Liability for Acid Mine/Rock Drainage in Australia. *In Proceedings of the Fourth International Conference on Acid Rock Drainage*, Vancouver May 31-June 6, Volume IV: 1905.
6. Gustafsson, H. 1997. A Summary of the Swedish AMD Experience. *In Proceedings of the Fourth International Conference on Acid Rock Drainage*, Vancouver May 31-June 6, Volume IV: 1897

7. Feasby, G., Jones, R. 1994. Report of Results of a Workshop on Mine Reclamation, Toronto, CANMET, March 10-11.
8. MEND Project 2.11.9. Design Guide for the Subaqueous Disposal of Reactive Tailings in Constructed Impoundments. 1998.
9. Wilson, G.W., Newman, L., Barbour, S.L. 1997. The Cover Research Program at Equity Silver Mine Ltd. *In Proceedings of the Fourth International Conference on Acid Rock Drainage*, Vancouver May 31 – June 6, Volume I: 197.
10. Tremblay, G. Using Tailings in Dry Covers. Recycling Technology Newsletter. CANMET. Vol 4, Issue 1, June 1999. (Mend Project 2.22.2).
11. MEND Project 2.22.4a. Construction and Instrumentation of a Multi-layer Cover at Les Terrains Aurifères. 1999.
12. MEND Project 2.22.2b. Field Performance of Les Terrains Aurifères Composite Dry Cover. 1999.
13. Lewis, B.A., Gallinger, R.D. 1999. Poirier Site – Reclamation Program. In Proceedings of Sudbury '99 – Mining and the Environment. Sudbury 1999 (in preparation).
14. Young and Wiltshire. 1996. Evaluation Study of the Mine Environment Neutral Drainage Program, MEND Project 5.9. October 1997.

New technologies and approaches

Tailings and Mine Waste'00 © *2000 Balkema, Rotterdam, ISBN 90 5809 126 0*

AMD treatment, it works but are we using the right equipment?

J.H. Smith III
SEPCO Incorporated, Fort Collins, Colo., USA

ABSTRACT: For the past 40 years various approaches have been developed to treat acid waters coming from abandoned as well as operating mining operations. System designs have evolved to meet increasingly stringent discharge permit limits for treated water, as well as to provide solids disposal within economic constraints.

A treatment system for remediation of acid mine drainage (AMD) or acid groundwater (AG) requires two main steps:

1. The addition of chemicals to precipitate dissolved metals contained in the waters, and if necessary, to coagulate the precipitated solids ahead of physical separation.

2. Physical separation of the precipitated solids from the water so the water can be lawfully discharged from the site.

Choosing the appropriate technology and equipment results in the most efficient plant design, the lowest capital outlay, and minimum operating cost. The goal of these plants is to discharge liquids and solids able to meet standards. The separation of solids from liquids can be achieved through various means, including gravity settling, flotation, mechanical dewatering, filtration and evaporation.

As important as the liquid solids separation unit operations are, they are driven by the chemistry of the water to be treated. The content of the dissolved solids will influence the quality and quantity of the solids produced during precipitation. Thus the two aspects must be integrated, with chemistry first, then mechanical engineering.

This presentation will provide an overview of a number of liquid solids separation tools currently being used to treat AMD-AG at several sites in the USA. It will also discuss how their operations are impacted by the chemistry of their particular acid water feeds. The tools used include clarifier-thickeners, solids contact clarifiers, dissolved air flotation, polishing filters, membrane filters, and mechanical dewatering devices (belt and filter presses, vacuum filters, and driers).

1 INTRODUCTION

During the past two years the author has visited or assisted in the startup of ten different AMD-AG sites. He was also involved with the design and startup of other systems dating back to the late 60's. The following observations are based on this first hand experience, and on information provided by the plant operators of these systems. Patterns observed serve as the basis for the conclusions drawn about the design of AMD-AG treatment systems and the operational problems common in the liquid solids separation equipment now in general use.

Each site produces acid flows which are unique to that site. The composition of flows at each site varies considerably over a 12 month period or longer as the AMD-AG system continues to "clean the area" of dissolved solids. In most cases water leaving the plant meets acceptable standards. When an exception occurs, it is often because of a breakthrough of a polishing filter, or the lack of a polishing filter. It should be noted that the permit limits vary from different agencies (BAT versus NPDES) and that in some cases the discharge waters, even though meeting permit limits, will attract attention because of a negative aesthetic impact on a receiving stream. Since the regulations for discharged waters continue to be adjusted downward, today's acceptable results may need to be improved.

There is some uniformity across various sites: the process flow sheet, the type of neutralizer and the types of liquid solids separation equipment used. However, since each inflow is different, each site produces a unique precipitated solid. This leads to differing operational problems with the liquid solids separation equipment installed.

It is the author's belief that the tools applied to date have been adequate, however improvements would reduce costs (including operating personnel) and allow the industry to meet tighter discharge requirements for both water and precipitated solids.

2 BACKGROUND

One of the oldest operating AMD operations in the intermountain area of the USA was started in the early 70's. The wastes at first came to an impounding area from three sources: mine drainage, surface water, and discharge from the minerals processing plants. These contained Fe, Pb, Zn, Cd, Hg, F, and SO_4. In some streams the suspended solids content was up to 100,000 ppm. Since the closing of the processing plant, water comes only from the first two sources. Initial laboratory tests utilizing lime treatment produced 2% solids by weight settled sludge. Utilizing a process which mixed a portion of precipitated settled solids from the clarifier-thickener with the lime solution in a reactor and then with the acid water increased settled sludge concentration to over 15%. The resulting precipitated solids have the ability to settle quickly in the clarifier-thickener allowing for a reduction in equipment size, and a marked reduction in the volume of sludge needing to be stored in an impounding area at the site.

Because of high iron content, the treatment system includes aeration. This has a positive impact on achieving the levels of sludge concentrations produced in the clarifier-thickener.

The plant was designed to treat 5,000 gpm of waste water in a clarifier-thickener fitted with a special energy dissipating feedwell designed to maximize settling of precipitated solids. The clarification area has 31,000 square feet of surface area (0.16 gpm/sq.ft.). At the present time the feed rate is about 2,400 gpm (0.08 gpm/sq. ft.). It should be noted that from the start of the operation, there were times when the overflow from the clarifier-thickener contained enough TSS to leave residue on the banks of the receiving stream and on rare occasions discharged waters contained some metals above permit levels. There are still times during the year when the TSS in the clarifier-thickener overflow is higher than desired. This occurs during peak run off periods when upper workings in the old mines are flooded and additional "clays" are discharge with the acid water.

We have gone into detail on this site because of its long operating history and the fact that the system has been able to meet discharge permit limits. It also is a design that has been used on most newer AMD-AG systems with similar operational problems. Such as:

1. Uncontrollable variations in raw water TDS and TSS contents. These impact clarifier-thickener overflow clarity, sludge concentrations, and sludge dewatering.

2. Inability to completely automate plant operation.

3. Potentially ineffective system for the blending of polymers with precipitated solids.

4. Lack of overflow polishing filter to ensure minimum TSS content in treated discharge water.

5. Lack of information by the operators about water chemistry, aside from that pertaining to the current discharge permit limits.

3 OVERVIEW

From this point on we will summarize liquid solids unit operations at a variety of AMD-AG treatment sites. Because of the importance of water chemistry on treatment system designs and operations, we will begin with acid waters coming to the treatment system.

3.1 Chemistry

In our view, AMD-GA should be thought of as solutions containing dissolved and suspended solids such as come from a solution mining operation except that the operator of the AMD-GA system has no control over, or in most cases does not have a current total analysis, of the water they are to treat. At a minimum the volume of water needing treatment will vary with spring run off, or wet/dry summers. In the case of old mine workings, the potential exists for blow outs of plugs or bulkheads which can impact the quantity and quality of waters needing treatment. In all cases noted during recent site visits, the operators said they know when something had changed because: 1. the incoming water "looks different", 2. the system is disturbed (more TSS in clarifier-thickener overflows), 3. the chemical addition rates need to be changed, or 4. the cakes from the sludge dewater device (when used) are different.

Changes in volume of the incoming waters are accompanied by changes in the quality of the wastewaters, measured as total dissolved solids (TDS). Higher volumes often result in lower ppm, but the content of the acid water may be affected by which surfaces the water contacts and by the age of the acid water. During dry weather conditions, longer acid water detention times in the workings can result in upward swings in TDS content.

In general the basis for the design of AMD-AG plants has been based on a complete water analysis taken at some point, the current permit limits for the site, and projections (guesses) of changes that will occur as operations continue. There is a goal that someday the water leaving the "workings" will be free of contaminants. But even if the site is never clear, there will be an eventual change in the water chemistry which will have an impact on the operation of the wastewater treatment system. This is the challenge faced by designers and, most importantly, by the plant operators. It is imperative to perform a complete water analysis as close as possible to the beginning of plant design and again near the end of the design effort. The latter will become the base line for future determinations of raw feed TDS changes.

It is our view that an operating plant should have at least two complete water analyses each year. The key to a successful AMD-AG remediation treatment system is knowing what the incoming feed water contains in the way of TDS. One could say "Why worry about anything but what is permitted?", however changes in the type of solids (TDS and TSS) in the incoming acid water will have an impact on many aspects of the operation. These include solids settling, floc dosage, sludge concentration, life of polishing filter media (if utilized), and the dewatering characteristic of settled sludge.

3.2 Equalization basins

As it is not possible to control the makeup of the incoming feed streams, it is beneficial to at least minimize the variations in TDS/TSS. This is best done in a basin as large as economically possible, installed on the site. This basin also serves as a buffer for large increases of water volume (heavy rains and or snow melts) which otherwise would have to be factored into the size of the equipment in the wastewater treatment facility.

The design of the equalization basin must take into account the potential need to remove accumulated settled solids from the basin and to provide a means to monitor the build up of sludge to avoid pumping of high concentrations of TSS to the wastewater treatment system.

We have direct knowledge of two sites which function without adequate equalization basins. The TDS and TSS to the plants vary greatly whenever rainfall flows down through different tailings piles or when underground mine bulkheads break and flood the system with acid water held there for long periods. These variations necessitate frequent adjustments in chemical additions and extra backwash cycles to maintain final effluent permit levels.

3.3 Neutralization

The selection of the specific chemical needed to change the pH of the acid water entering the plant will depend on three items: 1. the discharge permit limits of specific dissolved materials in the water, 2. the projected volume of the precipitated solids to be generated, and 3. the cost of the chemical (delivery, storage, and equipment). Of the systems reviewed, lime (either pebble or hydrated) was the predominate source for neutralization. The other was sodium hydroxide.

In general, when lime is used as the neutralizer, it results in the generation of larger volumes of precipitated solids than when sodium hydroxide is used. This impacts the sizing of the clarifier-thickener, the sludge pumps, the mechanical dewatering devices, and containment areas used for sludge disposal. The cost of transport is also impacted by increased solids generation when off site disposal of sludge is required. The use of lime is usually justified by its low cost, the higher concentration of sludge produced, and the fact that the sludge will eventually dewater to a higher concentration.

The use of sodium hydroxide as the neutralization chemical requires less makeup equipment than a lime system and generally produces fewer participated solids than lime. However in addition to high cost of sodium hydroxide, the sludge generated with it in a clarifier-thickener will generally be less concentrated and, if there is no calcium carbonate in the raw water, the precipitated solids can contain entrained water, which results in lower solids contents in cakes produced with mechanical dewatering devices.

In addition to the neutralizer chosen, other factors affecting sludge concentration are: 1. the way the clarifier-thickener is operated, 2. the neutralization process, and 3. the type of precipitated solids produced.

3.4 *Neutralization process*

Early AMD treatment systems (such as at eastern coal operations) utilized lime and added the material directly to the incoming acid waters going to treatment. The resultant precipitated solids settle slowly in open basins or gravity settlers (clarifier-thickeners). The settled sludges were low in solids content, and when dewatered with a mechanical device produced cakes with high water contents. In an effort to reduce the settling basin size, increase sludge concentrations, and produce drier filter cakes, a process was developed which involved the mixing of previously settled solids with a lime slurry prior to that mix being added to the incoming acid waters going to treatment. This system was utilized at many USA mining operations (both coal and metal) starting in the 60's.

Of the sites used as the basis for this paper, the majority of those that use lime as the neutralizer also use some variation of this sludge-to-lime recycle process. However this does not mean that settled sludge coming from all gravity settlers will be at the highest possible concentration. The operator will have other constraints (sludge storage, clarifier-thickener type, clarifier-thickener overflow TSS content, mechanical dewatering device, etc.).

It must be noted that the lime source is important. Some sites have experienced problems with lime supplied with high amounts of "grit"and low amounts of $CaCO_3$. For this reason, most operations utilize hydrated lime.

3.5 *Aeration*

When acid waters contain high amounts of ferrous iron, aeration is used to convert it to the ferric state, in which form iron is more readily removed by an AMD-AG treatment system. Two sites visited (and most systems handling coal mine wastewater) were aerating the water after neutralization. One site used a fixed surface aerator while the other introduced air through diffusers set in the bottom of a tank. The surface aerator lacked the ability to adjust oxygen input, created a great deal of splash, and caused a general housekeeping problem. It may be that had a few small floating surface aerators been used, oxygen input and splashing could have been controlled.

3.6 *Polymer addition*

Polymer addition is an art, involving decisions related to supplier, chemical properties, one or two types, purchased in dry or liquid form, dilute versus less dilute, single or multiple addition points, addition in line versus into a blending tank, high shear versus low or controlled shear. No two plants do it in the same way. What is constant is that by trail and error each operator has found an acceptable way to use it, frequently different from the system first installed because changes in the raw feed required adaptations in the polymer as well as in the operation of the clarifier-thickener.

If we use the generally accepted approaches given by the polymer suppliers, we note some constants regarding how to get the maximum utilization of polymer, defined as the smallest amount needed to create the particle that will settle in the shortest time.

1. Add polymers in as dilute a concentration as practical.

2. Add polymers in a number of locations prior to the start of gravity separation.

3. Use only enough mixing energy to blend the polymers with the neutralized slurry but not enough to shear the floc once formed.

4. Add only enough polymer to treat the TSS contained in the feed to the gravity settler. (Too high a polymer dosage will cause major negative changes in the solids mass contained in the gravity settler: doughnuts, clouds, sticky mess).

5. Conduct frequent on site jar tests to see how the precipitated solids are settling.

6. Ensure that there are enough TSS in the feed being treated with polymers to allow contact of particles as the floc forms.

In the ideal design, we would suggest a tank (fitted with a variable-speed paddle mixer) in which polymer and clarifier-thickener feed are blended. The dilute polymer would be added through a distribution pipe set across the surface of the blender tank. The tank would have a baffle which allows slurry from the bottom of the tank to exit the top and into the feed pipe (or launder) going to the clarifier-thickener.

A second polymer addition point would be at the outlet of the blender, and a third would be located at the centerwell of the clarifier-thickener. Even though this requires piping and valving, the approach allows the operator to change polymer addition rates and locations to suit changing conditions as they are encountered.

3.7 Gravity settlers

If the use and addition of polymers is an art, then the design and sizing of gravity settlers is a "black art." If there is one best design, then there are a dozen, ranging from conventional open top tanks, to units fitted with plates or tubes. Within this group there are units designed only to produce a "clear" overflow water (clarifiers) while others will clarify as well as thicken the settled solids (clarifier-thickeners). There are units with deep feedwells (high-rate settlers), dilute feed systems, ballasted flocculation, internal solids recycle, energy-dissipating feedwells, extended sludge collection rake blades, angle or pipe rake arm designs, two or four rake arms, and units with or without sludge rake arm lifting devices. The selection of any one design seems to be more a factor of client preference (based on past experience or on cost), design engineer's comfort level, and finally, the ability of the equipment salesperson to convince everyone that the units they sell are the best for the service.

We noted a majority of gravity settlers are open topped clarifier-thickeners without plates or tubes, but each has variations of some sort specific to feedwell designs, sizing criteria, tank depth, number of rake arms, and rake arm speeds (from a low of 32 minutes per revolution to a high of 7 minutes, with most being about 12 minutes per revolution) Two facilities use the clarifier-thickener tank for sludge storage ahead of a filter press. Two sites have units with tube settlers, another has a solids contact unit like those used in the drinking water treatment industry, and another utilizes a dissolved air flotation unit (DAF) for separation of precipitated solids.

The design basis for the clarifier-thickeners varied from 0.16 to 0.53 gallons per minute flow per square foot surface area (including backwash and recycle volumes). The current operational rates vary from 0.08 to 0.43. The solids contact clarifier was designed at 0.31 but operates at 0.23 most of the time. Even at these relatively low hydraulic loading rates, no site is able to continually produce overflows with less than 2 ppm (most up to 5 ppm) TSS. All (including the tube settler units) had increases in overflow TSS contents at various times during the year because of variations in the TDS/TSS content in the feed. Note that operations using NTU as a measurement often show less than 1 unit when discharging overflows containing +3 ppm TSS.

It should be kept in mind that with clarifier-thickener feed streams containing upwards of 10,000 ppm TSS, even a 10 ppm TSS content in the overflow equals a removal rate of 99.999 percent.

Settled sludge concentrations vary from 20% solids by weight to 4% (including the units with tube settlers). On average the underflow concentrations are between 8% and 10% solids by weight. There was one exception: that site operates with underflows always above 50% solids by weight, however the precipitated solids are primarily Manganese, Calcium, and Magnesium.

Those clarifier-thickeners units that are operated as sludge storage tanks ahead of a filter press operate with varying sludge levels within the tank. If not watched carefully, this could result in an increase in overflow TSS content if there is an increase in the TSS in the feed coming to the unit (as could occur when a polishing filter is backwashed).

Those that have a separate settled sludge storage tank ahead of a dewatering device operate with a more constant sludge level in the clarifier-thickener tank, yet still have the occasional problem with increases in overflow TSS due to an upward movement of the sludge level in the clarifier-thickener tank.

In observing the operation of the solids contact clarifier with a variable TSS content in the feed, it becomes obvious how difficult it is to maintain a constant sludge level even in this type of unit, and how quickly the overflow TSS content can deteriorate as the level either raises or drops below the outlet of the unit's special feedwell. As a result, the operators have to take extra care to maintain a set sludge level in the tank and to adjust the withdrawal rate of sludge to suit varying feed conditions.

When the tube settler units are closely monitored, the TSS content in the overflow is often less than 3.0 ppm. This requires that the TSS concentration in the feed to the clarification zone and the settled sludge inventory in the "thickening" zone of the unit be controlled. However even these "clarifiers" can produce overflows containing higher amounts of suspended solids with varying plant feed conditions.

It is interesting to note that almost all of the clarifier-thickeners have "floaters" in the upper zone of the tanks. These are visible particles which appear to hang in the water but with time migrate to the unit's overflow launder and report as TSS.

A recent study showed that under laboratory conditions it is possible to duplicate the formation of these "floaters" and to eliminate their formation by modifying the level of introduction of the feed slurry into the simulated clarifier-thickener. This procedure has been duplicated with other AMD waters and leads one to hypothesize that under certain conditions it is possible to break (or scrub) a portion of the formed floc from the main particle. When this fragment enters the zone of clear water in the clarifier-thickener tank there is little opportunity to contact other fragments needed to reform into a larger particle which would settle. As a result the particles leave the tank in the overflow and are reported as TSS. High shear occurring in the clarifier-thickener feedwell feed pipe can also break the formed floc and create fragments.

In addition to targeting clear overflow, it is desirable to obtain maximum concentrations in settled solids underflow to yield the smallest volume of material for either mechanical dewatering or backfilling. When the gravity settler is used to store concentrated settled solids ahead of a dewatering device (or in some cases to get the slurry thicker prior to pumping to disposal) the sludge level will rise in the tank. If not watched carefully, the sludge level can raise into the clear water zone of the unit and cause TSS to carry over into the overflow.

This potential conflict in system needs can be avoided by following the clarifier-thickener with a polishing filter on the overflow and a separate settled sludge storage tank between the settler and the dewatering or backfill system.

In an ideal operation, a clarifier-thickener with the following features would reduce some of the current shortcomings:

1. A variable-speed, sludge collection rake drive unit with at least a 100 percent speed adjustment range.

This would allow for adjustments of rake speed, which are needed if it is found that the sludge blanket is being "agitated" by the rotational speed of the rake arms. In the writer's view this is generally not the case.

2. A deep and large diameter feedwell which has a lower telescoping section which can be adjusted from the top of the unit.

This would reduce the shear of floc particles within the feedwell and allow the operator to fix the height of the lower outlet of the feedwell just above or within the sludge blanket in the tank to ensure floc particle contact.

3. A feed system within the feedwell to send the incoming feed flow across two separate plates fixed at different levels within the feedwell.

This would minimize feed inlet velocities to reduce floc breakage.

4. A polymer addition distributing system with at least two outlets across the top of the feedwell.

This would allow the operator to add polymer to the clarifier-thickener feed just prior to the material entering the settling zone of the unit.

5. A lifting device for the sludge collection rake arm lifting device with minimum 24" lift.

Should the rake arms ever stop due to a mechanical or power failure, leaving them in the zone of concentrated settled sludge could make restarting the rakes difficult and lead to the need to empty the tank. If the arms can be lifted from the zone of thick sludge within the tank and the rake drive restarted, it would save time and production.

6. A clarifier-thickener tank with a center sludge thickening pocket achieved by having a tank bottom with a center slope of 3" to 12", and a tank with at least a 15 foot side water depth.

A double sloped tank bottom will give additional sludge storage capacity without increasing the overall depth of the sludge level within the tank.

7. A separate settled sludge collection tank.

This is important if a filter press is used to dewater the settled sludge because the removal of sludge is not continuous. It avoids a varying level of sludge within the clarifier-thickener.

Even with the above listed design features, a clarifier-thickener may not be able to produce an overflow with less than 3 ppm TSS on a continuous basis. Considering that the existing operations (even with their different combinations of designs) are able to produce TSS of less than 10 ppm, one would want to be sure that the cost for the "ideal" design is warranted.

One plant visited utilizes a dissolved air flotation (DAF) unit rather than a gravity settler for the removal of precipitated TSS from the treated wastewaters. With this type of unit, small air bubbles attach to the precipitated solids and cause the solids to float to the surface of the tank. Both a polymer and a surfactant are added to ensure attachment.

In general the float concentrate in a DAF is not as high in solids content as the settled sludge produced in gravity settlers, and the TSS content of the overflow from the unit is higher than that from a gravity unit. A DAF unit is more sensitive to changes in feed TSS content, which can impact overflow clarity. On the other hand, the loading rate (gpm per square foot of surface area) on a DAF is higher, so for the same flow volume a DAF would be smaller than a gravity settler and, because of a smaller volume in the tank, the system can recover more quickly than a gravity settler.

3.8 *Polishing filters for gravity settler overflows*

A number of AMD sites include a granular media unit filter to remove TSS from the overflow coming from clarifiers. By adding these filters, the discharge TSS from the plant generally can be held to 1.0 ppm or less. This discharge water would be free of suspended solids that could cause stream bed discoloration. If treated water with less than 1 ppm TSS causes discoloration in a stream bed, this could be due to oxidation of other than the regulated TDS in the treated water.

The advantage of having a polishing filter is countered somewhat with the need to backwash the units. This recycles dirty water to the inlet of the treatment plant, in some cases up to 1/3 of the plant's daily capacity. This "extra" material has to be taken into account when designing the gravity settlers.

The majority of plant sites which utilize polishing filters use a round sand as the filtering medium.

It is best to backwash the filter often enough to ensure that the trapped solids do not penetrate too deeply into the filter media bed. This is to avoid the possibility of buildup of solids in the bed which would not be removed during the backwash cycle. If the bed becomes full of solids, a breakthrough of TSS into the filtrate could occur. TSS in the filter feed from the clarifier will dictate the frequency of the backwash sequence.

Filter media (sand or coal) need to be replaced in time because of a tendency of the media particles to grow in size, which can change the performance of the filter operation. This change is due either to buildup of TSS on the media particles (often associated with residue polymer content in the filter feed), or because of mineral deposits on the particles.

Some plants change filter media every two years, while others have operated without replacement for over 10 years. The under drain system which supports the filter media also needs replacement with time.

3.9 *Sludge disposal (impounding or mechanical dewatering)*

As stated throughout this paper, the primary function of the AMD-AG remediation systems is to produce a plant discharge water which meets the permit levels set by some regulatory agency, and that is generally accomplished in all cases. The other "product" generated from the treatment system is sludge, which occupies space and requires disposal, which costs money.

Many sites have the option of disposing of sludge on their own property (above ground in evaporation/drying ponds, or in underground workings.) Others are required to haul the material off site for disposal at private or public dumps.

Onsite disposal generally is done in one of two ways: into backfill areas, where "dry" material is preferred, or into impounding areas or underground workings, which require a slurry that can be pumped. The solids content in the "slurry" is set by the specific site conditions, but is generally from 5% to 15% solids by weight.

For off site disposal, the standard is for the "cake" to meet the paint filter test required by the dump operator. In some cases this may require that a "binder" (fly ash or cement) be added to the filter cake.

This no dripping liquid standard does not always relate to the total solids content of the cake; some sites produce cakes with as low as 16% solids by weight while others produce cakes with up to 40 % solids (70-80 lbs per cubic foot) and both meet the paint filter test. The key is in the physical nature of the participated solids and their ability to hold water. However, any water in the filter cake takes room and adds weight, leading to higher costs for hauling and disposal.

Most dewatering of sludges generated in AMD-GA treatment plants is done on filter presses, even though they operate in a batch mode, because they produce a cake with a high percent solids (even at 100 psi operating pressure). In some cases increasing the pressure of the dewatering device produces a drier cake. But in a recent test on a somewhat unusual sample from an operating plant, increasing the pressure to 2,000 psi caused only a 5 % increase in cake solids over that produced in a unit operating at 100 psi.

Filter cycle times range from under one hour to over four hours, dependent on the feed concentration and the physical nature of the particles being produced in the neutralization step of the wastewater treatment plant. A membrane filter press generally will operate with a shorter cycle time than a non membrane unit. This advantage is offset by higher initial equipment, installation, and maintenance costs.

Filter presses can be provided with a great deal of automation, including a device to stop the filter cycle when zero flow of filtrate is detected. However an operator is required during the time when the filter cakes are being removed from the plates, in order to ensure that there are no solids left on the surface of the filter cloth. Such residue could lead to incomplete closure of the plates, which could result in a leak of feed slurry during the next filling of the press.

Many operators of filter presses report that at times the unit will produce a "wet" cake or the filter cycle time will be "different". Some have assumed it had something to do with the press itself even though there were no changes in the mechanical operation of the unit. In all cases they "solved the problem" by making adjustments to the wastewater treatment process ahead of the filter press (neutralization, flocculation, or thickening).

When sizing a filter press it is important to match the filter cycle time to the number of shifts per 24 hours which an operator will be on site, to the sizing of the sludge thickening and filter feed storage units. In general it is better to have excess capacity for sludge dewatering rather than just enough because of the variations in volume and TDS content in the feed to the wastewater treatment plant.

Some sites dewater sludges with continuous belt presses which produce filter cake solids content within the average range of those achieved with filter presses but at cake thicknesses of under 1/4". (The impact on cake solids content versus cake thickness is also seen on filter presses). The need for polymer addition to the feed to the belt press and the large number of moving parts associated with the unit must be considered.

Continuous vacuum disc filters have been used to dewater sludges generated with some success but their use appears to be limited.

Those sites which require the highest filter cake solids (+ 60 %) are using driers. One type of drier is a modified filter press. The plates are designed as heat exchanges through which hot fluid is circulated when the filter press is in its "drying" sequence. Recent developments are allowing the "drying" sequence to be completed in about an hour. Aside from the equipment needed to deliver the hot fluid, the physical appearance and operational requirements of the units are identical to a standard membrane filter press.

3.10 *Instrumentation*

The degree of instrumentation varies from extensive systems to very basic. This does not appear to impact the ability of the treatment systems to produce permit-quality water. The number of operating people required appears the same for all plants regardless of the sophistication of instrumentation. We believe this is the case because:

1. There is no proven instrumentation for determining the amount of precipitable solids contained in the raw water coming to the treatment system. Although in many cases pH and water volume are measured within the treatment system, these do not relate to the amount of precipitable solids that would be produced at the specific pH setting used for neutralization.

2. There is no widely used continuous method to determine the amount of polymer needed to ensure adequate separation rates of precipitated solids going into the gravity settler.

3. There is no trouble-free continuous method to measure the depth of settled solids in a gravity settler.

4. There is no way to automatically ensure that all solids produced on a filter press have been discharged.

4 CONCLUSION

Even with problems, the current technology works, so why fix it? Here are a few answers:

1. There are too many steps involved.
2. The systems take too long to correct after an upset occurs.
3. They need polymer to operate.
4. There is no way to ensure 100 percent compliance with discharge requirements under the varying feed conditions inherent with AMD-AG treatment systems.
5. The systems take a lot of land space and have a high cost to design and build.
6. Operating costs (chemicals and operators) are high.
7. When a polishing filter is utilized, the system has to be designed to treat extra TSS and water volume associated with backwashing.
8. Lime, often generates additional precipitated solids over those produced when using sodium hydroxide for neutralization.
9. As the permit limits change for discharge water and or sludge disposal, operating capabilities of existing systems may be exceeded.
10. The systems cannot be fully automated.
11. There will always be floaters in the water coming from gravity settlers. Without polishing filters these pose potential water-discharge problems.

We believe it is time for careful consideration of different technologies, and membrane filtration may be one answer.

Recent improvements in the manufacturing and cost of operation of membranes have allowed the use of membrane filtration on a wide variety of industrial applications, including AMD treatment. Indications are that this approach would make the treatment of AMD-AG simpler and less costly than the approaches now utilized. As with current treatment technologies, there are some feed streams which can cause operating difficulties or require pretreatment of the membrane filtration feed streams.

Tailings and Mine Waste'00 © 2000 Balkema, Rotterdam, ISBN 90 5809 126 0

Using Envirobond™ ARD to prevent acid rock drainage

M. Gobla
US Bureau of Reclamation, Denver, Colo., USA

S. Schurman & A. Sogue
Rocky Mountain Remediation Services, L.L.C., Golden, Colo., USA

ABSTRACT: Acid Rock Drainage (ARD) is one of the most serious and persistent pollution problems facing today's mining industry. ARD is formed when pyrite (FeS_2) is exposed to natural weathering conditions. Pyrite weathering, formally the oxidation and hydrolysis of pyrite, leads to the generation of soluble, hydrous iron sulfo-salts and the production of acid in the form of H^+. The resulting effluent is water that is characterized by high levels of iron (and any other metal contained in the rock undergoing weathering) as well as high concentrations of sulfate and a low pH (acid).

Rocky Mountain Remediation Services, L.L.C., (RMRS) has developed a patents pending Envirobond™ product, Envirobond™ ARD, to treat acid rock drainage. Treating pyrite with Envirobond™ ARD inhibits the oxidation and hydrolysis of pyrite thus curtailing the pyrite oxidation cycle.

1. THE ENVIROBOND™ TREATMENT PROCESS

The only way to stop the oxidation of pyrite is to limit its exposure to weathering (oxidation and hydrolysis). Envirobond™ ARD effectively stops this process by combining Fe^{2+} with Envirobond™ ARD (EB-ARD) to form a stable, insoluble compound Fe(EB-ARD). The iron/Envirobond™ ARD compound renders the Fe^{2+} in the pyrite unavailable for oxidation by permanently coating all surfaces that it comes into contact with. Envirobond™ ARD also reacts with the oxidizing Fe^{3+} ion to form a second, equally stable Fe(EB-ARD) compound that also coats all surfaces that it comes into contact with.

The following equations show the chemistry of Acid Rock Drainage (ARD):

Step

$$FeS_2 + 7/2\ O_2 + H_2O \longrightarrow Fe^{2+} + 2\ SO_4^{2-} + 2H^+ \tag{1}$$

Oxidation of Fe^{2+} to Fe^{3+}

$$Fe^{2+} + 1/4\ O_2 + H^+ \longrightarrow Fe^{3+} + \frac{1}{2}\ H_2O \tag{2}$$

Hydrolysis of Fe^{3+}

$$Fe^{3+} + 3\ H_2O \longrightarrow Fe(OH)_3 + 3H^+ \tag{3}$$

Oxidation of FeS_2 by Fe^{3+}

$$FeS_2 + 14\ Fe^{3+} + 8H_2O \longrightarrow 15\ Fe^{2+} + 2\ SO_4^{2-} + 16\ H^+ \tag{4}$$

The next equation shows what happens to Acid Rock Drainage (ARD) chemistry when Envirobond™ ARD is added. As indicated, the acid generating process is curtailed when the Fe(EB-ARD) compound is formed and precipitates on the surface of the pyrite grains. Once the pyrite is coated with Fe(EB-ARD), pyrite oxidation cannot occur.

Step

$$FeS_2 + Envirobond^{TM}\ ARD \longrightarrow Fe\ (EB\text{-}ARD)\ pyrite\ grain\ coating\ is\ formed \quad (1)$$

2. ENVIROBOND™ ARD APPLICATION INFORMATION

RMRS has conducted a number of Envirobond™ ARD treatability studies on rocks from two mining Superfund projects, Leadville and Summitville, located in Colorado. Both treatability studies have been successful in stopping acid rock discharge and it is anticipated that Envirobond™ ARD will be deployed at both sites. Envirobond™ ARD will be sprayed in its liquid form onto mine dump material at the Leadville site and at the Summitville site it will be sprayed onto rock in the open pit highwall. On test plots or small applications, Envirobond™ ARD can be sprayed using a small mixing tank attached to a spray wand. During full-scale deployment of the liquid form of Envirobond™ ARD, a hydromulch spray cannon can be used for treating entire mine dumps or open pit highwalls.

At other projects where a deeper penetration of the Envirobond™ chemicals is desired, Envirobond™ ARD can be blended with ARD generating rock in its solid form to the desired depth. In this application Envirobond™ ARD can be disced, tilled or ripped into the surface of a mine dump to create a non-oxidizing rind over the ARD generating material that will stop the runoff of acid and metals contaminated water.

Once applied to the ARD generating rocks, the reaction of the Envirobond™ ARD with the pyrite should be complete within 24 to 48 hours and the pH and available ferrous iron in the system will begin their respective increase and decrease within that time frame. As long as the temperature of the surface of the rocks is above 32°F, temperature should have no effect on the reaction of the Envirobond™ ARD with the pyrite.

2.1 Summitville Mine Pit Highwall, Colorado, EPA Superfund Site

The Bureau of Reclamation submitted rocks from the Summitville open pit highwall to RMRS to determine if Envirobond™ ARD could be used to inhibit or stop the generation of acid and the discharge of metals from the surface of the Summitville highwall. The highwall rock samples were an argillically altered latite that contained from 3 to 5 percent pyrite and had an acidic pH of 3.30.

Of concern to the Bureau of Reclamation was:
- The pH of the discharge,
- The metal content of the discharge, and,
- The swelling of clays in the argillically altered latite.

Success was defined as chemically treating the highwall rocks to reduce or eliminate the acidic runoff that contained metals while having a minimum impact on the highwall clays.

To determine the ability of Envirobond™ ARD to treat the Summitville highwall rocks, a number of samples were placed in Buchner funnels with treatment chemicals and water was flowed through the ARD generating rocks on a daily basis. The leachate from these samples was collected on a daily basis for analysis and was to designed to test the following parameters:
- Sample 1 – Untreated Baseline Sample (Baseline on accompanying graph)
- Sample 2 – Treated with Envirobond ARD (EB Treated on accompanying graph)
- Sample 3 – Envirobond™ ARD and Strong Oxidizer (Oxidized/EB Treated on accompanying graph)
- Sample 4 – Treated with Sodium Phosphate (Phosphate Treated on accompanying graph)

In order to simulate the high precipitation conditions of Summitville (Summitville's annual precipitation rate varies from 36 to 42 inches) the samples were saturated on a daily basis with the equivalent of three inches of water. Daily leachates from the funnels were collected and analyzed for pH and ferrous ions (Figures 1 and 2).

As shown on Figure 1, pyritic latite rock samples treated with Envirobond™ ARD resist oxidation and acid generation even when treated with a supplemental oxidizer. This compares to the baseline sample where the acidity in the rock remains at a constant pH of about 3.2. The phosphate treated sample controls the formation of acid for a limited amount of time but eventually loses its ability to stop the formation of acid and the pH drops.

Figure 2 shows that even as the pH of the leachate becomes neutral with the addition of Envirobond™ ARD (Figure 1), the formation of ferrous iron (Fe^{2+}) is reduced over time until virtually no ferrous iron is present to initiate the pyrite oxidation cycle. Again, the ferrous iron content is reduced because it combines with Envirobond™ ARD to form the Fe(EB-ARD) compound which coats the pyrite (and any other surface that it comes into contact with) with a permanent, insoluble compound that curtails the oxidation of pyrite.

2.2 *Ponsardine Mine Dump, Leadville, Colorado*

The Ponsardine mine dump is located in the Leadville Mining District of central Colorado. It is located on the eastern edge of the town of Leadville and encompasses an area of approximately 300 feet x 200 feet or 60,000 ft^2. The Ponsardine shaft is located on the top of the mine dump and is caved. The elevation of the project area is approximately 10,560 feet above sea level. The climate is warm with occasional rainstorms in the summer with extreme cold, snow and wind in the winter.

Mine Dump Composition

The composition of the surface of the Ponsardine mine dump is generally fine grained (clayey) with larger fragments, minus 1", armoring the surface of the clayey material. The surface is primarily covered with sulfide waste but may be covered with as much as 40% oxide waste rock. Approximately half of the surface area is 30 to 45 degree side slopes with the other half being flat. Localized, extremely high concentrations of sulfide occur various points around the sides and on the top of the Ponsardine mine dump.

A physical examination of the Ponsardine mine waste indicates that much of the debris covering the surface of the mine dump contains fragments of pyrite and other sulfides that are generally less than ½" in size but can be as large as 1" in size. These sulfide fragments can be free or still intergrained with other rock particles. Water that has passed through a funnel containing a two inch bed of Ponsardine mine waste has an extremely acidic pH of 1.2 to 2.2.

3. ENVIROBOND™ ARD TREATMENT APPROACH

The challenge of the Ponsardine project is that RMRS must decrease the discharge of acid and metals from the surface of the mine dump but cannot excavate or change the color(s) of the dump material because of historical concerns. The EPA has asked RMRS to treat the Ponsardine dump with Envirobond™ ARD in fall 1999 so they can monitor the quality of the runoff from the mine dump this fall and spring. An improvement in water quality over the baseline sampling will determine if Envirobond™ ARD has been successful in reducing the oxidation of pyrite thereby reducing the discharge of metals and acid from the Ponsardine mine dump.

4. ENVIROBOND™ ARD TREATMENT

4.1 *Laboratory Testing*

RMRS conducted full-scale laboratory tests to optimize two formulations of Envirobond™

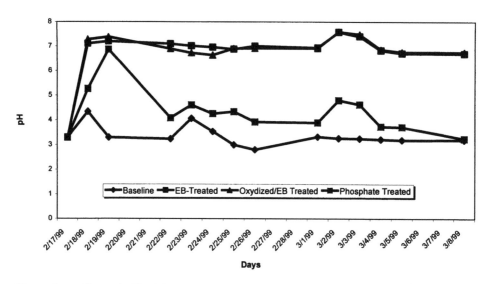

Figure 1: **Summitville ARD Study**
pH Variations with Time

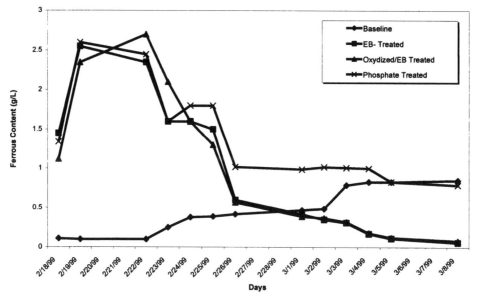

Figure 2: **Summitville ARD Study**
Available Ferrous Ions with Time

ARD for use on the Ponsardine mine waste pile. The first formulation, ARD-1, proved successful in treating ARD generating rocks from the Summitville Superfund site that had an initial pH of 3.2. However, RMRS needed to adjust the formula, ARD-2, to treat the more acidic conditions of the Ponsardine waste (pH 2.2). ARD-2 is an optimization of ARD-1 and is an attempt to substantially reduce ARD chemical treatment costs. The lab tests yielded excellent treatability results with both formulations and will result in lower ARD treatment cost.

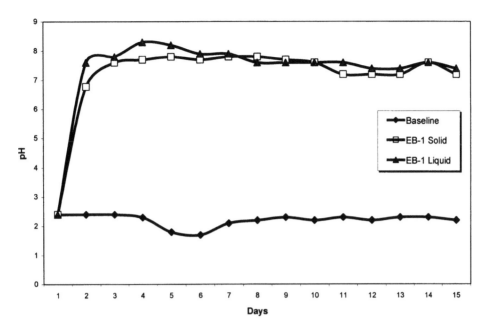

Figure 3: **Ponsardine ARD Study**
Envirobond ARD-1

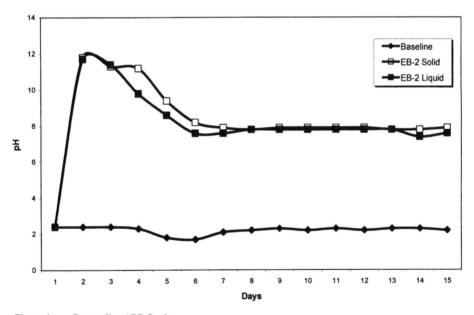

Figure 4: **Ponsardine ARD Study**
Envirobond ARD-2

Figures 3 and 4 show the Leadville treatability results for formulations ARD-1 and ARD-2. As shown, both liquid forms of Envirobond™ ARD formulations 1 and 2 increased the pH of the leachate from 2.2 to a pH of 7 and 12, respectively, in one day. After approximately one week, the pH of the leachate stabilized at the pH of the tap water flowing through the sample (pH 8).

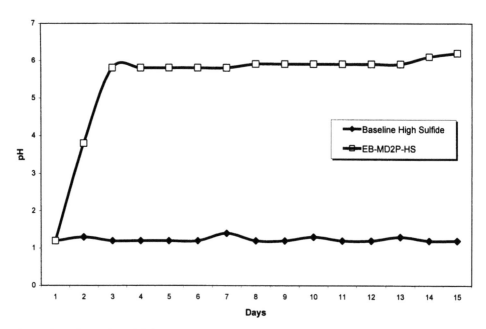

Figure 5: Ponsardine ARD Study
Envirobond ARD-2 (High Sulfide)

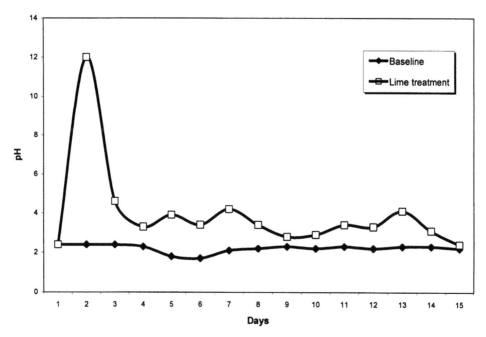

Figure 6: Ponsardine ARD Study
Baseline vs. Lime Treatment

A third sample with a very high sulfide content of approximately 40% was treated with Envirobond™ ARD in a similar fashion as samples ARD 1and 2. The initial pH of the leachate from this sample was 1.2 and after two days of treatment with Envirobond™ ARD achieved and maintained a pH of 6 (Figure 5).

434

Figure 4 shows an average Leadville ARD sample that was treated with lime. The purpose of this test was to show the long-term capability of lime to treat ARD generating rock. As shown, the lime increased the pH of the rock to 12 in one day but by the second day the pH had fallen to 4.5 before falling back to baseline pH. Obviously the lime was being consumed by the acid generated from the ARD rock and only offered a short-term solution to ARD remediation.

4.2 *Test Plots*

At the time that this paper is being prepared, RMRS is conducting tests on plots located on the top, flat portion of the Ponsardine mine dump. The purpose of these plots is to field test the optimized formulation of Envirobond™.

4.3 *Full Scale Deployment*

Following the evaluation of the data from the test plots, the optimized Envirobond™ ARD formula will be deployed on the entire surface of the Ponsardine mine dump. RMRS has constructed small catchment basins below the runoff points from the Ponsardine mine dump to directly sample the water quality of the runoff before it comes into contact with any other contaminant source. This sampling will be an excellent gauge of the effectiveness of Envirobond™ ARD in stopping the oxidation of pyrite on the Ponsardine mine dump.

Water quality samples will be collected from the catchment basins and from above and below the Ponsardine mine dump at the same location that the EPA has historically collected baseline water quality samples. Water quality samples will be collected in the fall and spring and at approximately the same time as the EPA baseline samples.

4.4 *Expected Results*

Based on the results of treatability tests conducted on sulfide rock samples from the Leadville and Summitville mining districts, it is expected that the pH of the leachate from the Leadville's treated dump rock and Summitville's open pit walls will increase and the available ferrous ions will slowly decrease.

5. CONCLUSIONS

The patents pending Envirobond™ ARD is a new generation approach to controlling acid and metal discharge from mine sites. Envirobond™ ARD can either use the available ferrous iron already in the oxidation system or it can liberate enough ferrous iron from pyrite to create a permanent, insoluble compound Fe(EB-ARD) that coats the pyrite and any other surface that it comes into contact with. The result is that when Envirobond™ ARD is applied to acid generating rock, the pH of the treated rock rises and the amount of ferrous iron decreases.

The use of Envirobond™ ARD will result in substantial savings in reclaiming projects with acid rock drainage. Because Envirobond™ ARD can be sprayed (or land farmed into) onto a mine dump to stop ARD generation, no expensive dirt work to remove or place the acid generating material in a repository will be required. Dirt work, lime treatment and repository construction are all expensive propositions when compared to an effective, in situ treatment of ARD generating rock with Envirobond™ ARD.

6. ENVIROBOND™ ARD APPLICATIONS

Because of the ability of Envirobond™ ARD to interrupt the pyrite oxidation cycle and stop the discharge of acid and metals to the environment, there are many applications that the mining sector can take advantage of. The following suggestions are a few of these applications:

Open Pit Highwall Treatment: Envirobond™ ARD can be sprayed onto open pit highwalls to stop acid water runoff.

Mine Waste Piles: Envirobond™ ARD can be sprayed onto or blended into ARD generating coal and hard rock mine dumps located throughout the United States to stop the discharge of acid and metals to ground and surface water.

Underground Mine Workings Treatment: Underground metal mine workings can be treated with Envirobond™ ARD to stop the oxidation of pyrite thereby reducing or eliminating acid and metals discharge from mine water pumped to the surface from the mine. Mines with water treatment plants would especially benefit from this application.

Mill Tailings: Mill tailings with pyrite can be treated with Envirobond™ ARD upon discharge from the mill to prevent the future discharge of ARD from the tailings impoundment.

Open Pit Backfill: Open pit mines backfilled with acid generating rock often results in acid-metal discharges to ground and surface water. Envirobond™ ARD can be applied to the surface of the backfill and allowed to migrate downward through the backfill eliminating ARD *in situ.*

Tailings and Mine Waste'00 © 2000 Balkema, Rotterdam, ISBN 90 5809 126 0

Treating metals contaminated sludges with Envirobond™

B. Littlepage
Leadville Mine Drainage Tunnel, US Bureau of Reclamation, Colo., USA

S. Schurman
Rocky Mountain Remediation Services, L.L.C., Golden, Colo., USA

D. Maloney
Kaiser-Hill, Rocky Flats Environmental Technology Site, Golden, Colo., USA

ABSTRACT: Rocky Mountain Remediation Services (RMRS) has developed the Envirobond™ chemical treatment system for hazardous metallic sludges. Metallic sludges are commonly generated by mine water treatment plants, as by-products of industrial and defense manufacturing, and in mine and industrial water evaporation ponds. The Environmental Protection Agency characterizes metallic sludges as hazardous unless they pass Toxicity Characteristic Leaching Procedure (TCLP) testing for the RCRA metals. Hazardous sludges are expensive to dispose of because of high transportation costs to properly licensed hazardous waste facilities and high disposal fees. In contrast, treated metallic sludges that pass the TCLP test for the RCRA metals can be disposed of as a non-hazardous, metal hydroxide sludge in any licensed clay lined landfill. When metallic sludges are treated with Envirobond™, they will pass the RCRA TCLP test, and consequently, may be more easily disposed.

1. INTRODUCTION

Rocky Mountain Remediation Services, L.L.C. (RMRS) has developed the Envirobond™ product to treat metals contaminated sludges from water treatment plants, evaporation ponds, and mining and milling operations. Typical metal contaminants in sludges include cadmium, lead, chromium, arsenic, aluminum and barium. The Envirobond™ treatment system will treat these metals to RCRA TCLP standards so they can be disposed of in less expensive, non-hazardous landfills. In contrast, sludges that do not pass RCRA TCLP leach tests must be disposed of in more expensive hazardous landfills.

2. HISTORICAL PERSPECTIVE

The development of sludge stabilization has undergone a number of steps over the years. Early remediation of sludges generally utilized cement or grout to stabilize the sludge. Unfortunately, this process substantially added to the weight and volume of the sludge and substantially increased disposal costs. Also, interfering species of metals and other compounds often inhibit the hardening or curing of the cement stabilizing media.

After cement, polymers came into vogue to stabilize sludge. With the use of polymers the process equipment is often both complicated and expensive to operate, waste stream volumes are increased, and the polymers generally encapsulate the metals which can lead to TCLP failure.

Vitrification, the process of stabilizing waste in a glassy matrix, was then developed to sequester sludge contaminants. Unfortunately, the vitrification process often creates secondary waste streams such as off-gases and slags that can be as expensive to treat as the original waste stream.

TABLE 1: Leadville Tunnel Sludge Hazardous vs. Non-Hazardous Disposal Costs*

20 Yards Hazardous Sludge	Annual Disposal Hazardous Sludge	20 Yards Non-Hazardous Sludge	Annual Disposal Non-Hazardous Sludge
$5,000/ Load	$250,000/Year	$1,000/Load	$50,000/Year

* These figures do not include the cost of transportation

The latest development in sludge treatment is the use of phosphate-based additives to stabilize the metals in contaminated sludges. There are many types of phosphate-based additives ranging from natural, apatite mineral based phosphates that are nominally effective with a limited range of metals to manmade phosphate products such as Envirobond™ that are extremely efficient in sequestering a wide range of metals.

3. METALLIC SLUDGE SOURCES AND PROBLEMS

Metallic sludge and sediment is present at 35% of all National Priority List (NPL) sites. Tremendous amounts of metallic sludges are also generated at mine water treatment plants, industrial evaporation ponds, and in municipal sludges that have too high a metal content for bio-solids land application. The problems with metals contaminated sludges are:
- Volume – a typical mine water treatment plant generates 1,000 yd^3 to 1,500 yd^3 of sludge per year;
- Cost of transportation – metals contaminated sludges must be disposed of in specially permitted, hazardous landfills that are often a great distance from the treatment plant; and,
- Cost of disposal – metallic sludges are generally classified as hazardous waste and can cost upward from $300/yd^3 to dispose.
Table 1 shows the actual cost comparison of hazardous vs. non-hazardous disposal costs for the Bureau of Reclamation's Leadville Tunnel Water Treatment Plant located in central Colorado.

4. ENVIROBOND™ SLUDGE TREATMENT PROCESS

Envirobond™ sludge treatment is an easy to use process that utilizes non-hazardous, phosphate based chemicals. Metallic sludge treatment with Envirobond™ is generally a one-step addition of a benign chemical that easily mixes with the waste. In a water treatment plant the Envirobond™ metals treatment formulation can be added to the sludge treatment process prior to the contaminated water entering the reactor vessel, into the low-density sludge holding tank or added directly to the high-density sludge that comes out of the filter press.

The addition of Envirobond™ to each of these points in a water treatment plant has its own advantages. When Envirobond™ is added to untreated water before the reactor vessel it will aid in the flocculation of difficult metals such as manganese and aluminum. When Envirobond™ is added to the low-density (5% solids) sludge holding tank, the metallic sludge that ultimately emerges from the sludge filter press will pass TCLP analysis. Finally, when Envirobond™ is blended with high-density sludge from the filter press the metals in the sludge will be stabilized and will pass TCLP analysis.

RMRS has also used Envirobond™ to successfully stabilize sludges from radioactive and heavy metal contaminated evaporation ponds at the Rocky Flats Environmental Technology site as wall as calcined radioactive fines derived from historic uranium processing operations at the Fluor Daniel Fernald DOE site. RMRS has also stabilized high selenium content sludges derived from glass manufacturing. Table 2 shows the results of treating various metals in sludges with Envirobond™.

438

5. ENVIROBOND™ PROCESS CHEMISTRY

The Envirobond™ metals treatment process utilizes a patented phosphate chain that sequesters a wide range of metal ions (including radionuclides) to form insoluble phosphate metal complexes. As shown in Figure 1, metal ions bond to the oxygen sites on the phosphate ligand. However, the unique Envirobond™ phosphate formulation allows the phosphate chain to freely rotate around the P–O–P bond allowing metal ions to be linked to more than one site on the phosphate chain. Envirobond's multiple site bonding capability results in metal phosphate complexes that have extremely low Ksp (solubility product) values and the ability to capture metals with 2^+ and 3^+ valences.

TABLE 2: Envirobond™ sludge treatability results for select metals

Metal	Pre-Treatment TCLP (ppm)	Post-Treatment TCLP (ppm)	Regulatory Levels	
			RCRA	UTS
Arsenic	17	0.13	5.0	5.0
Barium	34,083	0.03	100	21
Cadmium	147	0.41	1.0	0.11
Chromium	61	< 0.10	5.0	0.65
Lead	680	0.15	5.0	0.75
Mercury	500	0.07	0.2	0.025
Selenium	190	0.89	1	5.7
Vanadium	1.7	< 0.50	NS	1.6
Zinc	108	2.00	NS	4.3

NS = No Standard

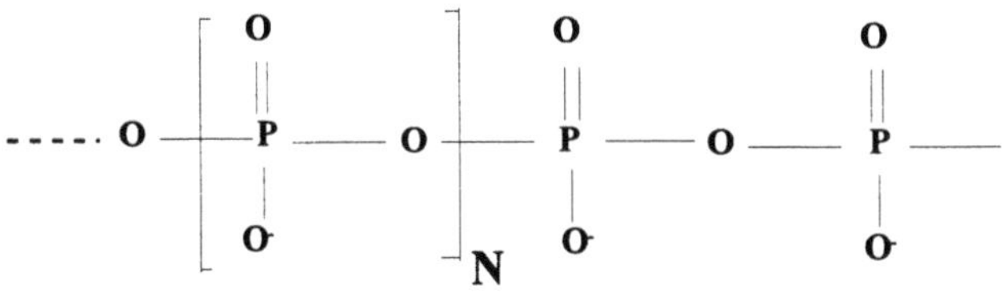

N = multiple units of phosphate structure

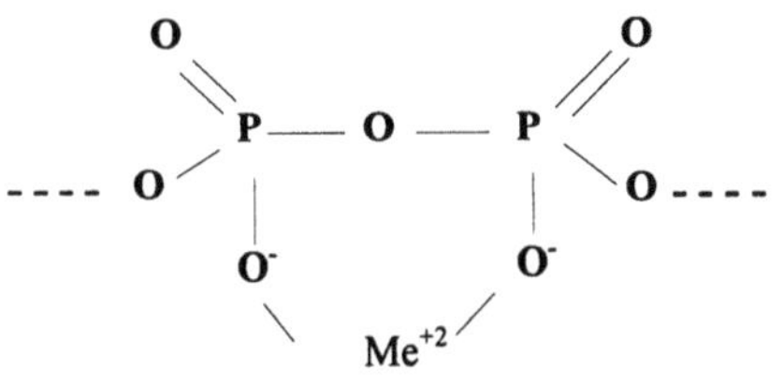

The 2^+ metal ion binds to the two O⁻ sites in this example of Envirobond™ metals chelation

FIGURE 1: Envirobond™ process chemistry

Tailings and Mine Waste'00 © 2000 Balkema, Rotterdam, ISBN 90 5809 126 0

Hyperspectral mapping of abandoned mine lands, drainage, and watershed impacts in Utah

F. B. Henderson III – *HENDCO Services, Nathrop, Colo., USA*

A. Selle & K. Wangerud – *US Environmental Protection Agency Region 8, Denver, Colo., USA*

W. Farrand – *Farr View Consulting, Westminster, Colo., USA*

P. L. Hauff – *Spectral International Incorporated, Arvada, Colo., USA*

D. C. Peters – *Peters Geosciences, Golden, Colo., USA*

R. Stewart – *Earth Search Science, Incorporated, McCall, Idaho, USA*

EXTENDED ABSTRACT: In 1997, U.S. Environmental Protection Agency (EPA) Region 8 initiated an interagency study on five Utah mining areas which is utilizing the Jet Propulsion Laboratory (JPL) AVIRIS (Advanced Visible-InfraRed Imaging Spectrometer) instrument. The mining areas included in the study are Park City-Alta (northeastern Utah), the Oquirrh Mountains (central Utah), the Tushar Mountains (south-central Utah), Leeds-Silver Reef (southwestern Utah), and the Tintic and East Tintic districts (central Utah). The purpose of the study is to determine how well AVIRIS hyperspectral data can identify and map acid-producing and potentially toxic mine and mill wastes and downstream watershed impacts, including effects on biota, riparian vegetation, and other ecosystem health indicators. The U.S. Geological Survey (USGS) Spectral Laboratory in Denver is performing data processing and information extraction on the AVIRIS data, including application of its "Tetracorder" algorithm for mineral and vegetation identification and classification of the hyperspectral data.

In addition, the commercial remote sensing community was challenged by EPA Region 8 to demonstrate acquisition of hyperspectral data (using commercial sensors) and data processing (of commercial and/or AVIRIS hyperspectral data) that can provide useful information on a cost-effective and operational basis. A challenge for the commercial participants is to do this on a pro bono basis. Earth Search Science's Probe I and G.A. Borstad Associates' CASI (Compact Airborne Spectrographic Imager) and SFSI (SWIR (Short-Wave InfraRed) Full Spectrum Imager) hyperspectral systems were flown over three mining districts covered by AVIRIS. Ten additional companies are processing the EPA-provided AVIRIS data and selected commercial data in an effort to demonstrate their unique data processing techniques. In spite of industry worries about this advance in the government's ability and potential to identify, target, and discover mining-related pollution and possible noncompliance with environmental regulations, seven mining and engineering companies participated in ground validation and interpretation of the data.

The Tintic and East Tintic mining districts were chosen by the commercial participants as common calibration, field data collection, and image processing areas to allow direct comparison between their different technologies and results. In particular, the Dragon Pit, an open-pit clay mine in the southern part of the Tintic District, was chosen as a site for detailed ground characterization and comparison among commercial participants and with the USGS.

A focus of this government and commercial collaborative study is to further validate hyperspectral data for identifying and mapping potentially acid-producing and toxic (containing heavy metals) minerals and materials in mine and mill wastes. Another focus is to evaluate what the data can identify with respect to impacts on watersheds, such as stressed vegetation, erosion-prone soils and wastes, transport and accumulation of stream sediments, impacted biota and habitats, and any other ecosystem health indicators.

This collaboration is helping to set common procedures for hyperspectral information extraction, validation and acceptance by both the regulatory and mining communities for better public and private environmental management decision making while minimizing litigation and discovery liability. Preliminary results of the USGS and commercial efforts were presented in several papers at the 1999 JPL AVIRIS Workshop in Pasadena, California (Green 1999) and in a recent paper by Hauff et al. (1999). Final results of the government and commercial studies will be presented at an EPA-sponsored conference on the project in Salt Lake City in the first half of 2000.

REFERENCES

Green, R.O. (ed.) 1999. *Summaries of the Eighth JPL Airborne Earth Science Workshop, February 9-11, 1999*: JPL Publication 99-17.

Hauff, P.L., D.C. Peters, G.A. Borstad, W. Peppin, N. Lindsay, L. Costick, R. Neville, & R. Glanzman 1999. Hyperspectral evaluation of mine waste and abandoned mine lands—NASA and EPA sponsored projects in Utah and Idaho, In *Proc. Int. Symp. on Spectral Sensing Res., 31 Oct.-4 Nov. 1999:* In Press.

Tailings and Mine Waste'00 © 2000 Balkema, Rotterdam, ISBN 90 5809 126 0

Increased accuracy in suction measurements using an improved thermal conductivity sensor

D.G. Fredlund, Fangsheng Shuai & Man Feng
Department of Civil Engineering, University of Saskatchewan, Sask., Canada

ABSTRACT: Suction measurements play an important role in mine reclamation and environmental protection. The thermal conductivity matric suction sensor has been proven to hold the great promise for the *in situ* measurement of soil suction. There were some limitations associated with previously developed versions of these sensors. An improved thermal conductivity sensor has been developed at the University of Saskatchewan, Saskatoon, Canada. The accuracy of the new sensor has been increased through use of a specially designed ceramic, enhanced electronic and improved interpretation technique for the data.

INTRODUCTION

Suction measurements play an important role in mine reclamation and environmental protection. Frequently, the most economical and feasible reclamation option is to use a soil cover over the waste rock and tailings. Evaluation of the performance of cover systems can be ensured through *in situ* measurements of soil suction. The rate of water movement and degree of storage of the cover are related to the suction in the soil.

One of the more common methods used for continuous soil suction measurement involves the use of thermal conductivity matric suction sensors. Comparing with other suction measurement systems (e.g., filter paper, TDR and psychrometer), thermal conductivity sensors produce a reasonably reliable measurement of soil suction over a relatively wide range and are essentially unaffected by the salt content of the soil (Lee and Fredlund, 1984 and Fredlund and Wong, 1989). These sensors also have the advantage of versatility and ability to be connected to a data acquisition system for automated recording. However, some difficulties have been encountered with previously developed thermal conductivity sensors. These difficulties relate to poor durability, low accuracy and reliability for geotechnical and environment monitoring.

In order to obtain a relative inexpensive and reliable sensor to measure soil suction, an improved thermal conductivity soil suction sensor has been developed. In this paper, the special design considerations associated with the new thermal conductivity sensor are described. The factors that may influence the accuracy of the suction measurement, such as heating voltage variation, stability of the output signal, interpretation of the data and the hysteresis properties of ceramic, are discussed. The techniques developed to eliminate each of these negative influences are described. Some soil suctions measured using the new thermal conductivity sensor are also presented.

AN IMPROVED THERMAL CONDUCTIVITY SENSOR

Thermal conductivity sensors have been used to measure soil suction for long time (Shaw and Baver, 1939 and Johnston, 1942). However, the use of early versions of these sensors have been experienced numerous difficulties. The difficulties with the use of thermal conductivity sensors has ranged from problem associated with: i.) ceramic tips that were soft, friable and

Figure 1 Thermal conductivity sensor developed at the University of Saskatchewan.

liable to crumble and crack during calibration or installation ii.) accuracy that was poor in the high suction range (i.e., greater than 200 kPa), and iii.) stability of the electronic signal from the sensor. In addition, the influence of factors such as heating voltage variation and the hysteresis upon wetting and drying of the ceramic, on the accuracy of the suction measurement needed to be studied. In order to resolve the difficulties associated with using thermal conductivity sensors, an improved thermal conductivity sensor was developed in the University of Saskatchewan, Saskatoon, Canada (Fig. 1).

The sensor developed at the University of Saskatchewan mainly consists of a specially designed ceramic tip, an integrated circuitry and a heating element (Fig. 2). The strength and durability for the new ceramic tip has been significantly improved over that of previous sensors. The compressive strength of the ceramic has been increased to approximately 2100 kPa. The tensile strength is about 600 kPa. The stronger sensor has a positive impact with respect to the prevention of cracking and crumbling during installation.

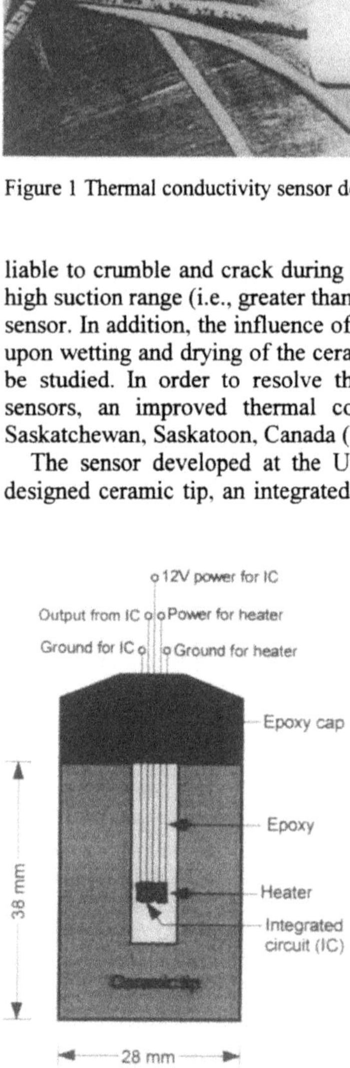

The new ceramic tip also has a high porosity and a wide range of pore sizes ranging from 0.05 mm to less than 0.005 mm. As a result, the soil-water characteristic curve for the new ceramic tip has been significantly improved ensuring a increased accuracy for suction measurements up to 1500 kPa.

With respect to the electronics, the quality of the output signal from the new sensor has been improved by using superior integrated circuitry for the temperature sensing device. Another advantage of using integrated circuitry for the temperature sensing device is that the

Figure 2 The physical layout of the thermal conductivity sensor developed at the University of Saskatchewan.

444

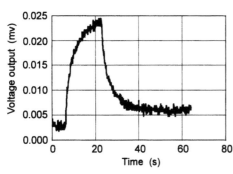 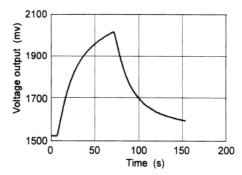

A. Measured heating curve before improvement   B. Measured heating curve after improvement

Figure 3 The heating curves measured before and after signal conditioning.

sensor can be used to measure the soil temperature as well as the soil suction. In the other words, the sensor is not only a soil suction sensor but also a soil temperature sensor. Soil suction can be measured as long as the temperature is above the freezing point of water while temperature can be measured over a wide range.

The advanced signal conditioning technique was also used to enhance the stability of the output signal from the sensor. The technique includes amplification, isolation and filtering. As a result, the stability of the output signal was significantly improved. Figure 3 shows the heating curves measured before and after signal conditioning.

With the above design considerations, the durability and accuracy of the new sensors have been significantly improved over that of earlier versions. The new thermal conductivity sensor has been found to be quite sensitive and accurate in measuring soil suction in the range from 5 to 1500 kPa with a coefficient of variation less than ±5%. The main technical specifications for the new sensor are listed in Table 1.

Table 4.1 Technical specification

Measurement parameters	Soil suction
	Soil temperature
Measurement range	Soil suction 5 to 1500 kPa
	Temperature -40 °C to 110 °C
Accuracy	Less than ± 5% for suction measurement
	± 0.5 °C for temperature measurement
Resolution (using CR10X)	0.33 mV
Soil types	Suitable for all soil types
Protection	Suitable for long-term burial
Temperature	0 to 40 °C for suction measurement (no
	damage when used in frozen soils, but
	suction reading will be incorrect)
Power supply	12V ~ 15V DC, 250 mA
Size	Diameter: 28 mm, Length: 38 mm
Cable length	Standard: 8m, Maximum: 100m

FACTORS WHICH INFLUENCE THE ACCURACY OF THE SUCTION MEASUREMENT

Some factors that may influence the accuracy of the suction measurement using a thermal conductivity sensor were investigated. These factors are the heating voltage variation, the interpretation of the data and the hysteresis of the ceramic.

445

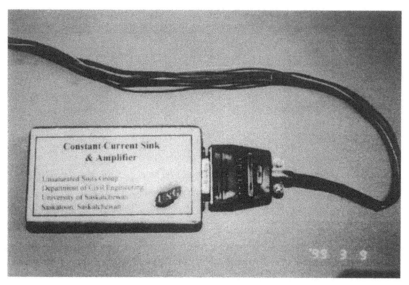

Figure 4 The Constant Current Sink & Amplifier device designed to be used with the data acquisition system.

The heating voltage variation

A precise controlled heating voltage across the heating element is mandatory if readings are to be reproducible. There are many factors that influence the heating voltage across the heating element, such as, cable length, environmental temperature and voltage vibration in the power source, to name only a few. In order to eliminate these influences, a constant current sink was designed and manufactured.

The device is able to maintain a constant current of 200 mA through the heater resistor. The constant current compensates for differences in resistance when different lengths of extension wires are used. It also compensates for small changes in the heating resistance caused by the change in environmental temperature. This constant current sink, along with the amplifier mentioned previously, were installed on the same PC board and called a constant current sink and amplifier. According to the test results, the constant current sink and amplifier is able to function under freezing condition (i.e., -40°C). A picture of this device is shown in Fig. 4.

Interpretation of the data

The calibration curve obtained from the calibration process is non-linear (Fig. 5). In order to use the calibration curve to calculate the soil suction from the output voltage of the sensor, it is important to have a reasonably accurate characterization of the calibration curve. The following equation is proposed to fit the relationship between output voltage, ΔV, and the soil suction, ψ.

$$\Delta V = \frac{ab + c\psi^d}{b + \psi^d}$$

[1]

where:

a = parameter designating the output voltage at saturated condition
c = parameter designating the output voltage under a total dry condition
d = parameter designating the slope of the calibration curve
b = parameter related to the inflection point on the calibration curve

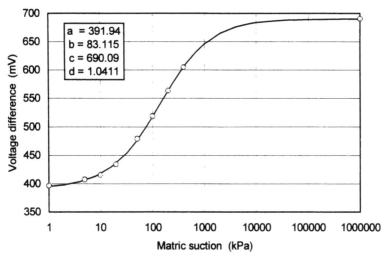

Figure 5 Typical calibration curve for a thermal conductivity sensor.

A typical calibration curve for a sensor is shown in Fig 5 along with its parameter values. Eq. 1 accurately fits the calibration data over the entire suction range. Since there are only four parameters in Eq. 1, only five calibration points are required to establish the calibration curve. As a result, the calibration process is simplified and the time required for calibration is significantly reduced. As well, using calibration Eq. 1 facilitates the calculation of the soil suction from the output voltage of the sensor and increases the accuracy of the suction measurement.

The hysteresis of the ceramic

The water content versus matric suction curves for any porous material during wetting and drying are generally not the same. The hysteresis in the soil-water characteristics of the ceramic may cause hysteresis in the sensor response upon wetting and drying. To-date, little research has been done on the hysteresis associated with the ceramic.

Figure 6 shows measured wetting and drying curves for the new ceramic. Some hysteresis is observed in the soil-water characteristic curve of the ceramic, along with some initial saturation condition. When the suction is lower than 20 kPa, there can be a significant hysteresis during the first drying and wetting cycle. The hysteresis becomes less significant during the second wetting and drying cycle. The reason for this may be that the saturation condition was not the same during the first and second drying and wetting cycle.

At beginning of the test, the sensor was saturated under a vacuum of 80 kPa to ensure 100% degree of saturation. However, at the end of the first drying and wetting cycle, the sensor was saturated under zero matric suction. Since no vacuum or back pressure was applied to the sensor, it is impossible to remove all of the small air bubble isolated inside the ceramic. Therefore, the water content of the ceramic at beginning of the second drying and wetting cycle is less than the water content at beginning of the first drying and wetting cycle. Further test results indicate that, using the sensor saturated under 0 kPa matric suction gives a hysteresis at the first drying and wetting cycle that is almost the same as that of the second drying and wetting cycle. This result indicates that, in order to accurately measure soil suction, the sensor should be saturated under 0 kPa matric suction instead of being saturated under a vacuum.

The hysteresis of the ceramic became less significant and quite reproducible when the measured matric suction is greater than 20 kPa. The research conducted by Feng (1999) indicated that the maximum possible relative error in the suction measurement caused by the hysteresis is about 30%. In the other words, the hysteresis of the ceramic should be considered during the measurement of soil suction. The suctions measured with or without consideration of

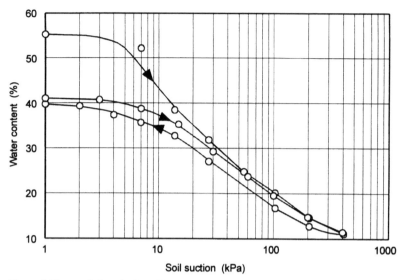

Figure 6 Hysteresis loop in the soil-water characteristic curve of the ceramic.

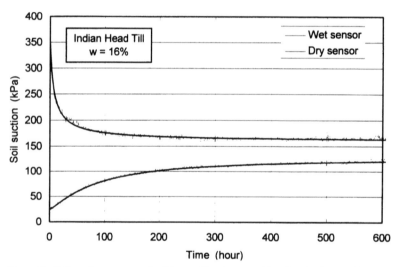

Figure 7 Soil suctions measured without taking into consideration the hysteresis of the ceramic

hysteresis in the ceramic are shown in Figs. 7 and 8. The suctions were measured by placing sensors with differing initial water contents into a soil specimen. A gap of 40 kPa was observed between the suctions measured with an initially dry sensor and with an initially wet sensor. The gap was eliminated by taking the hysteresis in the ceramic into consideration (Fig. 8).

The scanning curves between the drying and wetting curves were also investigated and are shown in Fig. 9. In order to improve the interpretation of the sensor data, a hysteresis model was proposed (Feng, 1999). This model can be used to modify data obtained in engineering practice according to the wetting or drying history of the sensor. As a result, it is possible to obtain greater accuracy in the assessment of suction. A comparison between the actual suction change and the predicted suction change using the hysteresis model is shown in Fig. 10. Good agreement was noticed between the predicted suction changes and actual suction changes.

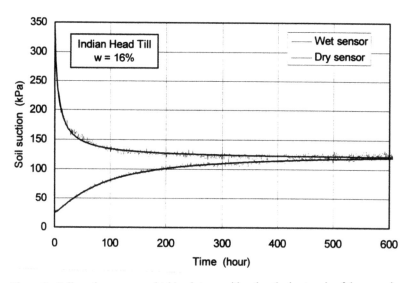

Figure 8   Soil suctions measured taking into consideration the hysteresis of the ceramic

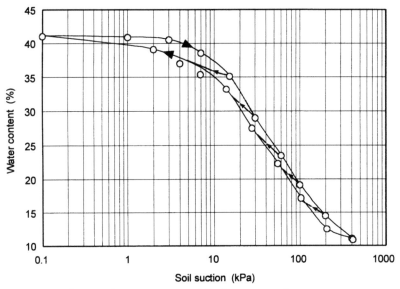

Figure 9   Wetting scanning curves measured for the new ceramic.

CONCLUSIONS

Currently available thermal conductivity matric suction sensors show promise for the *in situ* measurement of soil suction. An improved thermal conductivity soil suction sensor was developed to eliminate limitations associated with earlier sensors.

The durability and accuracy for the new thermal conductivity sensors have been significantly improved. The new thermal conductivity sensor has been found to be quite sensitive and accurate in measuring soil suction in the range from 5 to 1500 kPa with a coefficient of variation less than ±5% over the range.

A constant current sink was developed to ensure a constant heating voltage. The influence of the cable length on soil suction measurements was eliminated through the use of this constant current device.

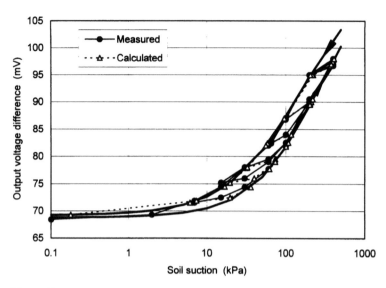

Figure 10   A comparison between the applied suction and the predicted suction using the hysteresis model.

A calibration equation was proposed to facilitate in the calculation of soil suction versus output voltage. There is an increase in the accuracy of soil suction measurements. It was found that the equation accurately fits calibration data over the entire suction range.

The hysteresis of the ceramic has an influence on soil suction measurement. A hysteresis model was proposed and can be used to interpret sensor data collected according to the drying or wetting history of the sensor.

ACKNOWLEDGMENTS

The authors wish to acknowledge the financial support from the Natural Sciences and Engineering Research Council of Canada (NSERC) and other seven organizations, Placer Dome Canada Ltd., Placer Dome Inc., Westmin Resources Ltd., Cominco Ltd., Cambior Inc., Rio Algom Ltd., and Saskatchewan Highways and Transportation.

REFERENTS

Feng, M. 1999. The effects of capillary hysteresis on the measurement of matric suction using thermal conductivity sensors, *M.Sc. thesis*, University of Saskatchewan, Saskatoon, Saskatchewan, Canada. 168p.

Fredlund, D.G. and Wong, D.K.H. 1989. Calibration of thermal conductivity sensors for measuring soil suction, *ASTM Geotechnical Testing Journal*, vol. 12, no. 3, pp. 188-194.

Johnston, L.N. 1942. Water permeable thermal radiators as indicators of field capacity and permanent wilting percentage in soils, Soil Sci., vol. 54, pp. 123-126.

Lee, R.K.C. and Fredlund, D.G. 1984. Measurement of soil suction using the MCS 6000 gauge, *in* Proc. 5th Int. Conf. Expansive Soils, Inst. of Eng., Adelaide, Australia, pp. 50-54.

Shaw, B. and Baver, L.D. 1939. An electrothermal method for following moisture changes of the soil insitu, Proc. Soil Sci. Soc. Amer., vol. 4, pp. 78-83.

Tailings and Mine Waste'00 © 2000 Balkema, Rotterdam, ISBN 90 5809 126 0

Tailings management using TDR technology

K. M. O'Connor
GeoTDR Incorporated, Apple Valley, Minn., USA

D. A. Poulter
The Glasgow Engineering Group, Littleton, Colo., USA

D. Znidarcic
Department of Civil, Environmental and Architectural Engineering, University of Colorado, Boulder, Colo., USA

ABSTRACT: This paper focuses on applications of TDR technology in the management of tailings. Case study summaries are presented which demonstrate applications of TDR for various aspects of tailings management including remote continuous monitoring of cover performance, ground water changes, contaminant transport, and slope movement. Recent studies have shown this technology may have the potential to provide a reliable means for real time monitoring of tailings consolidation and the impact on long term storage capacity. Remote monitoring is particularly applicable to post-closure monitoring where savings can be made in personnel costs and alarm systems can be implemented.

1 INTRODUCTION

Time Domain Reflectometry (TDR) was developed by the power and telecommunications industries to locate faults in cables. A cable tester launches a voltage pulse into a coaxial cable, parallel pair wire or twisted pair wire. Wherever there is a change in electrical properties, due to cable damage or water ingress, a portion of the voltage is reflected back to the tester which displays the ratio of reflected to transmitted voltage as a reflection coefficient. The waveform shape is a function of the type and magnitude of cable damage. The travel time is converted to distance by knowing the propagation velocity which is a property of the cable or wire. Consequently, it is possible to display all reflections and identify the type and location of cable damage.

2 MEASUREMENT OF VOLUMETRIC WATER CONTENT

Time Domain Reflectometry (TDR) was developed by the power and telecommunications industries to locate faults in cables. A cable tester launches a voltage pulse into a coaxial cable, parallel pair wire or twisted pair wire. If a probe consisting of two or more parallel rods is embedded in a porous medium and a voltage pulse is launched along this probe, a reflection is created at the top of the probe and a second reflection is created at the end of the probe so the travel time can be measured. The particular probe shown in Figures 1 and 2 is segmented such that reflections are created at the top and bottom of each segment (ESI, 1997). Since the segment lengths (L_p) are fixed, the pulse velocity along each segment can be computed as twice its length divided by the time (t) required for a pulse to travel along the segment and back. Typically, this velocity is normalized with respect to the speed of light, c (= 3×10^8 m/s), and expressed as a dimensionless propagation velocity,

$$V_p = 2L_p / ct .$$ (1)

Figure 1. Measuring water content of tailings using
segmented TDR probe

This ratio of velocities is approximately equivalent to the dielectric constant of the medium in which
the probe is embedded. Since the dielectric constant of water is about 81 while that of mineral soil
grains is 3-5, the measured dielectric constant is predominately a function of volumetric water
content,

$$\theta_v = V_w/V_t \tag{2}$$

where V_w is the volume of water and V_t is the total volume. The relationship between volumetric
water content and propagation velocity is linear (Herkelrath et al, 1991),

$$\theta_v = b\,(\,1/V_p\,) - a \tag{3}$$

as shown in Figure 3 for clay loam and nickel mine tailings. The slope, b, and offset, c, are functions
of the particular porous material, and they are determined by conducting laboratory calibrations
(Topp et al, 1994; Zegelin and White, 1994).

2.1 Performance of landfill covers

By installing an array of probes at various depths it is possible to monitor progression of a wetting
front through the vadose zone. This technique is being used to monitor the performance of landfill
covers by tracking changes in volumetric water content as water seeps through prototype landfill
covers which are being evaluated as viable alternatives to the designs required by the Federal
Resource Conservation and Recovery Act Subtitle D (Benson et al, 1994; Nyhan et al, 1994). The
segmented probes are installed in arrays of eight probes, four of which extend down to a depth of
1.2 m and four are buried extending from 0.6 to 1.8 m. All probes are connected to a remote data
acquisition system (ESI, 1997; Boehm and Scherbert, 1997). The system is currently programmed
to read the soil moisture probes once per hour. The internal data storage unit is downloaded
approximately weekly onto diskettes via a portable computer. The field data collection effort will
indicate a limiting depth for significant/measurable moisture fluctuations and thus provide real
evidence that percolation through the final cover is extremely low. Computer modeling is used to
calculculate the actual percolation and thus provide a means for extrapolating soil moisture behavior
over a 30-year post closure period.

2.2 Performance of mine waste covers

This same approach is being at a mine in the western U.S. There has been a problem with
performance of the clay cap placed over a stockpile of low grade ore. During closure of the site,

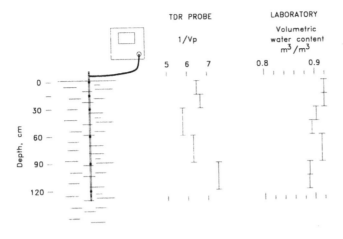

Figure 2. Schematic of segmented probe and depth profile obtained in phosphatic clay tailings.

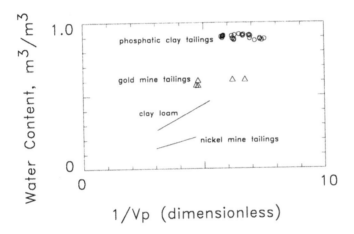

Figure 3. Relationship between volumetric water content and propagation velocity for a variety of materials; phosphatic clay and gold mine tailings from present studies; clay loam (Hook and Livingston, 1996); nickel tailings (Sun, 1996).

the low grade ore was isolated in a clay lined area then covered with a clay cap and overburden/growth media material. The clay cap over the face of the stockpile was poorly compacted which resulted in excess infiltration. Water infiltrated and collected in the base of the low grade ore stockpile which caused seepage and stability problems at the toe of the stockpile. During field exploration, it was found that moisture content of the clay cap and underlying stockpile material along the slope was greater than at the top of the stockpile. Therefore, repair of the clay cap was limited to the slope area.

The slope material was recompacted and the clay cap reconstructed under controlled conditions. Upon reconstruction of the stockpile and clay cap, moisture contents were taken through the clay cap over the slope and surface using the TDR profiling probe shown in Figures 1 and 2 (ESI, 1997). Following the Spring melt and runoff, a second set of readings will be taken to assess the occurrence of infiltration. If relatively high moisture contents are measured in the slope area, other remedial measures will be developed to prevent future infiltration.

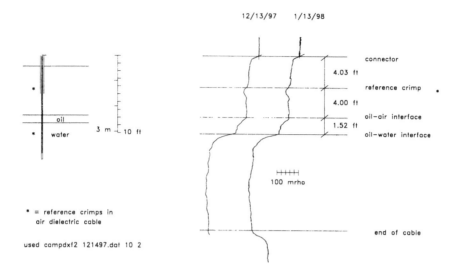

Figure 4. Waveforms acquired from air dielectric cable installed in monitoring well. It is possible to monitor changes in the water level as well as monitor changes in thickness of the oil resting on the ground water.

The use of TDR in this project allowed for quick and easy collection of field measurements at several locations over a remote area. The use of the 1.2 m long probe made it possible to obtain average moisture content values over continuous intervals which provided an assessment of the moisture profile. The hole created by the probe was backfilled with granular bentonite to protect the integrity of the clay liner.

3   GROUNDWATER AND CONTAMINANT MONITORING

A parallel pair wire, twisted pair wire, or air-dielectric coaxial cable can be placed in a well to monitor changes in water level as shown on the left side of Figure 4. When a voltage pulse is transmitted along the cable or wire, a large magnitude negative reflection occurs at the air-water interface (Dowding, Huang and McComb, 1996; Nicholson et al, 1997), and a large magnitude positive reflection occurs at the end of the cable or wire.

The waveforms shown in Figure 4 were obtained by monitoring an air-dielectric cable installed in a well at a site at which there had been an oil pipeline break. There are residual pools of oil resting on the ground water, and a TDR reflection is created at both the air-oil and oil-water interfaces. The time history of depth to oil and water surfaces are shown in Figure 5. This data is being evaluated to interpret changes on the oil pool thickness and to monitor the effectiveness of remediation as the oil is removed by pumping.

The TDR reflection at the end of the cable can be attenuated depending on the fluid conductivity. The reflected voltage magnitude at the bottom of the cable will decease if conductivity of the ground water increases (Dalton et al, 1984). It is this remarkable capability of TDR to measure both apparent permittivity (dielectric constant) and conductivity of fluids and porous media that makes it particularly suitable for monitoring contaminant transport. This technology has been used to monitor transport of sodium chloride, calcium chloride, kerosene, and tetrachloroethylene (Kachanoski and Ward, 1994; Redman and D^{c}Ryck, 1994). It has also been used to monitor heavy metal concentrations in mine tailings (Norland, 1994).

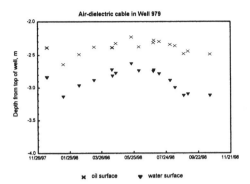

Figure 5. Time history of depth to top of oil and depth to top of water; note the seasonal change in oil pool thickness.

3.1 *Leak detection*

Coaxial cable is manufactured which includes a central electrical conductor, annular spacer of electrically insulating material around the central conductor and an outer sheath (PermAlert, 1995). The annular spacer is formed of a braid of electrically conductive wires which are coated with corrosion resistant material for protection against water, acids, alkalis, solvents or other liquids or environmental contaminants that may be present in the vicinity. By virtue of being braided, the annual spacer has void spaces into which fluid can enter. The cable is interrogated using TDR and reflections are created wherever fluids have penetrated the annular spacer. The outer sheath may have an uninsulated wire incorporated which makes it possible to detect conductive liquids. Consequently a leak detection system utilizing this cable can distinguish between leaks of conductive and non-conductive liquids.

4 SLOPE MONITORING

Inclinometers are the most common method used to locate failure zones in slopes and quantify the rate of movement. Similarly, a metallic coaxial cable can be placed in a drillhole and anchored to the surrounding soil or rock by tremie placement of an expansive cement grout. When movement along the failure zone is sufficient to fracture the grout, cable deformation occurs. If the cable is subjected to localized shearing the TDR reflection is a characteristic spike such as those shown in Figure 6. Cable faults can be located with an accuracy of ±2% of the distance from cable tester to fault. This accuracy can be improved by crimping a cable at some desired spacing. The reflections flagged with asterisks in Figure 6 are caused by crimps at known physical locations. These reflections provide control markers in the TDR records. Not only is it possible to locate movement but also to distinguish shearing from tensile deformation and to quantify the magnitude of deformation (Dowding et al, 1988; O'Connor and Dowding, 1999).

4.1 *Tailings embankments and highwalls*

TDR has been used to monitor movements in highwalls and tailings embankments. The example shown in Figure 6 involved installation of coaxial cable in the highwall slope of an oil sands mine (O'Connor et al, 1995). Slope failure is an operational problem at this mine and many kilometers of inclinometer casing have been installed to continuously monitor slope stability in the vicinity of the multimillion dollar dragline. Given the large commitment to man-hours and hardware

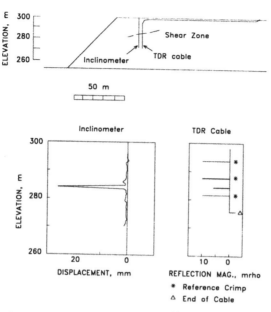

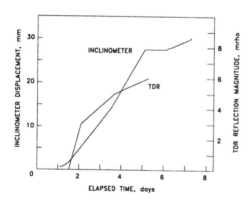

Figure 6. Monitoring highwall movement; *top*, location in slope; *bottom*, comparison of inclinometer profile and TDR reflections along coaxial cable. Note movement at a depth of 18 m.

Figure 7. Time history of inclinometer displacement and TDR reflection magnitude at depth of 18 m.

required for monitoring, the mining company investigated the potential of TDR for remote monitoring. At three locations in the mine, a coaxial cable was grouted into a hole located 30 m from an inclinometer and the results from one comparison are shown in Figure 5. Note the reflection which developed at a depth of 18 m is consistent with the shear zone location and inclinometer displacement profile. When the TDR reflection magnitude time history is compared with the inclinometer displacement time history as shown in Figure 7, it can be seen that they are consistent.

5 CURRENT DEVELOPMENTS - DIRECT MEASUREMENT OF CONSOLIDATION

As tailings consolidate, there is an increase in the volume and weight of solids,

456

$$\Delta W_s = \gamma_w \, G_s \, \Delta V_s \qquad (4)$$

where G_s is the specific gravity of the solids. If the tailings are saturated, and there is no change in the total volume, an increase in the solids volume is equal to a decrease in the volume of displaced water, $\Delta V_s = - \Delta V_w$, so

$$\Delta W_s = -\gamma_w \, G_s \, \Delta V_w . \qquad (5)$$

Combining with eqtn [2],

$$\Delta W_s = -\gamma_w \, G_s \, V_t \, \Delta \theta_v \qquad (6)$$

which states that as the tailings consolidate and solids displace water there is a decrease in the volumetric water content. This is measured using TDR technology as discussed earlier.

5.1 *Preliminary field trials*

Field trials have been performed in the beach area of a gold mine tailings impoundment in California and a phosphatic clay tailings impoundment in Florida using the segmented probe shown in Figures 1 and 2. This probe was simply pushed into the tailings and measurements made using the push-button TDR unit. For these initial trials, only near-surface profiles in the beach area along the embankment were obtained. Direct samples were also collected for laboratory determination of water content. The profiles in Figure 2 show that the TDR measurements reflect changes in moisture content with depth as did the oven-dried samples.

The compilation in Figure 3 illustrates that the relationship between water content and TDR travel time is materials-specific which is consistent with the experience of all users (Topp et al, 1994; Zegelin and White, 1994; Hook and Livingston, 1996). This is a limitation of all techniques used for in situ measurement of water content in clays and organic soils. Among the factors identified by researchers have been the influence of high specific surface, bound water, and the frequency content of the voltage pulse as it travels along the probe (O'Connor and Dowding, 1999)

Use of TDR in loam, sand, and gravel can produce accuracies of $\pm 0.03 \text{ m}^3/\text{m}^3$ and this can be improved with material-specific calibration. The field trials have highlighted the need for establishing a rigorous calibration protocol for various tailings to assure that the collected data is reliable. As current studies continue, it will be possible to determine the accuracy required for monitoring consolidation.

5.2 *Use with existing consolidation model*

A reliable testing technique to evaluate the highly nonlinear constitutive relations of soft, cohesive soils has been developed (Znidarcic et al, 1992; Abu-Hejleh et al, 1995). The technique is an enhanced seepage induced consolidation test which eliminates most of the limitations of the previously existing methods and the theory has been implemented in the CONDES finite element program (Yao and Znidarcic, 1997). However, topographic surveys and indirect measurements are still required for field validation, and one of the main shortcomings of this approach to tailings impoundment modeling is the reliance on laboratory testing of slurry samples to obtain consolidation characteristics for the analysis. While the developed laboratory testing methods are reliable they can never address the issue of material variability within a single impoundment. The amount of testing required to properly characterize a typical impoundment would be prohibitively expensive and time consuming. It is also important to note that prior to impoundment filling, there is no way of obtaining representative samples for the tailings. Material segregation is omnipresent during the filling operation and consolidation characteristics change dramatically as the slurry is sorted by grain

457

sizes. The only way of overcoming this major obstacle is to monitor and determine consolidation characteristics as tailings are being deposited in the field. TDR technology in combination with CONDES has the potential to effectively address this need.

6 REMOTE MONITORING AND INTEGRATED SYSTEMS

TDR cable testers are available that allow for serial communication via an RS232 module (Tektronix, 1989) and also have low power requirements (Dowding et al, 1996). The waveforms in Figure 4 were acquired using a remote modem, datalogger (Campbell Scientific, 1991) and cable tester to acquire and store data. Data is downloaded over a phone line. This capability allows for acquisition of digitized TDR waveforms which facilitates analysis and interpretation (Huang et al, 1993).

As shown in Figure 8, the datalogger and TDR cable tester can be interfaced with a multiplexer. This allows several cables and probes to be connected to a central data acquisition system. Consequently, TDR technology makes it possible to adopt a systems approach to tailings management. Integrated monitoring in mine tailings areas can include the components shown in Figure 9: monitoring consolidation during active operation, monitoring slope movement, monitoring ground water and contaminants, and monitoring hydrologic performance of the cap after closure.

7 SUMMARY

Brief summaries have been presented to demonstrate diverse applications of TDR technology that are pertinent to tailings management. These include monitoring the hydrologic performance of covers, monitoring movement along slopes, and monitoring changes in groundwater conditions. Furthermore, specially designed coaxial cables can be used for leak detection beneath liners. Another promising application of TDR technology is remote real time monitoring of tailings consolidation. All these applications can be integrated with remote data acquisition systems that are commercially available.

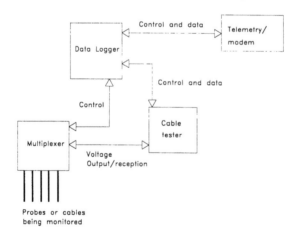

Figure 8. Schematic of data acquisition system used for remote monitoring of several cables and probes using a single TDR cable tester.

458

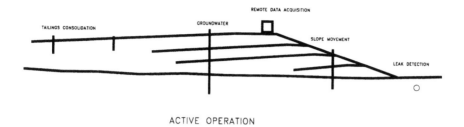

ACTIVE OPERATION

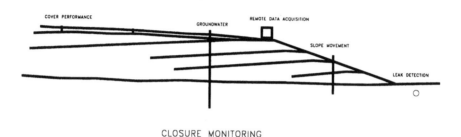

CLOSURE MONITORING

Figure 9. Potential applications of TDR for monitoring the tailings disposal area during active operation and after closure.

8 REFERENCES

Abu-Hejleh, A.N. and D.Znidarcic 1995. Desiccation theory for soft cohesive soils. *Journal of Geotechnical Engineering*, ASCE, Vol. 121, No. 6, pp. 493-502.

Benson,H.B., P.J.Bosscher, D.T.Lane, and R. J. Pliska 1994. Monitoring system for hydrologic evaluation of landfill covers. *Geotechnical Testing Journal*, GTJODJ, Vol. 17, No. 2, June, pp. 138-149.

Boehm,R.G. and G.S.Scherbert 1997. Development of the Glendale landfill test cover study. *Proceedings, Waste Tech '97*, Tempe, Arizona, February, pp. 95-120.

Campbell Scientific, Inc. 1991. *Time domain reflectometry for measurement of rock mass deformation.* Logan, UT.

Dalton,F.N., W.N.Herkelrath, D.S.Rawlins, and J.D.Rhoades 1984. Time-domain reflectometry: simultaneous measurement of soil water content and electrical conductivity with a single probe. *Science*, Vol 224, June 1, pp. 989-990.

Dowding, C.H., F.C.Huang, and P.S.McComb 1996. Groundwater pressure measurement with time domain reflectometry. *Geotechnical Testing Journal*, GTJODJ, Vol. 19. March, pp. 58-64.

Dowding,C.H., C.E.Pierce, G.A.Nicholson, P.A.Taylor, and A.Agostan 1996. Recent advancements in TDR monitoring of ground water levels. *Proceedings, 2nd North American Rock Mechanics Symposium: NARMS '96*, Montreal, 19-21 June, pp. 1375-1381.

Dowding,C.H., M.B.Su and K.M.O'Connor 1988. Principles of time domain reflectometry applied to measurement of rock mass deformation. *Int. J. Rock Mech. and Min. Sci.*, v. 25, No. 5, pp. 287-297.

ESI Environmental Sensors, Inc. 1997. *MoisturePoint landfill closure monitoring system.* Victoria, British Columbia.

Herkelrath,W.N., S.P.Hamburg, and F.Murphy 1991. Automatic, real-time monitoring of soil moisture in a remote field area with time domain reflectometry. *Water Resources Research*, Vol. 27, No. 5, pp. 857-864.

Hook,W.R. and N.J.Livingston 1996. Errors in converting time domain reflectometry measurements of propagation velocity to estimates of soil water content. *Soil Science Society of America Journal*, Vol. 60, No. 1, pp. 35-41.

Huang, F.C., K.M.O'Connor, D.M.Yurchak, and C.H.Dowding 1993. NUMOD and NUTSA: software for interactive acquisition and analysis of time domain reflectometry measurements. *BuMines IC 9346*, 42 pp.

Kachanoski, R.G. and A.L. Ward 1994. Measurement of subsurface chemical transport using time domain reflectometry. *Proc. Sym. on Time Domain Reflectometry in Envir., Infrastr., and Min. Appl.*, Evanston, IL, Sept., p. 171-182; NTIS PB95-105789.

Nicholson, G.A., J.F.Powell, and K.M.O'Connor 1997. Monitoring groundwater levels using a time domain reflectometry (TDR)pulser. *Final Report, CPAR-GL-97-2*, U.S. Army Corps of Engineers WES, Vicksburg.

Norland,M.R. 1994. TDR waveform analysis of Cd, Pb, and Zn in coaxial cells. *Proc. Sym. on Time Domain Reflectometry in Envir., Infrastr., and Min. Appl.*, Evanston, IL, Sept., p. 235; NTIS PB95-105789.

Nyhan,J.W., T.G.Schofield, and C.E.Martin 1994. Use of time domain reflectometry in hydrologic studies of multilayered landfill covers for closure of waste landfills at Los Alamos, New Mexico. *Proc. Sym. on Time Domain Reflectometry in Envir., Infrastr., and Min. Appl.*, Evanston, IL, Sept., pp. 193-206; NTIS PB95-105789.

O'Connor,K.M. and C.H. Dowding 1999. *GeoMeasurements by pulsing TDR cables and probes*. Boca Raton: CRC Press.

O'Connor,K.M., D.E.Peterson, and E.R.Lord 1995. Development of a highwall monitoring system using time domain reflectometry. *Proc., 35th U.S. Sym. Rock Mech.* Reno, Nevada, June, pp. 79-84.

PermAlert ESP 1995. *Leak detection/location systems*, Niles, IL.

Redman,J.D., and S.M.DeRyck 1994. Monitoring non-aqueous phase liquids in the subsurface with multilevel time domain reflectometry probes. *Proc. Sym. on Time Domain Reflectometry in Envir., Infrastr., and Min. Appl.*, Evanston, IL, Sept., pp. 207-214; NTIS PB95-105789.

Sun,Z.J. 1996. Feasibility studies for measuring water content in mine tailings using MP-917. ESI Environmental Sensors, Inc., Victoria, 2p.

Tektronix, Inc. 1989. *SP232 serial extended function module operator/service manual*. Redmond, OR.

Topp,G.C., S.J.Zegelin, and I.White 1994. Monitoring soil water content using TDR: an overview of progress. *Proc. Sym. on Time Domain Reflectometry in Envir., Infrastr. and Mining Appl.*, Evanston, IL, Sept., pp. 67-80; NTIS PB95-105789.

Yao,T.C.D. and D.Znidarcic 1997. Crust formation and desiccation characteristics for phosphatic clays, users' manual for computer program CONDES0. Florida Institute of Phosphate Research, Publication No. 04-055-134, 115p.

Zegelin,S.J. and I.White 1994. Calibration of TDR for applications in mining, grains, and fruit storage and handling. *Proc., Symp. on Time Domain Reflectometry in Environmental, Infrastructure, and Mining Applications*, Evanston, Illinois, Sept 7-9, U.S. Bureau of Mines, Special Publication SP 19-94, NTIS PB95-105789, pp. 115-129

Znidarcic,D., A.N.Abu-Hejleh, T.Fairbanks, and A.Robertson 1992. Seepage induced consolidation test, equipment description, and users' manual. *Report for Florida Institute of Phosphate Research*, Dept. of Civil Eng., Univ of Colorado, Boulder.

Tailings and Mine Waste'00 © *2000 Balkema, Rotterdam, ISBN 90 5809 126 0*

Use of risk assessment to evaluate effects and plan remediation of abandoned mines

Terence P. Boyle
US Geological Survey, Midcontinent Ecological Science Center, Colorado State University,
Fort Collins, Colo., USA

ABSTRACT: A framework of risk assessment is elaborated for the evaluation of the effects of abandoned mines and mills. Steps in this process include environmental description, identification and characterization of sources, assessment of exposure, assessment of effects, risk characterization, and risk management or remediation. The development and use of ecological endpoints for remediation is discussed in terms of the chemical constituents, toxicity tests and the biological community.

1 INTRODUCTION

Mining waste from historic hardrock mining is one of the unfortunate legacies of development in the Western United States. Drainage from old mines and mills is a major water quality problem in most states in the Rocky Mountain West. Remediation of these impacts has become a major regional concern (USEPA, 1995). Approaches for how to develop information and to plan for mitigation activities are in the process of being developed in a number of state and federal programs (USEPA, 1997). This paper presents a prospectus for the use of the procedures developed in ecological risk assessment as an overall strategy to develop information to determine effects of abandoned mine and mills, to plan for remediation activities, and to determine when remediation has been completed. Risk analysis has been adapted by the U.S. Environmental Protection Agency as the overall paradigm for addressing threats to human health and environmental quality from toxic substances (USEPA 1992, 1998).

2 RISK ASSESSMENT

Risk assessment is most frequently used in anticipating and predicting the effects of planned or occurring activities: risk of developing cancer and death from smoking, risk of environmental damage from pesticide use, local health risks of building a nuclear power plant, etc. However, the principles of risk assessment can also be used in analyzing the risk of existing effects of hazardous waste sites, chemical spills, chemicals in current use, etc. (USEPA, 1992, 1998). In these cases risk assessment becomes adaptive in that the point of departure can be variable:
1) Source driven, where there is a known source of contamination with unknown exposure and unknown effects;
2) Effects driven, with biological effects with unknown exposure and unknown source; and;
3) Exposure-driven, with known exposure from a hazardous substance with unknown source and unknown effects.

Collectively these strategies for analyses for developing information are known as retrospective ecological risk assessment (Suter, 1993). This is the framework that will be developed for analyzing risk and development of remediation plans for abandoned mine and mill sites. This is a source driven assessment in that it can be undertaken without prior knowledge of exposure or effects.

Steps in retrospective ecological risk assessment of abandoned mines and mills.

2.1 Hazard assessment

It requires the identification of the bounds of the mine-impacted area including draining, adits, tailings and receiving waters, lakes, streams and wetlands. Critical environmental descriptions are essential in defining the spatial boundary of expected impact and gathering of relevant physical, chemical, and biological data for a preliminary assessment of the impact of mining activities. There may be some legal considerations in determining the area to be included as well as what type of biological data to be considered. This step should include identification of important representative species, threatened and endangered species, consideration of chemical and biological endpoints to evaluate acute and chronic toxicity, reproductive effects, bioaccumulation/biomagnification, and community level changes in structure and function. This step should employ a biological survey and analysis of impacts on the biological community (algae, fish, and invertebrates) using appropriate ecological indicators.

2.2 Source evaluation

Identification and appraisal of the various sources of environmental contaminants involves measurement of concentration and discharge from adits, characterization and leachability from tailing piles and mill deposits. Some detail in the estimation of the relative importance of individual sources in complex mine waste situations is important in establishing the level of remediation work and the priority of sequence of remediation efforts. Estimate of release of toxic chemicals from adits, tailings, and other origins should be made in order to evaluate the likelihood of their need for remediation. Characterization of the possibility of transport by water or wind should be determined.

2.3 Exposure assessment

For each of the sources identified above an assessment of release and environmental exposure should be made. This effort should include estimations of release rate, chemical characteristics of the contaminant, chemical characteristics of the environment affecting the transport and fate, measured and expected bioavailable concentration in the water and sediment, and comparison of these concentrations with known toxicity information and water quality standards. These concentrations should be calculated in both adequate temporal and spatial scales to consider their exposure to biological organisms in the boundary considered from (2.1) above. Models for the determination of exposure can be generated at this point to give estimate fate and transport of the contaminants within the environment (Bartell, et al 1992).

2.4 Effects assessment

Most effects assessment is based on toxicity testing and/or evaluation of the biological community using ecological indicators. The effects assessment should consider an appropriate range of receptor species and their surrogates which are representative of those organisms found in the area receiving the effects of mine waste. This information can come from the literature, toxicity based water quality standards, or actual testing of media from the environment in question and from the hazard assessment information from (2.1) as above. These tests should consider exposure response at both acute and chronic temporal scales based on the information generated by the exposure assessment step. If there is more than one contaminant expected, then special efforts should be made to consider their combined effects by deriving toxicity units to combine their expected effects and in the testing of the actual mixtures. Endpoints should be derived from these tests that not only reflect their toxicity but concentrations at which no effects in the environment are expected. Assessment should also consider affects from sources through sediment and through the food chain if warranted by the exposure assessment.

2.5 *Risk characterization*

This is essentially a combination of exposure assessment and effects assessment. Integration of these two steps will yield an estimation of the effects from the calculated exposure. This information can include concentrations of contaminants to which the biological community is exposed, extent of exposure, portion of community affected and measured, and an estimation of the magnitude of the impact in time and space. To plan for regulation of the chemical, plan mitigation or plan remediation risk managers require this information. Integration of these two steps will yield an estimation of the effects from the calculated exposure. This information can include concentrations of contaminants to which the biological community is exposed, extent of exposure, portion of community affected and measured, and as estimation of the magnitude of the impact in time and space. In order to plan remediation of the mine waste impacts risk assessment should specifically format the information from risk characterization for use by risk managers.

3 ENDPOINTS FOR REMEDIATION

Before measures have been undertaken for the remediation of mine and mills sources, specific ecological endpoints should be formulated that represent environmental quality goals for these efforts. Remediation endpoints can be defined as specific values for variables to be attained as a result of measures enacted to mitigate the impacts of mine waste. These variables can be specific concentrations of constituent chemicals, no effect level in specific toxicity tests of ambient receiving or discharge waters, and the values of a set of ecological indicators for fish or benthic macroinvertebrate communities. Not only should the value of these endpoints be determined, but also some consideration should be given to their persistence before remediation can be declared successful. In complex economic, political, or social situations, stakeholder groups may be involved in the formulation of endpoints for remediation that reflect a consensus of valuation on the ecosystem in question (Pittinger, 1998).

REFERENCES

Bartell, S.M., R.H. Gardner, and R.V. O'Neill. 1992. *Ecological Risk Estimation.* Lewis Publishers. Boca Raton. pp 252.

Pittinger, C.A.(ed) 1998. A Multi-Stakeholder Framework for Ecological Risk Management: Summary from a SETAC Technical Workshop *Environmental Toxicology and Chemistry* 18 (2): 1-22

Suter, G.W. 1993. *Ecological Risk Assessment.* Lewis Publishers. Boca Raton. pp 358.

U.S. Environmental Protection Agency. 1992. *Framework for Ecological Risk Assessment*. Washington DC: Risk Assessment Forum EPA/630/R-92-001.

U.S. Environmental Protection Agency. 1995. *Historic Hardrock Mining: The West's Toxic Legacy*. U.S. Environmental Protection Agency Region 8. 10 pp.

U.S. Environmental Protection Agency. 1997. *EPA's Hardrock Mining Framework*. EPA 833-B-97-003.

U.S. Environmental Protection Agency. 1998. *Guidelines for Ecological Risk Assessment*. Washington DC: Risk Assessment Forum EPA/630/R-95/001F.

Case histories

Tailings and Mine Waste'00 © 2000 Balkema, Rotterdam, ISBN 90 5809 126 0

Stream morphology and habitat restoration of Pinto Creek, Gila County, Arizona

Scott C. Bourcy
AGRA Earth and Environmental Incorporated, Portland, Oreg., USA

Ralph E. Weeks
Golder Associates, Phoenix, Ariz., USA

ABSTRACT:

The partial failure of a permitted tailings impoundment at the Pinto Valley Operations copper mine resulted in the deposition of approximately 370,000 cubic yards of mine waste rock and saturated tailings (Debris) into Pinto Creek, an interrupted stream located in Gila County, Arizona. Resultant damage to the environment included: disruption of the physical character of the creek, loss of riparian and terrestrial vegetation, and temporary damage to ecological health and function. Following the bulk removal of the Debris from the canyon, repairs to the stream morphology and revegetation efforts were designed and implemented. Photogeologic interpretations, mapping, direct measurements of stream discharges, channel geometry, bankfull dimensions, and the characterization of native vegetation were completed. Applied restoration techniques, based on geomorphologic principles, were then utilized to repair the impacted terrains, and included the construction of specialized stream features (rock vanes, sills and weirs). Monitoring to assess recovery of the ecosystem is on-going.

INTRODUCTION:

On October 22, 1997, the partial failure of a waste rock/tailings impoundment complex at the Pinto Valley Operations copper mine located in central Arizona, resulted in the deposition of approximately 370,000 cubic yards of saturated tailings and entrained waste rock (Debris) into a portion of the streambed and along the canyon sideslopes of Pinto Creek (referred to as the "Incident"). The Debris flowed about 2,000 feet along a steep arroyo, dropping in elevation approximately 400 feet to the streambed of Pinto Creek, much of which is situated on public lands managed by the USDA Forest Service (Forest Service). Upon entering the creek, the Debris flowed approximately 900 feet upstream and 2,500 feet downstream. The mass ranged in thickness from a few feet at either terminus of the flow to a maximum of 42 feet, and extended across the entire canyon bottom. The Debris impacted approximately 25 acres of terrain, including the creek bottom, its flanking riparian corridor, and the terrestrial upland slopes. Some materials originating from the Debris were subsequently remobilized and deposited downstream, extending an additional 4,500 feet down the streambed. The remobilization of Debris was the result of pore water seepage emanating from the downstream terminus of the Debris (North Toe), and surface runoff from winter rains.

This paper provides a summary of reconstruction activities completed to restore the stream morphology beneath the footprint area of the former Debris mass, repair the upland areas, and revegetate the impacted area. Long-term monitoring parameters utilized to assess the recovery of the ecosystem are also discussed. Repairs to morphological characteristics of the impacted area were initiated immediately following the incident, continuing through two phases, culminating in June, 1998.

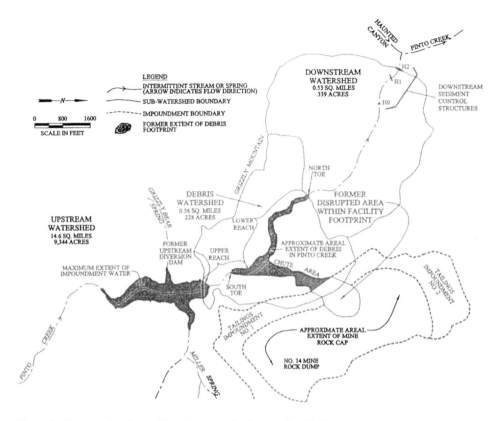

Figure 1. Site map showing incident features and sub-watershed divisions.

SITE DESCRIPTION AND REMOVAL ACTION SUMMARY

The Pinto Creek Watershed encompasses an area of 175.4 square miles which empties into Roosevelt Lake, a primary drinking water source for the Phoenix metropolitan area. Stream flow in the dendritic-shaped Pinto Creek drainage basin contains ephemeral, intermittent and perennial flow regimes (interrupted stream). The incident occurred below a 14-square mile segment of the Pinto Creek watershed and approximately 21 miles above Roosevelt Lake. The impacted portion of Pinto Creek was divided into three separate sub-watersheds, identified as the upstream, debris (upper and lower reaches) and the downstream watersheds, see Figure 1. The rationale for segmenting the watershed was based upon geomorphologic distinctions, the characteristics and magnitude of impacts to each watershed area, and the relative position of the constructed sediment/water control structures. Logistical factors were also considered, such as equipment access limitations into the rugged mountainous terrain.

During Phase I activities, both upstream and downstream containment and diversion structures were constructed to prevent further impacts to surface water quality, stream morphology, riparian vegetation and related biological habitat. Phase I work also involved the stabilization of the breached tailings impoundment and construction of a buttress to mitigate the potential of future failures. These activities were conducted concurrently with mechanical/hydraulic removal of Debris in the debris watershed and the hydraulic/manual removal of the remobilized debris in the downstream watershed.

During Phase II, dismantling of the sediment/water containment structures and restoration of the physical setting (morphology and vegetation) of the impacted area was completed. Several Phase II alternatives regarding morphological rehabilitation of the creek, riparian and upland areas were considered for those reaches or areas altered as a result of the debris removal and/or subsequent activities completed during Phase I. Four main areas underwent channel reconstruction during Phase II, the former footprint area of an upstream diversion dam, the upper and lower reaches of the debris watershed and the site of former downstream sediment/water control structures (gabion area) located in the downstream watershed. The entire stream channel in the impacted area in the upstream, debris and downstream watersheds were repaired where full-scale reconstruction was not required. This paper focuses on the debris watershed area (footprint of debris) which was subjected to the greatest impacts and required the most intensive channel reconstruction and re-introduction of vegetation.

As part of the restoration process, the entire impacted area of Pinto Creek was classified in accordance with the Rosgen system (1996) and general geomorphological and fluvial processes principles/concepts developed by Schumm (1977). This classification system and associated restoration techniques were used as a guideline to repair or improve conditions in affected areas, and to better understand the pre-incident stream conditions. The upper reach of the debris watershed and the gabion area underwent full-scale channel reconstruction, with the intent of improving on and exceeding pre-incident stream channel conditions. Stream morphology improvement in these areas involved the construction of rock vanes, sills and a rock weir, coupled with alluvial bar enhancement and riparian habitat improvements. All reaches were, at a minimum, restored to pre-incident morphologic conditions.

Following the repair of stream morphology, flanking valley side slopes, and portions of riparian habitat in the impacted area were seeded, and containerized plants (deergrass), trees (sycamore and ash) and tree pole cuttings (willow & cottonwood) were introduced in selected locations within the riparian corridor. This effort was completed to enhance the revegetation process and minimize erosion of newly repaired areas along the streambed. The upland valley slopes in the former dam footprint and flanking the stream channel in the debris watershed were repaired, contoured, hydro-mulched and seeded with a USDA Forest Service (Forest Service) approved seed mix following the completion of Debris removal.

REHABILITATING STREAM MORPHOLOGY

Channel reconstruction and stream repair was accomplished in stages, with the process guided by the morphologic characteristics within specific, classified reaches of the stream. Prior to initiating stream rehabilitation, numerous reconnaissance surveys through, above and below the impacted area of Pinto Creek were conducted. The purpose of these stream reconnaissance surveys was to measure bankfull dimensions, and interpret the fluvial processes and morphologic conditions characteristic of relatively un-disrupted portions of the creek. The results of the surveys were then used to compare these characteristics to conditions which probably existed in the debris watershed. At a minimum, an understanding of bankfull channel dimensions, geologic conditions and dynamic fluvial processes unique to Pinto Creek was required in order to return the stream to a balanced, stable fluvial system.

Pre-incident aerial photographs (August 1993 & others) were utilized as a template for channel restoration. An aerial photography review coupled with field mapping activities were conducted to identify and map areas of bedrock exposures and the position of riffles and pools, measure stream gradients and bankfull dimensions, and identify depositional channel features such as alluvial bars and floodplain areas. These studies also evaluated the relative abundance of alluvium. Following the completion of these tasks, the data was interpreted and the Level I/Level II stream classification system developed by Rosgen (1996) was used to identify specific geomorphological reaches in the impacted area of Pinto Creek. The Rosgen approach provides a system to evaluate the following factors catagorize overall stream types: channel gradient, dominant bed materials, entrenchment, sinuosity, channel width/depth ratio and meander width ratio. This classification assigns an alphabetic notation (A thru G) for the overall channel morphology category. The classifications range from an A-type stream flowing in a steep, narrow, deeply incised confined canyon typical of the headwaters of a watershed, to a G-type stream which flows in a moderately steep, fluvial dissected alluvial or colluvial valley. In addition, and a numeric notation is assigned for dominant channel material particle size (1 thru 6), ranging from bedrock (1), boulders (2), cobbles (3), gravels (4), sands (5),and silt and clay-size materials (6). Stream channels of lower gradients, higher sinuosity, and developed floodplains which generally occur in broad alluvial valleys are typical of C and E stream types.

The excursion and subsequent inundation by Debris affected the physical setting of the stream channel, flanking riparian corridor and upland slopes in the debris watershed. Due to the presence of bedrock in the canyon walls and channel floor, the overall physical setting was not drastically altered. The channel characteristic affected most by the removal action was the alluvial sediments. The debris watershed contains a bedrock-controlled channel, with an alluvial bedload of limited thickness. The incident and removal action resulted in some of the limited alluvial material being entrained with Debris, which was subsequently excavated and transported from the creek bed.

The channel morphology in the lower segment of the debris watershed is controlled by the presence of Paleozoic limestone bedrock formations (Martin, Escabrosa and Naco Formations), minor karst features, and faults. These conditions result in a stream which is moderately incised into a steep, narrow, confined canyon. The stream types transition from a B-type stream at the downstream end (in the vicinity of the north toe) to an A-type in the lower to middle portion, then back into a B-type stream at the upstream portion of the lower reach. The stream morphology in the B-type stream reaches are characterized by riffle/step pool morphology, slightly lower channel gradients, wider channel, development of floodplain/alluvial bar areas and higher quantities of alluvial bed load than the A-type stream section located in the lower half of the lower reach. The low-flow channel configuration is defined by scour surfaces incised on the surface of the limestone bedrock. The presence and spacing of the pools in this reach are irregular, and coincide with the scour holes on the bedrock surface. Along portions of this reach a gradual transition from stream channel to terrace to valley slope is lacking, with the slope often grading from stream channel to vertical or near-vertical limestone ledges.

The stream morphology in the lower reach basically remained intact, with the exception of the area beneath the north toe where a sediment control structure (north toe dike) was constructed at the downstream terminus of the debris. The north toe dike was keyed into the streambed alluvium at this location. Large boulders were strewn throughout the length of the channel and represent rock-fall from the flanking slopes and outcrops. During removal operations, many boulders were dislodged. Most of the original stream bank remained intact, as demonstrated by the presence of former tree roots/trunk fringing the top of the boulder-lined bank on the northeast side of the channel. Approximate bankfull widths were measured from the horizon of the boulder lined bank where the tree trunk is present, then projected across the stream to a similar height on the exposed bedrock. Bankfull width/depth measured in this reach were approximately 20 feet wide by 1.5 to 2.0 feet deep (30-40 square feet). The flood-prone width measured in this zone was as much as 50 feet. Stream rehabilitation in this reach utilized the measured bankfull dimensions and pre-incident aerial photographs as a guideline for channel reconstruction.

Restoration of the lower reach consisted of the final removal of Debris/plating rock, temporary removal of channel boulders (to remove debris entrained between boulders and alluvium), the placement of boulders, hydraulic washing of the limestone outcrops and streambed materials, jetting/cleaning of the pools and the reshaping and fortification of the limited alluvial bar/floodplain features. Based on the pre-existing data reviewed, the alluvial bedload in this reach was thin prior to the incident. However, Debris removal resulted in the further reduction of available alluvium. Remarkably, the channel features and boulder positions noted on the 1993 aerial photographs remained relatively un-altered. The bulk removal of Debris and plating rock was accomplished mechanically, followed by manual removal of Debris by shovel/bucket methods and finishing with pressure washing utilizing gravity-fed well water. Due to the fractures and undulating bedrock surfaces, much of the Debris entrained within the features were manually removed by shovels, picks and brooms. The majority of the Debris was removed while the sediment was dry. This procedure was done to avoid mixing Debris into the limited alluvium prior to washing. The alluvium consisted mainly of sand and gravel which partially fill the scour pools: however, gravels, cobbles and boulders are present along the riffle/run sections. The pool, riffles, rans and streambed alluvium were hydraulically cleaned out by jetting methods. Following completion of the manual cleanup, the limestone cliff faces, ledges, scour surfaces, fractures and boulders were pressure washed. This effort was done to remove the Debris coating on the rock surfaces and Debris entrained on bedrock ledges, in fractures and solution voids. The pools were utilized as slurry-water transfer pump stations during removal operations, which resulted in the settling/mixing of some Debris materials into the alluvium. The pools were jetted to extract the entrained residual Debris from the alluvium. The jetting process forced the silt-fraction of the Debris into suspension, while the sand and gravel fraction of the alluvium settled out. As the pools were jetted, pumps were operating simultaneously to remove the turbid water containing the suspended Debris.

The stream channel in this reach were basically restored to the pre-existing conditions, with the construction of two rock vane features for habitat improvement. Numerous large boulders were originally situated within the low-flow channel. Some of these boulders were removed from the channel to improve flow conditions, and utilized for vane structures on the alluvial bar/floodplain feature located near the North Toe. Two rock vanes were constructed on both the upstream and downstream portion of the floodplain/bar feature. The vanes serve several functions which enable the bankfull channel to function efficiently. During high flows the vane helps force water back into the main channel and away from terrace and valley margins, thereby minimizing lateral erosion. The vanes also enhance sediment aggradation on its downstream side, which further builds and maintains the terrace bar feature. Additionally the vane maximizes shear stress, which allows sediment accumulation on the floodplain terrace. The fine-fraction of sediment deposits on the floodplain terrace, which in turn, enhances the riparian habitat. This occurs during periods of high flow, when the vane creates a back-eddy on the downstream side of the boulder vane. The back-eddy dissipates flow energy, facilitates sediment transport and helps maintain sediment load. Maintaining sediment load and sediment transport enables the channel dimensions to be sustained.

The rock vanes were keyed into the bedrock on the valley slope and into either the uppermost surface of the limestone contact, or 2-3 feet into the alluvium of the terrace. The largest boulder (footer boulder) was keyed into the terrace-valley slope transition. From the footer boulder, extending into the channel bank, boulders were placed ranked in size from largest to smallest. The boulders tapered down from the valley slope (largest boulder) to the right bank of the channel (smallest boulder). The vane structure points upstream and is positioned at approximately a 30 degree angle towards the creek.

Two additional alluvial bar features, located in the lower reach, remained relatively undisturbed during the incident and removal actions. The bars were re-contoured and boulders were placed on top of and fringing portions of the alluvial bar features. Both bar locations coincide with the inside meander bends of the stream. Excess boulders within the stream channel were removed and either placed on the alluvial terrace bars, along channel/valley floor slope contacts or set in pockets up on the flanking colluvial valley slopes. The purpose of the boulder placement on and fringing the alluvial bars is to help stabilize the bar, prevent bar erosion and migration until vegetation on the bars has fully recovered.

STREAM MORPHOLOGY REPAIR IN THE DEBRIS WATERSHED - UPPER REACH

The upper reach of the debris watershed is characterized by a wider channel, higher sediment supply, riffle/pool morphology and some developed floodplain/alluvial bar features. The stream morphology (based upon the 1993 aerial photographs) were characterized by multiple-thread (braided) patterns with the presence of point bars, relatively unstable banks and historic impacts to the stream related to grazing, road construction and mining related activities. During Phase II, a workgroup was formed to address channel re-configuration for this reach. It was decided that this reach would be rehabilitated to channel conditions that improved pre-incident conditions and exceeded removal action objectives. Prior to channel reconstruction, the Debris and plating material were removed from the streambed and flanking valley slopes and the former earthen diversion dam structure was dismantled. The streambed was then hydraulically washed with well water using high pressure hoses.

The bedrock in this reach consists predominantly of the Whitetail Assemblage (Mid-Tertiary Age) which is exposed along many portions of the creek bottom, and is characterized as a fanglomerate. The uppermost surface exposures on the channel floor are weathered and friable. Due to the lack of significant alluvium in this reach, the low-flow channel was scoured into and exposed on top of the bedrock surface. The alluvium was depleted as a result of flooding which occurred in 1993 and additionally from the removal action activities. Alluvium obtained from an approved borrow source were introduced to this reach, for purposes of bar enhancement, channel reconstruction and habitat improvement.

Following the completion of the Debris and plating rock removal, the channel centerline profile,

471

cross-sectional areas, alluvial bedload thickness and extent, valley-slope profiles and transition zones were surveyed and mapped. This information was used as a baseline to design a channel configuration which improved on the pre-incident stream morphology conditions. Shear stress determinations for riffle/run areas were calculated based upon the survey and mapping data collected. The shear stress values calculated for this reach were critical because as channel gradients change, bankfull depths or dimensions were adjusted accordingly. Modifications were made to the conceptual design based upon field conditions encountered during the reconstruction effort. The overall objective was to construct the channel to a single thread, C-type stream with riffle/pool morphology. This was accomplished through the development of gradient controls features and the enhancement of pool, alluvial bar and floodplain terrace features.

Field activities commenced at the downstream portion (near the chute confluence) and progressed upstream to the former dam footprint. The first task was to backfill and compact a former sump (used for water management purposes) located at the north end of an existing gravel/cobble bar. During removal activities, attempts were made to minimize the impact to this alluvial bar feature. Attempts were successful, in fact, some remnants of the pre-existing woody, rooted, riparian vegetation still remains on the bar surface. This bar feature was enhanced by enlarging, re-contouring and reshaping the alluvial bar. Based upon a review of historic aerial photographs, the original channel split around the point bar. A trackhoe was used to cut a low-flow swale on the east-side or outside meander bend of the point-bar. The intent was to convert what had been a central point bar into a floodplain terrace feature and to modify the stream pattern from braided to a single-thread channel. The basis for improvement to the bar-floodplain feature was to situate the terrace on the inside or depositional portion of the meander bend. This should allow for continued formation and building of the terrace-bar feature through sediment deposition, especially during higher flow events.

A boulder sill was constructed for the purpose of dissipating high flow energy away from the terrace toward the main channel, enhancing sediment aggradation onto the bar, thereby aiding in stabilizing the alluvial bar. The sill is located on the upstream portion of the bar. The boulder sill was keyed two to four feet into and below the top of the bar. The sill was then covered and compacted with alluvium. Boulders were also placed in small clusters on top of and fringing the bar, to aid in sediment/bar stabilization.

The outside meander bend of the creek adjacent to the terrace-bar feature is in contact with the limestone bedrock and forms the geologic transition which divides the upper and lower reach of the debris watershed. Several pools are located in the lower section of the upper reach and coincide with the upstream and downstream portion of the bar-floodplain terrace. Pool spacing is approximately 200 feet, with the pools separated by riffles and runs, many of which flow over exposed bedrock. Boulders were strategically placed along portions of the bank-valley slope transitions, especially along the outside of meander bends. On the east side of the creek the streambed-valley slope transition is fairly abrupt. Beneath the former dam footprint and portions of the lower valley slopes were denuded of vegetation as a result of the incident and removal action. The combination of moderately steep slopes and lack of vegetative cover, greatly increases the potential for erosion. Boulders were placed into and backfilled along portions of the channel banks, scattered along the colluvial valley slopes and on the upstream toe of the bar feature. The primary function of the boulders is to dissipate water energy and act as an erosion inhibitor along the stream channel and on slope faces.

Progressing upstream, a low-flow channel thalweg was excavated. Surface water was flowing during channel reconstruction, and actually facilitated in establishing stream gradient during restoration of the low-flow channel. Slope contouring was completed on the east side of the stream. A more gradual slope break transition was completed for this area. On the west side of the channel, several natural seeps are present at about the level of the floodplain/slope transition. Small potholes were excavated on the downhill side of the seeps. Boulders were placed into the pothole trenches and then backfilled with native alluvium/colluvium. The rationale for the boulder lining was to aid in seep water retention which will provide a source of shallow, subsurface water for riparian vegetation on the floodplain. A gravel, cobble and boulder alluvial bar-floodplain terrace is located directly below these natural seeps. This terrace was selected as a target area for the planting and re-establishment of woody riparian plants.

Directly upstream of this alluvial bar, a gradient control feature (modified rock weir or vane) was constructed. The function of the rock weir is to emulate a natural rock outcrop or bench in the creek bottom. Several natural occurring bedrock benches are located in the streambed in the upper segment of the lower reach. A comparison can be made in the function and geometry between the natural and constructed features. A V-shaped trench (plan view) was excavated to a depth of approximately 3-4 feet below the channel floor. The rock weir was keyed approximately two feet into the underlying bedrock. The "V" notch extends from slightly above the bankfull level on each side of the channel, with the apex of the "V" pointing upstream. The largest boulders (footer rocks) that form the weir were set on each bank side of the channel then subsequently smaller boulders were placed in succession, so that the weir tapered down to the smallest boulders at the apex of the "V". Once the boulders were placed, coarse alluvium was used to backfill around the boulders.

The purpose of the weir is to simulate a natural bedrock outcrop in the stream channel which acts as a gradient control, enhances pool formation and habitat, coalesces flow over the structure which in turn facilitates sediment transport by increasing shear stress as water flows over the weir. The shear stress dissipates as water flows into the scour pool which forms on the downstream side of the weir. The water cascading over the weir into the scour pool creates a back-eddy, allowing some sediment being transported to be deposited while facilitating the transport and deposition of alluvial materials further downstream. The weir is anticipated to aid in building and maintaining bankfull channel dimensions. Maintaining bankfull dimensions enables the channel to maintain stream gradient, *i.e:* equilibrium. The weir will prevent the pool from migrating up or downstream by assisting in sediment transport and deposition. Pool migration inhibits the channel from maintaining bankfull dimensions, which typically results in channel instability causing head-cutting, lateral shifting of the channel (avulsion), and degradation of bedforms such as bar and bank features.

Following the removal of the upstream diversion dam and subsequent sideslope re-contouring, channel reconstruction was performed. Stream repair consisted of the following elements: re-creation of the low-flow channel, construction of a floodplain terrace, armoring of the former dam key trench with coarse alluvium, introduction of additional alluvium to the stream bottom and channel banks, and the introduction of boulders which were strategically placed along portions of the channel banks and terrace margins to assist in minimizing erosion potential. The purpose of the boulder placement was to aid in initial channel/bank stabilization, to dissipate flow energy in the form of water deflection, to improve biologic and riparian habitat, and for aesthetic qualities. Prior to the re-creation of the low-flow channel, an approximate bankfull channel dimension of 30-40 square feet (20 feet wide by 1.5-2.0 feet deep) was measured for this zone.

In summary, the upper reach of the debris watershed was reconstructed to a C-type stream, which consists of a single-thread channel with riffle/pool morphology and developed floodplain terraces. Upon completion of channel reconstruction in the debris watershed, retained water from a nearby spring-fed pond was released into the streambed of Pinto Creek. The water was initially released in June, 1998 and has since flowed continuously. The rationale for releasing water into the creek was to augment growth of riparian vegetation, aquatic macro-invertebrate recovery, initiate substrate armoring and to allow the creek to continue restoration of a low-flow channel configuration.

REINTRODUCTION OF RIPARIAN VEGETATION

Following the completion of the stream morphology repair activities, the reintroduction of riparian vegetation was initiated. Approximately 6.09 acres of riparian-floodplain habitat were impacted in the upstream, debris and downstream watersheds. The debris watershed and the adjacent former upstream diversion dam footprint were the only areas for which extensive riparian rehabilitation were prescribed. The following rehabilitation prescription was adopted: herbaceous understory plantings, tree pole cuttings, tree plantings, and seeding; and the inoculation of algae into the stream channel. Seeding and understory plantings were prescribed to improve visual quality and the potential for short-term soil stabilization. Tree pole cuttings and plantings will increase the likelihood that these native species will more quickly thrive and minimize or exclude tamarisk.

The lower reach is narrow, bedrock-controlled, lacks significant quantities of alluvial sediments and is scour prone; whereas the upper reach is comparatively wider, more alluvial in nature, has more alluvial sediments and has rehabilitated floodplain areas that are suitable habitat and could sustain riparian plant growth. Consequently, the introduction of tree pole cuttings and tree plantings were planted in greater amounts in the upper reach, with lesser amounts planted in the lower reach. This phase of the rehabilitative effort was completed in early Spring 1999.

Seeding and partial understory planting were completed in the upper reach, and in selected areas in the lower reach of the debris watershed. Final species selection was based on several factors: (1) whether or not the species is native to the impacted area; (2) the potential for the species to grow rapidly and stabilize sediments along the stream channel; and (3) the availability of nursery stock. The two riparian habitats targeted during rehabilitative efforts are wet areas at seeps and along the existing watercourse, and drier terrace areas above the creek channel. Approximately 3,000 plant starts including: softstem bulrush, alkali bulrush, spike rush and knotgrass and approximately 290 one-gallon-size deergrass were planted in the upper reach and in selected areas of the lower reach where alluvial sediments are present. Floodplain terrace areas in the upper and lower reach of the debris watershed were seeded with vine mesquite and with the standard seed mix applied to Pinto Creek side slopes.

In 1999, tree pole cuttings (approximately 100 individuals) and tree plantings (approximately 40 individuals) were established during the period January through mid-March when the trees are dormant. Pole cuttings consists of Goodding willow and Fremont cottonwood which were collected locally. Tree plantings consisting of Arizona sycamore and velvet ash obtained from local nursery stock. All of these species are native to the Pinto Creek drainage near the debris watershed. Poles and plantings were placed in wet areas at seeps, along the stream channel/banks and riparian areas throughout the debris watershed.

REPAIR OF UPLAND SLOPE AREAS AND REINTRODUCTION OF VEGETATION

Debris removal on the colluvial valley side slopes in the debris watershed was accomplished by mechanical means. In areas where bedrock cliffs and ledges are exposed, Debris was removed by a combination of mechanical, physical and hydraulic processes. Following the completion of Debris removal, mechanical slope preparation was conducted and included: dimpling, roughing, notching, cross-slope furrowing, and the placement of colluvial boulders were implemented on colluvial slopes to increase water retention and decrease erosion potential. No mechanical slope preparation was needed on cliffs or rocky ledges, however, the limestone cliffs and ledges were pressure washed utilizing well water to remove the residual Debris. Soil samples were collected from the colluvial slopes and analyzed to evaluate potential deficiencies in soil chemistry, nutrient and organic content, or soil organisms (e.g., fungi and bacteria) to support revegetation efforts. Test results indicated that the colluvial soils are highly variable with respect to the biological and chemical parameters evaluated, and therefore, the addition of liquid humus (15 gallons/acre) and compost (6 cubic yards/acre) to the hydromulch mixture might benefit the success of plant growth.

ASPECTS OF THE LONG-TERM MONITORING PROGRAM

The objectives of the currently on-going monitoring program are to evaluate and quantify the process of the ecosystem's recovery along the impacted portions of Pinto Creek and upland areas. The monitoring program is designed to periodically (quarterly) evaluate the impacted area as a whole on the reconnaissance level, coupled with more detailed data collection appraisals (quantitative) at selected monitoring sites on an annual basis, which will result in a time series of direct observations. Collectively, these observations will monitor the rate and character of physical changes occurring in the ecosystem.

Seven permanently benchmarked stream transect sites have been established above, within and below the impacted portion of Pinto Creek. Stream morphology conditions, processes or hazards defined by Schumm and Chorley (1983) that are being evaluated and data collection parameters include: erosion (degradation, nickpoint formation), deposition (aggradation), channel pattern changes, metamorphosis, channel stability evaluation criteria developed by Pfankuck (1975), and changes in bankfull dimensions. Sediment characterization including size and distribution (both bed and bank), pebble count procedures (Bevenger and King, 1995), the degree of embeddedness (Burns and Edwards, 1985), and the role of vegetation. Stations were established for photo-documentation purposes, and mapping of the stream geometry, cross sectional area/profiles and channel gradients are being performed at each transect location. In addition, a yearly bioassessment monitoring program is being conducted at the same stream morphology monitoring locations, to collect data regarding the aquatic faunal communities health and recovery. Vegetation monitoring is being conducted in a parallel manner and frequency, for both the riparian and upland communities/corridors and includes numerous riparian line-point transect and upland/sideslope quadrant plot data collection sites.

REFERENCES

Bevenger, G.S., and King, R.M., 1995. A Pebble Count Procedure for Assessing Watershed Cumulative Effects: USDA Forest Service, 17p.

Burns, D.C., and Edwards, R.E., 1985. Embeddedness of Salmonoid Habitat of Selected Streams on the Payette National Forest, Unpublished Report, Payette National Forest, McCall, Idaho.

Pkankuch, D.J., 1975. Stream Reach Inventory and Channel Stability Evaluations, USDA Forest Service, R1-75-002, Government Printing Office #696-260/200, Washington, D.C.

Rosgen, Dave, 1996. Applied River Morphology. Wildland Hydrology. Pagosa Springs, Colorado.

Schumm, S.A. and Chorley, R.J., 1983. Geomorphic Controls on the Management of Nuclear Wastes, Publication of the U.S. Nuclear Regulatory Commission No. NUREG/CR-3276.

Schumm, S.A., 1977. The Fluvial System. John Wiley and Sons, Inc. New York, NY.

Tailings and Mine Waste'00 © *2000 Balkema, Rotterdam, ISBN 90 5809 126 0*

Stabilization of the Pinto Valley tailings impoundment slide

L. A. Hansen & N. J. LaFronz
AGRA Earth and Environmental Incorporated, Phoenix, Ariz., USA

M. B. Yasin
BHP Copper Incorporated, Miami, Ariz., USA

ABSTRACT: The 1997 failure of an approximate 900-foot long section of the No. 14 Mine Rock Dump at the BHP Copper, Inc. - Pinto Valley Operations (PVO) copper mine resulted in the release of 370,000 cubic yards of tailings and non-mineralized mine rock. The released materials flowed about 2,000 feet down a steeply sloping dry arroyo tributary to the dry Pinto Creek bed, traversing an elevation difference of about 400 feet. This paper initially discusses the monitoring and investigative programs established, and then discusses the analyses completed to evaluate the stability of the slide mass and the failed area. Assessing the stability of the slide mass, particularly within the dry wash and at the former perimeter of the impoundment, was necessary to permit safe removal of the slide debris in the dry arroyo and within Pinto Creek.

The paper then describes the design and construction of a stabilization buttress and drainage system at the perimeter of the former tailings impoundment. Because of limited capability to fully investigate conditions within this area within a reasonable time frame, and access restrictions imposed by the uncertain conditions, a conservative preliminary design was developed to facilitate the initiation of construction. The design was then modified as construction proceeded, based on actual conditions encountered, including seepage flows and unstable zones of tailings and debris. This design-build process and the final constructed design of the buttress and drainage system are presented. The recontouring of the failure area is discussed, including control of surface runoff, soil cover, erosion protection and revegetation.

1 INTRODUCTION

On October 22, 1997, BHP Copper, Inc.- Pinto Valley Operations (PVO) experienced a failure of an approximate 900-foot long section of the west-facing slope of the No. 14 Mine Rock Dump. The facility was the prior location of the Nos. 1 and 2 Tailings Impoundments, which were being converted to a waste rock dump by placing lifts of waste rock on top of the existing tailings. An initial 50-foot thick life of waste rock had been placed and the second 50-foot lift was being placed when the incident occurred. The failure resulted in the release of materials consisting of non-mineralized waste rock and saturated tailings, which flowed in a viscous mass down the west-facing valley slope of a small, dry arroyo tributary (the Chute) to Pinto Creek. The flow distance to Pinto Creek was about 2,000 feet, with a drop in elevation of about 400 feet.

The failure resulted in the formation of a roughly circular depression that encompassed about 14.3 acres inside the original trace of the facility footprint (Figure 1). The failure zone above the steeply sloping arroyo was characterized by several large masses of dumped mine waste rock that had dropped in elevation as much as 30 feet and tilted to the east at an angle of about 35 degrees. The blocks had an en echelon configuration, and were oriented parallel to the circular boundary of the depression. Three PVO vehicles, including a Caterpillar D-10 dozer, a

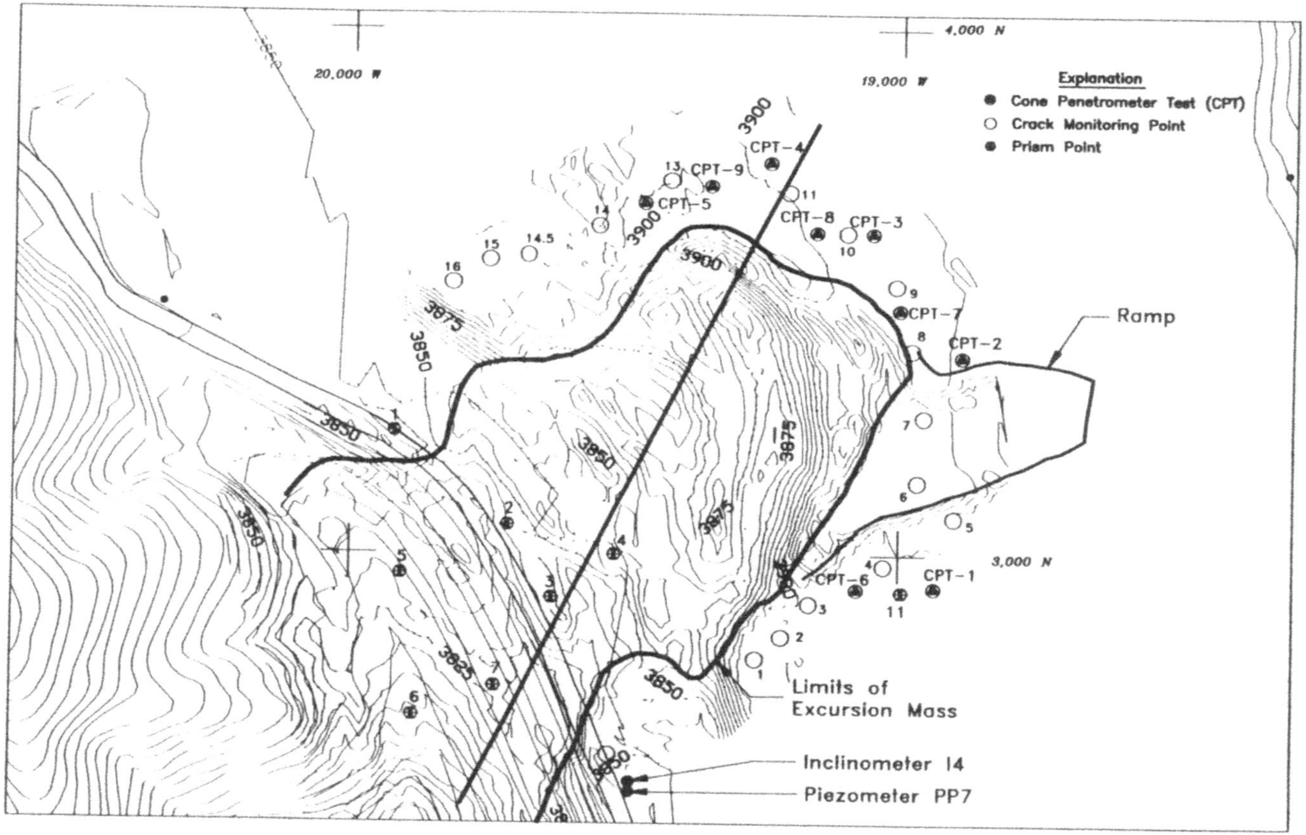

FIGURE 1 : INITIAL MONITORING PROGRAM

478

pickup truck and a 180-ton mine haul truck, were captured in the incident and partially buried, but no injuries occurred to mine employees.

The incident impacted public lands under the stewardship of the USDA Forest Service, as administered by the Tonto National Forest. Due to the imminent possibility of further significant impacts, a rapid response program of abatement and restoration was established by the Forest Service, with the overall goals being to remedy the consequences of the incident to the environment, prevent further damage and restore the impacted habitat to pre-incident conditions. The site safety and health considerations were under the jurisdiction of thé US Mine Safety and Health Administration (MSHA), particularly the efforts to remove the debris within Pinto Creek and the Chute, and the design and construction of the buttress.

The initial step in the removal action was the removal of the slide debris within the dry bed of Pinto Creek. Because of the potential for further movement of the slide mass downslope in the Chute to Pinto Creek, an assessment of the stability of slide mass was required in the short term, and design and construction of a buttress across the failure area was required in the long term. Completion of this assessment enabled MSHA to allow removal of debris within Pinto Creek to continue beyond its confluence with the Chute, and removal of debris within the Chute, since the assessment established that it was safe for personnel and equipment to work in these areas. It also was necessary to establish safe working protocols for removal of slide debris, particularly tailings, within the area where the buttress was to be constructed.

2 OBSERVATION, MONITORING AND INVESTIGATION

Following the incident, a detailed mapping effort was completed to characterize the geometry of the tailings flow in order to understand the sequence of events comprising the incident and evaluate the probability of a secondary or follow-on event. This effort included a review of descriptions of historic tailings impoundment failures available in the public domain. As noted above, it appeared that the tailings flowed beyond the perimeter of the impoundment and that the waste rock dropped as blocks into the space previously occupied by the tailings. Within the tailings remaining in the failure zone, and in the Chute area, many sand boils were observed, with obvious flows of water. Within about one week, the sand boils dried up and the many water streams had coalesced, eroding a predominate channel within the Chute area. The total quantity of flow was estimated to be 80 gallons per minute (gpm).

Within two days following the incident, an initial monitoring system had been installed within the slide mass and areas adjacent to and back of the scarps which defined the limits of the failure zone. The monitoring system initially included seven total station survey prisms and 17 specific monitoring points around the perimeter of the failure zone (Figure 1). Six of the prisms were located within the slide mass and one (No. 1) was located on the north side of the failure zone on a road which previously traversed the slide area. The monitoring points were established at the most distant location from the perimeter of the main failure scarp where cracking within the waste rock was observed; monitoring included measuring the width of the crack that had formed and whether an additional crack located further from the main scarp had developed. During the first month following the incident, the prisms were surveyed on at least a daily frequency, as were the monitoring points. Neither appreciable widening of the cracks within the waste rock, nor development of new cracks was observed.

The new monitoring equipment augmented the many pizometers and the few inclinometers that were part of the ongoing monitoring program for the No. 14 Mine Rock Dump. However, most of these devices were located at considerable distance from the slide area, typically within the No. 1 and No. 2 Tailings Dams which formed the majority of the perimeter of the facility. One inclinometer (I-7) and one pneumatic piezometer (PP-7) were installed from the surface of the existing road at the south end of the slide mass. Material for these installations was readily available since it had been purchased for a different project. The initial reading of the piezometer did not indicate significant excess pore water pressures, and by the end of November 1997, the inclinometer detected movements to the northwest of about 1/4 inch.

By November 24, 1997, approximately one month after the incident, monitoring of the survey prisms indicated very small, continuing downward movements of prism Nos. 2 through

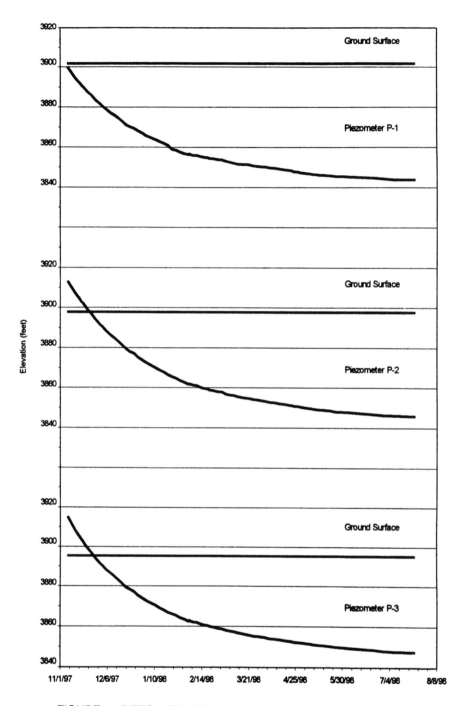

FIGURE 2. PIEZOMETRIC SURFACE ELEVATIONS VERSUS TIME.

5, likely as a result of continuing consolidation of the slide debris. The most important secondary component of the movement was toward rather than away from the failure area. Review of the original topography in the area determined that the natural ground surface in the area of the slide sloped away from a ridge above the area of the Chute, thus establishing down gradient to be toward the failure area. Maximum vertical movements were about 1.8 to 2.3 feet, monitored by prism Nos. 3 and 4 located within the slide mass. Total lateral movements of the same prisms were about 0.5 to 1.0 foot to the north-northeast. The downward movement of prism Nos. 6 and 7, located west of the slide mass within the upper area of the Chute;. had essentially ceased, after total movements of 0.3 to 0.8 foot had been recorded.

Between October 31 and November 6, 1997, nine piezocone soundings were completed at locations about 100 to 200 feet back of the main scarp of the failure zone (Figure 1). The soundings required casing of borings 35 to 75 feet through waste rock, and extended to total depths of 62 to 219 feet below the surface of the waste rock. The investigation included pore pressure dissipation tests performed at frequent depths by recording pore pressure decay versus time. The shear wave velocity of the encountered tailings also was determined using an accelerometer mounted internally in the cone. At many locations, the shear wave velocity of the tailings (normalized to effective overburden pressure) was determined to be 500 feet per second.

Vibrating wire piezometers were installed in piezocone soundings CPT-7, 8, and 9 (Figure 1) in order to monitor the dissipation of pore water pressure within the tailings underlying the remaining waste rock at the perimeter of the failure zone. This was considered very important to evaluating the stability of the remaining tailings, particularly since placement of piezometers within the failure zone was not deemed practical. Initial readings of the piezometers on November 7, 1997, determined piezometric surfaces at elevations varying from elevation 3900 to 3915 feet. Since the elevation of the surface of the tailings below the waste rock was 3850 feet, the measured piezometric surface elevations indicated excess pore water pressures within the tailings (at the time of their initial measurement) of 50 to 65 feet. Pore water pressure dissipation tests performed during the piezocone soundings indicated pore pressure values 15 to 25 feet higher than the tailings surface.

Most important to the assessment of the stability of the area of the failure was the dissipation of pore water pressures with time. The piezometers were monitored at least daily, with the monitoring frequency increased during storm events. Dissipation curves (Figure 2) for the three piezometers are very similar, indicating a significant decrease over time that approaches the surface of the tailings at elevation 3850 feet; the tailings underlying the waste rock were known to be essentially saturated prior to the placement of waste rock. By November 26, 1997, about one month after the incident, pore pressure decreases of 16 to 20 feet had been recorded. By the end of December, 1997, decreases of 25 to 38 feet had been recorded, indicating a somewhat diminishing rate of decrease as the tailings responded to the initial load applied by the waste rock.

3 STABILITY ASSESSMENT

In support of the design of the buttress to stabilize the failure zone, a complete review of the monitoring data and of the underlying topography of the ground surface prior to tailings impoundment construction was completed. A detailed exploratory investigation of the general area of the buttress was not completed, primarily because of access restrictions and time constraints. The bases for the assessment were the results of the piezocone soundings and available topographic mapping, from which typical cross sections extending through the failure area and within the Chute were developed. In particular, the significant decrease in pore water pressure and the pattern and rate of movements of survey monuments indicated the stability of the failure area had been enhanced significantly by the end of November 1997, with continuing improvement anticipated.

Several cross sections through the failure area were developed, based on contours of the area prior to construction of the tailings impoundment (circa 1970) with a contour interval of 5 feet, topographic mapping from 1993 with a contour interval of 10 feet showing impoundment construction completed in 1987 prior to the placement of waste rock, and topographic mapping with a contour interval of 2 feet based on aerial photography from October 25, 1997. The most critical cross section (Figure 3) extended through the slide mass in a southwesterly direction to the ridge forming the former western boundary of the tailings impoundment. This section was considered for stability assessment after consideration of the

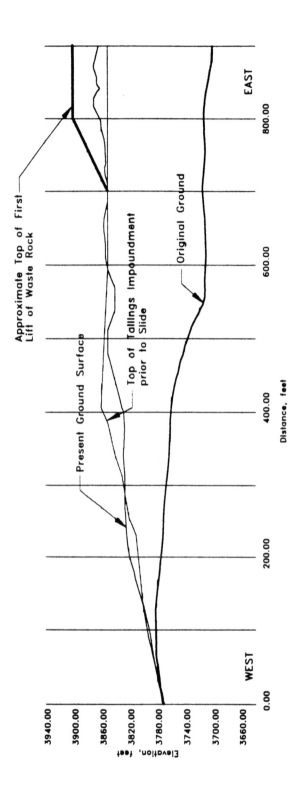

FIGURE 3 : CROSS SECTION THROUGH FAILURE AREA

482

impoundment layout, estimated thickness of the remaining tailings, orientation of natural drainages and slope orientation to maximize the driving shear stresses. It appeared that the overlying waste rock had displaced the tailings, leaving a new tailings/waste rock interface 40 to 50 feet below the pre-incident interface. Though a linear interface was conservatively assumed for the analysis, the existence of such an interface was considered unlikely. Cross sections oriented in a more east to west direction were buttressed by the hills located west of the slide area.

The tailings were analyzed as undrained, with a low steady-state shear strength, which is appropriate for post-liquefaction stability analysis. The ratio of steady-state strength to effective overburden stress was assumed to be equal to 0.08, resulting in an undrained shear strength on the order of 400 pounds per square foot (psf) for the tailings located in the area where the piezocone soundings were completed and large pore water pressures were measured. This strength also was assumed for the tailings located near the western boundary of the slide area, which was unlikely given the drainage that was occurring. The calculated safety factor for a deep seated, sliding block failure surface was 1.09; safety factors for other potential failure surfaces, including surfaces located near the toe of the remaining slide mass, exceeded 1.2. Based on these and other evaluations considered, it was determined that the slide debris within the failure area above elevation 3780 feet had an adequate degree of safety to allow continued removal of debis from Pinto Creek.

The thickness of the slide debris within the Chute was estimated, based on topography prior to construction of the tailings impoundment and subsequent to the slide. The depth of the tailings on the approximate 20 percent slope above Pinto Creek was estimated to be 2 to 3 feet over a significant portion of the slope, with the tailings having infilled low areas in the natural topography. At many locations, the tailings appeared to be present as only a thin veneer. Other than a lobe of tailings with a maximum depth of about 14 feet located near the top of the Chute on the south side of the failure area, the depth of the tailings was determined to be less than 6 to 8 feet. Considering the natural development of preferential drainage channels within the tailings on the slope, resulting in continued drainage of the tailings, it was concluded that there was only a very low risk of continued movement of the slide debris within the Chute, allowing removal of the debris to proceed.

4 STABILIZATION BUTTRESS DESIGN & CONSTRUCTION

Development of alternatives considered for the stabilization of the failure area included evaluation of the following issues:

The types of equipment which could efficiently perform debris removal, including conventional earth moving equipment,

Seepage control during excavation, including local dewatering as necessary,

Stability of temporary excavation slopes in the tailings and debris, particularly along the eastern (upslope) side, including adjustment of slopes as required in response to material changes and seepage conditions,

Safety considerations, particularly the uncertainty which existed at the time regarding the stability of saturated tailings with intermittent blocks of mine waste rock,

Material availability, with the intent of developing a simplified barrier design using readily available materials and conventional construction procedures, and considering the required height of the structure and the need for permanent seepage and drainage controls at the eastern toe and beneath the barrier, and

Developing a design that would allow for future remedial construction within the area disturbed by the slide.

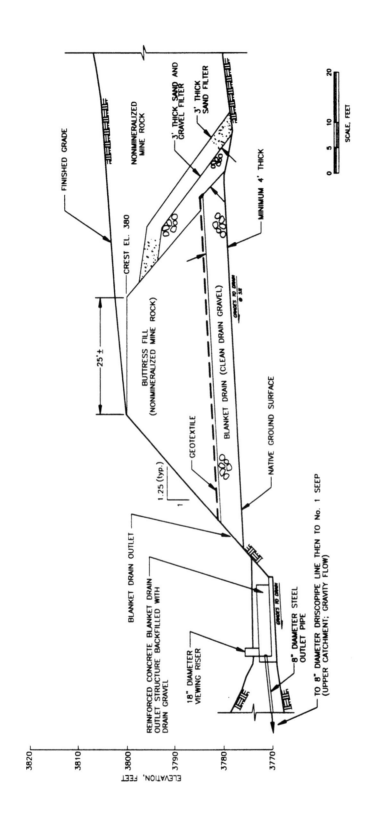

FIGURE 4 : TYPICAL CROSS SECTION THROUGH STABILIZATION BUTTRESS

The timing of the development of the preliminary design, and its acceptance by MSHA, was critical to the removal action. By the end of November 1997, separate haul roads had been established to Pinto Creek to allow removal of the debris in this area, and debris removal had extended downstream to about the confluence of the Chute and Pinto Creek, an area characterized by near vertical bluffs about 50 feet in height. Clarification of the stabilization plan and assessment of the stability of the slide mass were essential to MSHA allowing continued removal of the debris in Pinto Creek.

The basis for the design was the topographic mapping previously discussed, from which typical cross sections along the alignment of the buttress were developed. The buttress was designed to be approximately 420 feet in length, extending from a tie-in to competent bedrock at the north end of the slide area to the intact No. 1 Tailings Dam at the south end. After consideration of several alternative alignments, the buttress was located to take advantage of an underlying natural topographic ridge and swale, a relatively flat area with an elevation of about 3780 feet. This location also had the advantage of minimizing the amount of slide debris to be removed, with estimated depths at the centerline of the buttress varying from 20 to 25 feet.

The location of the western toe of the buttress controlled its alignment, since for purposes of stability the toe was not to extend onto the steep slope at the top of the Chute. The preliminary design included a 20-foot high buttress extending to about elevation 3800 feet with a crest width of about 40 feet. The buttress was to be constructed using non-mineralized waste rock readily available on the No. 14 Mine Rock Dump. The design also incorporated a section of drain rock placed at the eastern toe of the buttress to intercept seepage flows from the slide debris. This drain was then to be connected to three blanket drains extending through the buttress, from which flows would be collected and routed to an existing seepage collection pond (the No. 1 Seep) upstream of the Chute.

Access to the buttress area was provided by haul roads located at the north and south sides of the slide area that extended to the top of the waste rock surface. The haul roads ultimately allowed an efficient, single direction flow of truck traffic. The north haul road required extensive removal of saturated tailings and slide debris, and the south haul road had to be cut into the face of the No. 1 Tailings Dam. Both haul roads traversed the existing tailings surface at locations far removed from the slide area, requiring placement of as much as 50 feet of waste rock to provide shallow grades for the haul trucks to travel to the top of the waste rock at elevation 3900 feet.

Construction of the buttress was initiated at its south end, which was more readily accessible to equipment. Removal of tailings and slide debris generally progressed to the north, exposing the weathered bedrock and stiff soils forming the ridge, until such time that the north haul road was completed. Difficulties encountered during construction included removal of very wet to saturated tailings that slumped toward the excavation and seepage flows of several tens of gpm which initially ponded on the native bed rock and soils, and then against the east toe of the buttress as it was being constructed. Continuous pumping from temporary collection ponds and routing of seepage flows was required. Experience with the removal of tailings, however, had been gained during the ongoing removal activities in Pinto Creek. The tailings were removed with track-mounted excavators and bulldozers, and hauled to designated disposal sites located on the top of the No. 14 Mine Rock Dump.

The buttress was constructed using non-mineralized rockfill borrowed from the mine open-pit and minor amounts of mine-run rock that previously had been placed on the No. 14 Mine Rock Dump. The mine-run rock was placed within the buttress during the middle of the construction period, such that these materials were largely encapsulated within the finer grained rockfill. The rockfill was placed by end-dumping from haul trucks, dozing into nominal one-foot thick lifts and compacted using bulldozers and segmented rollers. The rockfill utilized was restricted to a maximum particle size of one foot; larger particles were winnowed to the edges of the structure. The side slopes of the buttress were constructed at angle-of-repose for the rockfill (about 1.5H:1V or slightly steeper).

A typical cross section through the buttress is shown in Figure 4. The buttress included three blanket drains, as considered in the preliminary design, comprised of commercially produced coarse drain rock covered with geotextile, which cross beneath the full width of the buttress at selected locations. The blanket drains are connected to a toe drain (which also consists of coarse drain rock covered with geotextile) which extends along the eastern toe of the buttress. In this regard, Figure 4 presents somewhat schematic details, since construction of the toe drain, in particular, required placing drain rock adjacent to very wet to saturated tailings and essentially "chasing" ponded water in order to collect seepage flows. The initial geotextile encapsulated toe drain become blinded by the tailings and was replaced with the separate sand and gravel drains shown, allowing complete collection.

The toe drain and blanket drains route seepage from the tailings and debris upstream of the buttress by transporting the collected flow beneath the structure. Subsequent to completion of the buttress, additional gravel-encapsulated drains were extended to the north to collect seepage flows that had historically occurred down slope from the impoundment and apparently from a pond adjacent to another mine facility. Reinforced concrete outlet structures, which were not considered in the preliminary design and which were fitted with risers to allow viewing the flow through the structures, were constructed at the down gradient ends of the three blanket drains. The outlet structures, in turn, were connected to 8-inch diameter HDPE pipe via 8-inch diameter steel pipe. The HDPE pipe transports the collected flow to an existing seepage collection pond (the No. 1 Seep), from which the flow is then routed back to the PVO process circuit. The volume of collected flow can be readily monitored at the No. 1 Seep, since no additional flows are collected by the system.

5 REHABILITATION OF FAILURE AREA

Subsequent to completion of the stabilization buttress, and recovery of the partially buried PVO vehicles, the failure area up slope of the buttress and extending to the limit of the intact No. 14 Mine Rock Dump was graded and contoured to provide safe, stable slopes. The regraded slopes generally were constructed from the head scarp areas toward the buttress and from the buttress toward the head scarp areas. The regraded slopes range from about 8 percent or less just up slope from the buttress to about 25 percent within the major portion of the area. The slopes were somewhat steepened at the north and south edges of the failure zone to match the existing mine rock dump slope of about 1.5H:1V. Regrading also included capping the stabilization buttress. Regrading required removal of additional tailings and debris, particularly just up slope from the buttress, and capping these areas with non-mineralized waste rock. Following regrading the area was revegetated using a seed mixture specifically formulated for the environment.

The regrading established a drainage swale that extends from the eastern most section of the failure zone toward the southwest. Where this contoured drainage intersected the hill located northwest of the buttress, it was turned to the south to connect with the original drainage within the Chute. The surface drainage was covered with coarse rock riprap (erosion protection) to protect the capping material. Construction of the drainage swale, and the regrading, was completed in accordance with a conceptual grading plan, but as with the construction of the stabilization buttress, modifications were made as construction proceeded in response to specific conditions. In general, rehabilitation of the area upslope from the buttress was completed in a manner consistent with the removal action objectives in that it served to minimize or eliminate the potential for future impacted runoff or seepage from exiting the area in an uncontrolled manner.

6 CLOSING COMMENTS

Construction of the stabilization buttress in a timely manner required the extraordinary commitment and daily communication of BHP personnel, regulatory agencies and all other personnel involved. Design of the stabilization buttress was initiated in late October 1997, shortly after the incident occurred, and completed for all practical purposes by mid-January 1998. Detailed plans were not prepared and only general specifications were provided for the components comprising the buttress and its associated drainage system, requiring almost daily decision making by a partnership of designers and construction personnel. The stabilization buttress was only one element, but an important element, that allowed of 99.98 percent of the debris discharged into Pinto Creek to be removed by the summer of 1998. Long-term monitoring of the previous failure area, the stabilization buttress and its drainage system, including additional instrumentation installed after completion of construction, indicates the general stability of the area continues to be enhanced by ongoing drainage of the tailings and debris underlying the area.

Evaluation of mine tailings impact to the environment in mineralized terrains

R.A. Mongrain, B.E. Lary & T.O. Looff
AGRA Earth and Environmental Incorporated, Phoenix, Ariz., USA

E.L.J. Bingham
BHP Copper Incorporated, Miami, Ariz., USA

ABSTRACT: A tailings impoundment failure resulted in mine tailings and waste rock being deposited into Pinto Creek in Arizona. Resulting impacts were quantified using comprehensive characterization programs. These programs included collection and analysis of data to evaluate the tailings debris, mineralized terrain, effects of historic mining operations, surface water, and alluvial groundwater. Potential effects to water quality, riparian and upland flora and fauna were also evaluated. Geochemical models then integrated data from solid media (Debris, remobilized Debris, residual Debris, Earlier Tailings Residue, and Natural Stream Sediments), and aqueous media (pore water, surface water and alluvial groundwater).

Characterization and modeling efforts: 1) identified Potential Constituents of Concern; 2) evaluated geographic distribution of impacts; 3) assessed acid-generation potential of residual Debris and the buffering capacity of alluvial sediments; 4) evaluated adequacy of the Removal Action; 5) assessed impacts and interaction of surface/groundwater quality; and 6) facilitated ecological hazard and human health risk assessments.

1 INTRODUCTION

On October 22, 1997, BHP Copper Inc.- Pinto Valley Operations (PVO) experienced a failure of a portion of the slope of the No. 14 Dump and No. 1-2 Tailings Impoundment (the Incident). The failure resulted in the release of materials consisting of tailings, un-mineralized mine rock and pore water (Debris) from the tailings impoundment into the dry streambed of Pinto Creek, in east-central Arizona (Figure 1). Slope failure of the No. 14 Dump and the underlying mill tailings resulted in a 600-foot-wide breach in the west-facing No. 1-2 tailings embankment. The release of Debris through this breach resulted in the formation of a roughly circular depression that encompasses about 14.3 acres inside the original trace of the facility footprint (Figure 1). Three PVO vehicles and two fuel-powered light stands were captured in the slope failure incident. Although the vehicles were partially buried within the Debris slide during the incident, no injuries occurred to mine employees.

Upon exiting the facility, the Debris flowed in a viscous mass down the west-facing valley slope of a small, dry arroyo tributary (the Chute), and eventually settled into the dry streambed of Pinto Creek. The direction of flow was southerly, dropping in elevation approximately 400 feet over a distance of approximately 2,000 feet. Upon reaching the valley bottom, the Debris flowed both upstream and downstream along the stream bed of Pinto Creek. The approximate extent of this migration was about 900 feet upstream and 2,500 feet downstream. The ribbon of Debris along and within the drainage was estimated to contain approximately 370,000 cubic yards of wet, sandy to silty material. The total areal extent of impact within the confines of Pinto Creek was about 8.1 acres, with the Debris obtaining a maximum thickness of

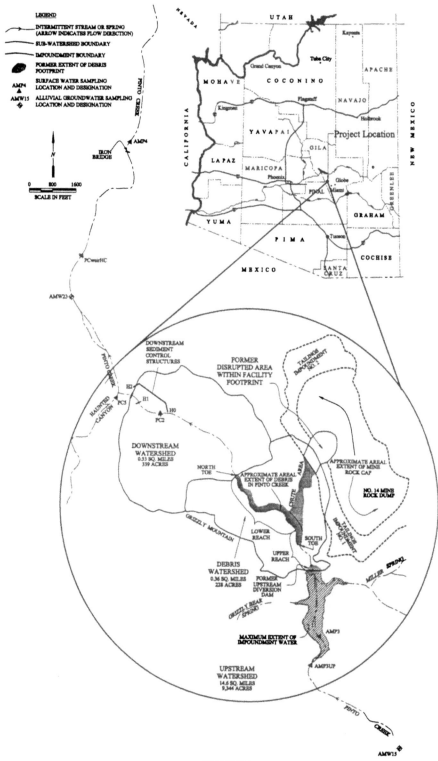

FIGURE 1
VICINITY MAP AND PROJECT AREA DETAIL

about 42 feet. This maximum thickness occurred near the region where the Debris entered the drainage bottom, with most of the mass averaging about 20 feet in thickness.

A 14.7-square mile portion of the Pinto Creek watershed lies above the southernmost extent of the Debris mass. Below this point, to the confluence of Pinto Creek with Haunted Canyon, about 4,000 feet northwest and downstream of the Debris mass, watershed contributions originate from a 0.9-square mile area. In order to minimize downstream impacts from the Incident associated with the upcoming winter rainy season, water and Debris management structures both upstream (Upstream Diversion Dam) and downstream (Gabion Sediment Control Structures [H0, H1, and H2]) were constructed. In this manner, impacts resulting from the Incident and efforts associated with removal of the Debris and restoration of Pinto Creek (the Removal Action) could remain focused on a more confined area upgradient of the Haunted Canyon confluence.

By July 1998, the construction activities associated with the Removal Action were complete. A total of 99.98% of the Debris had been removed, the water and sediment control structures were demolished, and reconstruction of the creek morphology and revegetation efforts had been completed. Also by July 1998, the characterization programs described herein had been completed and an evaluation of data to determine the adequacy of the Removal Action had been conducted. An evaluation of the impacts associated with the Incident and potential future effects were completed during the Fall of 1998. A Long-Term Monitoring program was developed and initiated to evaluate the recovery of affected media in Pinto Creek. The following sections describe the efforts completed to characterize and evaluate impacts to the environment as a result of the Incident.

2 CHARACTERIZATION OF DEBRIS MATERIALS & AFFECTED MEDIA

Certain of the impacts to the environment, specifically in regard to water quality, risk to human health, and risk to other ecological receptors, could only be understood and quantified if the constituents of the Debris and those of various affected media were known. To this end, a sampling and analysis program was conducted to characterize the physical and chemical properties of the Debris, the natural alluvium, earlier tailings residue material, and groundwater and surface water, both upstream and downstream of the Debris. Additionally, the surface water quality sampling was performed to comply with the PVO National Pollutant Discharge Elimination System (NPDES) permit requirements. The characterization program also evaluated flora and fauna present in the area to provide an understanding of the number and type of these resources that may have been affected as a result of the Incident.

The characterization data provided an understanding of conditions prior to and following the Incident in the area impacted by the Debris and selected un-impacted upstream and downstream locations. The overall intent of the characterization program was to identify potential constituents of concern (PCOC) which might originate from the Debris, support an evaluation of potential impacts to the environment and ecosystem related to the PCOC identified and evaluate the adequacy of the Debris removal effort.

A geographic distribution of sample sites in both solid and liquid media provided the basis for an analysis of changes in parameter values from upstream to downstream. An appropriate distribution of upstream sample sites was incorporated in order to establish pre-Incident conditions for evaluating temporal changes in downstream data. The characterization program was also designed to address variability in the geology and the degree of impacts along the Pinto Creek channel. As such, the program divided the Creek into three watershed segments, the Upstream Watershed, the Debris Watershed, and the Downstream Watershed. The watershed area extending from the headwaters (origin) of Pinto Creek downstream to the Upstream Diversion Dam is termed the Upstream Watershed. Water chemistry in and downstream of this watershed had previously been impacted by the local presence of naturally occurring mineralized terrain. The watershed area extending from the Upstream Diversion Dam (just above the southern extent [or South Toe]) of the Debris mass downstream to the northern extent of the Debris (North Toe) in and adjacent to the Pinto Creek stream channel is termed the Debris

Watershed. This area contained the initial Debris mass. The watershed area extending from the North Toe to the crest of the H2 gabion structure located just upgradient of the Haunted Canyon confluence (Figure 1) in and adjacent to the Pinto Creek channel is termed the Downstream Watershed. This area was characterized by the presence of fine sediment carried downstream from the Debris mass.

2.1 Characterization of solid media

Three types of Debris materials are referenced throughout the project: the original Debris, the remobilized Debris, and the residual Debris. The term "Debris" refers to all materials initially released during the October 22, 1997 Incident. This material is a mixture of mine rock, tailings, pore water, soil and other materials deposited outside of the Tailings Impoundment, onto the Chute Area, and into the Pinto Creek channel (Figure 1). Tailings and pore water made up the bulk of the material and influenced the manner of movement and transport of the material into Pinto Creek. The remobilized Debris represents fine-grained sediment that was mobilized from the Debris mass by pore water outflow and surface runoff flowing downstream from the North Toe of the Debris. Residual Debris refers to the small quantity (0.02%) of Debris material that remained in the creek following the completion of the cleanup efforts.

The alluvial sediments placed naturally in the Pinto Creek channel were designated as Natural Stream Sediment (NSS) or natural alluvium. Tailings material in the Pinto Creek channel left from historic tailings excursions (primarily 1940s and 1993 incidents) that were deposited prior to the Incident are referred to as Earlier Tailings Residue (ETR).

A statistically significant set of samples was collected from the original Debris material between the South Toe and North Toe. A second set of samples was collected to characterize the remobilized Debris in the creek bottom in the Downstream Watershed, between the North Toe and the gabion sediment control structures. The NSS and ETR materials were characterized in like manner. The sample sets were analyzed for physical characteristics (grain size, permeability, moisture content and porosity [calculated]), acid-base accounting and an extensive suite of chemical constituents by completing a whole rock analysis (Figure 2). Additionally, the solid media samples were subjected to extraction and analysis for leaching potential in accordance with the Nevada Meteoric Water Mobility Test (MWM) and the Synthetic Precipitation Leachate Procedure (SPLP).

2.2 Characterization of liquid media

Surface water and shallow alluvial groundwater in the Pinto Creek drainage were characterized to evaluate changes in water quality as a result of the Incident. These media were characterized concurrently due to the integrated nature of the flow dynamics between surface water and the shallow alluvial groundwater throughout the Pinto Creek Watershed. A literature search identified field studies conducted prior to the Incident that reported four different groundwater/surface water relationships along Pinto Creek (Rivers West, 1991). Three of these relationships described losing stream conditions, where surface water is infiltrating and recharging the alluvium. The fourth is an interaction in which the stream is gaining water from the alluvial aquifer, resulting in increased surface water flow. Previous field studies also indicated that the vast difference in permeability between the shallow alluvium and bedrock would preclude measurable impacts to the bedrock aquifer, and therefore, groundwater quality in bedrock was not characterized as part of the project.

A network of monitoring locations was developed to characterize the quality of surface water and groundwater following the Incident. The location of sample sites was predicated upon several criteria, including: 1) site access during ideal and less than ideal conditions; 2) consistent exposure of the media to be sampled; and 3) general compatibility between existing data (e.g. PVO NPDES ambient monitoring point historic data), data acquired during the characterization and interim monitoring activities, and data collected during the course of long-term

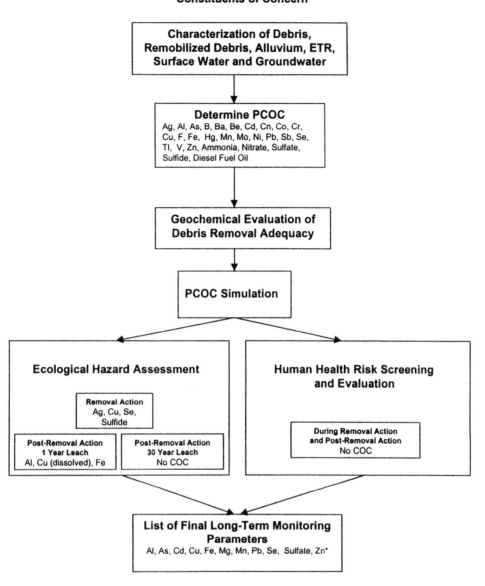

Flow Chart of Work Tasks to Determine the
Constituents of Concern

**Characterization of Debris,
Remobilized Debris, Alluvium, ETR,
Surface Water and Groundwater**

Determine PCOC
Ag, Al, As, B, Ba, Be, Cd, Cn, Co, Cr,
Cu, F, Fe, Hg, Mn, Mo, Ni, Pb, Sb, Se,
Tl, V, Zn, Ammonia, Nitrate, Sulfate,
Sulfide, Diesel Fuel Oil

**Geochemical Evaluation of
Debris Removal Adequacy**

PCOC Simulation

Ecological Hazard Assessment

Removal Action
Ag, Cu, Se,
Sulfide

**Post-Removal Action
1 Year Leach**
Al, Cu (dissolved), Fe

**Post-Removal Action
30 Year Leach**
No COC

**Human Health Risk Screening
and Evaluation**

**During Removal Action
and Post-Removal Action**
No COC

**List of Final Long-Term Monitoring
Parameters**
Al, As, Cd, Cu, Fe, Mg, Mn, Pb, Se, Sulfate, Zn*

*Including constituents from site NPDES permit

Figure 2

monitoring of the site and adjoining areas. Water quality data was compared to applicable standards for surface and groundwater quality and evaluated spatially and temporally to monitor changes following the Incident. Surface water and groundwater were analyzed for total and dissolved concentrations of the same chemical constituents as those evaluated for solid media.

Review of existing literature and data and reconnaissance level surveys in the field provided information on pre-Incident environmental conditions to characterize media and natural attributes potentially affected by the Incident. Information on geomorphology, native flora and fauna was obtained from a review of published literature, interpretations of aerial photographs provided by PVO and field surveys. Supplemental information was obtained from PVO and from hydrogeologic investigations performed as part of the associated Environmental Impact Statement (EIS) for the adjacent Carlota Copper Mine project (USDA Forest Service, 1997). Historic surface water flow and climatic data were compiled from previous reports, and real-time flow data and weather information, critical to decision-making, were obtained through the United States Geological Survey, the Salt River Project and PVO.

3   STATISTICAL ANALYSIS OF DATA

Data collected during the program underwent a strict quality control review. Quality assurance/quality control (QA/QC) criteria address the following data attributes: accuracy, precision, completeness, representativeness, and comparability. Factors that relate to the quality of data included all aspects of the data collection process, starting with sample collection methods through laboratory analysis and database review. A rigorous standardization protocol was established early on so that all field data were reported in an identical manner regardless of the site, sample location, or operator.

The data compiled from characterization and monitoring efforts were subject to statistical validation in order to accurately assess contamination and demonstrate regulatory compliance. This required statistical validation of data significance and statistical evaluation of the significance of a detected exceedance. The sampling program conducted for the Removal Action was designed to accommodate a 95% confidence level, which adequately addressed the prospect of assuming an invalid conclusion based upon the sample data. A 95% confidence level implies that 5 sample events in 100 will reflect aberrant or otherwise inconsistent results.

In accordance with U.S. Environmental Protection Agency (EPA) guidance documentation (EPA, 1992), a 19-sample protocol establishes statistical validity (at the 95% confidence level) for a non-parametric statistical analysis. This program adopted a 21-sample protocol in order to accommodate the possibility of sampling mishaps.

The original Debris mass was sampled at 19 randomly selected locations (plus two replicates) within the Pinto Creek channel. In similar fashion, twenty-one samples each of the Remobilized Debris, ETR, and natural Alluvium (19 samples plus two replicates) were collected to evaluate physical and chemical properties of these materials. Each surface water location was sampled on 21 consecutive days during December 1997. Groundwater monitoring sites were sampled on a weekly basis starting in December 1997 for 21 weeks. In addition to characterizing the specific medium the sampling strategy was designed to support a monitoring schedule which encompassed the full duration of the Removal Action project and provide a basis for continuity through post-removal monitoring.

A series of statistical data evaluations were then performed to refine the data package for use in geochemical modeling efforts. Standard statistical parameters were defined, including identification of data outliers and non-useable data. In the event a certain parameter was not detected during the characterization event, it was determined not to be a PCOC, and therefore, an evaluation regarding removal of that parameter from future monitoring was completed.

4   GEOCHEMICAL APPRAISAL OF DEBRIS REMOVAL ADEQUACY

Although initial geochemical information was used to establish target cleanup levels, the nature of the Removal Action allowed the functional removal of most of the Debris. The geochemical information provided documentation that the removal of 99.98% of the Debris was sufficient to protect human health and the environment. This adequacy appraisal included an evaluation of geochemical issues associated with potential impacts to human health and the potential long-term impacts to the environment possible from the small volume of residual Debris remaining in Pinto Creek following completion of the Removal Action.

Two primary geochemical issues were considered in this appraisal:

1. The potential for persistent acidification, which could be due to the oxidation of residual sulfides which, in turn, may result in an accelerated release of metallic constituents from the Debris; and

2. The potential that deleterious constituents may be released from the Debris by short- and/or long-term weathering or local acidification related to natural processes.

This appraisal was based upon the 21 sample characterization data for Debris and remobilized Debris, additional analyses of the residual Debris, and direct observation of the amount and location of residual Debris, sulfide material, ETR, and NSS (natural alluvium) within and downstream from the Debris Watershed. The Debris, as well as pre-existing Pinto Creek channel materials such as ETR and Alluvium were evaluated in order to assess the influence of each material on the overall chemistry observed in soils, surface water and groundwater. For example, the Debris could locally generate an acidic seepage that the natural Pinto Creek Alluvium has the capacity to neutralize. Additionally, one source material may release a constituent which is absorbed by another material, thus decreasing the overall observed concentrations for that particular constituent.

4.1 *Analysis of acid generation/neutralization potential*

The initial focus of this appraisal was to quantify the potential for residual Debris to generate acidic conditions within the natural environment. Acidic conditions may increase the mobility of certain metals present in the Debris. Therefore, the acid generation potential of the Debris and residual Debris, and the sulfide content of the Debris was determined. The Acid-Base Account (ABA) value for each material was determined where the reported value represented the potential that a given material would generate or neutralize acid. The method used evaluated total sulfur, as well as the amount of sulfur present as sulfide, which may oxidize to form an acid, and the amount of sulfur present as sulfate, which would not generate acid.

The adequacy of Debris removal was evaluated using the acid generation potential of the original Debris and acid neutralization potential of existing materials in morphologically similar sections throughout the Downstream and Debris Watersheds. This approach became the basis for the determination of adequacy for each particular section as the Removal Action progressed. The quantity of residual Debris was measured by reconnaissance and mapping in each section, and that quantity evaluated against the quantity of alluvial materials to determine overall acid generation/neutralization potential. The evaluation of characterization results and the results of the ABA analysis indicated that all stream sections have sufficient acid neutralization capacity to offset any acid generation potential for the small amount of residual Debris present in a particular section. Follow-up sampling and analysis of residual Debris in the Downstream Watershed near the end of Debris removal activities provided analytical results which support this conclusion.

4.2 *Analysis of short-term (acute) or localized constituent leaching potential*

The MWM and SPLP extractions and analyses were used to determine potential short-term (acute) and localized impacts to surface water quality, due to the leaching of PCOC from the Debris, respectively. The MWM test yields information regarding the amount of readily leachable PCOC in ambient (local) water conditions, while the SPLP analysis is used to evaluate readily leachable PCOC in acidic conditions, due to the localized oxidation of sulfides.

The purpose of these evaluations was to determine if the quantities following the Incident would contribute to exceedances of applicable surface water quality standards (SWQS) given the concentrations of these constituents already present in Pinto Creek due to the naturally occurring mineralized terrain and historic mining activity. The constituent concentrations leached from the Debris as a result of MWM and SPLP testing were added to existing constituent concentrations present in the system to complete the analysis. The evaluations indicated that exceedances of SWQS were more likely due to occur from upstream water quality entering the system or contributions from the natural Alluvium because of the mineralized terrain present in the area than from the Debris or the cumulative effect of the Debris and existing conditions.

4.3   *Analysis of Long-term (chronic) constituent leaching potential*

A quantitative whole rock (QWR) analysis was completed as part of the chemical characterization of Debris materials (e.g., Debris, Remobilized Debris, Alluvium, ETR). QWR analysis provided the chemical characterization required to evaluate PCOC release rates for various weathering scenarios and the long-term leachability of PCOC.

The results of QWR analysis were used to evaluate the potential concentrations of PCOC which might be generated by long-term weathering of residual Debris. The rate of weathering for the different materials varies according to geochemical environment, concentrations of the constituent in associated media and synergistic relationships with other commingled materials. The release of specific constituents depends largely on the weathering rates of the individual minerals in which these occur. Weathering rates of 1, 5, 10, 25, 30, and 50 years were simulated for each PCOC in order to provide a comprehensive evaluation. In each of these simulations the total PCOC content of the "weathered" material was released, at an even rate, over the time period specified. It should be noted that each such simulation was designed to evaluate the specified period of time only. The results of a ten-year simulation cannot be evaluated as the culmination of ten, individual one-year intervals. The volume of surface water flow applied in all simulations to calculate concentration units was based on the typical annual run-off from the Upstream Watershed. The results of the simulations, expressed in terms of PCOC concentration, were then compared to SWQS and aquifer water quality standards (AWQS) values to determine if a specific PCOC posed a potential long-term (chronic) risk of impact to water quality.

In even the most conservative scenarios, the results indicate that exceedances of SWQS or AWQS are unlikely to occur as a result of the Incident. In the Debris Watershed, SWQS or aquifer water quality standards (AWQS) were not exceeded by any of the PCOC included in the QWR simulations, including the most conservative one-year weathering scenario. Downstream Watershed simulations indicated that copper concentrations could exceed the dissolved SWQS if all of the copper content of the residual Debris could be weathered (released) in less than a five-year period. However, existing conditions in Pinto Creek (i.e., ETR chemistry) indicate that the prospect of a complete weathering of materials to release copper is highly unlikely in this time period. The total copper content of the ETR may be viewed as remnant copper remaining in the earlier residue following as much as 30 to 40 years of weathering. ETR QWR results suggested that the copper minerals within the residual Debris would weather under natural conditions in tens to hundreds of years. Overall, this geochemical appraisal indicated that the Debris removal activities were successfully completed.

5   HUMAN HEALTH RISK SCREENING EVALUATION

Empirical chemical data was available to assess potential risk and be protective of human health during the Removal Action. However, geochemical modeling was used to assess potential long-term exposure scenarios. The Post-Removal Action water exposure point concentration was defined as the total QWR concentration of PCOC that may leach from residual Debris into water originating at or above the Impacted Area (Figure 1). A standard geochemical model was used to estimate Post-Removal Action water exposure point concentrations. The QWR PCOC concentration was determined by the input concentration for Incident-related PCOC contribution, background PCOC contribution, and the cumulative contributions of background and the Incident. Naturally occurring alluvium, mineralized upstream terrain, and ETR are likely the primary sources contributing to concentrations of PCOC in background. Risk associated with future leaching of residual Debris was calibrated to a 30-year period for purposes of the human health risk screening evaluation. Future cumulative PCOC contribution combines the background contribution with the modeled residual Debris contribution to estimate the overall cumulative risk of PCOC concentrations at specific potential receptor locations during a hypothetical 30-year period following the Removal Action.

The human health risk screening evaluation indicated that there was no unacceptable risk to human health which may be related to past, present, or future effects as a result of the Incident or Removal Action. The geochemical analysis of the Debris and water quality analyses resulted in a list of dissolved and total metals identified as PCOC present in Pinto Creek Watershed as a result of the Incident (Figure 2). The exposure assessment determined that the human receptors most likely to be exposed to these PCOC were recreational users, ranchers, residents downstream at Roosevelt Lake and the population in the Phoenix metropolitan area using Roosevelt Lake water as a drinking source. In the toxicity assessment, published information regarding the potential toxicological effects of PCOC to the receptors was compiled. The risk characterization combined the results of each of the three prior components to determine if an unacceptable health threat was posed to human receptors. The results of the evaluation indicated that the Incident did not cause unacceptable risk to human health. Therefore, no Constituents of Concern (COC) were identified and site-specific target levels which would need to be met in order to be protective of human health were unnecessary.

6   ECOLOGICAL HAZARD ASSESSMENT

Ecological risk during the Removal Action was assessed using the empirical data collected and analyzed from the characterization sampling. Risk to aquatic life and wildlife following the Removal Action was analyzed by comparing predicted risk at selected background or reference sites to the cumulative risk at these same locations. This cumulative risk considers the effects of both the background water chemistry and the contributions from long-term weathering of the residual Debris that remains in the creek channel following the Removal Action. The approach for estimating Post-Removal Action PCOC concentrations at all sites was addressed in the geochemical models.

The results of the Ecological Hazard Assessment indicate negligible risk for wild life both during the Removal Action as well as under Post-Removal Action conditions. An acute (short-term) as well as chronic (long-term) risk for aquatic life existed for certain PCOC during the Removal Action. However, the short duration of the project and removal of the Debris material negated risk for those PCOC.

Two weathering or leaching scenarios were evaluated as part of the Ecological Hazard Assessment. The 1-year Leach Scenario was the most conservative assumption where all PCOC in the residual Debris leached in 1 year. This scenario indicated potential risk existed for aquatic life from selected PCOC. However, the more realistic 30-year scenario indicated no significant long-term impacts or risk to aquatic life within Pinto Creek.

7.   CONCLUSIONS

The sampling and analysis program created for this project provided an extensive compilation of chemical and physical characteristics of existing conditions in Pinto Creek and a thorough understanding of characteristics of the emplaced Debris materials. Data sets provided a background, or Pre-Incident representation that could be compared both spatially and temporally with conditions following the Incident, and evaluate changes which occurred as the Removal Action progressed. The project team assessed the data using geochemical models to provide real-time information to regulatory agencies, engineers and construction crews as removal of the Debris from the creek was completed and restoration of the natural attributes of the creek conducted. Geochemical appraisals were used to evaluate potential acute and chronic risks to human health and the environment associated with the Incident. The extensive QA/QC procedures and sound use of technical models demonstrates the Removal Action achieved its environmental objectives. Use of scientifically sound procedures of data collection and interpretation provides the basis for long-term monitoring which will document the recovery of the Pinto Creek Impacted Area.

REFERENCES

Rivers West, Inc. and Water and Environmental Systems Technology, Inc.1991. *Pinto Creek In stream Flow Assessment*. Tonto National Forest, Gila County, Arizona, Contract No. 43-8180-0-198.

United States Department of Agriculture Forest Service. 1997. *Final Environmental Impact Statement - Carlota Copper Project*. USDA Forest Service, Southwest Region, Volumes I-III.

United States Environmental Protection Agency. 1992. *Risk Assessment Guidance for Super fund, Volume 1: Human Health Evaluation Manual, Part A, Interim Final*. EPA/540/1-89/002. Office of Solid Waste and Emergency Response, NTIS: PB90-155581.

Tailings and Mine Waste'00 © 2000 Balkema, Rotterdam, ISBN 90 5809 126 0

Water management during the Pinto Valley removal action

R. E. Weeks
Golder Associates, Phoenix, Ariz., USA

R. Krohn
BHP Copper, Miami, Ariz., USA

T. H. Walker
AGRA Earth and Environmental Incorporation, Phoenix, Ariz., USA

ABSTRACT: The 1997 release of 370,000 cubic yards of material from the Pinto Valley Operations (PVO) copper mine No. 14 waste rock/tailings impoundment complex resulted in the blockage of Pinto Creek. The incident occurred in mountainous terrain below a 14-square mile segment of the Pinto Creek watershed above Roosevelt Lake, a source of drinking water for the Phoenix metropolitan area. As part of a CERCLA Time-Critical Removal Action, PVO implemented a strategy to prevent further water quality impacts. This strategy included managerial elements to facilitate rapid planning and execution, coupled with the design, construction and operation of both upstream and downstream containment and diversion systems. The criteria utilized to size containment, diversion and spillway capacities were developed by evaluating watershed and climatic characteristics, calculating various probabilities of exceedance, and considering both time constraints and other site limitations. Within 8 months, the tailings were removed from the impacted area, with no release of affected surface water below the downstream control system.

1 INTRODUCTION

The partial breach of the No. 14 waste rock and tailings disposal facility at the BHP Pinto Valley Operations (PVO) copper mine in Arizona on October 22, 1997 (the Incident) resulted in the excursion of 370,000 cubic yards of saturated tailings and entrained waste rock (Debris). This debris entered a steep tributary arroyo of Pinto Creek, subsequently flowing as a viscous mass into the main stream channel. With the normally wet winter season approaching in an anticipated El Niño year, the incident posed a distinct threat of further impacts to surface water quality. Pinto Creek flows to Roosevelt Lake about 21 miles downstream of the incident. Roosevelt Lake supplies drinking water for the City of Phoenix. The large debris mass blocked the Pinto Creek channel with a 42-foot high plug of loose, sandy material. The most urgent potential consequence was the risk of rapid downstream migration of dissolved inorganic constituents and affected sediment should runoff from the upstream 14.3 square-mile watershed accumulate behind and subsequently overtop the debris. Such an occurrence would have transported the unconsolidated sandy tailings and impacted surface water down the Pinto Creek channel, potentially affecting the water quality downstream.

The Incident impacted public lands under the stewardship of the USDA Forest Service (Forest Service), as administered by the Tonto National Forest. Due to the imminent possibility of further significant impacts, a rapid response program of abatement and restoration was established by the Forest Service under the protocols and requirements of a CERCLA Time-Critical Removal Action. The overall goals of the Removal Action were to remedy the consequences of the incident upon the environment, prevent further damage, and restore the impacted habitat to pre-incident conditions. Meeting all these goals required a focus on isolating the debris from storm water contributions of the surrounding and upstream terrain, and capturing any runoff or seepage having previous contact with the tailings. This effort required immediate attention at the outset of the rapid response, then continued diligence during the removal of debris, ensuing repairs to the streambed and adjacent terrain, and subsequent restoration of the terrestrial and aquatic habitats.

With the participation of the mine operator, the Forest Service, and other designated stakeholders, the Removal Action began with a strong definition of project objectives and responsibilities. With regard to managing the surface water, the physical constraints were initially evaluated, data needs were defined, and possible control strategies were developed. The methods and procedures used to select, design, construct and operate each control system were continuously analyzed, with the decision to proceed based on an interdisciplinary evaluation of alternatives. Each strategic decision was weighed in light of its effect upon other components of the Removal Action, and the current and anticipated short-term runoff conditions in the watershed.

Due to the urgency needed to prevent further impacts, the hydrologic aspect of the Removal Action was comprised of two related, but distinct components. The first objective was to establish appropriate design criteria for the sizing of water retention or diversion structures. The second task was to continuously evaluate both weather and watershed conditions to predict and monitor runoff during and following storm events. These predictions and the monitoring of storm inflows were used to operate the control structures, thereby providing sufficient retention and diversion capacity to prevent further downstream discharges of impacted surface water.

2 CONSTRAINTS AND OPPORTUNITIES

The excursion of PVO tailings into Pinto Creek resulted in a 3500-foot long ribbon of debris in a narrow and mountainous segment of the stream channel. With the exception of the upstream terminus of the debris, and near the location where the debris entered the streambed, no immediate equipment access along Pinto Creek was available. To further complicate the pending removal of the debris, an additional 4500 linear feet of stream bottom was present between the downstream terminus of the tailings (the North Toe) and the first point along the creek accessible to heavy equipment. This 4500-foot reach of Pinto Creek was also impacted by limited discharges of surface water runoff, fine tailings sediment and seepage originating from the debris. The possibility of mobilizing equipment from this point up the creek bottom was considered, as was the construction of additional access roads. These alternatives and the feasibility of constructing the primary surface water controls immediately adjacent to the debris were eliminated from further consideration. Unacceptable environmental damage would have occurred in the riparian corridor and within the canyon uplands flanking the stream channel.

Due to the proximity of the active mining and milling operations, sufficient electrical power to operate high output pumps was available within 900 feet of the upstream extent of the debris mass. As the Removal Action progressed, electric power service was improved and extended to the area near the upstream terminus of the debris, and high-power portable generators were mobilized to the confluence of Haunted Canyon with Pinto Creek, about 4500 feet downstream from the North Toe.

An additional factor associated with the mining operation clearly influenced the selection of preferred action alternatives and the efficiency of ensuing construction and operation of the surface water control facilities. This factor was the mine personnel's familiarity and experience with large-scale earthmoving, and the installation and operation of pumping systems. For example, through the immediate action of PVO, the existing access roads were widened and modified, and an unconventional but effective haulage pattern was improvised.

Removal of the debris from Pinto Creek clearly involved difficult excavation and haulage in the narrow confines of the stream channel, requiring expert and innovative construction management. Designing a workable system for controlling the flow of surface water was dependent upon a full recognition of these constraints of access and debris removal. It became evident early in the Removal Action that the primary upstream and downstream control structures would need to be located at sites somewhat removed from the debris mass. In addition, detention basins or other structures in the area directly impacted would inherently interfere with debris removal due to the narrow configuration of the affected stream channel. However, earthen materials suitable for embankment construction were readily available on the mine property, and the project team possessed the expertise necessary to quickly process and transport these borrow materials.

Although the threat of significant runoff from the upstream 14.3 square-mile watershed was ever-present throughout the course of the Removal Action, the incident deposited tailings in a predominantly dry streambed. Pinto Creek is typical of streams in the upland, semiarid Southwest. Over much of the year, only discrete, relatively short segments of this portion of Pinto Creek are perennial, with these sustainable flows commonly well under 1 cubic foot per second. The thickness of the stream alluvium is quite limited and subflow is often interrupted by bedrock exposures along the channel bottom. Substantial streamflow occurs only as a direct response to precipitation events. Fortunately, the environmental issue at the outset of the Removal Action was the management of pending runoff events, not the immediate containment of large volumes of affected water caused by large base flows or alluvial subflow.

3 PINTO CREEK WATERSHED

For the purpose of selecting, designing and operating both upstream and downstream control structures, that portion of the Pinto Creek watershed capable of impacting the debris removal activities were divided into three segments. The first of these segments was the 14.3 square miles of contributing watershed above the debris blocking the creek channel. The second subdivision encompassed the limited, flanking terrain adjacent to the debris, consisting of a total of 0.36 square miles. The third and final segment was the 0.34 square mile sub-watershed between the downstream terminus of the debris and the point of nearest equipment access downstream of the impacts. This downstream point was immediately above the confluence of Pinto Creek with Haunted Canyon, a tributary with a substantial watershed of 12.6 square miles. Containing the sediment and water quality impacts above this confluence was a primary focus of the Removal Action.

That portion of the Pinto Creek watershed with the potential to contribute runoff to the impacted area ranges in elevation from about 3430 feet near the debris deposit to greater than 6650 feet to the southeast in the Pinal Mountains. Stream gradients, for tributaries as well as for the main-thread channel, are generally steep. The channel gradient of Pinto Creek proper in the reach upstream of the disturbed streambed averages about 180 feet per mile; gradients in Pinto Creek in higher-elevation areas near the watershed boundaries are greater than 300 feet/mile. The occurrence of prominent and laterally extensive outcrops of competent bedrock are common throughout the watershed, with the more gentle slopes mantled with rocky colluvium. Soil development is quite limited, and the vegetation is typical of the Arizona high Sonoran desert.

4 DEVELOPING RAPID RESPONSE CRITERIA

As the Removal Action Plan proceeded immediately following the incident, several key issues were identified. To prevent storm runoff from contacting and mobilizing the debris in the Pinto Creek channel, the rapid design and construction of a stormwater management system, including diversion structures, impoundments, pumping systems, and other components, were required. This in turn required establishment of hydrologic design criteria for the system components.

It was estimated that about 6 months to 1 year would be required for the completion of the Removal Action. The selection of the storm event to be used in design of the components of the management system would determine the estimated level of risk of occurrence of an event exceeding the design event. For example, a design storm with a 2-year return period corresponds to a probability of 50 percent that the event would be exceeded during any one year. (The annual exceedance probability is the inverse of the return period.) In addition to determining this acceptable level of risk, other issues to be resolved included the selection of appropriate analytical and modeling procedures and which data sets and other parameters should be used in the determinations. During the Removal Action, several variations in hydrologic design criteria were considered as part of an evolving, culminating in a process a final set of criteria which were utilized in the design of the system ultimately constructed.

501

The first proposed hydrologic criterion was based on the daily precipitation record from a rain gauge operated for about 25 years on the mine property, near the incident site (the PVO gauge). At the outset of the Removal Action, with the records from this rain gauge readily available, a 2-year/24-hour storm with a precipitation depth of 1.83 inches was initially considered for use in capacity design of the control structures. This preliminary design storm was based on a statistical analysis of PVO precipitation data for the 6-month period of November through April from 1972 through the Fall of 1997.

The next refinement in the estimation of peak flows and runoff volumes utilized regionalized regression relationships for flood frequency for the southwestern U.S. developed by the U.S. Geological Survey (Thomas et. al., 1994). Using these relationships, the 2-year return-period peak runoff on the upper 14.3-square mile watershed was estimated to be 222 cubic feet per second. By definition, use of the 2-year storm recurrence interval implies a probability of exceedence of 50 percent in any given year. With the anticipated time necessary to complete the Removal Action between 6 months and 1 year, this probability of exceedance was deemed unacceptable, and the selection of the design storm event was revisited.

The use of a design storm with a 5-year return period, corresponding to an annual exceedence probability of 20 percent, was then proposed. The 5-year/24-hour storm precipitation depth for the upstream watershed was determined in accordance with procedures presented in the NOAA Atlas 2 for Arizona (Miller et al., 1973) to be 3.56 inches. Discussions also took place at this time as to the appropriate SCS Curve Number (CN) and the temporal distribution of the design storm rainfall. While summer thunderstorms in Arizona tend to be intense but short in duration, winter storms tend be less intense, but last longer. Because the incident occurred in autumn, most of the Removal Action program would be completed during late autumn and winter, so a winter storm distribution was adopted. Rainfall-runoff modeling, using the HEC-1 computer package (U.S. Army Corps of Engineers, 1991), was performed for the 14.3 square mile watershed upstream of the debris, yielding an estimated peak discharge of 1030 cubic feet per second and a total runoff volume of 950 acre-feet.

However, because winter precipitation in Central Arizona often occurs in multiple-day storm sequences, it was concluded that a single 5-year/24-hour storm event did not adequately model the most likely scenario. To better reflect this situation, a storm sequence consisting of a 5-year/24-hour storm of 3.56 inches on the first day followed by a 2-year/24-hour storm of 2.76 inches on the second day, was utilized in rainfall-runoff modeling. Both storms were distributed according to the winter storm pattern adopted earlier. The model utilizing this storm sequence yielded an estimated peak discharge of 1030 cubic feet per second and an estimated runoff volume of 1600 acre-feet. These results were ultimately used to design the protective, upstream components of the surface water control system.

In evaluating the available data, it became apparent that the PVO rain gauge, located near the incident site at a relatively low elevation, did not properly represent the precipitation regime for the higher-elevation regions in the upper watershed. The use of PVO gauge data was accordingly curtailed early in the assessment of applicable criteria for the upper, higher watershed segment. However, data from the PVO gauge was subsequently applied to the problem of deriving realistic peak flow and runoff volume estimates for the lower subwatersheds flanking and downstream of the debris. The 5-year/24-hour storm estimated from a statistical analysis of the PVO gauge record, with an SCS curve number of 81 for Antecedent Moisture Condition (AMC) II, were applied in analyses utilized in the design of the downstream controls.

5 DIVERTING PINTO CREEK INFLOWS

After an in-depth consideration of an array of alternatives, the project team concluded that the most expedient and feasible means of managing Pinto Creek inflows from the upstream watershed involved the construction of a diversion dam and ancillary pumping system. The dam was subsequently sited about 500 feet upstream from the debris, and electrical service was routed to the stream bottom. Construction began on a homogenous earthen embankment, ultimately reaching a maximum height of 53 feet with a final crest width of 15 feet. The structure was constructed with a downstream slope of 2H:1V and an upstream slope of 3H:1V. The ultimate length of the dam was 470 feet, with an impoundment capacity of 685 acre-feet with 3 feet of embankment freeboard.

The embankment was comprised of compacted, low-permeability clayey borrow derived from a soft conglomerate exposed in the PVO open pit mine. This material was screened to remove its cobble fraction, then placed in lifts and compacted on grade. A centerline key trench was constructed in the dam foundation floor. The foundation materials consisted of a moderately-indurated and largely unfractured conglomerate. A downstream toe and blanket drain were also constructed, consisting of clean gravel encapsulated in two layers of nonwoven geotextile.

At an invert elevation that would allow discharge 35 feet below the crest, three 36-inch diameter steel pipe outlets were constructed through the embankment. All three of the low-level outlet pipes were equipped with manually-operated valves at the dam. One of these outlet pipes was connected to a 36-inch diameter high-density polyethylene (HDPE) pipeline placed on the ground surface and extending over and beyond the entire debris mass in the streambed. This piping provided for the contingency of gravity discharging up to 63,000 gallons per minute from the diversion reservoir. The dam was also fitted with a full-width, over-the-crest spillway designed to pass a 100-year/24-hour storm, assuming the reservoir at full capacity prior to the storm event.

As depicted on Figure 1, the dam and reservoir facility was then augmented with a dual pumping system, consisting of an array of on-line pumps, booster stations and pipelines. The combined capacity of the pumping system was about 11,400 gallons per minute, returning the water to Pinto Creek beyond and downstream of the impacted area.

6 CONTAINING DOWNSTREAM IMPACTS

To prevent further downstream transport of affected water and sediment, a means of intercepting stream flow was needed immediately upstream of the confluence of Haunted Canyon with Pinto Creek. After employing several minor temporary measures, an effective system of retention basins, pumps and pipelines were constructed to capture and route sediment-laden water back to the PVO mine property for appropriate disposal. Two low embankments were constructed using wire-basket gabions filled with cobbles borrowed from the stream bottom, with a woven geotextile as a filter layer within the baskets. These structures were placed across the Pinto Creek channel, creating small impoundments, which were fitted with pumps with a combined capacity of about 2,250 gallons per minute. The total capacity of the constructed basins was about 43 acre-feet. These basins received runoff from the affected channel bottom and flanking hillsides, as well as intentional discharges used to hydraulically clean the impacted alluvial streambed.

Due to the location and limited amount of the debris materials remobilized below the North Toe, repeated attempts were made to isolate the original debris mass from the downstream watershed. This action would also have allowed the downstream area between the North Toe and the sediment basins to be cleaned early in the process, in the event that discharges through the large 36-inch pipeline were needed to maintain capacity in the upstream reservoir. However, due to the extremely difficult access to the North Toe over the debris in the channel, sediment and water from the original debris mass not contained behind the working removal face was routed to the lower sediment basins and removed. Later in the project, access was achieved and a dike was eventually placed at the North Toe, isolating discharges from the original debris area and preventing them from reaching the downstream segment intercepted by the sediment basins.

As the debris removal proceeded, a small retention basin was placed in the upper reaches of the impacted area, downstream of the diversion dam. Water that collected behind the working debris face was pumped to this small basin using portable pump units; these small volumes were eventually routed to permitted containment basins on the PVO mine property.

As the mechanical removal of the tailings proceeded north using excavators and haul packs, hydraulic cleaning of the sideslopes and channel bottom began in a staged fashion. Most of the water used to perform this hydraulic washing originated from the upstream reservoir using the 36-inch HDPE pipe as a working conduit. The outfall end of the 36-inch HDPE pipe was fitted with a energy dissipater, and large surges of clean water were periodically discharged to the area between the North Toe and the downstream sediment basins. Augmented with considerable

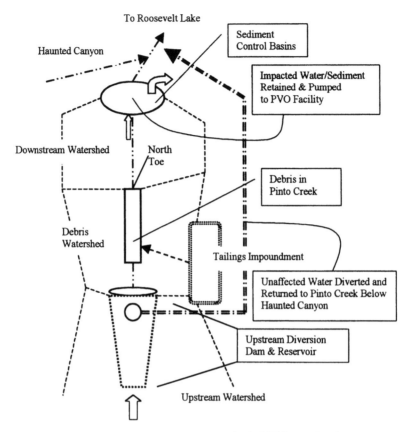

To Roosevelt Lake

Haunted Canyon

Sediment
Control Basins

Impacted Water/Sediment
Retained & Pumped
to PVO Facility

Downstream Watershed

North
Toe

Debris in
Pinto Creek

Debris
Watershed

Tailings Impoundment

Unaffected Water Diverted and
Returned to Pinto Creek Below
Haunted Canyon

Upstream Diversion
Dam & Reservoir

Upstream Watershed

Figure 1: Water Management System for the PVO Removal Action

manual labor, this action served to clean the residual sediment previously mobilized from the principal debris mass.

7 OPERATING THE SYSTEM

The guiding principle in operating the Pinto Creek surface water control system was to maximize storage capacity, within both the upstream diversion reservoir and the downstream sediment basins. This objective called for the continual maintenance and 24-hour operation of the 11,400 gallon-per-minute pumping system. Over the 7 months that the system was operated, this system pumped about 3,500 acre-feet of detained runoff around the impacted area. Only scheduled flows of water passed through the 36-inch outlet structure, and the water surface in the reservoir never reached the spillway elevation. On one occasion during a significant storm sequence, the reservoir was at near capacity, with runoff contributions well in excess of the pumping rate during peak flows.

Maintaining sufficient capacity in the downstream sediment control basins proved to be the most difficult to achieve. The upper of the two basins was utilized as a collector for the mobilized sediment and affected water used for cleaning of the stream channel. Capacity steadily decreased as sediment accumulated, but the system proved adequate throughout its use as a functional and protective component of the overall system.

Far exceeding the expectations of the project team was the usefulness of on-site predictions of approaching storm fronts, and the estimates of upcoming precipitation amounts and duration. These reports and estimates, combined with daily reports of stream inflow rates and available capacities in detention basins, were discussed with the project team. This information largely dictated whether preparations should be made to gain additional storage capacity in both the upstream and downstream facilities.

Key personnel were assigned the responsibility of monitoring National Weather Service (NWS) and related data accessible through the Internet, augmented by discussions with Salt River Project personnel in Phoenix, Arizona. Short-range forecasts received special attention, in addition to real-time precipitation data at stations to the west. Data collected and reported by the San Diego NWS proved very useful, with precipitation amounts at similar elevations roughly equivalent to amounts received at PVO some 12 to 24 hours later. This information was incorporated into an interactive model which gave onsite personnel information to make hour by hour decisions on how much to pump or whether critical valves should be open or closed.

U.S. Geological Survey stream gaging data was also available for a station downstream of the impacted area, on Pinto Creek just downstream of the confluence with Haunted Canyon. The project team also installed a continuous stage recorder in the Pinto Creek channel above the diversion dam reservoir. These stage records were combined with manually acquired measurements of stream flow velocity in a measured cross-section upstream of the reservoir, to produce estimates of discharge. During storm events, reservoir elevations were measured on an hourly basis if an upward trend was detected, and reserve capacities were periodically calculated.

8 REACHING EXPECTATIONS

Functional diversion and containment systems are normally the product of a protracted design process, supported by an equally detailed development of design criteria and parameters. The critical nature of the PVO incident inherently foreshortened the normal design development process. The project team was forced to rely more on their judgement and past experience, and less on the findings of detailed analytical solutions or simulations. By necessity, the system required active management, demanding constant attention and adjustment. Such characteristics are likely typical of a management system that evolves from a crisis, in that the variables that determine its ultimate course may initially be poorly understood and inappropriately prioritized.

In the case of the PVO incident and ensuing Removal Action, the success experienced in operating the installed surface water controls related as much to the effects of daily strategic decisions as to the criteria that dictated the capacity and operational flexibility of the system proper. Each complication was met with a multi-disciplinary assessment of potential consequences, and the system or its operation was subsequently adjusted to mitigate the problem. These appraisals were accomplished by individuals working together as a team, representing a diverse array of expertise, including the mine operator, system designers, contractors, environmental investigators, and regulators. The team was repeatedly successful in making effective decisions due to a clear set of overall project objectives, and a commitment to overcome an equally distinct and common adversary, the debris in Pinto Creek.

As delineated at the outset of the Removal Action, the primary goals of the remedial response were to prevent further environmental degradation and restore the impacted habitat. These goals were realized through the design, construction and integrated operation of the management system described herein. By the Summer of 1998, 99.98% of Debris discharged into Pinto Creek had been removed. All the control facilities have been removed, and their sites have been reclaimed. Throughout the entire course of the Removal Action, no discharges of impacted surface water or sediment passed beyond the downstream controls located immediately upstream from the Pinto Creek/Haunted Canyon confluence. The surface water quality of Pinto Creek currently meets applicable standards, and both the stream morphology and the riparian habitat are in recovery.

REFERENCES

Miller, J.F., Frederick, R.H. & Tracey, J.F. 1973. NOAA Atlas 2, Precipitation-frequency atlas of the western United States, Vol. VIII, Arizona. Silver Springs, MD: US Department of Commerce, National Oceanic and Atmospheric Administration, National Weather Service.

Thomas, B.E., Hjalmarson, R.H. & Waltemeyer, S.D. 1994. Methods for estimating magnitude and frequency of floods in the southwestern United States. US Geological Survey Open-File Report 93-419. Tucson, AZ.

US Army Corps of Engineers 1991. Flood hydrograph package (HEC-1) and user's manual, Version 4.01.1E. Davis, CA: Hydrologic Engineering Center.

Tailings and Mine Waste'00 © 2000 Balkema, Rotterdam, ISBN 90 5809 126 0

Mine waste management at Ok Tedi mine, Papua New Guinea: A case history

L. Murray & M. Thompson
Klohn-Crippen Consultants Limited, Vancouver, B.C., Canada & Brisbane, Qld, Australia

K. Voigt & J. Jeffery
Ok Tedi Mining Limited (OTML), Tabubil, Papua New Guinea

ABSTRACT: The Ok Tedi mine commenced production in 1983 and is currently milling ore at a nominal rate of 83,000 tpd. The mine is located in unstable mountainous terrain which experiences extremely high rainfall of over 10 m per annum, and where conventional approaches to storage of waste rock and tailings are unacceptable from a risk-cost-benefit perspective. Apart from a brief period at start up, the mine adopted riverine tailings disposal until mid 1998 when, due to increasing concerns over environmental effects, trial storage of some of the waste, predominantly sand, was commenced.

Mine derived sand is trapped within a dredged slot and delivered to the storage site at a nominal rate of 20 mtpa via a conventional cutter suction dredge which operates in the Ok Tedi River near the village of Bige about 100 km downstream of the mine site.

An extensive study of all practical sites in the mountainous terrain near the mine was made prior to selection of the floodplain storage site. The chosen site has challenging foundation conditions ranging from lowland swamp to weathered tropical clay.

The paper discusses the background to selection of the site and the construction of the dredged sand storage scheme in the context of a high rainfall, weak foundation, seismically active environment. Results of performance monitoring of the sand storage area are presented.

1 INTRODUCTION

The Ok Tedi mine is located in the Star Mountains in the interior of the western Province of Papua New Guinea adjacent to the border with Irian Jaya, see Figure 1.

The mine produces predominantly copper concentrate and mills ore at a nominal rate of 83,000 tpd. In addition to this, the mining of waste rock is carried out at a current rate of 152,000 tpd. All mill tailings and waste rock (mine waste) is currently placed into the headwaters of the Ok Tedi system.

There is some retention predominantly of the coarser fractions of the waste rock in the headwater river valleys, however, the majority of the tailings and the sand fraction of the waste rock is transported through the upper and middle Ok Tedi system to the lower Ok Tedi and Fly River.

It is estimated that about 60 mt per annum of mine waste, valley wall erosion at the mine and natural erosion, passes through the lower Ok Tedi system at the current trial sand storage site near the village of Bige.

In early 1994, Ok Tedi commenced a study of all identified possible waste storage options which could help mitigate the environmental impact of riverine waste disposal. This paper overviews some of the waste storage options and discusses the current trial scheme to store predominantly sand tailings on the lower Ok Tedi flood plain.

2 WASTE STORAGE OPTIONS STUDY

Numerous waste storage studies have been undertaken for the Ok Tedi mine, particularly since the ill fated Ok Ma tailings dam failed during construction in 1983. The current mine waste management study, which commenced in 1995, identified over 46 potential candidate storage sites, most of which had been previously studied by Klohn-Crippen or others. Of these, 33 potential sites were identified for tailings storage, 8 for waste rock and 5 for combined tailings and waste rock storage.

All 46 previous schemes were assessed based on potential environmental benefit, risk of failure, storage life, and cost/storage ratio.

The majority of the candidate schemes were rejected during the study due to small storage volume resulting from the rugged terrain, or high risk due to poor geotechnical conditions, severe water handling problems or high earthquake hazard. A total of 7 options were identified as having the potential to store sufficient material to provide an environmental benefit. These options were evaluated at feasibility level based on a common set of design criteria. The location of the 7 potential schemes plus an outline of the Ok Tedi system are shown on Figure 2.

Engineering studies indicated that the most cost effective schemes for tailings storage were the Lukwi scheme or schemes involving dredging in the lower reaches of the Ok Tedi.

The Lukwi scheme involves construction of a 90 m high impervious core rockfill dam on a karstic limestone and weak shale foundation. The Lukwi site is downstream of the Ok Ma dam which failed during construction in 1983. Although the Lukwi dam was considered to be technically feasible to construct and operate, the results of a risk assessment identified failure of the dam as a low probability but high consequence risk, which was not acceptable to OTML. The remoteness of the site combined with high rainfall, erosion and seismic events etc. would make long-term care and maintenance of the dam site very difficult and increase the probability of long-term unplanned release of tailings.

Although, perceived to be more costly to construct than the Lukwi option, the dredge schemes in the sand deposition reach of the lower Ok Tedi appeared to provide a relatively low risk option for storage of either sand or sand and tailings. Tailings storage would require construction of a roughly 110 km long tailings pipeline, mostly along the Kiunga-Tabubil mine access road.

3 OUTLINE OF DREDGE OPTIONS

3.1 *Sand Storage*

The objective of the sand storage is to remove up to 20 mtpa of sand from the lower Ok Tedi river near Bige and to store the material on the river bank a minimum of 500 m back from the existing Ok Tedi channel.

Figure 3 shows the proposed sand storage development concept which would cover an area of about 6 km² if implemented to the end of mine life.

3.2 *Sand and Tailings Storage*

Under this scheme, a series of perimeter dykes would be constructed using dredged sand and subsequently tailings would be stored in a series of cells confined by dredged sand perimeter dykes and internal hydraulic tailings splitter dykes, see Figure 4. The sand perimeter dykes would take about 5 years to construct, tailings storage could start about 2 years after commitment to construct the 110 km long tailings pipeline.

Both the sand tailings and sand storage options require the ability to dredge large volumes of sand from the Ok Tedi and the ability to construct stable slopes on the river flood plain. Consequently, OTML decided to initiate a trial dredging program which commenced in March 1998. The intent of the trial program is to prove the viability of dredging, prove the feasibility of constructing competent hydraulic fill structures and monitor the environmental benefit of the dredging on the lower Ok Tedi and Fly River systems.

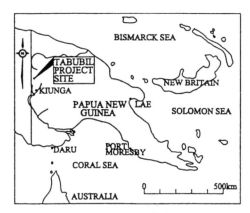

Figure 1. General Site Plan

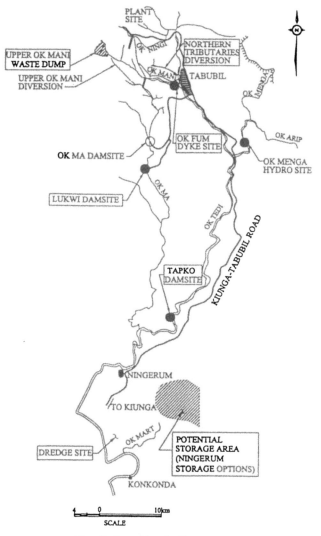

Figure 2. General Location Plan

509

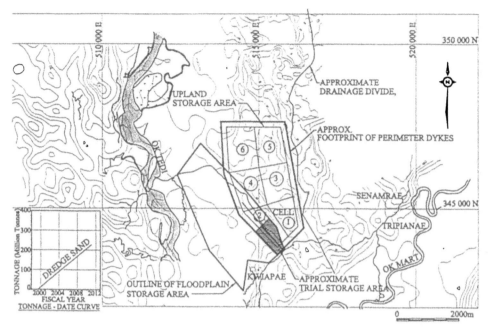

Figure 3. Sand Storage - Site Plan

The following sections describe the engineering aspects of the dredging and hydraulic fill operations.

4 SITE DESCRIPTION

The site of the trial dredge operation is located within the lower reaches of the Ok Tedi just north of the Ok Mart confluence near the villages of Bige and Konkonda, see Figure 2. The area is characterized by lowland forest, however extensive die back and progressive conversion to grassland has occurred at the site. At this site, the large and fast flowing Ok Tedi is located within a relatively straight channel, some 300 m wide, which is incised to El. 14 m to 20 m into the alluvial flood plain. The regularly innundated flood plain is about 2 km wide and lies at EL 21 m to 24 m and is flanked by gentle rolling topography, designated as the upland area, which rises to about El. 35 m. The selected site is in a reach of the Ok Tedi where sand deposition commences and hence is at the upstream end of the zone of sand aggradation in the river channel and associated overbank deposition.

The river channel and the general area is underlain by very weak sandstone, conglomerate and siltstone of the young Awin formation. Above the Awin, on the flood plain, extensive gravel deposits occur along with soft silt/peat infill in cutoff river meanders. On the higher ground on the flood plain and on the upland areas, alluvial deposits and Awin have been extensively weathered in situ to a red mottled cream tropical soil.

The red tropical soil forms a stiff, roughly 3 m thick, dessicated crust at the surface, although the crust quickly softens if wetted and remoulded. Beneath the crust, the clay becomes softer and locally sandy with depth. The soft layer has a "jelly" like consistency and often releases free water when sequenced by hand. A summary of the weathered red tropical soil index and engineering properties, based on laboratory testing, is presented on Table 1.

One field permeability test was completed on the weathered soil from 3 m to 5 m depth. The test indicated a permeability of 6×10^{-5} cm/s and although there was concern over the accuracy of the test, the result is in reasonable agreement with a value of 10^{-5} cm/s calculated for the consolidation tests on sandy clay.

510

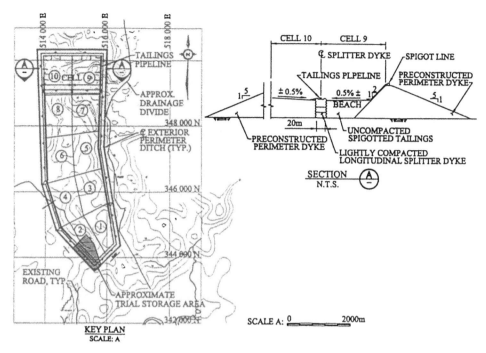

Figure 4. Proposed Tailings Storage

Table 1 Weathered Tropical Soil Engineering and Index Properties

Parameter	Unit	Range
Moisture Content	%	40 to 80
Liquid Limit	%	120 to 38
Plastic Limit	%	55 to 25
Direct Shear		
Peak Cohesion	kPa	0 to 50
Peak Friction	degrees	36 to 23
Residual Cohesion	kPa	0 to 20
Residual Friction	degrees	12 to 29
Triaxial		
Effective Peak Cohesion	kPa	0 to 21
Effective Peak Friction	degree	28 to 36
Effective Extended Strain Friction	degrees	30-34
Total Stress Extended Strain Friction	degrees	12
Consolidation (oedometer)		
Sandy Clay		
Preconsolidation Pressure	kPa	80 & 225
eo (Void ratio)		1.9 & 2.1
Cc (Coefficient of Consolidation)		1 & 0.75
Mv (0-375 kPa) m2/kN		2 to 5
Cv m2/yr		40 to 50
Clay		
Preconsolidation Pressure	kPa	160 & 125
eo		1.9 & 2.0
Cc		0.63 & 0.75
Mv (0-375 kPa) m2/kN		1.53 to 5
Cv m2/yr		0.75 to 5.0
Peak Undrained Strength Clay (pocket penetrometer and field vane)	kPa	18 to 155
Post Peak Undrained Strength Clay	kPa	10 to 70
Pore Pressure Response (undrained condition)	B	
Total (saturated) Unit Weight	kN/m^3	

511

4.1 Dredging

The dredge commenced mobilization from Taiwan in May 1997 and proceeded to Cairns where a major refit was carried out. It was then stranded in the Fly river due to a severe drought and finally commenced dredging in the Ok Tedi in March 1998. The dredge and support vessels are operated by the prime contractor, Dredeco, and consists of the following:
- Cutter Suction Dredge "Cap Martin" a 58 m long, 15 m wide vessel with a maximum 2.97 m draught capable of over 20 mtpa dredged sand production;
- Support vessels including:
  - One multicat work boat;
  - 150 t crane barge "Asian 15";
  - Survey and crew boat "Dome" (10 m long); and
  - 360 t motor push tug "Bige".

The dredging commenced on March 29, 1998 with excavation of a slot in the river channel for sand settlement. The slot development required dredging of a mixture of recent sandy mine sediments and old river bed gravels. The river gravels were stored on the East or West bank close to the dredge slot. The slot is currently about 800 m long and 250 m wide and excavated down to El. 15. The slot captures sand during flood flow events and can store about 20 days of material for dredging.

Sand which settles into the dredge slot is pumped to the storage area on the East Bank by the cutter suction dredge through a 900 mm diameter steel pipeline with upto 400 m long floating line then via a 900 mm diameter main land line 2300m long. From this main line, branch lines 250 m to 1200 m long, run to fill in the various areas. The pipe is 16 mm to 19 mm wall thickness and is installed in 6 m or 12 m lengths.

4.2 Sand Storage Trial Area

The area selected for the sand storage trial was the south-west corner of the potential tailings and sand storage area, see Figure 4.

As at the 21st of August 1999, a total of 10,500,000 m^3 of material has been stored in the area.

The east half of the trial area, which would potentially form part of the tailings dam embankment, was instrumented by installing electrical piezometers and settlement posts into the tropical soil foundation. Settlement posts consist of a steel plate placed on the surface of the soil with a 100 mm I.D. steel pipe riser, the posts are raised periodically by adding sections of threaded 100 mm I.D. steel pipe.

Electrical piezometers were installed in 8 locations adjacent to the settlement posts with three piezometers placed at each location at nominal depths of 3 m, 8 m and 18 m below ground level.

Initially, a series of inclinometers were installed, but these proved difficult to service and raise and have largely been abandoned.

Site development, sand placement and material handling utilizes the following equipment:
- 2 bulldozers Cat D7 LPG
- 2 excavators Komatsu PC 300
- 1 wheel loader Komatsu WA 500

The bulldozers keep dredge materials clear of the pipe discharge area direct the flow and compact the sand within the final embankment perimeter zone.

The excavators and bulldozers compact, shape and level the dredge material to the desired lines and grades.

Where possible and prior to placement of sand, the footprint of the dyke perimeter is stripped of organic material which is windrowed at the dyke toe for future reclamation. A starter dyke is constructed using either local tropical soil or dredged sand. The sand storage area is then raised in nominal 4 m lifts to produce final overall maximum 5H:1V side slopes in areas underlain by tropical clay soil. The side slopes are flattened in areas of swamp or across infilled old river meanders to 10H:1V or even 20H:1V.

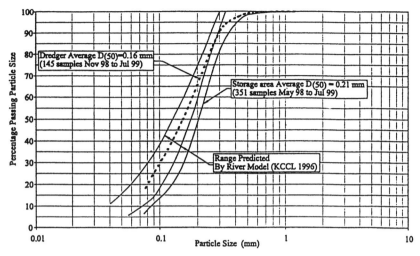

Figure 5. OTML Lower Ok Tedi Dredge Trial Project - Particle Size Distribtuion

4.3 Geotechnical Considerations

The geotechnical appraisal of the work has concentrated on assessing the in place density achieved within the sand and monitoring the performance of settlement posts and piezometers. At the time of preparing this paper, the fill height had reached about El. 34 m with perimeter dykes constructed to El. 38 m ready to receive the next lift. Thus the sand fill has reached a maximum height of about 10 m and has roughly reached the height of land in the upland region.

4.4 Sand Gradation

The results of some 500 particle size distribution tests undertaken on both the sand recovered from the dredge slot and in the storage area dykes during the period November 1998 to August 1999 is shown on Figure 5 as average gradation curves. These results indicate the material trapped within the slot has an average D(50) of 0.16 mm. The D(50) has been relatively variable to date probably associated with the effects of progressive slot extension, sampling error, large variations in river flow and flood frequency, etc. The sand stored within the perimeter dykes has an average D(50) of 0.21 mm and has remained relatively consistent mostly due to placement methods and loss of fines to lower less accessable regions. Figure 5 also shows the predicted gradation of the material trapped in the slot based on modelling done by Klohn-Crippen in 1996.

4.5 Sand Density

The results of 105 laboratory compaction tests on the sand from the storage area dykes indicate an average Standard Maximum Proctor dry density of about 1.61 t/m³ at an optimum moisture content of 20%. The optimum moisture seems anomolously high, but agrees quite well with the insitu sand moisture content which is also high at an average 16.4 %. The results of 102 field density tests within the dyke footprint indicate the in-situ density averages 96% of maximum Standard Proctor density with about 29% of results falling below the design target 95% Standard Proctor density (although only 8% of test results are below 90%). The lower density results are generally concentrated at points remote from the dredge discharge although some lower densities occur under the perimeter slopes particularly within areas of lower elevation

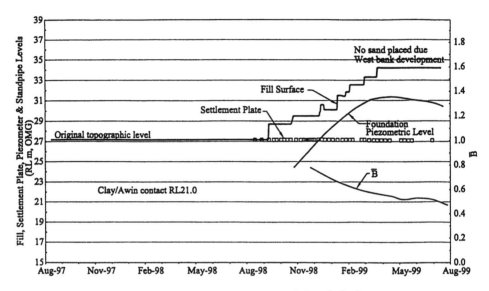

Figure 6. Trial Dredge Embankment - Typical Monitoring Data

which are saturated or periodically inundated. Because the sand storage site will ultimately be closed without ponded water, these isolated low density zones in the perimeter are not considered to be a significant concern for long-term stability or earthquake deformation. It is anticipated that after the area is completed to full height, about El. 50 m, exterior slopes will be tested by drilling or CPT to measure final in place density.

4.6 Pore Pressure Response and Foundation Collapse

The vibrating wire piezometers have operated successfully with about a 25% loss of units to date. A typical pore pressure response of the foundation to loading is shown on Figure 6. The plot shows measured piezometric response and also the calculated parameter $\overline{B}$ which is defined as follows:

$$\overline{B} = \frac{hw \cdot \gamma w}{hs \cdot \gamma s} \qquad (1)$$

where: hw, γw are the height and unit weight of water above the piezometer, and hs, γs are the height and unit weight of new fill placed above the piezometer tip.

A $\overline{B}$ value of about 0.75 was anticipated during design and current values are lower at about 0.6.

Studies of weak weathered tropical soils (Wallace, 1973) indicate that in some cases, soils can suddenly collapse due to failure of the internal brittle soil skeleton, see Figure 7. The open skeleton of weakly cemented minerals is responsible for the very high water content of these tropical soils. A sudden increase in $\overline{B}$ is a possible indicator of collapse. Although problems are not expected, collapse of the tropical clay soil could occur as the sand storage area load exceeds the apparent pre-consolidation pressure shown in Table 1, that is around the current pile height of 10 m. Sudden collapse would be followed by a loss of undrained strength in the foundation and possible embankment failure. The potential for collapse was a significant major identified risk for the tailings storage option, since a perimeter dyke breach would result in tailings discharge.

514

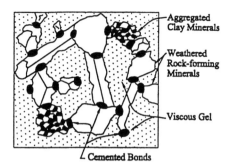

Figure 7. Idealized residual soil structure

4.7 Settlement Posts

A series of 8 settlement posts have been installed adjacent to the piezometers. The posts duplicate as stand pipes to measure the water level within the dredged sand. Currently, the water level in the sand is close to the fill surface indicating that drainage is slow possibly as a combined result of the high rainfall, relatively low sand storage elevations to date. To date, the maximum settlement range is 40 to 70 mm or about 1% of the current sand fill height.

4.8 Surface Water Handling and Erosion Protection

The Bige/Konkonda area receives about 6 m of rainfall per year, often in the form of tropical thunder storms. The estimate runoff from the area for the 1 in 1000 year rainfall is 58 m^3/s/km^2, the 1 in 2 year runoff is estimated at 8 m^3/s/km^2.

The trial embankment area will be about 1.5 km^2, therefore, allowing for some pond storage on the fill surface, and outflow of 12 m^3/s due to rainfall must be handled during construction. This outflow is in addition to the approximately 4 m^3/s delivered by the dredge.

During construction, the combined flow is handled by 'water boxes'. The water boxes vary in dimension from 2 m × 2 m × 1 m to the size of a shipping container. The boxes are fitted with grooves for stoplogs, placed in the required position in the dyke and sealed with dredge material, plastic sheeting and sandbags. The water boxes are relocated between each lift.

Ultimately, water from 1 in 1000 year rainfall will be discharged from the pile surface by means of a series of gabion lined spillways. Erosion of the exterior slopes is intended to be controlled by covering with stripped topsoil, windrowed material and revegetation. To date, efforts to grow vegetation on trial plots have been undertaken with some success. Revegetation trials of actual sand slopes commenced in the second quarter of 1999.

5 SUMMARY AND CONCLUSION

The dredging trial has been underway since March 1998, the dredge slot has been developed to some 800 m long on the lower Ok Tedi and 9.1 million m^3 of sand and 1.5 million m^3 of gravel has been dredged and stored on land to date.

The hydraulic sand fill has generally achieved the design density beneath the perimeter embankments, and the results of field piezometer and settlement monitoring indicate that the tropical clay foundation materials are behaving as anticipated in the design. The sand fill is currently approaching a critical height when potential collapse of the soil structure in the tropical weathered soil could occur.

The ability to dredge large volumes of sand from the fast flowing river has been proven but the environmental benefits of long-term sand removal have not yet been demonstrated.

Tailings and Mine Waste'00 © 2000 Balkema, Rotterdam, ISBN 90 5809 126 0

Acid In Situ Leach uranium mining – 1. USA and Australia

Gavin M. Mudd

School of the Built Environment, Victoria University, Melbourne, Vic., Australia

ABSTRACT: The technique of In Situ Leach (ISL) uranium mining is well established in the USA, as well as being used extensively in Eastern Europe and the former Soviet Union. The method is being proposed and tested on uranium deposits in Australia, with sulphuric acid chemistry and no restoration of groundwater following mining. The history and problems of acid ISL sites in the USA and Australia is presented.

1 BACKGROUND

The unconventional mining technique of In Situ Leach (ISL) is now the primary producer of refined uranium in the United States, with a market share of around 95% in the mid 1990's (DoE, 1999). ISL mines appear set to assume a greater role in Australia's uranium industry.

It is perhaps an historical curiosity as to where the conceptual processes for ISL (as applied to uranium) were first conceived and applied. The Chinese were apparently the first to use solution mining to produce copper as early as 907 AD, with references to the technology dating back to 177 BC (Morris, 1984). In the 1890's, the Frasch process for mining elemental sulphur was invented, and ISL mining of gold was first suggested by Russians (Morris, 1984). The first trials of uranium ISL were developed in the USA and Soviet Union in the early 1960's. It is uncertain who developed the concepts or if they were developed separately (Mudd, 1998).

By the mid 1970's, there were ISL mines across the world as an alternative, low cost method (Mudd, 1998). In the USA, ISL mines generally used alkaline chemistry with only a few sites trialling acid chemistry. In contrast, Soviet mines generally used acid with only a few sites using alkaline reagents. In Australia in the 1980's, two ISL projects in South Australia proposed acid, while a third in Western Australia trialled alkaline chemistry (Mudd, 1998).

The environmental regulation of mining generally requires the restoration of affected groundwater to be returned to its pre-mining quality or use category. In countries controlled by the Soviet block, the need for restoration of contaminated groundwater following mining was ignored during operation, and the problems and magnitude of groundwater contamination now coming to light in the 1990's can only be described as extreme (cf. Mudd, 1998 & 2000).

Indeed, the use of alkaline chemistry in the USA has been partly related to the need to restore affected groundwater and that alkaline mine sites are recognised to be technically easier to restore (Mudd, 1998; Tweeton & Peterson, 1981). In direct contrast, Australian mines - historically and currently - proposed not to restore affected groundwater after acid ISL mining.

The resurfacing of the Australian acid ISL uranium mine proposals in 1996 (due to changes in federal government uranium policy), the lack of acid ISL mines in the USA, the research coming to light through the International Atomic Energy Agency (IAEA) and others of the extent of impacts from acid ISL mines in the Soviet block (cf. Mudd, 2000), led to a detailed review of ISL uranium mining by the author, completed in 1998 (Mudd, 1998).

2   ACID ISL IN THE USA

2.1   Brief History of ISL

The initial development of ISL mining in the USA occurred in Wyoming at the Shirley Basin uranium project from 1961-63, by the Utah Construction and Mining Company (UCMC; now Pathfinder Mines Corp.) (Larson, 1981). They experimented with 5 generations of wellfield design and over 100 patterns, using sulphuric acid chemistry (Underhill, 1992). The Shirley Basin ISL project operated on a small scale from 1963-70 to produce 577 MTU, however, the ISL mine was closed in 1970 and converted to an open cut operation (Underhill, 1992).

The late 1960's to mid 1970's witnessed rapid development and promise in ISL mining, principally in Texas, Wyoming, New Mexico and Colorado (Kasper et al., 1979). By May 1980, a total of 18 commercial and 9 pilot scale projects were either in operation or under active development (Larson, 1981). Virtually all of these sites utilised alkaline reagents such as ammonia or sodium carbonate/bicarbonate. The difficulty of restoring ammonia-based sites saw a quick shift in emphasis to sodium- or carbon dioxide-based leaching chemistry by the early 1980's (Tweeton & Peterson, 1981). Despite years of lower production in the late 1980's, ISL mines have gradually increased their share of the uranium market in the USA from about 1.2% in 1975 (Underhill, 1992) to greater than 90% during the mid-1990's (DoE, 1999).

By 1991, a total of 62 ISL projects had been developed, although only 24 of these sites were commercialised (Underhill, 1992), indicating more unsuccessful than successful projects (Mays, 1984). There has been no development of a commercial ISL mine since Shirley Basin using acid chemistry (Mays, 1984). Further detail on all ISL mines is in Mudd (1998). There were several sites in Texas, New Mexico and Wyoming which underwent pilot scale testing of acid ISL, although most were poorly documented in public literature (Mudd, 1998).

The best documented acid ISL project is Nine Mile Lake, near Casper, Wyoming. The project was developed by Rocky Mountain Energy Co. (RMEC) in association with research by the U.S. Bureau of Mines. The landmark study was reported in detail by Nigbor et al. (1982). RMEC's Reno Ranch trial in Wyoming was the second documented acid ISL trial, reported by (Staub et al., 1986). Further acid ISL trial sites, however, have not been reported widely in the literature. Acid systems were generally considered unsuitable for Texan deposits.

2.2   Nine Mile Lake, Wyoming

The geology and hydrogeology of the Nine Mile Lake (NML) site is given by Nigbor et al. (1982) and Staub et al. (1986). The following discussion is adapted from these references.

It is situated on the southwest flank of the Powder River Basin. The roll-front type uranium mineralization occurs in the Teapot Sandstone within the Mesaverde Formation. The ore body extends over a strike length of 6,100 m in a north-northwest direction and ranges between 15 to 900 m in width, consisting of upper and lower zones. The site is at an elevation of 1,600 m.

The uranium ore at NML was precipitated at the interface of oxidation-reduction boundaries in the Teapot sandstone, due to the presence of carbonaceous material and pyrite. The principal uranium mineral was uraninite, with minor quantities of coffinite. Vanadium was associated

Table 1 - Pilot Scale ISL Mines Using Acid Leaching Chemistry [1]

Project / Site	Company	Time Period
Nine Mile Lake (NML), WY	Rocky Mountain Energy Co. (RMEC)	Mining : Nov. 1976 to Nov. 1980 Restoration Suspended : Feb. 1982
Reno Ranch[2], WY	RMEC	Mining : Feb. 1979 to Nov. 1979 Restoration Suspended : March 1981
Irigary, WY	Wyoming Minerals	Unclear - acid trial referred to by Kasper et al. (1979)
Jackpile Paguate[3], NM	Anaconda	Early 1970 trial, 2 wellfields, with 2 injection bores & 18 extraction bores, upgraded to 29. Project discontinued.
Dunderstadt, TX	Cities Service	Trial operated between 1969-71. No reports.
Besar Creek, TX	RMEC	Early 1970's ?, details unknown (plant used at NML).

Notes : [1] - Mudd (1998), Underhill (1992) & Staub et al. (1986); [2] - also Reno Creek; [3] - North Windup Project.

with the mineralization (about 1.3%) and was proposed to be extracted from a commercial facility. The ore contained less than 0.1% carbonate, although total carbon content was higher at 0.2-2.0%. The major clay mineral present was kaolinite (2-5%), with minor montmorillonite, although this had a low cation exchange capacity at about 5 meq/100 g.

Due to the low carbonate content of the ore body and the low cost of sulphuric acid, NML was considered an ideal site for sulphuric acid ISL mining. Extensive laboratory tests on core samples suggested that savings in chemical costs would result from the use of acid. A total of four wellfield patterns underwent testing and development at the Nine Mile Lake project site. The chronology and detail for each pattern is summarised in Table 2.

Pattern 1, completed in the upper ore zone, experienced several problems leading to poor operational performance. These included problems with the PVC well casing, cement baskets and pumps. A buildup of gypsum scale on the injection well screens, possibly related to the degaradtion of the casing cement by the acid, contributed to poor injectivity. Potential channelling and poor injectivity led to disappointing overall uranium recovery.

Pattern 2, completed in the lower ore zone, with a detailed assessment provided by Nigbor et al. (1982), was generally considered a good success. Injectivity was good, although plugging problems due to "fungus growth" and gypsum precipitation were encountered in April 1978. No evidence was provided to substantiate the conclusion for "fungus growth".

The two injection bores of Pattern 3 were completed in both the upper and lower ore zones to test the feasbility of simultaneously leaching both zones. The extraction bores were completed independently in each ore zone. The pattern experienced sporadic problems with well plugging, frozen lines and equipment failures, leading to poor operational performance. Further problems were encountered in controlling lixiviant distribution to the two ore zones.

Pattern 4, using alkaline chemistry, was intended to give a comparison of alkaline and acid leaching on the same ore body. However, the results of the trial are not available, although RMEC described the test as "disappointing". Thus no comparison can be made of the respective advantages and disadvantages of acid versus alkaline for the same deposit.

There were 5 horizontal excursions (ore zone aquifer) detected at NML during the testing phase, with three in Pattern 3 and two in Pattern 4. All excursions were bought under control by increasing the extraction rate. No monitoring of overlying and underlying bores was undertaken, and determination of any vertical excursions is impossible. This potential exists at almost every ISL site, due to casing failures and abandoned exploration bores (Staub et al., 1986; Marlowe, 1984). The risk increases with the total number of bores and age of a site (Marlowe, 1984).

The restoration of each pattern was undertaken immediately after mining, followed by the regulatory period of stabilization. Post-restoration monitoring is critical in understanding the effectiveness of restoration efforts and long term impacts on water quality at NML. The available baseline, leaching phase and restoration groundwater quality data for each pattern is compiled in Tables 3 & 4, adapted from Nigbor et al. (1982) and Staub et al. (1986).

The restoration data is averaged from observation and extraction bores, due to the tendency of injection bores to reflect the quality of injected solutions rather than groundwater after mixing. The high sulphate levels of the ore zone were thought to be related to influx from Nine Mile Lake itself, 1.6 km to the south, which is naturally high in sulphate.

Table 2 - Research and Development Details for Nine Mile Lake

Pattern & Type	Lixiviant Chemistry	Period of Testing[4]	PV[5]
1	7-spot,	4 g/l H_2SO_4 (pH 1.7), 0.5 g/l H_2O_2,	Mining : Nov. 1976 to Aug. 1977	7
	15 m radius	0.15 g/l $FeSO_4$, flow ~2.5 L/s	Restoration : Sep. 1977 to Oct. 1978	12
2	5-spot,	3-5 g/l H_2SO_4 (pH 1.8), 1 g/l H_2O_2,	Mining : Dec. 1977 to Sep. 1978	13
	15 m radius	flow ~2.6 L/s	Restoration : Sep. 1978 to Aug. 1979	12
3	8-spot[6],	H_2SO_4, H_2SO_5 or O_2, flow ~3.8 L/s	Mining : Sep. 1979 to April 1980	5.6
	18 m radius		Restoration : Aug. 1981 to Jan. 1982	6
4	5-spot,	Na_2CO_3 / $NaHCO_3$ with $CO_{2(g)}$	Mining : June 1980 to Nov. 1980	?
	15 m radius	(pH ~ 7.5), 0.5 g/l H_2O_2 (later) O_2	Restoration : Nov. 1980 to Aug. 1981	?

Notes : [4] - Restoration refers to initial phase only; [5] - Aquifer Pore Volumes reached during testing; [6] - included 2 central injection and 6 extraction bores (effectively, one 3-spot pattern for each ore zone aquifer).

Table 3 - Baseline and Restoration Groundwater Quality, Patterns 1 & 3, Nine Mile Lake (all units mg/L, except for pH; EC in mmhos/cm) (Staub *et al.*, 1986)

Pattern & Phase		TDS	EC	pH	Cl	SO$_4$	Ca	U (U$_3$O$_8$)	V
1	Baseline	2,483	3.16	6.9	3.3	1,240	87	0.384	0.1
	Restoration	7,750	12.0	6.9	93	5,140	300	0.289	0.073
3	Baseline	2,034	2.38	6.9	35	1,244	74	0.060	0.18
	Restoration	1,450	2.50	7.1	26	920	61	0.126	0.57

Table 4 - Average Baseline, Leaching Phase and Restoration Groundwater Quality, with Standard Deviation, Pattern 2, Nine Mile Lake (mg/L, except for pH, EC in mmhos/cm, Redox in mV, As, Mo and Se in µg/L, and Ra^{226} & Th^{230} in pCi/L) (Nigbor *et al.*, 1982)

	TDS	EC	pH	Redox	DO[7]	Cl	SO$_4$
Baseline	4,300 ± 550	4.10 ± 0.51	6.7 ± 0.3	-120 ± 200	1	46 ± 4.3	2,510 ± 244
Leaching		10.0-20.0	1.5-2.0				up to 8,000
Restoration[8]	3,000 (2,390)	3.25 (3.08)	6.1 (6.9)	-22 to 120	<0.1	29 (37)	1,585 (1,584)

	HCO$_3$	F	Ca	Mg	Na	K	Al
Baseline	290 ± 30	0.77 ± 0.25	207 ± 43	92 ± 31	830 ± 145	14 ± 3.6	0.13 ± 0.05
Leaching			260				
Restoration[8]		0.6	805 (102)	42	485	6.2	

	As	B	Cr	Cu	Fe	Mn	Hg	Mo	P	Se
Baseline	40	0.67 ± 0.40	0.01	0.01	1.07 ± 0.4	0.31 ± 0.18	0.01	8 ± 18	0.2	2
Leaching					up to 200					
Restoration[8]	24				6.8	0.24			<0.1	13

	Si	U	V	Zn	Ra^{226}	Th^{230}
Baseline	4.2 ± 4.0	0.23 ± 0.10	0.5 ± 0.2	0.02 ± 0.02	510 ± 29	0.084 ± 0.005
Leaching		80-150	up to 800		10,000 ± 170	49,000 ± 3,200
Restoration[8]	14.8	1.05 (0.132)	11.1 (0.986)	1.97		

Notes : [7] - Dissolved Oxygen; [8] - includes additional restoration work undertaken in 1981-82 in brackets. Many elements buildup over time, while others approach a stable concentration - leaching values indicative only.

The methods for restoring each pattern differed slightly. Pattern 1 was restored using a groundwater sweep, while Pattern 3 involved a groundwater sweep combined with reverse osmosis treatment and mixed with "clean" formation water before reinjection into the ore zone.

The post-restoration monitoring of Pattern 1 from early 1978 to 1981 indicated substantial deterioration of water quality, due to gypsum dissolution increasing salinity levels. Reverse osmosis treatment of approximately 2.5 pore volumes of recirculated groundwater was undertaken in 1981, although later monitoring again showed a deterioration and stabilization at a high salinity level. The water quality, with salinity 4 times higher at 7,750 mg/L and SO$_4$ 3 times higher at 5,140 mg/L, is now unsuitable for stock purposes - it's pre-mining use category.

The restoration of Pattern 2, however, proved to be more recalcitrant. The first phase of restoration involved four months of a modified groundwater sweep with reinjection of process water and barren production fluid. Restoration using reverse osmosis treatment was then undertaken for a month. From May to mid-August 1979, a high pH, sodium hydroxide solution was injected to promote ion exchange and speed restoration. Clean water recycling with reverse osmosis continued for the next three weeks, by which stage nearly all major parameters were restored to pre-mining ranges, and active restoration ceased.

Post-restoration monitoring of Pattern 2 during late 1979 and early 1980 detected scattered areas of contaminated groundwater around the pattern interior, migrating slowly down gradient. Pumping resumed in August 1980, with the groundwater being treated with a lime/barium chloride precipitation process, and reinjected into the wellfield. The total quantity was about 3.5 aquifer pore volumes. Little improvement was apparent and by May 1981, water quality was again deteriorating. As of June 1984, V, ^{226}Ra and TDS were above pre-mining levels.

The restoration of Pattern 3 returned most parameters to baseline values or better but failed to restore U, V and ^{226}Ra to pre-mining levels. The lack of published data for Pattern 4 precludes a direct comparison of the efficacies of an acid to an alkaline leached site.

The Nine Mile Lake acid ISL trial demonstrated that acid was indeed an effective alternative leaching reagent to the alkaline chemistry prevailing at the time, albeit non-selective. However, other issues raised by the trial include the difficulty in scaling laboratory test results to the field.

The column leaching tests performed on NML core samples suggested significantly lower reagent consumption than required in the field. The restoration of the laboratory columns indicated that about 13 pore volumes would be required to restore the water quality, whereas in the field it was closer to 20 and still experienced deterioration following treatment efforts. Nigbor et al. (1982) concluded that, due to greater reagent consumption and the difficulty and expense of restoration, that acid leaching was no more cost effective than alkaline leaching.

The expansion of the NML site to commercial scale by RMEC proposed to use a 7-spot production pattern with a radius of 21 m. The lixiviant was 3-5 g/L H_2SO_4 and 1 g/L H_2O_2, with recovery of the vanadium by-product. As of June 1984, the RMEC had no plans for commercialisation, and to the best of the authors' knowledge, the site is yet be developed.

2.3 Reno Ranch, Wyoming

The Reno Ranch (Reno Creek) uranium deposit in Wyoming, although lesser known than the Nine Mile Lake site, underwent trials of acid ISL about the same time period. An alkaline 5-spot trial was also developed. However, unlike NML, the geology, hydrogeology and information on the ISL trials at Reno Ranch was published by Staub et al. (1986).

Reno Ranch is situated on the eastern flank of the Powder River Basin. The roll-front type uranium mineralization occurs in the Wasatch Formation, consisting of fluvial sandstones, siltstones, shales, claystones and coal seams. The ore zone contains high quantities of carbonate minerals, although quantitative data is unavailable. The site is at an elevation of 1,590 m.

Two wellfield patterns were developed and tested, the first being a conventional 5-spot pattern and the second being a 6-spot pattern with 2 injection and 4 extraction bores. Pattern 1 was leached with acid while Pattern 2 with alkaline reagents, details are in Table 5.

Although this paper is primarily concerned with the use of acid in ISL mining, it is of great value in comparing the acid and alkaline trial at Reno Ranch. The groundwater quality from each trial is compiled from Staub et al. (1986). The results from Pattern 1 are in Table 6, although space in this paper does not allow full data from Pattern 2 (see Staub et al., 1986).

After mining was initiated in Pattern 1, problems with gypsum precipitation and "fungus growth" reducing the efficiency of well field circulation. No evidence was provided to substantiate the conclusion for "fungus growth". The uranium recovery rates were low and the carbonate minerals in the host sandstone consumed high quantities of acid. Leaching was terminated prematurely and restoration began immediately, consisting of water treatment by ion exchange, groundwater sweeping, and treatment with potassium carbonate to raise the pH and facilitate further removal of calcium, heavy metals and radionuclides. The restoration sequence, although aggressive compared to other efforts at ISL mine sites, encountered many difficulties.

The ongoing restoration efforts of Pattern 1 failed to reduce free acidity, SO_4 and Ra levels. The RMEC suggested the use of a high salinity solution to displace the hydrogen ions from clay lattice structures, enabling these to be removed during the restoration process. The regulatory agencies refused this technique due to uncertainties and possible adverse effects on the aquifer.

Very little post-restoration water quality improvement has occurred at pattern 1. During the first quarter of 1983, groundwater monitoring indicated that : 1) pH levels in the aquifers have not changed significantly; 2) Ca and SO_4 concentrations have not changed significantly (270 and 1,500 mg/L, respectively); 3) U levels have decreased marginally to less than 1.0 mg/L; & 4) TDS (~ 2,650 mg/L) remains almost twice that before mining.

Pattern 2, leached with alkaline reagents, proved less problematic from an operational and restoration perspective, however, post-restoration monitoring indicated a significant increase in

Table 5 - Research and Development Details for Reno Ranch

Pattern & Type		Lixiviant Chemistry	Period of Testing	PV
1	5-spot,	5 g/L H_2SO_4 (pH 1.8), H_2O_2,	Mining : Feb. 1979 to Nov. 1979	?
	12 m radius	flow ~2.5 L/s	Restoration : Nov. 1979 to Oct. 1981	?
2	6-spot[9],	Na_2CO_3 / $NaHCO_3$, H_2O_2,	Mining : Sep. 1980 to Dec. 1980	?
	15 m radius	flow ~1.6 L/s	Restoration : Dec. 1980 to April 1981	6.5

Notes : [9] - included 2 central injection and 4 extraction bores.

Table 6 - Average Baseline, Injection, Extraction and Restoration Groundwater Quality, Ore Zone, Pattern 1 (acid), Reno Ranch (mg/L, except for pH, and EC in mmhos/cm)

	TDS	EC	pH	HCO$_3$	CO$_3$[10]	Alkalinity[10]	SO$_4$
Baseline	970-1,566	1,220-2,000	6.4-11.2	ND-190	ND-48	ND-225	486-1,006

	Cl	F	NH$_4$	NO$_3$	Ca	Mg	Na	K	Fe
Baseline	6 - 62	ND-0.57	ND-0.74	ND-7	72-182	9-51	145-323	7-25	0-3

	As	B	Mn	Se	SiO$_2$	U	V
Baseline	ND-0.03	ND-2.6	ND-0.22	ND-0.05	ND-8.7	0.007-0.27	0-8

	TDS	pH	SO$_4$	Ca	Fe	U	V
Restoration[11]		4.8	1,385	230	29.1	1.1	0.2
Post-Rest.[12]	1.267	9.3	764	102		0.059	
Post-Rest.[13]	2,551	5.3	1,551	263		0.64	

Notes : [10] - as CO$_3$; Alkalinity as CaCO$_3$; [11] - groundwater quality at the cessation of restoration efforts, February 1981; [12] - Post-Restoration groundwater quality, March 1983; [13] - Post-Restoration groundwater quality of extraction bores, March 1983. ND - Not detectable.

U levels to around 3.7 mg/L. This pattern of increasing uranium after restoration has been noted at many former ISL sites, although the mechanism remained unclear without more research.

There were no reported excursions at the Reno Ranch site, although it was questionable whether the control limits were sensitive enough to detect such an event, especially for a vertical excursion. As with Nine Mile Lake, the Reno Ranch site is yet to be commercialised, and new interest in the developing the deposit has been recently abandoned.

2.4 Acid ISL in the USA - Summary

The experience with acid In Situ Leach uranium mining at Nine Mile Lake and Reno Ranch has shown that it can be an alternative, albeit non-selective, to alkaline process. However, the choice presents two major potential problems : 1) precipitation of gypsum on well screens and within the aquifer during mining, plugging wells and reducing the formation permeability (critical for economic operation); and 2) gradual dissolution of the precipitated gypsum following restoration, leading to increased salinity and sulphate levels in groundwater. The further effects of the release of heavy metals and radionuclides, especially Ra, that were co-precipitated with the gypsum have not been assessed or quantified. A critical issue is that acid leaching was not found to be more cost effective than alkaline, when taking restoration into consideration. No commercial acid ISL uranium mine has been approved nor developed in the USA.

3   ACID ISL IN AUSTRALIA

3.1 Brief History

The history of the technique of ISL uranium mining in Australia coincides with the litmus paper of public concern regarding the environment, nuclear issues and indigenous land rights. There has been no commercial ISL uranium mine in Australia by 1999. Only three sites have had pilot scale testing - two with acid, at Beverley and Honeymoon, South Australia, and one with alkaline chemistry at Manyingee, Western Australia (see Mudd, 1998).

There has never been a commercial acid ISL copper mine, although several sites near Mt Isa, Queensland, have undergone trials (mostly in the late 1960's) and more recently at the Gunpowder (Mammoth) mine. A small experimental acid ISL copper project was trialled at the old Mutooroo mine, 100 km south of Honeymoon, during 1981-82. All projects proved difficult and sub-economic, and thus acid ISL copper mining is yet to be.

A different site of note was the western world's first proposed ISL gold mine at Eastville, Victoria, in the early 1980's. The regulators and community, however, were not convinced about the safety of cyanide leaching and expressed grave concerns about groundwater contamination problems in the rural farming area. The project was quickly abandoned by (then) CRA Ltd.

The Beverley and Honeymoon uranium deposits, situated in the Lake Frome Embayment east of the Gammon Ranges in northeastern South Australia, were both discovered in the early

1970's, at a time when the prospects for nuclear power and uranium mining seemed endless. The deposits were actively being developed towards commercial scale in the late 1970's.

The Beverley deposit was originally planned as an open cut operation, but with the rapid drop in uranium prices in the mid-1970's, the project was shelved by 1974. The Honeymoon deposit, however, was recognised to be uneconomic by conventional mining from the outset, and by the late 1970's, ISL was being investigated as a possible economic alternative.

The Joint Venture developers of the Honeymoon deposit first conducted alkaline push-pull tests in 1977 using ammonia-bicarbonate solutions, however the results were discouraging (Mudd, 1998). A second push-pull test using sulphuric acid was undertaken in 1979 with positive results, and the partners committed to commercial development (Mudd, 1998).

The environmental impact assessment (EIA) process was undertaken (MINAD, 1980, 1981) with federal government approval being obtained in late 1981 for pilot testing before commercial scale operations could proceed (Mudd, 1998). A semi-commercial scale pilot plant, with a capacity of about 115 t/yr U_3O_8, was built and operated briefly in 1982, but was plagued with severe operational problems due to jarosite precipitation and other issues (Mudd, 1998).

The joint venture partners developing the Beverley deposit, first investigated the use of ISL in about 1980, releasing their Draft Environmental Impact Statement (EIS) in 1982 (cf. SAUC, 1982). The EIA process was not completed, however, and final approvals were not given.

Both projects proposed not to restore affected groundwater following operations at each site.

In March 1983, the recently elected government of South Australia refused to issue mining leases for commercial operations at Beverley and Honeymoon, citing these reasons (Mudd, 1998) : 1) many of the economic, social, biological, genetic, safety and environmental problems associated with the nuclear industry were unresolved; 2) endorsement of the Government's position by a wide range of community organisations; 3) commitment to the Roxby Downs (Olympic Dam) project; and 4) community disquiet at the nature of the ISL process.

The later introduction of the "Three Mines Uranium Policy" by the federal government in 1984 saw no further development until the election of a new federal government in 1996 and the removal of the (infamous) policy (Mudd, 1998). With new owners, fresh plans for their development are now being actively pursued. The Honeymoon site was joined with all nearby deposits, including Gould's Dam 75 km northwest, to form the "Honeymoon Project".

The geology of the Lake Frome region is given in Curtis *et al.* (1990), Morris (1984) and Brunt (1978). The following discusion uses these references, except where noted.

3.2 Beverley ISL Project

The Beverley deposit was originally discovered by the OTP Group of companies in 1969, with further drilling in 1970 confirming economic uranium grade (Mudd, 1998). After nearly three decades and two unsuccessful attempts, the Beverley uranium deposit finally began development towards commercial operation in 1996 through new owner Heathgate Resources Pty Ltd (HR), a wholly-owned Australian subsidiary of US-based General Atomics Corporation.

The geology and hydrogeology of Beverley is given in HR (1998a) and SAUC (1982) . The deposit consists of three ore zones - North, Central and South, each with increasingly higher salinity, respectively. The total size is about 21,000 t U_3O_8 (HR, 1998a).

A series of new bores were constructed in 1996 and 1997, including two 5-spot patterns, and hydrogeological pump testing was completed. HR applied for operation of an acid Field Leach Trial (FLT) in late 1997 with no proposed restoration of the pilot patterns. Approvals were quickly forthcoming from the South Australian government. The trial began on January 2, 1998 - before public release of a revised Draft EIS for the project. The trial was to leach the 5-spot patterns in the Northern and Central ore zones each for about 6 months. Due to the unexpected success of the Northern pattern, the FLT apparently used this pattern until the end of 1998.

After the release of the revised EIS in mid-1998, further studies were required by the federal and state governments to address significant inadequacies in the EIS, such as the degree of isolation of the Beverley mineralised aquifer, long term impacts on groundwater quality, and especially the potential to contaminate surrounding groundwater systems. HR still proposed not to restore affected groundwater following current and future mining operations at Beverley. With completion of these extra studies, final government approval was received in April 1999.

523

Table 7 - Beverley Groundwater Quality (average) : North (N), Central (C) and South (S) Ore Zones, Northern Field Leach Trial data (Injection - I, Extraction - E; averages March to July 1998) and Retention Pond (P; July 1998) (units as noted; m - mg/L; b - µg/L; na - not available) (adapted from HR, 1998a & b, SAUC, 1982)

	pH	TDS	S	SO_4	Cl	F	Na	K	Ca	Mg	U	Ra^{226}	Rn^{222}
	units	g/L	g/L	g/L	g/L	m	g/L	m	m	m	m	Bq/L	Bq/L
N	7.3	3-6	na	1.6	2	0.85	1.2	42	380	198	0.076	22-967	500-2,000
C	7	6-10	na	2.1	na	na	na	na	610	na	1.91	1.2-3,100	5-32,140
S	6.8	11-13	na	2.6	na	na	na	na	850	na	0.70	13-111	20-585
I	1.93	11.5	1.6	4.79	2.0	7.67	1.43	59	610	337	2.9	8414	na
E	1.97	11.7	1.6	4.84	2.0	7.33	1.43	59	600	337	162	9881	na
P	2.10	62.1	9.8	29.5	6.1	5.50	15.1	105	460	369	272	1713	na

	Al	B	Ba	Cd	Co	Cr	Cu	Fe	Mn	Ni	Pb	Se	Si	SiO_2	V
	m	m	b	b	m	b	b	m	m	m	b	b	m	m	b
N	0.2	1.6	53	0.2	0.1	20	30	0.7	0.2	0.004	40	1	48	na	1
I	91	1.0	37	117	20	100	200	109	0.7	8.47	160	410	138	294	1,100
E	91	1.1	39	116	20	580	200	105	0.8	8.33	790	410	133	283	1,130
P	39	3.4	76	49	6.6	260	180	39	0.9	2.48	70	310	99	211	780

Note - This is a compilation only, many parameters display a buildup over time. Complete trial data unavailable. No measuered Rn^{222} analyses available from the Field Leach Trial.

It is worth pointing out some significant outcomes of the approvals process for Beverley : 1) it is the western world's first commercial acid ISL uranium mine; 2) it proposes to re-inject all liquid wastes back into the mineralised aquifer rather than deep re-injection (»1 km; as per some US-sites) or evaporation (as per most US-sites); 3) the extent of the palaeochannel system is underexplored beyond the surrounds of the three ore zones; and 4) it is the first mining project in modern Australian history not required to restore the majority of it's environmental impacts after cessation of operations (that is, groundwater contamination).

HR (1998a & b) argue that following mining, the levels of radionuclides, heavy metals and pH will return to pre-mining conditions given several years; no mechanism is provided. This deserves critical assessment. The ore contains low sulphide (0.13%), organic carbon (0.05%), carbonate (0.06%), Fe, Mn and clay content (HR, 1998a). Buma (1981) argued that natural geochemical processes within aquifers can restore ISL-contaminated groundwater, thereby saving valuable chemical, energy and financial resources. The processes include precipitation of reduced compounds; scavenging of of heavy metals by pyrite, organic matter, calcite and ferric oxyhydroxides; adsorption by quartz, feldspars and clays. The key was for active reductants to be present. The conditions at Beverley, therefore, fail to provide any geochemical mechanism for natural restoration following acid ISL mining. The current trial, now two years old, if the data were to be released publicly, might be able to shed important light on such behaviour.

Of further significance is that Morris (1984) stated clearly that "reliance on this process (natural restoration) has never been tested". The time and rates at which natural processes could attenuate such levels of pollution are yet to be established. The extreme levels of groundwater contamination wrought at acid ISL uranium mines across the Former Soviet Union suggests natural restoration appears to be spurious at worst, ineffective at best (Mudd, 2000).

The potential for excursions due to abandoned exploration bores (when an open cut was intended) still remains, as well as excursions due to well casing failures (Marlowe, 1984). Curiously, final approvals for Beverley included provisions that liquid waste reinjection only occurr in the Northern zone - the zone of least exploration drilling and, importantly, the region of the best quality groundwater. This zone has similar water quality to pastoral use in the region (excluding radionuclide content), although gold mines in Western Australia often operate with much more saline groundwater (TDS up to 250 g/L).

The high Ca and SO_4 levels of the Beverley ore zones, especially the Central and Southern ore zones, create the potential for gypsum precipitation. A geochemical saturation analysis of the data in Table 7 can demonstrate this. This creates potential problems, similar to Nine Mile Lake and Reno Ranch, both operationally and for post-mining geochemical conditions. By August 1999, HR had apparently begun leaching of the Central trial pattern, although the full results from the Northern trial are yet to be publicly released, nor are they likely to be. This is in contrast to the USA regulatory process, where the results and restoration of a pilot scale facility form the permit basis of a commercial mine (Mudd, 1998). The construction of the commercial

Table 8 - Honeymoon Groundwater Quality : Upper (U), Middle (M) and Basal (B) (ore) Sands and predicted Lixiviant (L) composition (Fe in mg/L for B, g/L for L) (adapted from MINAD, 1980, 1981; Morris, 1984)

	pH	TDS	SO$_4$	HCO$_3$	Cl	F	Na	Ca	Mg	Fe	U$_3$O$_8$	Ra^{226}
	units	g/L	g/L	mg/L	g/L	g/L	g/L	g/L	mg/L	-	mg/L	Bq/L
U	-	10	1.4	-	5	1-2.5	-	0.5	280	-	-	0.13-1.5
M	-	12-15	1.6	-	6	2.8	-	0.6	320	-	-	2-100
B	6.8-7	16-20	1.8-2	135	8-10	0-2	3.8-5	0.95	410	<0.5	0.1	90-445
L	1.8-2.5	16-36	6-20	-	7-10	-	-	1.05	450	1-5	150	740-3,400

operation at Beverley is proceeding rapidly during 1999, presumably to avoid potential changes in government policy to reflect community opposition to uranium mining in Australia.

3.3 Honeymoon ISL Project

The Honeymoon deposit was Australia's first attempt at developing an ISL uranium mine, and had it succeeded in the early 1980's, would have become the first commercial ISL mine.

The deposit is located within the Yarramnba palaeochannel, which consists of three distinct aquifer sand layers, separated by thin, discontinuous clay layers. The upper aquifer is used by pastoralists in the region, while the lower sand contains the uranium deposit.

The deposit has several unique features related to the use of ISL, including pyrite content at 5-15%, compared to less than 2% in USA deposits; higher salinity; low organic content (0.3%); high background radon activities in the ore zone (Rn^{222} at 6,000 Bq/L); and direct hydraulic connections between the three aquifers in the palaeochannel due to gaps in the clay confining layers. The leaky nature is confirmed by the three aquifer pressures rising to the same level. The high Rn, in disequilibrium with Ra, is anomalous but may be related to basement features.

By 1982, the pilot plant (using solvent extraction) and four 5-spot patterns drilled had been consutrcted, although the fourth pattern intersected silt lenses with little mineralisation, and a field leach trial staretd using sulphuric acid and ferric sulphate. The trial encountered significant operational failure, due principally due to jarosite precipitation. The details have never been published, although it is known that jarosite was difficult to control.

A new trial at Honeymoon was approved in March 1998, relying mainly on previous approvals, with work beginning in April 1998. The information from both the 1982 and new trials are presumably to be incorporated in the new EIS for the project, which SCR are due to release in late 1999. The new trial is evidently trialling oxygen, which should avoid jarosite formation, although ferric sulphate is apparently still being used.

The approvals for Beverley has important precedents for ISL in Australia that have critical implications for the Honeymoon project : 1) the project proposes to re-inject all liquid wastes into the palaeochannel distant from theore zone, but still into the lower aquifer which is known to be hydraulically connected to important aquifers used by pastoralists; 2) the potential for "natural restoration" is questionable, although this depends on the extent of pyrite remaining after mining; and 3) the Yarramba palaeochannel is the only groundwater resource in the region (the velocity is about 18 m/year). The potential for post-mining impacts on groundwater are quite significant, especially if restoration is again not required by government regulators.

4   DISCUSSION AND CONCLUSIONS

The use of acid ISL in the USA was considered problematic and has never been approved or used on a commercial scale, despite the lengthy research at Nine Mile Lake and Reno Ranch, Wyoming. If Beverley and Honeymoon succeed where they previously failed, Australia will be forging a new, more profitable method of ISL - acid leaching with no restoration of groundwater. This is more akin to practices in Eastern Europe and the former Soviet Union than the demonstrable experience in the USA (cf. Mudd, 1998 & 2000). This is not considered an acceptable approach for an arid region that is almost entirely dependent on groundwater.

5 ACKNOWLEDGEMENTS

There are a great many people who have contributed to the comprehensiveness of this paper (see Mudd, 2000). This is independent and voluntary research undertaken by the author, and thus cannot be taken to represent the views of any other person, entity or organisation except those of the author. It is trusted this paper makes a valuable contribution on the topic.

6 REFERENCES

Brunt, D. A., 1978, *Uranium in Tertiary Stream Channels, Lake Frome Area, South Australia*. AusIMM Proc., 266, pp 79-90.

Buma, G., 1979, *Geochemical Arguments for Natural Stabilization Following In-Place Leaching of Uranium*. In "In Situ Uranium Mining and Ground Water Restoration", Chap. 8, New Orleans Symposium, Society of Mining Engineers, AIME, Feb. 19, 1979, pp 113-124.

Curtis, J. L, Brunt, D. A. & Binks, P. J., 1990, *Tertiary Palaeochannel Uranium Deposits of South Australia*. AusIMM, Monograph 14, pp 1631-1636.

DoE, 1999, *Uranium Industry Annual 1998*. U.S. Department of Energy, Washington DC, USA, April 1999, 69 p.

HR (Heathgate Resources Pty Ltd), 1998a, *Beverley Uranium Mine : Draft Environmental Impact Statement*. June 29, 1998, 405 p.

HR (Heathgate Resources Pty Ltd), 1998b, *Beverley Uranium Mine : Response Document and Supplement to the Environmental Impact Statement*. Oct. 3, 1998, 165 p.

Kasper, D. R., Martin, H. W., Munsey, L. D., Bhappu, R. B. & Chase, C. K., 1979, *Environmental Assessment of In Situ Mining*. U.S. Bureau of Mines, OFR 101-80, Dec. 1979.

Larson, W. C., 1981, *In Situ Leach Mining - Current Operations and Production Statistics*. In "In Situ Mining Research", U.S. Bureau of Mines, Information Circular 8852, pp 3-7.

Marlowe, J. I., 1984, *An Environmental Overview of Unconventional Extraction of Uranium*.U.S. Environmental Protection Agency, EPA-600/7-84-006, Jan. 1984, 130 p.

Mays, W. M., 1984, *In Situ Leach Mining - A Decade of Experience*. In "AIF Uranium Seminar", Keystone, CO, October 1984.

MINAD (Mines Administration Pty Ltd), 1980, *Honeymoon Project : Draft Environmental Impact Statement*. Prepared by Gutteridge Haskins & Davey Pty Ltd, Nov. 1980, 74 p+.

MINAD (Mines Administration Pty Ltd), 1981, *Honeymoon Project : Final Environmental Impact Statement*. Prepared by Gutteridge Haskins & Davey Pty Ltd, March 1981.

Morris, L. J., 1984, *Solution Mining*. In "8TH Australian Groundwater School", Vol. 2, Chapter 14, Australian Mineral Foundation, Adelaide, SA, Aug. 27-Sep. 7, 1984, 131 p.

Mudd, G. M., 1998, *An Environmental Critique of In Situ Leach Uranium Mining : The Case Against Uranium Solution Mining*. Research Report, Melbourne, VIC, Australia, July 1998, 154 p. (http://www.sea-us.org.au/isl/)

Mudd, G. M., 2000, *Acid In Situ Leach Uranium Mining - 2 : Soviet Block and Asia*. These Proceedings.

Nigbor, M. T., Engelmann, W. H. & Tweeton, D. R., 1982, *Case History of a Pilot-Scale Acidic In Situ Uranium Leaching Experiment*. U.S. Bureau of Mines, Rep. of Investigations 8652.

SAUC (South Australian Uranium Corporation), 1982, *Beverley Project : Draft Environmental Impact Statement*. July 1982, 329 p.

Staub, W. P. and others, 1986, *An Analysis of Excursions at Selected In Situ Uranium Mines in Wyoming and Texas*. U.S. Nuclear Regulatory Commission, July 1986, 294 p.

Tweeton, D. R. & Peterson, K. A., 1981, *Selection of Lixiviants for In Situ Leach Mining*. In "In Situ Mining Research", U.S. Bureau of Mines, Information Circular 8852, pp 17-24.

Underhill, D. H., 1992, *In-Situ Leach Uranium Mining in the United States of America : Past, Present and Future*. IAEA TECDOC-720, pp 19-42.

Tailings and Mine Waste'00 © 2000 Balkema, Rotterdam, ISBN 90 5809 126 0

Acid In Situ Leach uranium mining – 2. Soviet Block and Asia

Gavin M. Mudd

School of the Built Environment, Victoria University, Melbourne, Vic., Australia

ABSTRACT: The technique of In Situ Leach (ISL) uranium mining is well established in the USA, as well as being used extensively in Eastern Europe and the former Soviet Union. The method is being proposed and tested on uranium deposits in Australia, with sulphuric acid chemistry and no restoration of groundwater following mining. The history and problems of acid ISL sites in countries of the Former Soviet Union and across Asia is presented.

1 BACKGROUND

The unconventional mining technique of In Situ Leach (ISL) is now the primary producer of refined uranium in the United States, with a market share of around 95% in the mid 1990's (DoE, 1999). ISL mines appear set to assume a greater role in Australia's uranium industry. The commercial ISL uranium mines in the USA use alkaline chemistry, compared to the proposed projects in Australia which are based on the use of acid (Mudd, 1998). In contrast, the ISL uranium mines of the former Soviet Union and Eastern Europe have primarily used sulphuric acid, with apparently little consideration given to environmental concerns during operation.

The first trials of uranium ISL were both developed in the USA and the Soviet Union in the early 1960's. It is uncertain who developed the concepts or if they were developed separately (Mudd, 1998). The use of ISL uranium mining continued to expand until the collapse in the late 1980's. It is worth noting that an unnamed Russian first suggested In Situ Leaching of gold by as early as 1896 (Mineev & Shutov, 1979).

The majority of countries with uranium mining under the influence or control of the Former Soviet Union has undertaken ISL projects, although different countries had contrasting success, from an operational perspective. Bulgaria, for example, experienced a major shift in uranium production from conventional to ISL mines, dramatically reducing the workforce and exacerbating already recalcitrant environmental problems.

Since the reunification of Germany and the collapse of the Soviet Union, the extent of the contamination of groundwater is beginning to come to light, mainly through co-operative programs of the International Atomic Energy Agency (IAEA), although other agencies are are becoming involved. There is also increasing interest in the use of ISL methods for the mining of low grade uranium ores in other parts of Asia, most notably China and Pakistan.

The resurfacing of the Australian acid ISL uranium mine proposals in 1996, the lack of acid ISL mines in the USA, the research coming to light through the IAEA concerning the extent of impacts from acid mines in the Soviet block, led to a wide ranging review of ISL uranium mining by the author, completed in 1998 (cf. Mudd, 1998).

2 BULGARIA

This review is based on IAEA (1999), Nedyalkov (1996), Dimitrov & Vapirev (1994), Vapirev *et al.* (1993), Kuzmanov *et al.* (1992) and Tabakov (1992), more detail is given in Mudd (1998).

Table 1 - Typical Lixiviant Components (mg/L) (summarised from 13 ISL sites)

pH	TDS	Na	K	Ca	Mg	SO$_4$	V
1.4-2.0	15,000-20,000	30-900	30-200	140-600	140-330	10,000-12,000	1.0-18
Al	Fe	Mn	Zn	HSiO$_4^-$	U	Ra (Bq/L)	
310-840	700-2,200	6-61	2.1-7.3	210-350	5-30	1-2	

The ISL technique was first applied in 1967 to low-grade deposits (ranging from 0.006 % to 0.03 %) at Orlov Dol and Selishte (a former conventional mine). Given the success of these sites, a revolution was perceived whereby many previously uneconomic deposits were exploitable. By 1990, the share of uranium production from ISL was 70%.

All uranium mining and milling in Bulgaria was closed down by government decree on August 20, 1992. Activity since has been aimed at cleaning up and rehabilitating the numerous mine sites. The total contaminated area due to all uranium industry activity is approximately 20 km^2, including 6 km^2 from ISL mining and 4 km^2 of contaminated forest.

There has been a total of 19 sites where ISL has been applied, and a further 11 sites where the ISL technique was applied within an underground mine. Most of these sites began operation in the late 1960's to early 1970's, although poor results from initial trials meant some sites were not continued. These are all concentrated in the southern and western regions of Bulgaria (refer to Figure 21). The deposits contain high amounts of organic matter, iron and sulphides.

The ISL mines had a dramatic impact on the workforce in conventional mines, falling from 5,000 workers between 1965 and 1970 to approximately 500 in 1988. The initial chemistry used was sulphuric acid, although this was later switched to sodium-carbonate and ammonium-carbonate leaching chemistry in deposits with a high carbonate content.

The siting and operation of many uranium mining operations across Bulgaria were often neglected to enable fast tracking of projects and minimise the costs involved in establishing a project. Only one mine was closed due to contamination of drinking water.

Almost no preventive measures or counter measures were implemented during the whole period of mining for the environmental protection of water, soil and air from mechanical, chemical and radioactive pollution. The secrecy of the uranium and nuclear industry was identified as a key reason behind this philosophy.

The leaching of uranium was generally progressed in three stages - first, acid was introduced at levels up to 10 g/L (lasting for a few months); second, the base period of acid leaching at 4-6 g/L (lasting for over two years); and third, the closeout period at 0.5-1 g/L. The overall time for a wellfield was between 3 to 5 years and the recovered uranium was about 60% to 80%. The Ra levels of leaching solutions was generally low.

The pre-mining quality of groundwater in the ore zone aquifers was typically 500-2,000 mg/L TDS, pH of 6.3-8.8, iron 0.1-648 mg/L, and sulphate 24-758 mg/L; good quality water.

There has been significant contamination of groundwater at most ISL sites, with major concerns arising from chemical, radiological and bacterial contamination. For the combined underground-ISL sites, Fe, Cu, Co, Ni, Mo, As and some rare earth elements are several times higher than allowable limits. In the Deveti septemvri ISL mine, Mo reaches 13.4 mg/L (regulatory limit 0.5 mg/L, 27 times higher), and Mn reaches values up to 13 mg/L in groundwater and 4.2 mg/L in the retention pond; quantities of B and Hg have been detected.

At the Orlov Dol site, after six years of monitoring, the acidity of the ore zone aquifer was declining from 1,300 mg/L to 10 mg/L, although the groundwater still contained elevated levels of uranium despite the associated small increase in pH.

The concentration of sulphate can be very high in surface waters and even in water supply wells of private owners as a result of accidental spilling of solutions at ISL sites. The average chemistry of contaminated groundwaters ranges in TDS from 15-20 g/L, SO$_4$ from 10-12 g/L, U between 5-20 mg/L, other salts and the presence of heavy metals and rare earths.

For the Cheshmata (Haskovo) site, in the valley downstream, SO$_4$ concentration is 1,400 mg/L (limit 300 mg/L), free sulphuric acid 392 mg/L and the pH is 2.2 (over 1,000 times more acidic than the surrounding aquifer). The private wells of residents of the area have also been affected with significantly high concentrations of SO$_4$ being noted, demonstrating that the leaching solutions have migrated into drinking water supplies.

A similar case has been recorded in Navusen, where in a valley the SO_4 concentration is 13,362 mg/L and almost 5 g/L free sulphuric acid, indicating the water is actually lixiviant or leaching solution. The groundwater quality of such sites has a TDS (salinity) level of greater than 20 g/L, of which SO_4 is 12-15 g/L. Heavy and rare earth elements were detected in some cases, such as V, W, Mo and La, due to recycling of the solution.

There were also noted problems due to bacterial contamination, although their exact effects were not able to be predicted and were not studied. It was thought that they were beneficial in the leaching process.

There remains concern that solutions at the various ISL sites could contaminate the deeper groundwater systems, as well as the shallow systems. For the deep systems, which contain the ore being mined, the U content can reach 20-30 mg/L and Ra 1-2 Bq/L, with U content in shallow aquifers around 3-4 mg/L and Ra 0.5 Bq/L, despite dilution effects during migration through the aquifer.

At some sites, where there were surface spills due to failure of distribution pipes, the U and Ra content of soils is 10 and 2-3 times the background level, respectively.

At most of the ISL sites undergoing restoration, the solutions were continually recycled through the mined aquifer without adding acid, and this led to deposition of salts within the pipes. These salts contained increased and significant levels of radioactivity. This process has been now been stopped.

A principal problem of the restoration work currently in progress is that the environmental requirements are quite strict, making uranium production unprofitable. It was argued that achieving an acceptable level of environmental protection required that preventative measures, planning and funds are set aside during the early stages of a project, and during the operational phase of a particular project. As this was not done, the necessary funds are not available and restoration work is thus significantly impaired.

Only one third of the land used by ISL operations has been remediated, and as the land will be returned to the original owners for agricultural purposes, there are grave concerns for public health and environmental safety. Some of the ISL sites (such as Bolyarovo, Tenevo, Okop and Gorna Trakiiska nizina) are close to areas where potable quality groundwater is extracted by local communities or the groundwater is considered to be an important future water resource.

3 CZECH REPUBLIC

This review is based on IAEA (1999), Tomas (1996), Andêl & Pribán (1994), Andêl & Pribán (1993), Fiedler & Slezák (1992), Khün (1992) and Benes (1992), more detail is in Mudd (1998).

The Bohemian Massif mining district in the Czech Republic has been an important source of uranium for the Russian military and nuclear power programs, with uranium ore from the Jáchymov mine being used to manufacture the first Soviet atomic bomb. A total of more than one hundred uranium mines were developed, including shallow investigation mines.

The Stráz Pod Ralskem district consists of sandstone-type uranium deposits, and acid ISL has been applied as the mining method since 1968 after successful trials during 1967. The associated mineralisation is also unusual, with zircon, titanium and phosphorous present.

The Stráz region is characterised by complex and unfavourable hydrogeological and biological conditions that make the application and success of ISL extremely difficult. The dissolution rates of uranium are quite slow. This causes two principal problems - firstly, large doses of chemicals are required (sulphuric acid at about 5% with nitric acid and nitrate as the main oxidants); and secondly the leaching periods are very long, ranging from 15 to 25 years. The total uranium production by ISL was 16,470 t U_3O_8.

The Hamr deposit was developed at the Hamr and Luzice mines with both underground and combined ISL techniques. The Hamr mine is only 5 km from the Stráz mine, exacerbating the technical problems at both sites, leading to higher production costs and greater environmental impacts. A principal problem for many of the sites is the density of population across the Czech Republic, with 40,000 people living near the Stráz mine, for example.

After detailed evaluation of the negative impact of uranium mining and milling, a progressive program of declining production from uranium mining has been adopted and an extensive remediation programme implemented by the Czech Government.

3.1 Stráz Pod Ralskem

The hydrogeology of the Stráz region is complex, but can be thought of as two distinct aquifers - the Cenomanian and the Turonian. The Cenomanian is a deep, confined and artesian aquifer, and the Turonian lies above this, separated by up to 100 m of thick low permeability clays and siltstones. The Turonian is designated as an important high quality drinking water reserve with a calcium bicarbonate ($Ca-HCO_3$) type of water quality, and is known to discharge to the Ploucnice River at about 40 L/s. The Cenomanian was known to contain elevated levels of Ra.

For the Stráz deposit, every tonne of uranium (t U) produced :

- 274 t of sulphuric acid injected;
- 7.9 t of ammonia injected;
- 19.3 t of nitric acid injected;
- 1.79 t of hydrofluoric acid injected;
- 7,260,000 L of contaminated groundwater in the Cenomanian aquifer;
- 1,500,000 L of contaminated groundwater in the Turonian aquifer.

- 0.95 t of sulphuric acid released to the air;
- 1.18 t of nitrous oxides released to the air;
- 53 GBq of radium released to the air;

By contrast to experience in the USA, the Stráz ore deposit required 50-70 g/L of sulphuric acid and a leaching period of 15-20 years to reach a yield of 60-80% of the uranium. This was due to the lower permeability of the aquifer materials.

By 1994, a total of 32 ISL sites had been commissioned covering a total of 6 km^2 consisting of 7,000 wells. The Stráz mining district, has seen approximately 3,800,000 t of sulphuric acid, 270,000 t of nitric acid, 103,000 t of ammonia and 25,000 t of hydrofluoric acid injected into the wellfields. The interactions between the leaching solutions and aquifer sediments are not well defined, and the speciation of many heavy metals and radionuclides remain unstudied. The Stráz site ceased producing uranium on April 1, 1996.

The leaching solutions from the Stráz wellfields were not operated with a bleed system to maintain a cone of depression around active wellfields, and this led to solutions being dispersed widely through the Cenomanian aquifer in the area, as well vertically into the Turonian aquifer. The excursions occurred mainly through production bores, but significant excursions also occurred at liquid waste disposal bores.

The contaminated water in the Turonian aquifer alone is spread over 245 hectares (43% of the area of the wellfields). A total of 200 billion L of groundwater has been affected, covering a total area of 6 km^2 and the volume of aquifer material affected is thought to 720 billion L. Approximately 50% of the contaminated water is thought to be residual leaching solutions, with sulphate higher than 20 g/L and salinities between 35-70 g/L. The remaining 50% is thought to be dispersed solutions, formed by migrating leaching solutions mixing with native groundwater, with a salinity level of 4.5 g/L.

Table 2 - Typical Lixiviant Composition at Stráz Pod Ralskem (mg/L)

Free Acid (H_2SO_4)	SO$_4$		NH$_4$	NO$_3$	F	P	SiO$_2$
15,000-38,000	40,000-65,000		1,000-2,000	200-800	100-300	50-150	100-200
Na	K	Ca	Mg	Al	Fe	Cr	Ni	U	V
10-15	40-70	200-300	20-30	4,000-6,000	500-1,500	5-15	20-30	20-500	10-15

Table 3 - Groundwater Quality Before and After ISL mining at Stráz Pod Ralskem

	pH	TDS	SO$_4$	NO$_3$	F	U	Ra	H_2SO_4
	units	g/L	g/L	mg/L	mg/L	mg/L	Bq/L	g/L
Lixiviant	0.5	50-100	33-80	600-1,400	150-250	1-30	50-90	15-20
Cen. Before	6.7	0.14	0.033	<1	<1	0.02	8.74	NA
Cen. Affected	1.8-2.8	5-20	3.3-13	5-100	5-50	0-15	30-70	0.5-5
Tur. Before	6.7	0.1	0.035	5.2	<1	0.01	0.07	NA
Tur. Affected	2.5-7.0	0.5-5.5	0.05-3.3	5-1,000	0.5-25	<1	0.1-1.0	<0.5

The urgent need for restoration is governed by the extremely high concentrations of radionuclides and heavy metals in the various solutions and the large volumes of contaminated water involved. The most critical factor is that the Cenomanian aquifer is artesian, and the pressure difference between the Turonian and Cenomanian aquifers will always ensure groundwater flow is vertically upwards, as was the case before ISL mining began.

The presence of known excursions through boreholes highlights the above problem, and if the bores are not effectively sealed during restoration, there will remain a pressure gradient for new excursions of contaminated groundwater from the Cenomanian into the Turonian aquifer.

The contaminated groundwater in the Cenomanian aquifer is approaching the sanitary protection zone of the Mimon water supply (70 L/s). The contaminated groundwater in the Turonian is within 1.2-1.5 km of the sanitary protection zone of the Dolánky water supply (200 L/s). The region presently utilizes 1,500 L/s of groundwater for drinking supplies.

The restoration of the groundwater is proving a difficult task, with 1 pore volume of groundwater only removing about 70% of the contaminated groundwater and 5 pore volumes required for 90% removal. This equates to about 940 billion L of water. While regulators in the USA require a proven pilot-scale test to demonstrate effective groundwater quality restoration, the Stráz mine only received approvals for liquid waste disposal and restoration requirements in the mid 1990's after three decades of operation.

Currently, restoration programs are aimed at determining the optimal strategy for long term remediation of groundwater quality. The technology being used for restoration involves pre-treatment, reverse osmosis, volume reduction by evaporation, crystallisation and processing of the concentrated saline solutions or brines. Some components are re-utilised, such as sulphuric acid (H_2SO_4), aluminium oxide (Al_2O_3), ammonia (NH_3), and gypsum ($CaSO_4$). The presence and removal of Ra and other metals of concern is a significant barrier to these programs.

Further options for groundwater quality restoration are being investigated with a view to a compromise between environmental demands and economic feasibility. It is intended to return the Turonian aquifer to it's original quality (as much as possible). Recent modelling studies indicate that restoring the Cenomanian groundwater to a salinity of 3 g/L can cause undesirable impacts on the Turonian aquifer, due mainly to structural and tectonic conditions and the instability of the groundwater regime after returning to natural flows.

For the Cenomanian aquifer, though, it appears impossible to achieve restoration to it's original good quality. The philosophy being adopted is to ensure that any escape of Cenomanian groundwater will disperse to an appropriate quality and not impact on potable or surface waters.

The region, once covered by pine forests, underwent deforestation for mining purposes. This was undertaken hastily with many trunks left in the ground, dead and rotting. The surface soils, devoid of tree cover, were therefore exposed to accelerated rates of weathering, sheet erosion and wash-down of the poorly cohesive sandstones and deep furrows. This was exacerbated by the movement of heavy machinery across the site. In the low-lying areas near the Ploucnice River, the alteration of surface drainage patterns, together with the removal of vegetation, led to a gradual rise in the water table and the formation of lagoons and wetlands.

In the areas where pine trees had been left, to try and preserve some of the remnant forest, it was found that the forest was weakened and unsustainable since it was no longer continuous. This led to increased exposures of the wells and piping systems and high incidences of dead, falling trees. In the hill areas, wells and piping systems were often built partly on benches and partly on platforms. Together with the spills of solutions from pipes and surface runoff, the siting of these parallel to slopes led to significant rates of erosion and the prevention of further vegetation growth due to the lack of suitable soil. Attempts were made from the mid 1980's to address these problems, such as hydromulching, different seed species and other techniques, but they were of varying short duration and thus limited success.

Due to the intransigence of the chemical and physical changes caused by ISL mining at the Stráz site, the restoration efforts are anticipated to last several decades, or even centuries.

4   GERMANY

This review is based on Biehler & Falck (1999), Diehl (1999), Ettenhuber (1996) and Hähne & Altmann (1992).

The rich pitchblende uranium deposits of East Germany were one of the former Soviet Union's first targets for supplying uranium during the late 1940's and 1950's for weapons and nuclear power programmes. Simultaneous work was carried out on all types of uranium deposits in the area. Most uranium mines and mills in East Germany were underground, although Königstein and to a lesser extent Ronneburg, also had underground ISL applied for the extraction of uranium. Underground uranium leaching began in 1968 to take advantage of low grade ores and increasing conventional costs, since in-situ leaching costs were only 60-70% of conventional methods.

4.1 Ronneburg

Although a less prolific producer of uranium by ISL, repeated attempts were made to increase uranium recovery from 1970. The leaching solutions used included sulphuric acid or alkaline reagents. Between 40-70% of the uranium reserves were recovered at concentrations of sulphuric acid from 3-10 g/L, pH of approximately 1.5-2.5 and uranium content between 20-100 mg/L. A total of 3,203 t of U_3O_8 was produced by heap and waste pile leaching, with 106 t of U_3O_8 produced by underground ISL techniques.

4.2 Königstein

From 1971 both underground and in situ leach mining was being used, until ISL mining took over in 1984. The extraction of uranium with ISL operated until 1990, with the total uranium production from the life of the mine being 22,711 t of U_3O_8, 6,517 t by sulphuric acid ISL.

The aquifers at Königstein generally contain excellent quality water, with salinity less than 200 mg/L and pH near neutral. The heavy metal content is typically quite low. The uranium mineralisation at Königstein is found within the fourth aquifer of a regional groundwater system. The clay layer separating the third and fourth aquifers was intersected by the underground mine workings. The third aquifer is used by residents of the region for their water supply, as well discharging into the Elbe River 600 m east of the mine site. The dewatering of the fourth aquifer for the mine led to a decrease in water level of the third aquifer.

One of the most difficult problems associated with remediating the contaminated groundwater is that at the time of closure a new underground block had just been prepared for leaching. The prevailing unsaturated conditions allowed the pyrite to oxidise, generating significant quantities of sulphuric acid, further mobilising heavy metals, uranium and radionuclides and adding to the contaminant load to remediate.

The average concentrations of leaching solutions was 2-3 g/L sulphuric acid, pH 1.5-1.8, salinity (TDS) 10-14 g/L and uranium 10-150 mg/L. The uranium recovery was generally about 65-75% within three years. A total of 100,000 t of sulphuric acid was injected into the mine. The leaching process has chemically affected more then 55 million m^3 of rock and aquifer, while approximately 1.8 billion L containing 1.2-1.7 g/L sulphuric acid and more than 30 mg/L uranium remains circulating or trapped in the pore space of the rocks. A further 850 million L are circulating between the leaching zone and the recovery plant. Expressed as multiples of applicable German drinking water standards, the trapped liquids have levels 400 times higher in Cd, 280 times higher in As, 130 times higher in Ni and 83 times higher in U.

The principal concerns for restoration of the site are centred around the flooding of the underground mine workings that will occur after the mine is closed down. There is potential for contamination of surrounding groundwater and surface water streams with U, Ra, SO_4, Fe and heavy metals. Although small scale flooding trials are currently being conducted, restoration is still not complete and the mine still represents a threat to the surrounding aquifer, an important potable groundwater resource for the region.

5   FORMER SOVIET UNION

A general overview of the use of ISL in the Former Soviet Union is presented. A more detailed review of each new republic (the "stans"), however, will be treated separately. This review is based on Skorovarov & Fazlullin (1992) and Skorovarov et al. (1987).

Table 4 - Typical Acid and Alkaline Leaching Solution Composition (mg/L; Ra in pCi/L)

	Cl	SO_4 (g/L)	Na + K	Ca	Mg	Al	Fe^{2+}	Fe^{3+}	Ra
Acid	400-600	17-25	100-200	400-600	300-500	500-800	800-1,500	400-1,000	100

	HCO_3	NH_3	SO_4	Cl	Na + K	Ca	Mg
Alkaline	500-2,500	400-600	2,000-3,000	500-1,200	500-1,000	700-800	100-300

There has been active development of ISL-type mines across former Soviet block countries such as Ukraine, Uzbekistan, Kazakhstan and Russia, since the early 1960's. The ISL technique was typically applied to low grade deposits between 0.03-0.05%. Sulphuric acid was the more popular leaching agent, although alkaline carbonate-bicarbonate agents were used at some sites, depending mainly on the carbonate content of the ore. The wellfield patterns used were quite variable, including 10x10 m, 10x20 m, 25x50 m and up to 10x100 m spacings.

The concentration of sulphuric acid ranged from 2-5 g/L, with the stronger the acidity the greater the recovery of uranium and shorter the period of leaching. The average acid consumption per 1 kg of uranium recovered as an end product varied widely from 18-150 kg. The recovery rate of uranium was generally between 70 and 90%. No oxidant was needed to ensure dissolution of the uranium. Associated metals were also thought to be extractable, such as V, Re, Se, Mo, Sc, Yt and rare earths.

Some of the main problems of using sulphuric acid was the necessity to use acid-resistant materials and equipment, deterioration of the ore zone permeability due to chemical and gaseous plugging, and the very high salinity levels during mining (ranging from 15-25 g/L).

For ores with a carbonate content higher than 1.5-2.5%, alkaline solutions were used consisting of ammonium bicarbonate. The concentrations generally varied from 500-5,000 mg/L. On some sites sodium bicarbonate was also applied. The alkaline ISL sites used oxygen as the oxidant. The recovery rate of uranium was generally between 50 and 60%. The use of alkaline agents also tended to show much smaller increases of salinity during mining. The main recognised problems of alkaline ISL were the high degree of solutions escaping outside the mining zone (often due to gaseous oxygen plugs forming), compulsory pre-treatment to soften the water and restoration difficulties following completion of ISL.

Numerous techniques were being trialled to restore the quality of the groundwater, including lime pulp treatment, hyperfiltration and electrosorption. The success of these technologies on restoring contaminated groundwater is not known. The production costs of ISL were 40-45% of conventional costs, with significantly lower energy and capital costs and reagent consumption.

6   KAZAKHSTAN

This review is based on Catchpole (1997) and Carroll (1997).

The uranium resources of Kazakhstan are considerable, of which a large proportion are amenable to ISL extraction. As with many former Soviet-controlled states, the use of ISL in Kazakhstan began in 1970 and continued to increase in importance, centred around the large amenable deposits in southern Kazakhstan. The large scale ISL mines began in 1978. All ISL mines utilise sulphuric acid leaching chemistry.

By 1990, ISL technology had displaced conventional mines as the predominant uranium production method. The large ISL-amenable resources are seen as the future of the Kazakhstan uranium industry. There are several operating ISL projects, although an accurate assessment of current and prospective projects is not an easy task. Some operating sites include Stepnoye, Centralia and Chiili. Further sites being assessed and/or developed are the Inkai, Mynkuduk and Moynkum sites. The environmental impacts and operational issues are yet to be published, although given the history of the nuclear industry in Kazakhstan, it is likely to be similar to other parts of the Former Soviet Union.

7   UKRAINE

This review is based on Chernov (1998), Rudy (1996) and Molchanov (1995). In-situ leach uranium mining was carried out on the Deviadovskoye, Bratskoye and Safonovka deposits.

The Deviadovskoye ISL mine operated from 1966 to 1983, using sulphuric and nitric acids. The surface area of the mine is 12 ha, with the ore body about 218 ha and the area for underground storage is 120 ha. As a result of ISL mining, groundwater was contaminated at a depth of 80 m. The residual solutions are distributed a distance of 1.7 km along the flowpath and for 0.35 km against the upstream gradient.

The nearest settlement down gradient is 4 km only distant. The volume of residual solutions after the ISL mining of uranium in the Buchak aquifer is 7.09 billion L. The volume of tailing ponds water is 1 billion L with contaminated silt in ponds-collectors about 40,000 m^3. Leakage from pipelines has contaminated surface soils, totalling about 50,000 m^3.

The Bratskoye ISL mine site operated from 1971-84, using sulphuric and nitric acids. The orebody area is 95.5 ha. At the end of mining, the leaching solutions within the orebody were simply abandoned. The 5.2 billion L of contaminated groundwater is distributed 3 km down gradient and 1.2 km up gradient, to a depth of 50 m.

The Safonovka ISL site was mined from 1982-93. The surface area of the mine site covered 5 ha, although no further information is available on the extent of groundwater impacts and future management and remedial programs.

The severe lack of financial resources has led to the freezing of restoration activities in 1996.

8   UZBEKISTAN

This review is based on Venatovskij (1992). The state of Uzbekistan has numerous uranium deposits that host operating or potential ISL mines, concentrated within the large Central-Kizilkum province. They are generally around 300 m in depth and contain uranium ore in several distinct layers. The ore grades vary from 0.03-0.70%. Many contain low carbonate content less than 2.5% although some deposits are rich in carbonaceous matter (higher than 5%). Various leaching agents are used, with sulphuric acid being the preferred acid. Information on the extent of operational and environmental impacts is not presently available.

9   CHINA

This review is based on Xu et al. (1998) and Jian & Ning (1992), except where noted.

Approximately 61% of Chinese uranium is contained within deposits smaller than 3,000 t U_3O_8, mostly below a 0.2% grade. The In Situ Leach technique, for underground mines and the more traditional solution mines, has been viewed as the preferential method for economically extracting uranium since the early 1980's, with trials on all types of mineralisation being conducted. Two main ISL projects are currently being actively developed or operated at Tengchong and Yining (Diehl, 1999).

The Tengchong uranium deposit is hosted in sandstone with gangue minerals including pyrite and carbonaceous matter. A trial of sulphuric acid ISL lasting 42 days was undertaken on a pattern of 31 wells, with the uranium content reaching a maximum of 150 mg/L and an effective yield of 62%. The deposit is being developed as a commercial facility.

In Situ Leaching at Yining, also known as Deposit No. 512, began in 1994 uranium using sulphuric acid and a hydrogen peroxide oxidant. The deposit is hosted in sandstone being up to 20 m thick and is 0.011-0.17% in grade. Sulphuric acid levels were initially injected at 2% (20 g/L) and gradually increased to 8% (80 g/L) with hydrogen peroxide concentrations up to 0.55 g/L. The acid was later reduced to 4-6 g/L. The ISL trial ran for 92 days, with the injection of 9.8 million L of lixiviant, 41.59 t of sulphuric acid , 2.11 t of hydrogen peroxide, and the extraction of 11.7 million L of solutions (18.7% bleed rate) with uranium at 40-75 mg/L.

Table 5 - Typical Lixiviant Composition at Deposit No. 512 (Yining) (mg/L; Eh in mV)

pH	Eh	SO_4	S^{2-}	Cl	CO_2^1	PO_4	F	Na	K	Ca	Mg	Fe^{2+}	Fe^{3+}	Al^{3+}	U
1.26	652	22,800	2.7	762	0.99	22	5.2	172	28	17	140	377	440	323	75
As	Cd	Cu	Cr^{6+}	Cr_T^2	Mo	Mn	Ni	Pb	Re	Sb	Sc	Se	Ti	V	Zn
0.1	0.04	0.5	0.1	0.74	2.4	8.4	1.3	0.67	3.1	0.4	1.86	0.001	<1	4.8	0.88

Notes : [1] - Free CO_2; [2] - T is Total Cr.

10   DISCUSSION AND CONCLUSIONS

The experience of acid In Situ Leach uranium mining in areas controlled by the Former Soviet Union provides a stark contrast to experiences in America and Australia. In most applications of the technique, there has been extreme occurrences of groundwater contamination. At some sites, this contamination has migrated considerable distances to impact on potable drinking water supplies. For other sites, the potential for contamination to reach an undesirable receptor remains significant. The problems at these sites were severely exacerbated by the prevailing paradigm of uranium production without regard for environmental damage. Apart from Asian nations such as Kazakhstan, Uzbekistan and China, all countries are in the process of closing down and developing remedial action programs. The restoration of groundwater is proving difficult, both technically but also due to a lack of financial resources within these countries.

Morris (1984) noted that reliance on natural attenuation processes has never been tested for restoration of ISL. The former ISL sites across Eastern Europe and the Former Soviet Union allow some insight into the use of "natural restoration" as a remedial technique.

Buma (1981) argued that natural geochemical processes within aquifers can restore contaminated groundwater from ISL mines, thereby saving valuable chemical, energy and financial resources. The processes he outlines include precipitation of reduced compounds; scavenging of of heavy metals by pyrite, organic matter, ferric oxyhydroxides and calcite; adsorption by quartz, feldspars and clays. The key was for active reductants to be present.

At many ISL mines outlined above, there was high organic, iron or sulphide content, such as Bulgaria and the Czech Republic. The contamination at these sites, including the high concentrations of major ions, heavy metals and radionuclides, has not attenuated significantly over time, and instead migrates away from the mine sites, up to several kilometres in some instances. The geochemical mechanisms controlling this migration are unclear, although co-precipitation, which may give rise to higher solubilities for species such as $(Ca.Ra)SO_4$, and the complete oxidation of reducing agents during ISL mining with no active agents remaining after mining, are likely to be significant, key issues.

It would appear, therefore, that "natural restoration" is not a desirable approach in the slightest, even given the complex hydrogeochemical conditions known to exist at some sites across Eastern Europe and the former Soviet Union.

11   ACKNOWLEDGEMENTS

There are a great many people who have contributed to the comprehensiveness of this paper. Of special note is Peter Diehl (Germany), David Noonan and Dennis Matthews (South Australia), Paul Robinson (New Mexico, USA) and several supportive colleagues from around Australia (eternal thanks). This is independent and voluntary research undertaken by the author, and thus cannot be taken to represent the views of any other person, entity or organisation except those of the author. It is trusted this paper makes a valuable contribution on the topic.

REFERENCES

Andĕl, P. & Pribán, V., 1993, *Environmental Restoration of Uranium Mines and Mills in the Czech Republic*. IAEA TECDOC-865, Vol. 1, pp 113-135.

Andĕl, P. & Pribán, V., 1994, *Planning for Environmental Restoration of Contaminated Sites of the Uranium Industry in the Czech Republic*. IAEA TECDOC-865, Vol. 2, pp 81-95.

Benes, S., 1992, *In Situ Leaching of Uranium in North Bohemia*. IAEA TECDOC-720, pp 55-63.

Biehler, D. & Falck, W. E., 1999, *Simulation of the Effects of Geochemical Reactions on Groundwater Quality During the Planned Flooding of the Königstein Uranium Mine, Saxony, Germany*. Hydrogeology Journal, 7 (3), pp 284-293.

Buma, G., 1979, *Geochemical Arguments for Natural Stabilization Following In-Place Leaching of Uranium*. In "In Situ Uranium Mining and Ground Water Restoration", Chap. 8, New Orleans Symposium, Society of Mining Engineers, AIME, Feb. 19, 1979, pp 113-124.

Carroll, P. A., 1997, *The Reconstruction of the Uranium Industry in Kazakstan*. Uranium Institute (London), 22ND Annual Symposium, Sep. 5-7, 1997, London, UK, 8 p.

Catchpole, G., 1997, *The Inkai ISL Uranium Project in Kazakstan*. Uranium Institute (London), 22ND Annual Symposium, Sep. 5-7, 1997, London, UK, 5 p.

Chernov, A., 1998, *Uranium Production Plans and Developments in the Nuclear Fuel Industries of Ukraine*. Uranium Institute (London), 23RD Annual Symposium, Sep. 10-11, 1998, London, UK, 6 p.

Diehl, P., 1999, *Impacts of Uranium In Situ Leaching*. WISE Uranium Project, Germany, Jan. 1999, 7 p. (http://antenna.nl/~wise/uranium/uisl.html).

Dimitrov, M. & Vapirev, E. I., 1994, *Uranium Industry in Bulgaria and the Environment : Problems and Specific Features of the Period of the Technical Close-Out and Remediation of the Negative Consequences*. IAEA TECDOC-865, Vol. 2, pp 43-53.

Ettenhuber, E., 1996, *Environmental Restoration Plans and Activities in Germany*. IAEA TECDOC-982, pp 85-101.

IAEA, 1999, *Technical Options for the Remediation of Contaminated Groundwater*. IAEA, Vienna, Austria, June 1999, IAEA TECDOC-1088, 135 p.

Fiedler, J. & Slezák, J., 1992, *Experience With the Coexistence of Classical Deep Mining and In-Situ Leaching of Uranium in Northern Bohemia*. IAEA TECDOC-720, pp 115-128.

Hähne, R. & Altmann, G., 1992, *Principles and Results of Twenty Years of Block Leaching of Uranium Ores by Wismut GMBH Germany*. IAEA TECDOC-720, pp 43-54.

Jian, W. & Ning, D. Y., 1992, *In-Situ Leaching of Uranium in China*. IAEA TECDOC-720, pp 129-132.

Khün, P., 1992, *Environmental Aspects of the Operation and Sanation of the ISL Uranium Mining at Stráz pod Ralskem Czechoslovakia*. IAEA TECDOC-720, pp 181-190.

Kuzmanov, L., Simov, S. D., Valkov, T. & Vasilev, D., 1992, *In-Situ Leaching of Uranium in Bulgaria : Geological, Technological and Ecological Considerations*. IAEA TECDOC-720, pp 65-73.

Mineev, G. G. & Shutov, A. M., 1979, *Feasibility of Mining Gold by Underground Placer Leaching*. Soviet Min. Sci. (Engl. Transl.), 15, pp 400-404.

Morris, L. J., 1984, *Solution Mining*. In "8TH Australian Groundwater School", Vol. 2, Chap. 14, Australian Mineral Foundation, Adelaide, SA, Aug. 27-Sep. 7, 1984, 131 p.

Mudd, G. M., 1998, *An Environmental Critique of In Situ Leach Uranium Mining : The Case Against Uranium Solution Mining*. Research Report, Melbourne, VIC, Australia, July 1998, 154 p. (http://www.sea-us.org.au/isl/)

Mudd, G. M., 2000, *Acid In Situ Leach Uranium Mining - 2 : USA and Australia*. These Proceedings.

Nedyalkov, K., 1996, *Plans for Environmental Restoration of Uranium Mining and Milling Sites in Bulgaria*. IAEA TECDOC-982, pp 21-34.

Rudy, C., 1996, *Environmental Restoration in Regions of Uranium Mining and Milling in Ukraine : Progress, Problems and Perspectives*. IAEA TECDOC-982, pp 189-198.

Skorovarov, J. I. & Fazlullin, M. I., 1992, *Underground Leaching of Uranium in the Commonwealth of Independent States*. IAEA TECDOC-720, pp 75-80.

Skorovarov, J. I., Sadykov, R. H. & Nosov, V. D., 1987, *In Situ Leaching of Uranium in the USSR*. IAEA TECDOC-492, pp 143-152.

Tabakov, B., 1992, *Complete Mining of Uranium Deposits in Bulgaria by In-Situ Leaching Mining Systems Used After Conventional Systems*. IAEA TECDOC-720, pp 105-114.

Tomas, J., 1996, *Planning Environmental Restoration in the North Bohemian Uranium District, Czech Republic : Progress Report 1996*. IAEA TECDOC-982, pp 49-60.

Vapirev, E. I., Dimitrov, M., Minev, L., Boshkova, T., Pressyanov, D. & Guelev, M. G., 1993, *Radioactively Contaminated Sites in Bulgaria*. IAEA TECDOC-865, Vol. 1, pp 43-65.

Venatovskij, I. V., 1992, *Study of the Geotechnical Conditions of Uranium Deposits in Uzbekistan During Exploration Work*. IAEA TECDOC-720, pp 95-103.

Xu, L., Wang, Y.X., Wang, S., Peng, D. & Huang, A., 1998, *Estimating H_2SO_4 Balance and Optimal pH and SO_4^{2-} Values For H_2SO_4 In-Situ Leaching of Uranium*. In "Uranium Mining and Hydrogeology II", Freiberg, Germany, Sep. 1998, Vol. 1, pp 613-622.

Tailings and Mine Waste'00 © 2000 Balkema, Rotterdam, ISBN 90 5809 126 0

The Shikano North pervious tailings dam

Howard D. Plewes
Klohn-Crippen Consultants Limited, Richmond, B.C., Canada

Timothy D. Pitman
Klohn-Crippen Consultants Limited, Calgary, Alb., Canada

Kent Christensen
Quintette Operating Corporation, Tumbler Ridge, B.C., Canada

ABSTRACT: This paper describes the design, construction and performance of the Shikano North pervious rockfill tailings dam at the Quintette coal mine in north-eastern British Columbia. The dam provides containment for fine coal tailings produced by Quintette's coal washing plant. The key design features of the dam are an internal filter zone to control the rate of seepage from the tailings pond and prevent tailings from piping through the rockfill, and a seepage interception system at the downstream toe of the dam to collect the seepage water and convey it to a sedimentation pond for secondary clarification. The seepage performance of the dam is monitored by piezometers installed in the dam, and by flow measurements and water quality sampling of the dam discharge. The results to date show that the filtration provided by the internal filter zone has met the design expectations.

1 INTRODUCTION

This paper describes the design, construction and performance of the Shikano North pervious rockfill tailings dam at the Quintette coal mine in north-eastern British Columbia. The dam was built to create the Shikano North In-Pit Tailings Impoundment. This facility provides storage for fine coal tailings produced by Quintette's coal washing plant and has an operating life of 5 to 7 years, depending on coal production rates. Tailings deposition into the facility commenced in February of 1997.

2 SHIKANO NORTH TAILINGS FACILITY

The layout of the mined-out Shikano North Pit, shown on Figures 1 to 3, comprises a 1200 m long trench cut across the outcropping coal seams in the Shikano deposit. The depth of the trench is 25 m at the north-west end and increases to 200 m at the south-east end. The Shikano North Tailings Impoundment is formed by the tailings dam constructed across the outlet of the trench at the north-west end.

Tailings are discharged over the face of the pitwall at the south-east end of the pit. The deposited tailings solids form a flat beach sloping at 0.5% towards the tailings dam. The tailings slurry water which runs off the beach surface pools in a shallow pond against the upstream face of the dam. Because of the continuous flow of water across the narrow tailings beach, problems of wind erosion and dusting on exposed tailings beach surfaces are minimized.

The tailings dam is designed as a pervious structure to continuously decant the tailings pond by seepage through the dam. This eliminated the need for a reclaim water pumping system to transfer the tailings water from the impoundment. The tailings dam itself is a compacted rockfill embankment with a pervious upstream filter zone to control the rate of seepage and prevent the tailings from migrating through the rockfill. The filter zone consists of processed coarse

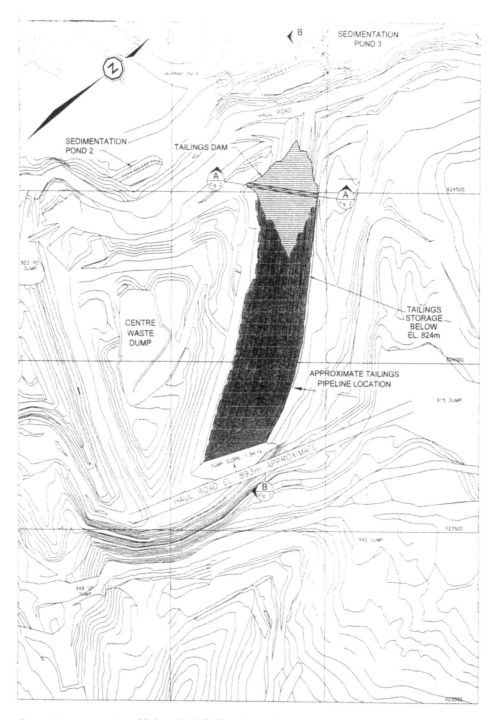

Figure 1. Arrangement plan of Shikano North Tailings Impoundment.

and fine filter material covered by a geotextile filter fabric on the upstream side. A system of seepage collection pipes is installed at the downstream toe of the dam to collect the seepage water and convey it to the pre-existing Sedimentation Pond No. 3 which is located below the toe of the Shikano North Pit. The sedimentation pond provides a contingency for further re-

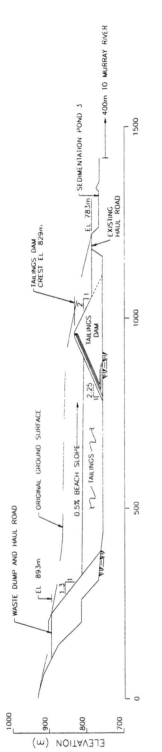

Figure 2. Section through tailings impoundment.

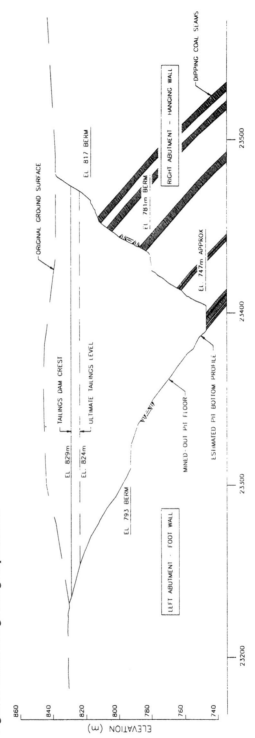

Figure 3. Centreline section through tailings dam

duction of any suspended solids in the seepage water prior to discharge to the environment. The dam is designed to decant an average daily inflow of 0.09 m³/s of tailings process water, with a maximum discharge of 0.33 m³/s during a 200-year runoff event from the impoundment catchment of 61 Ha.

The dam was constructed to El. 810 m in 1996 to provide an initial 2.5 years of tailings storage capacity. The dam was raised to its ultimate crest elevation of 829 m during the winter of 1998-1999. These two stages of construction are referred to as Stage 1 and Stage 2 in the remainder of this paper.

3 TAILINGS DAM DESIGN AND CONSTRUCTION

3.1 General Site Conditions

The Shikano North Pit is founded in the Gates Formation which consists of an interbedded, coal bearing sequence of carbonaceous shales, siltstones, sandstones and conglomerates. The stratigraphy is locally folded into a northwest-southeast trending syncline-anticline fold pair that plunges 15° to the southeast. As shown on Figure 3, the Shikano North Pit cuts across the outcropping coal seams in the "Shikano" Anticline which dips at about 40° into the pit. As a consequence of the mining, the left dam abutment consists of a relatively smooth, planar surface formed by the dipping bedrock bedding planes. The right dam abutment comprises a transverse cut across the exposed bedding planes which dip downward at 40° into the pit wall. The surface of the pit slope is quite rough because of the varying differences in rock type and competency.

Geotechnical site investigations of the pitwalls were carried out for the design of the tailings dam. The work included a review of existing mine geology core holes, drilling of new core holes, in-situ hydraulic conductivity tests, point load tests on selected rock core samples,, and installation of piezometers. Two inclinometers were also installed in the pitwall on the left abutment to monitor for potential movement along weak bedding planes parallel to the pitwall surface. Review of the results of this work is out of the scope of this paper. However, Table 1 provides a summary of the point load tests on the various rock types. This data is also representative of the mine waste rock used to construct the tailings dam as discussed below.

Table 1 Summary of Point Load Tests

Lithology	No. of Tests	Test Type	Estimated UCS			Rock Index(1)	
			Minimum (MPa)	Maximum (MPa)	Average (MPa)	Grade	Description
Siltstone	8	-	61	130	82	R4	Strong
Conglomerate	15	-	37	121	83	R4	Strong
Sandstone	17	with bedding	0	81	32	R3	Medium Strong
	14	across bedding	30	>96	70	R4	Strong

(1) ISRM (1981)

3.2 Dam Design Overview

Figures 4 and 5 show the plan and typical section of the tailings dam. The key design feature of the dam is the pervious filter zone beneath the upstream slope of the dam to control the rate of seepage from the pond and to prevent the tailings from piping through the rockfill. This filter zone consists of processed coarse and fine filter materials covered by a geotextile filter fabric on the upstream side. The geotextile filter fabric is extended to elevation 827 m, which corresponds to the maximum operating pond water level. The coarse filter is 6 m wide (horizontal) from the base of the dam to elevation 821.5 m and then narrows to 2 m wide above this elevation. Similarly, the fine filter is 4 m wide from the base of the dam to elevation 823.5 m and then narrows to 2 m wide.

Geotechnical site investigations showed that the permeability of the bedrock is higher (up to 10^{-5} m/s) in the upper 5 m to 10 m below the surface of the pit walls. The higher permeability is attributed to the opening of fractures and joints from stress relief, and the effects of distur-

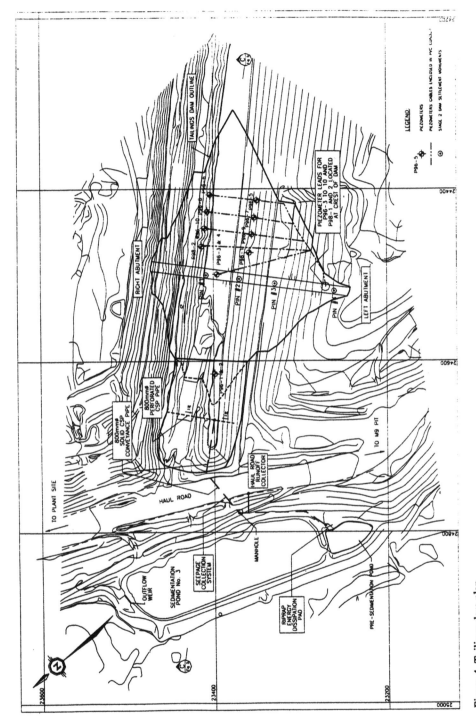

Figure 4. Tailings dam plan.

541

bance from blasting. Because the permeability of the dam materials is more than 10 times higher than the bedrock, the potential for high seepage gradients through the bedrock causing excessive seepage or piping was considered to be low. Nevertheless, the fine and coarse filter zones were extended in the downstream direction along the contact of the dam with the pit wall as indicated on Figure 5. The lengths of these filter zone extensions were 6 m to 20 m on the left (smooth wall) abutment and 6 m to 60 m on the right (rough wall) abutment, with the largest lengths occurring at lower dam elevations. The extended filter zones serve as a filter blanket to mitigate against possible piping of the bedrock materials or piping of tailings through fractures and crevasses in the bedrock below the filter zones.

The tailings dam was constructed by the downstream method, using mine waste rock to support the sloping filter zones in the upstream half of the dam. The upstream side of the filter zones is covered by an outer shell of waste rock to protect the filter zones from erosion and frost penetration, and to provide a conduit for the tailings water to flow into the dam.

The downstream dam slope and the upstream slope of the filter zone is 2H:1V (2 horizontal to 1 vertical), whereas the upstream slope of the outer waste rock is 2.25H:1V. The dam was built to elevation 810 m in the first construction stage in 1996 and to 829 m during the winter of 1998-1999. The completed dam has a maximum height of 80 m and a crest length of 225 m. The total volume of structural fill in the dam is 1.15 million m^3, with mine waste comprising 91% of the total.

3.3 Material Zonations

The outer dam shells and filter materials are comprised of Cretaceous sedimentary sandstones, siltstones and carbonaceous shales which are waste products from active coal mining. The specifications and placement methods for the various zones in the dam were as follows:

1. Zones A1 to A3 consist of clean, competent waste rock produced from nearby open pits. The maximum size of the rock was typically 1.0 m or less. In Zone A3, the waste rock was placed in maximum loose lift thicknesses of 1.0 m, whereas in Zone A1 and A2, the rock was placed in 2.0 m thick lifts. All rockfill was compacted by the passage of equipment traffic. The upstream 20 m of Zone A3 was also compacted by three passes of a 10 tonne vibratory roller.

2. Zone B consisted of a coarse filter zone consisting of 200 mm minus crushed mine waste rock. The filter was placed in maximum loose lift thicknesses of 0.6 m and compacted by four complete passes of a vibratory roller.

3. Zone C consisted of a 38 mm screened fine filter material produced from crushed waste rock. The filter material was placed in maximum loose lift thicknesses of 0.3 m and compacted by four complete passes of a vibratory roller.

4. Zone D consisted of clean, competent waste rock with a maximum particle size of 1.0 m. The waste rock was placed in maximum loose lift thicknesses of 2.0 m and compacted by the passage of equipment traffic.

5. Heavy duty Nilex 4553 non-woven geotextile filter cloth (apparent opening size = 0.150 mm) was placed on the upstream side of the fine filter zone. Typical overlaps at the seams of the filter cloth were in excess of 0.6 m, with no overlap less than 0.3 m. Over the filter cloth, 0.3 m of fine filter was placed to protect the integrity of the cloth during the placement of Zone D rockfill.

The average gradations for Zones B and C are shown on Figure 6. For Stage 1 construction, both Zones B and C were continuously watered during placement and compaction using a water hose (600 l/min). In Stage 2, the watering was omitted in the winter weather conditions. All overhanging rocks were scaled from the abutment contacts below Zones B and C so that no continuous overhangs greater than 1 m in length remained. Because there were no significant open fractures noted in the abutment contacts, treatment with slush grout or high slump concrete was not required. When placing the rockfill in Zone A3, well-graded rockfill with a maximum particle size of 300 mm was placed against the bedrock abutments.

The sandstone and siltstone waste rock materials are moderately resistant to weathering, but site observations indicate the potential for breakdown of the carbonaceous shales when exposed

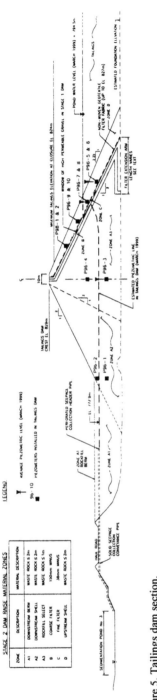

Figure 5. Tailings dam section.

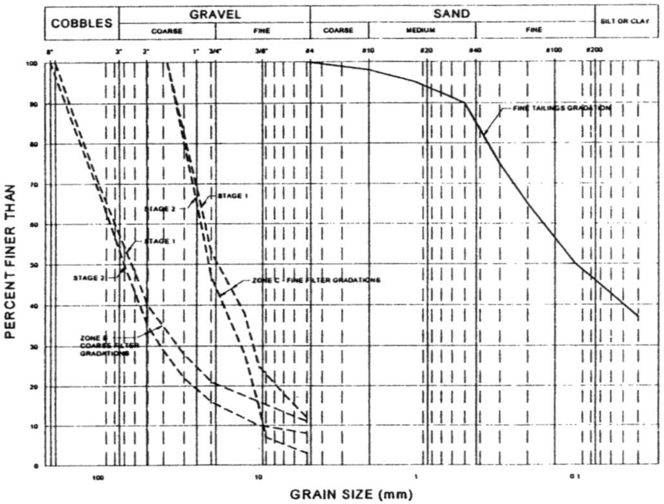

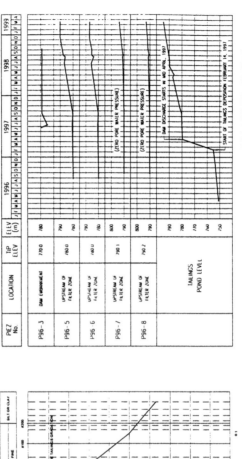

Figure 7. Piezometer readings.

Figure 6. Average coarse and fine filter crushing gradation summary.

Table 2 Summary of Quality Control Test Results for Filter Materials

Test	Design Criteria	Average Results For Stage 1 Dam Construction	Average Results For Stage 2 Dam Construction
Specific Gravity ASTM C127	2.6 minimum	2.56	2.68
Los Angeles Abrasion ASTM C535	40% maximum loss after 500 revolutions	23	24
Slake Durability ASTM D4644	Greater than 95% of original weight retained	95	
Deleterious Substances	Less than 5% by weight of coal or carbonaceous shales	-	3.4
Permeability of Fine Filter Material	0.1 cm/s minimum for 19 mm minus fraction	0.2 - 0.4	0.1 - 0.4

to the atmosphere. This potential for breakdown of the shales was a concern for the performance of the filter zones and efforts to minimize the amount of shales was made in selecting the source material for filter crushing. Material quality was also monitored by a quality control program consisting of tests for specific gravity, Los Angeles Abrasion, slake durability, deleterious substances and permeability. Table 2 summarizes the quality control test results and the design criteria.

The filter material in Stage 1 marginally failed the specific gravity requirement of 2.60. However, because the filter material met criteria for Los Angeles Abrasion, slake durability and permeability, the lower average specific gravity was judged acceptable.

In the early part of the Stage 1 construction, there were concerns that the permeability of the fine filter produced from crushed waste rock may be less than the design value of 10^{-1} cm/s. To compensate for this possibility, three 1 m high "windows" of high permeability crushed gravel were added in the fine filter zone to enhance the seepage flow-through capacity of the dam. These "windows" were added at elevations 781 m, 792 m and 800 m as shown on Figure 5. Subsequent permeability testing showed that these "windows" may not have been required.

3.4 Seepage Collection System

Because of the steeply sloping pit walls, the seepage water passing through the upstream filter zones tends to drain towards the middle of the dam. The water then flows through the downstream rockfill shell where it is intercepted by a seepage collection pipe and conveyed to Sedimentation Pond No. 3 below the dam.

As shown on Figures 4 and 5, the seepage collection system consists of a 800 mm diameter perforated corrugated steel (CSP) pipe passing across the downstream toe of the dam. The perforated pipe slopes at 1% slope to the middle of the dam, where the invert elevation of the pipe is 777.9 m. Above and below the pipe, "clean" well-graded waste rock was placed to enhance drainage into the perforated pipe. The perforated pipe itself was surrounded with 0.3 m of highly pervious 20 mm diameter drain gravel.

From the perforated pipe, a 800 mm diameter solid CSP pipe laid at a slope of about 2% conveys the water under the haul road and into a 1600 mm diameter CSP manhole drop structure. From the drop structure, water is then conveyed in another 800 mm diameter CSP pipe down the valley slope below the haul road to Sedimentation Pond No. 3. The pipe along the valley slope is covered by a minimum of 1.0 m of coarse coal refuse for frost protection. At the outlet, the flow is discharged into a channel lined with large rocks that act as energy dissipators.

3.5 Instrumentation

The seepage performance of the dam is monitored by 12 pneumatic and vibrating wire piezometers installed in the dam structure (Figures 4 and 5). Four piezometers (P96-1 to P96-4) are installed in the downstream rockfill shell along the mid-point of the dam to confirm the expected low piezometric levels within the dam. The remaining 8 piezometers (P96-5 to P96-10 and P98-1 to P98-2) are installed in the upstream waste rock shell to measure the pore pressure levels on the upstream side of the filter zone. These pore pressure provide information regarding the elevations at which seepage is passing through the filter zone and where potential clogging of the filter zone may have occurred.

To monitor post-construction settlements of the tailings dam, three or four settlement monuments were installed along the centreline of the crest of the dam after the end of the Stage 1 and Stage 2 construction. The settlement monuments consisted of steel pins with plates attached to their base, buried in 1 m below the surface of the rockfill. A steel pin was installed in the right dam abutment to use as a reference point for the settlement surveys.

As discussed earlier, other piezometers and inclinometers were installed in the bedrock in both dam abutments to monitor for water level changes and potential movements in the bedrock. These instruments and their results are not discussed in this paper. However, it can be reported that the water levels changes to date meet design expectations and no significant movement in the left dam abutment has been measured.

4 TAILINGS DAM PERFORMANCE TO DATE

4.1 *Seepage Performance*

The piezometric levels for selected dam piezometers are shown on Figure 7. To date, Piezometer P96-3 has remained relatively constant at 779 m, just above the invert elevation of 777.9 m for the seepage collection pipe system. Upstream of the filter zone, Piezometers P96-5 and P96-6 installed at elevation 780 m, began to respond as the pond level rose above 780 m in the fall of 1997. The current water levels are 1 m to 3 m lower than the tailings pond level, indicating a downward seepage gradient in Zone D. Piezometers P96-7 and P96-8 installed at elevation 790 m also started to respond as the pond level rose above 790 m in August 1998. The piezometers tips of P96-9 and P96-10 and P98-1 and P98-2 are above the tailings pond water level and therefore have shown no response to date.

The average depth of water against the dam in September 1998 was measured to be 0.8 m. This shallow water depth suggests that clogging of the internal filter zones caused by the migration of the coal tailings through the coarse rockfill of the upstream dam shell (Zone D) is probably minimal. Calculations during the dam design indicated that greater pond depths of 3 m to 5 m would form if complete clogging of the filter zone below the elevation of the tailings beach were to occur.

Figure 5 shows the piezometric surface in the tailings dam that has been interpreted from the water levels recorded in March, 1999. The low water level in the downstream rockfill shell is evidence that the downstream shell is well drained and water levels are being controlled by the seepage collection system at the toe of the dam. Upstream of the filter zone, the pressures observed in P96-5 to P96-8 indicate the filter zone does restrict the flow of seepage through the dam. Further readings as the pond level rises will provide data to assess whether clogging of the filter zone with the fine coal tailings is a factor.

4.2 *Water Quality*

Concentrations of total suspended solids (TSS) are measured regularly from samples taken from the supernatant tailings pond, at the discharge from the seepage collection pipe, and at the discharge from Sedimentation Pond No. 3.

The TSS in the supernatant tailings pond is typically in the range of 30 to 200 mg/l. In comparison, the TSS at the seepage collection pipe discharge is generally 4 mg/l or less and the TSS in the water discharged from Sedimentation Pond No. 3 to the environment has typically ranged from 2 to 10 mg/l. The suspended solids levels in the final discharge water are well below the maximum allowable of 50 mg/l set by regulatory agencies.

The source of the suspended solids in the seepage outflow from the dam is attributed to fines washed from the rockfill. No evidence of coal tailings in the outflow has been detected to date. The absence of tailings in the outflow suggests that the filter zone is successfully preventing tailings from piping through the dam.

4.3 *Dam Settlements*

After completion of the first stage of dam construction in October 1996, the maximum settlement at the centre of the dam was 254 mm after 1 year and 350 mm after 2 years. Settlements decreased toward the abutments. The maximum settlement of 350 mm is 0.6% of the maximum Stage 1 dam height of 58.5 m. This was within the maximum post-construction dam settlement of 1% predicted in the dam design. Settlements following the recent Stage 2 construction will be recorded over the operating life of the impoundment.

5 IMPOUNDMENT CLOSURE SCENARIO

At closure, surface runoff water will be the only inflow into the Shikano North Impoundment basin. Under normal conditions, it is expected that all surface runoff water will quickly drain from the impoundment by seepage through the dam and long-term water levels in the dam will drop below the operating levels, particularly upstream of the filter zone. Water may temporarily accumulate in the pond during the spring freshet or following major rainstorms. The maximum flood level in the tailings impoundment will be regulated by a spillway located in the left abutment of the dam. The spillway will discharge into Sedimentation Pond No. 3.

The tailings exposed in the impoundment will be vegetated to stabilize the surface against water and wind erosion. The slopes of the tailings dam will also be vegetated. If required, the crest of the dam will be raised with additional fill to compensate for loss of freeboard caused by long-term settlements.

6 CONCLUSIONS

The following are the main conclusions drawn from the construction and performance data for the Shikano North Tailings Dam:
1. The dam body and filter materials are comprised of Cretaceous sedimentary sandstones, siltstones and carbonaceous shales which are waste products from active coal mining. The sandstones and siltstones materials are moderately resistant to weathering, but site observations indicated the potential for breakdown of the carbonaceous shales when exposed to the atmosphere. This potential for breakdown of the shales was a concern for the performance of the filter zones and efforts to minimize the amount of shales was made in selecting the source material for filter crushing. In addition, material quality was carefully monitored by a quality control program consisting of tests for gradation, specific gravity, Los Angeles Abrasion, slake durability, deleterious substances and permeability.
2. The piezometric levels in the tailings dam are increasing as the tailings pond level rises. The current water levels are consistent with expected seepage patterns.
3. The water quality of the seepage outflow from the dam shows that filtration provided by the internal filter zone is meeting design expectations.
4. The piezometer data and pond water depth upstream of the dam suggests that full clogging of the filter zone with fine coal tailings has not occurred.
5. The settlement monuments indicate post-construction crest settlements of up to 0.6%, which is less than the 1% maximum expected settlement.

The tailings dam will continue to be monitored over the operating life of the impoundment and additional performance data for this dam will be presented at a future date.

ACKNOWLEDGEMENTS

The authors wish to thank Quintette Operating Corporation for permission to publish the information contained within this paper.

REFERENCES

International Society for Rock Mechanics (1981). "Rock Characterization, Testing and Monitoring - ISRM Suggested Methods", Oxford: Pergamon.

Klohn-Crippen Consultants Ltd. 1996a. *Shikano North Tailings Impoundment - Pervious Tailings Dam Option*. Report submitted to Quintette Operating Corporation. February 9, 1996.

Klohn-Crippen Consultants Ltd. 1996b. *Shikano North Tailings Impoundment - 1995 Geotechnical Investigations*. Report submitted to Quintette Operating Corporation. February 9, 1996.

Klohn-Crippen Consultants Ltd. 1997. *Shikano North Tailings Impoundment – 1996 As-Built Report*. Report submitted to Quintette Operating Corporation. March 14, 1997.

Klohn-Crippen Consultants Ltd. 1998. *Shikano North Tailings Impoundment – 1997 Annual Review Report*. Report submitted to Quintette Operating Corporation. February 20, 1998.

Klohn-Crippen Consultants Ltd. 1999. *Shikano North Tailings Impoundment – 1998 Annual Review Report*. Report submitted to Quintette Operating Corporation. March 5, 1999

Tailings and Mine Waste'00 © 2000 Balkema, Rotterdam, ISBN 90 5809 126 0

Operational results of a 1,200-gpm passive bioreactor for metal mine drainage, West Fork, Missouri

J.Gusek
Knight Piésold and Company, Denver, Colo., USA

T.Wildeman
Colorado School of Mines, and Knight and Piésold and Company, Colorado, Denver, Colo., USA

C.Mann & D.Murphy
The Doe Run Company, Viburnum, Mo., USA

ABSTRACT: An active underground lead mine produces water having a pH of 8.0 with 0.4 to 0.6 mg/L of Pb and 0.36 mg/L of Zn. This water is pumped at the rate of 1,200 gpm (0.076 m³/s) into a five-cell, bioreactor system covering about 5 acres (2 hectares). The gravity flow system is composed of a settling basin followed by two anaerobic bioreactors arranged in parallel which discharge into a rock filter polishing cell that is followed by a final aeration polishing pond. The primary lead removal mechanism is sulfate reduction/sulfide precipitation. The discharge has met stringent in-stream water quality requirements since its commissioning in 1996. However, there have been startup and operational difficulties. The system was designed to last about 12 years, but estimates suggest a much longer life based on anticipated carbon consumption in the anaerobic cells.

1 INTRODUCTION

The West Fork Unit is an underground lead-zinc mine purchased by the Doe Run Company from Asarco in 1998 that discharges water from mine drainage to the West Fork of the Black River (West Fork) under an existing NPDES permit. The West Fork Unit is located in Reynolds County in central Missouri, in the New Missouri Lead Belt, about three hours from St. Louis.

Flow rates in West Fork vary from about 20 cubic feet per second (cfs) to more than 40 cfs (0.56 to 1.13 m³/s). Water quality is relatively good, despite being located in an area with naturally high background levels of lead due to the bedrock geology. The mine discharges about 1,200 gpm (2.7 cfs or 0.076 m³/s) on the average or about 10 percent of the total flow in West Fork.

The adoption of water quality-based discharge limits, in its NPDES permit issued in October 1991, prompted Asarco to evaluate treatment methods for metal removal. Evaluations of alternative treatment processes determined that biotreatment methods were feasible and cost less than half as much as active sulfide precipitation. The goal of the water treatment project was to ensure that the stringent water quality-based limits in the permit would be consistently met.

Figure 1 - Site Location

Since 1987, a group from Knight Piésold and Co. and the Colorado School of Mines has been active in developing passive treatment methods for metal-mine drainages. The primary treatment method is through the generation of hydroxides and sulfides through microbial metabolism. The biogeochemical principles are summarized in Wildeman, et al. (1995), and Wildeman and Updegraff (1998). The design principles are explained in Wildeman, Brodie, and Gusek (1993). In the case of the West Fork Unit, biotreatment consists of two stages:

1. An anaerobic unit that generates sulfide through sulfate reduction and is responsible for the lead removal.

2. An aerobic unit that is a rock filter/wetland. This unit is responsible for removing dissolved organic matter and excess sulfide from the effluent from the anaerobic cell. The aerobic unit also reoxygenates and polishes the water before it enters the river.

Extensive laboratory, bench-scale, and pilot scale tests were made on the anaerobic unit. These are described in Wildeman, et al. (1997), and Gusek, et al. (1998). The design and permitting of the system are also discussed in Gusek, et al. (1998), and Wildeman, et al. (1999). This paper concentrates on the operation of the full-scale system since its start in 1996.

2 SYSTEM DESCRIPTION

The system was designed based on the performance of the pilot-scale reactor and the interim bench scale studies. The large-scale system was estimated to cost approximately $500,000 and required about three months of construction time. Operational costs include water quality monitoring as mandated by law. No additional costs for reagents are incurred; since the system uses gravity flow, moving parts are few and include valves, minor flow controls, and monitoring devices. Based on carbon depletion rates observed in the pilot system, the anaerobic cell substrate life was projected to be greater than 30 years; the full-scale biotreatment system should be virtually maintenance-free. Should mine water quality deteriorate, the full-scale design included a 50-percent safety factor.

The biotreatment system is composed of five major parts: a settling pond, two anaerobic cells, a rock filter, and an aeration pond (Knight Piésold, 1997). The system is fully lined. The design was also integrated into the mine's pre-existing fluid management system.

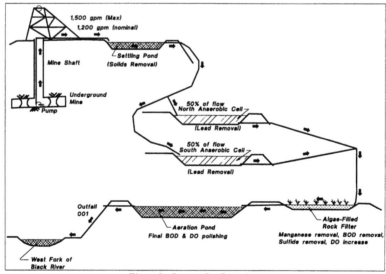

Figure 2 - System Configuration

A rectangular-shaped, 40-mil HDPE-lined settling pond has a top surface area of 32,626 ft² (3,030 m²) and a bottom surface area of 20,762 ft² (1,930 m²). The sides have slopes of 2 horizontal to 1 vertical (2H:1V). The settling pond is nominally 9.8 ft (3 m) deep. It discharges through valves and parshall flumes into the two anaerobic cells.

Two anaerobic cells are used, each with a total bottom area of about 14,935 ft² (1,390 m²) and a top area of about 20,600 ft² (1,930 m²). Each cell is lined with 40-mil HDPE and was fitted with four sets of fluid distribution pipes and three sets of fluid collection pipes, which were subsequently modified (see Start Up discussion). The distribution/collection pipes were connected to commonly shared layers of perforated HDPE pipe and geonet materials sandwiched between layers of geofabric. This feature of the design was intended to allow control of sulfide production in hot weather by decreasing the retention time in the cell through intentional short circuiting.

The spaces between the fluid distribution layers were filled with a mixture of composted cow manure, sawdust, inert limestone, and alfalfa, referred to hereafter as "substrate." The total thickness of substrate, piping, geonet, and geofabric was about 6 feet (2 m). The surface of the anaerobic cells was covered with a layer of crushed limestone. Water treated in the anaerobic cells flows by gravity to a compartmentalized concrete mixing vault and thereafter to a rock filter cell. The gravity-driven flows can be directed upward or downward.

The rock filter is an internally bermed, clay-lined shallow cell with a bottom area of about 63,000 ft² (5,900 m²) and a nominal depth of one foot (30 cm). It is constructed on compacted fill that was systematically placed on the west side of a pre-existing mine water settling pond. Limestone cobbles line the bottom of the cell, and the cell is compartmentalized by limestone cobble berms. The discharge from the rock filter flows through a drop pipe spillway and buried pipe into a 40-mil HDPE-lined aeration pond. The aeration pond surface covers approximately 85,920 ft² (8,000 m²). The aeration pond discharges through twin 12-inch (30-cm) HDPE pipes into a short channel that leads to monitoring outfall 001 and thence into West Fork.

After the water pumped from the underground mine enters the settling pond, all flows are by gravity.

3 START-UP EXPERIENCE

Bench-scale test results suggested that the anaerobic cells be incubated with settled mine water for about 36 hours or less before fresh mine water was introduced at full flow to minimize initial levels

Figure 3 - Aerial View

551

of BOD, fecal coliform, color, and manganese. For about two weeks, pumps recycled the water within the two anaerobic cells. Based on data collected in field, and subsequent laboratory confirmation, the water from the anaerobic cells was routed to the tailings pond for temporary storage and later treatment and release. At that point, the rock filter and aeration ponds were brought on-line. In the meantime, the mine discharged according to plan through an overflow pipe from the settling pond as it had during construction of the other components.

After about six weeks of full-scale operation, the apparent permeability of the substrate was found to be lower than expected and the system was operating nearly at capacity. The system had been designed so that either of the two anaerobic cells could accept the full flow amount on a temporary basis in case maintenance work required a complete cell shutdown.

Research found that H₂S gas, generated by the sulfate-reducing bacteria, was being retained in the substrate in the anaerobic cells; this created a gas-lock situation that prevented full design flow. A temporary solution was obtained by periodic "burping" of the cells using the control valves. However, the "burping" had to be performed at 24-hour intervals, and it was determined that this solution was too labor-intensive.

The sulfide gas lock problem was investigated in December 1996 by installing vent wells in the substrate and measuring the gas pressures. Observations indicated that the gas was a factor in apparent short circuiting of the water passing through the cell. The layered geotextiles (geonet and geofabric), originally intended to promote horizontal flow, appeared to be trapping the sulfide gas beneath them and vertical flow was being restricted. The permeability of the substrate itself was for the most part unaffected. However, construction practices in the south anaerobic cell could have contributed to the situation. Here, a low ground bearing bulldozer was used to place substrate in nominal 6-inch (15-cm) lifts. This could have created a layering effect that may have trapped gas as well. Substrate layers in the north anaerobic cell were placed in a single lift, and no layering effect was observed during subsequent excavation. It is noteworthy that the mid-cell geotextiles had not been a feature of the pilot test cell design.

The first phase of a permanent solution was implemented with a trenching machine that ripped through the geonet/geofabric layers in the south anaerobic cell. This disrupted the gas-trapping situation. Subsequently, the substrate from the entire south anaerobic cell was excavated and the cell refilled without the geotextiles in June 1997. Identical action was taken on the north anaerobic cell in September 1997. These actions have solved the gas lock problem.

4 MAINTENANCE EXPERIENCE

Although this is technically a passive treatment system, when one considers trying to direct the flow of 1,200 gpm (0.076 m³/s) through approximately 3,930 yd³ (3,000 m³) of material there is certain to be some hydraulic problems. In addition, the design of the anaerobic cells made provisions for the water to bypass portions of the cells during the summer to eliminate excess buildup of sulfide in the cell effluent. In the summer of 1997 and 1998, operation of the system included by-passing some portions of the cell to maintain lower sulfide concentrations. However, when this was tried, short-circuiting within the cells and plugging of the substrate made maintenance during the summer more extensive than during the winter.

Perhaps the most troublesome maintenance issue was that a combination of sediment in the mine water along with algae buildup on the cell surfaces would block the infiltration of water into the cells. This would necessitate periodically draining the cells and rototilling the top of the substrate so as to break up the accumulation cake. Often at the same time as a cell was tilled, water would be back-flushed through the discharge pipes to dislodge precipitate accumulation. When such maintenance was done, the rock filter would still receive discharge. It has proved to be an effective buffer between the cells and the discharge pond. This maintenance cycle of tilling and back-flushing had to be done almost once a month during the summer of 1998. During the winter, buildup was not as extensive and maintenance of the cell surfaces was less frequent. Currently, schemes are being

Table 1 - West Fork Mine Water Quality Data

Parameter	Typical Average Influent Water Quality in mg/L	Range of Water Quality Discharge in mg/L (June - November 1997)
Pb	0.4	0.027 - 0.050
Zn	0.36	0.055 - 0.088
Cd	0.003	<0.002
Cu	0.037	<0.008
Oil and Grease	--	<5.0
H$_2$S	--	0.011 - 0.025
Total Phosphorus	--	<0.05 - 0.058
Ammonia as N	0.52	<0.050 - 0.37
Nitrate and Nitrite	2	<0.050 - 1.7
True Color	--	10 - 15
BOD	1.7	<1 - 3
Fecal Coliform	—	<1 - 2
pH	7.94	6.63 - 7.77
TSS	—	<1 - 4.2

investigated to try a drastic reconditioning of the cells to permanently increase the hydraulic conductivity of the anaerobic cells.

Other than repairing a bubble that appeared under the liner of the aeration pond, there has been no maintenance needed on the rock filter and the aeration pond.

5 OPERATIONAL RESULTS

5.1 *The Anaerobic Cells*

The average influent water quality can be compared with discharge water quality (Table 1) during the June through November 1997 period. Discharge levels of Pb and other metals were reduced substantially from average influent levels. For Pb, the level was reduced from a typical average of 0.40 mg/L to between 0.027 and 0.050 mg/L. Zn, Cd, and Cu effluent concentrations were also reduced.

More extensive analysis of the operational data from June 1997 through June 1999 has shown some interesting results. The plumbing system in the anaerobic cells was designed to run the cells upflow or downflow, to use a portion of the cell when sulfide production became too high, and to be back-flushed in case precipitation occurred in the discharge line. All three features have been used. The cells have been run in the upflow direction during the first winter so that the substrate compaction that occurred during the summer could be relieved. The three levels of discharge pipes are routinely monitored for sulfide production, and the valves are adjusted accordingly to eliminate excess sulfide. In the summer, these adjustments become more difficult as attempts are made to only use portions of the cells. In addition, the cells are routinely back-flushed to maintain good circulation of mine water through the cells.

By operating the anaerobic cells in this fashion, over four seasons from July 1997 to July 1998, the average concentration of 40 analyses of total Pb in the water entering the cells is 0.45 and the average concentration of Pb in the water exiting the cells is 0.085. Results for zinc are not as extensive.

From March 1998 to November 1998, the average concentration of 10 analyses of total Zn in the water entering the cells is 0.44 and the average concentration of Zn in the water exiting the cells is 0.102.

Within the anaerobic cells, production of enough sulfide has never been a problem. During the summers of 1997 and 1998, sulfide concentration in discharges from some portions of the cells routinely exceeded 12.0 mg/L, the upper quantitation limit of the analytical procedure. This correlates with the pilot cell results where, during the two summers in which it operated, sulfide concentrations reached 20 mg/L. According to Wildeman, et al. (1997), at this level of sulfide concentration, the production of sulfide in the anaerobic cells is about 2 moles sulfide per cubic meter per day. As expected, during the winter, concentrations of sulfide in the cell effluent are lower. However, even during the months of December, January, and February, sulfide concentrations in the discharge from some portions of the cell were between 2.0 and 7.7 mg/L. These concentrations have been higher than the average of 0.3 mg/L of sulfide that was found during the winter the pilot cell operated (Wildeman, et al., 1997).

5.2 *The Rock Filter*

Of the five parts of the system, the operation of the rock filter has been the most interesting. It operates as a natural wetland where water of a depth of 1 to 2 feet (30 to 60 cm) meanders through the limestone cobbles. Flora and fauna have thrived in this ecosystem. It has served the important function of cleansing the excess sulfide in the water that is leaving the anaerobic cells. From July 1997 to September 1998, the average of 55 analyses of sulfide concentration in the water entering the rock filter is 3.3 mg/L. In 55 analyses of sulfide in the rock filter effluent, sulfide was detected in the water 20 times and none of these were above 0.25 mg/L.

Because the water entering the rock filter contains a significant concentration of sulfide, a unique ecosystem of algae and bacteria have developed in this area. In the summer of 1997, red algae/bacteria started to develop in this influent area and have persisted. In addition, a white scum has developed in this area. Indeed, the rock-filter influent area looks like a pool of the primordial soup. During the summer of 1997, when high levels of sulfide were entering the rock filter, the water would develop a milky white colloidal suspension that would persist throughout the wetland system. This milky suspension had diurnal characteristics. It would be more persistent in the morning and sometimes clear up during the day. In the summer of 1998, this milky suspension was not as evident even though the concentrations of sulfide entering the rock filter were sometimes higher. Vegetation in the rock filter was much more lush in the second summer. The speculation is that this milky suspension is colloidal sulfur. If it is, then this form of wetland ecosystem removes it.

Besides removing sulfide from the water, the rock filter also plays a significant role in further reducing the concentration of lead in the water. Over four seasons from July 1997 to July 1998, the average concentration of 40 analyses of total Pb in the water entering the rock filter is 0.085 and the average concentration of Pb in the water exiting the rock filter is 0.050. The mechanism for lead removal in the rock filter is not known.

6 CONCLUSIONS

In the introduction to this paper it was stated that the biotreament system should be virtually maintenance free. That has not been the case with the anaerobic cells. Keeping these cells from clogging has required periodic rototilling and back-flushing. Because attempts were made during the summer to use only a portion of the two cells, maintenance has been more extensive at this time than during the winter. Nevertheless, these cells have performed according to design and have been effective at removing lead from the mine water. Because of this necessary maintenance, the design of the plumbing system to include back-flushing, upflow and downflow, and use of only a portion of the cell has been particularly advantageous.

The need for the rock filter has been found to be essential. Its operation has shown some surprises. The presence of sulfide in the water has caused a unique ecosystem that effectively removes this constituent from the water. The removal of sulfide is more important in the summer. The rock filter also removes a significant amount of lead. The removal mechanism is unknown.

7 ACKNOWLEDGMENTS

The foresight and subsequent commitment of ASARCO and the Doe Run Co. to this type of treatment is most appreciated.

8 REFERENCES

Gusek, J.J., Wildeman, T.R., Miller, A., and Frickem, J., 1998. The challenges of designing, permitting, and building a 1,200 gpm passive bioreactor for metal mine drainage, West Fork, Missouri. In: *Proceedings of 15th Annual Meeting of American Society for Surface Mining and Reclamation*, pp. 203-212.

Wildeman, T.R., Gusek, J.J., Cevaal, J., Whiting, K., and Scheuring, J., 1995. Biotreatment of acid rock drainage at a gold-mining operation. In: *Bioremediation of Inorganics*, R.E. Hinchee, J.L. Means, and D.R. Burris, Eds., Battelle Press, Columbus Ohio, pp. 141-148.

Wildeman, T.R., Gusek, J.J., Miller, A., and Frickem, J., 1997. Metals, sulfur, and carbon balance in a pilot reactor treating lead in water. In: *In Situ and On-Site Bioremediation*, Volume 3. Battelle Press, Columbus, OH, pp. 401-406.

Wildeman, T.R., and Updegraff, D., 1998. Passive bioremediation of metals and inorganic contaminants. In: *Perspectives in Environmental Chemistry*, D.L. Macalady, Ed., Oxford University Press, New York, pp. 473-495.

Tailings and Mine Waste'00 © 2000 Balkema, Rotterdam, ISBN 90 5809 126 0

The Bradley Tailing Diversion and Reclamation Project

Lisa R. Yenne, William S. Eaton & Christopher N. Hatton
URS Greiner Woodward Clyde, Denver, Colo., USA

Dan E. Burnham
Mobil Business Resources Corporation, Fairfax, Va., USA

ABSTRACT: The Bradley Tailing Diversion and Reclamation Project was a highly visible, fast track regulatory driven project. The main project goal was to prevent tailing discharge into a partially constructed new Meadow Creek diversion channel. Design and construction were initiated and completed in 1998 under demanding site conditions in an isolated mountain valley located in Central Idaho. The Bradley Tailing Impoundment is located within an ecologically sensitive waterway. The design consisted of the construction of new diversion channel through and around the tailing impoundment while preventing the discharge of any tailing into the headwaters of the Salmon River. A portion of the channel also serves as a tailing filter. Other key project components included lining of a flowing channel, capping of fluid tailing, and state-of-the art revegetation. The Bradley Tailing Diversion and Reclamation Project illustrates the complex balance of engineering, biological and ecological sciences required to successfully achieve rigorous project goals.

1 SITE HISTORY

The Stibnite Mine (Site) is about 22.5 kilometers (14 miles) east of the town of Yellow Pine in Valley County, Idaho, deep in the Rocky Mountains of central Idaho. The project location is shown in Figure 1. The Site includes Boise National Forest land and other lands located within the exterior boundaries and adjacent to the Boise National Forest. The Site was first developed in the late 1800's by prospectors and developers mining antimony, tungsten, gold and other minerals.

A remnant of the historic Meadow Creek Mine is the 0.4 square kilometers (100-acre) Bradley Tailing Impoundment where tailing from the milling of antimony and tungsten were deposited in the 1940's and 1950's. The impoundment is located on the Site. The embankment for the impoundment was constructed of fine tailing using the upstream method of deposition by header and spigot. The fine tailing grain size required that the dam be raised with a 10 to 1 (horizontal to vertical) upstream slope to maintain dam stability.

The impoundment was inactive until the early 1980's when after cyanization and neuturalization of ore on nearby leach facilities, the spent ore was hauled and spread over approximately 0.24 square kilometers (60 acres) of the Bradley Tailing Impoundment. The spent ore ranges in thickness from about 23 to 30.5 meters (75 to 100 feet).

In 1981 a small diversion channel along the hillside south and east of the Bradley tailing was constructed to divert Meadow Creek away from the impoundment. Meadow Creek is a tributary creek to the East Fork of the South Fork of the Salmon River. This diversion channel was undersized for larger storm events and was reported to be unstable. Actions were taken in 1997 to stabilize the diversion and impoundment by others by relocating Meadow Creek from the existing diversion to a larger channel located at the south edge of the tailing impoundment. Construction of this new channel began during the summer of 1997 and substantial additional work and a re-

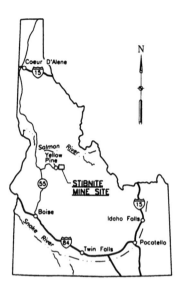

IDAHO STATE MAP
NOT TO SCALE

Figure 1. Site Location.

design was required in 1998 to complete the project since seepage from the tailing impoundment was eroding into the newly constructed channel area.

2 PROJECT BACKGROUND

URS Greiner Woodward Clyde (URSGWC) was contacted by Mobil Oil Corporation (Mobil) in late February 1998 to evaluate the feasibility of constructing agency (EPA and US Forest Service) proposed designs to stabilize the tailing and prevent tailing inflow to the new diversion channel. After careful review an alternative design was proposed by URSGWC. Given the project logistics, the alternative design was simpler and easier to construct, more cost effective and met agency project goals. The main project objective was to minimize the inflow of tailing from the tailing/spent ore pile to the new diversion channel. Negotiations proceeded between the EPA and Mobil eventually resulting in Mobil entering into an Administrative Order on Consent (AOC).

A preliminary design report, construction plans, technical specifications, and contractor selection were prepared in an accelerated two month period beginning in April and ending June 1, 1998. The preliminary design report included preparation of eight plans for Reclamation, Operations, Site Health and Safety, Construction Quality Assurance, Sampling Quality Assurance/Quality Control, Water Management, Alternative Design, and a Spill Prevention, Control and Countermeasure. During this two month period the preliminary design report and drawings were reviewed several times by multiple regulatory agencies with construction mobilization beginning June 1, 1998.

The project team, comprised of multi-disciplinary professionals, conformed to stringent multi-agency regulatory requirements. This project team also involved a unique client, agencies and engineer teamwork relationship to complete the project in an accelerated nine-month schedule. URSGWC working on behalf of Mobil served as the project engineer and contractor. Earthmoving activities were subcontracted to a local Idaho contractor. Other subcontractors

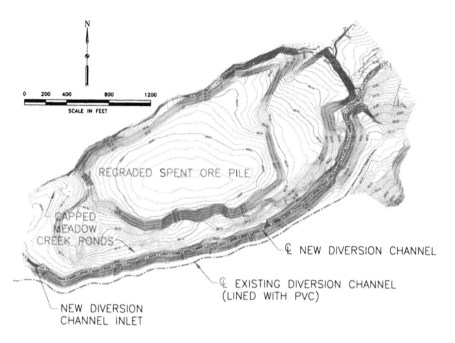

Figure 2. Site Map of Project Features.

were also utilized to provide the liner, cement and revegetation services. A system of promoting and maintaining teamwork between the agencies, URSGWC, and Mobil was developed which allowed design changes to be approved quickly and construction to proceed unimpeded.

The project key components discussed below, were lining of a flowing channel, unique channel design and construction, capping of fluid tailing and state-of-the art revegetation. Figure 2 shows the project site and key project feature locations.

3 LINING OF A FLOWING CHANNEL

The success of the project hinged on the ability to control local ground water and surface water infiltration into the new diversion channel construction area. During the 1997 channel excavation the contractor reported difficult construction conditions due to the quantity of water entering the channel excavation. Several alternatives were evaluated including the use of a traditional system of pumps and pipes to divert water around the site. However, due to the large peak flow early in the construction season and the potential for large storm runoff during construction the pumping costs were prohibitive.

Therefore a PVC flexible membrane liner was used to line the existing diversion channel, refer to Figure 2, located about 7.6 meters (25 feet) above the new channel and within 19 meters (50 feet) of the work area. Approximately 1,006 linear meters (3,300 linear feet) of a 30-mil PVC liner up to 12.2 meters (40 feet) wide was installed in the channel flowing 1.4 cubic meters per second (50 cubic feet per second). Installation of the liner without diverting the flow was an innovative approach, time saving, and to our knowledge, the first installation over such a distance and with such a large flow.

The liner panels were solvent welded prior to placement in the channel allowing the liner to be installed in one single 1,006 meter (3,300 feet) long piece. The entire liner was deployed and floated on the water surface, anchored with boulders and ropes above the channel side slopes. The liner also had a nylon rope hem on the edge to allow the liner to be tied to nearby trees in steep areas. Figure 3 shows a photograph of the liner outlet floating on the water surface. The upstream end was held above the water surface using the bucket of a backhoe. After the liner was deployed and properly anchored the upstream end was lowered and anchored with sandbags

Figure 3. Photograph showing liner "floating" on water surface.

and boulders into a small excavated trench at the channel inlet. The channel water flowing over the liner was then used to push the remainder of liner to the channel bottom.

Once installed, this liner cut off flow from the existing diversion channel and reduced groundwater inflow to the construction area by about 75 percent. After construction was complete, the liner was removed and a portion of the liner was reused. The liner saved the client over $300,000 when compared with pumping.

4 CHANNEL DESIGN AND CONSTRUCTION

The greatest challenge of the project was the design and construction of approximately a 1.6 kilometer-long (1 mile-long), rock lined diversion channel through and around the Bradley Tailing Impoundment, material prone to liquefaction and in difficult field conditions. The new diversion channel alignment can be seen in Figure 2.

4.1 *Channel Design*

After careful review of the agency proposed design, URSGWC proposed an alternative design to stabilize the tailing and prevent piping of the materials without requiring excessive tailing disturbance or deep excavation. The alternative design not only diverts surface water but also serves as a 610 meters (2,000 foot) long tailing filter. URSGWC proposed constructing a graded sand filter between the channel and the exposed tailing. The filter was designed to minimize tailing migration into the new diversion channel while allowing seepage flow. The typical sand filter channel section is shown on Figure 4. The alternative design also incorporated work previously completed to date.

4.1.1 *Filter Design*

The use of a filter system is a traditional engineering practice and has been used successfully for over 50 years both in water retention and tailing structures. However, the use of a filter system for this project was a unique application. The intent of the filter is to "capture" small particles that could erode from the base material (in this case, tailing) while allowing continued seepage

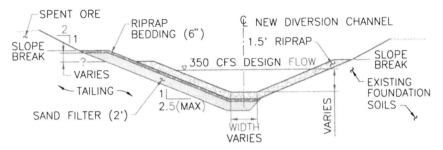

TYPICAL SAND FILTER TRANSITION SECTION

Figure 4. Typical Channel Section.

flow. The commonly accepted filter criteria methodology from Sherard (Sherard & Dunnigan 1989) was used for the initial filter evaluation and design. Sherard developed filter criteria for four soil groups separated based on the quantity of fines in the base soil.

Bulk samples of the tailing and the on-site sand filter borrow area were collected from the site in April 1998. The tailing samples were selected specifically to represent the coarse and fine tailing gradation limits.

Index property testing was performed on the tailing and on-site borrow area samples. The fine tailing material was classified as a low plasticity clay (USCS classification CL) and low plasticity silt (ML) with approximately 80 percent material passing the No. 200 sieve. Atterberg limits showed that the fine tailing had a plasticity index of 17 and a liquid limit of 42. The coarse tailing material classified as a sandy silt (SM) with approximately 27 percent material passing the No. 200 sieve. Fines in the coarse tailing sample were non-plastic.

A gradation analysis on the sample collected from the borrow area classified as a poorly graded sand (SP) and contained 41 percent gravel, 56 percent sand, and 3 percent finer than the No. 200 sieve.

The Sherard filter criteria was applied for the following project constructed material interfaces: 1) Tailing (base to the graded sand filter (filter); 2) Graded sand filter (base) to the riprap bedding (filter); 3) Riprap bedding (base) to the riprap (filter). The graded sand filter was manufactured on site by screening material from an on-site borrow area to gradation requirements similar to the ASTM C-33 sand. The riprap bedding and riprap material were also manufactured on site by screening material from the on-site borrow area.

Laboratory filter tests were performed on the tailing to the sand filter material interface for the following purposes: 1) to confirm the filter criteria methodology could be relied on to establish the sand filter gradation range specifications; and 2) to confirm this design would successfully prevent migration of the tailing particles. Two large-scale laboratory filter tests were performed at the U.S. Bureau of Reclamation laboratory in Denver, Colorado. One large-scale filter test was performed with fine tailing as the base soil and second with the coarse tailing as the base soil. For both tests, the ASTM C-33 sand manufactured from the borrow area material was used as the graded sand filter.

Laboratory testing confirmed that the Sherard filter criteria provided a strong basis for the design of the sand filter material. Laboratory filter test results indicated that the critical combination of materials, fine tailing and the sand filter, showed no evidence of tailing migration and the material interfaces meet the filter criteria function to filter the tailing.

4.1.2 Channel Criteria

The channel design criteria, as established in the AOC, required the channel to safely pass approximately a 500 year design flow of 9.9 cubic meters per second (350 cubic feet per second) while maintaining 0.6 meters (2 feet) of freeboard. The channel side slopes are 2.5 horizontal to 1 vertical with a longitudinal slope from 0.2 percent to 8 percent and the channel bottom width varies from 1.8 meters (6 feet) to 6.1 meters (20 feet). The design channel flow depth varies from 0.76 meters (2.5 feet) to 1.5 meters (5 feet). The channel was lined with about 1.5 feet of erosion resistant rock riprap obtained from the borrow area. The riprap was underlain by either natural soils or, where required, by riprap bedding and sand filter. The sand filter was placed where tailing was exposed during construction.

Even though the channel was an engineered structure, it was designed with meanders providing a more natural river appearance. In the interest of enhancing aquatic habitat and aesthetic character of the new diversion channel, some simple stream restoration elements were incorporated into the design. The use of sub-angular to sub-rounded, instead of angular rock was used to provide a channel bottom that promotes a natural stream bed. Stream restoration features such as resting pools and boulders, arranged to promote sand bar formation, were also included in the design to create fish habitat and promote plant development.

4.2 Construction

As mentioned previously, the sand filter, riprap bedding, and riprap were manufactured from an on site borrow area. The sand filter material was screened from the borrow area without washing. The sand filter gradation was tested frequently and adjustments in mining practices were made as necessary.

Channel construction provided another challenge. Soft channel foundation and side slopes susceptible to liquefaction triggered by heavy equipment traffic prevented the use of traditional channel excavation and material placement methodologies. A specialized excavator was mobilized to the site to remove the soft soils and place filter and rock in unstable channel areas. Figure 5 shows a photograph of the specialized long reach trackhoe placing sand filter. Approximately 72,000 cubic meters (94,000 cubic yards) of material were excavated to complete the new channel. A photograph of a completed channel section is shown in Figure 6.

5 CAPPING OF FLUID TAILING

A portion of the design involved capping approximately 40,500 square meters (10 acres) of exposed tailing, identified as the Meadow Creek Ponds on Figure 2. The majority of the exposed tailing was covered using traditional earthwork techniques such as pushing fill over a geotextile

Figure 5. Specialized equipment used for channel construction.

Figure 6. Portion of the completed channel.

with a dozer and brute force of pushing fill over the pond material. However, the final 8,100 square meters (2 acres) of cover required ingenuity and fast thinking to complete. Late in the season and one snowfall from shutting down the project for the season, an innovative combination of cement amendment, chemical additive and geofabrics were required to stabilize fluid tailing and complete the project within days of being snowed out.

The tailing contained within the final 8,100 square meters (two acres) of cover area contained a large fraction of suspended clay particles. The material had a natural water content in excess of 140 percent and was more liquid than a solid. An innovative technique of mixing cement into the upper 0.9 meters (3 feet) of the tailing material was used to solidify the tailing. Bulk Type I and Type III cement were applied at average cement ratio of 5 percent or about 65 kilograms per cubic meter (133 lbs per cubic yard). Calcium chloride was added as cement accelerator and to alter the clay cation exchange. This combination of materials resulted in partial solidification of the tailing surface creating a firm platform on which a heavy woven geofabric was placed and overlain by a soil cover. Figure 7 shows a photograph of the pond in the process of being capped.

6 STATE-OF-THE ART REVEGETATION

The project reclamation employed state-of-the art revegetation techniques utilizing site specific geochemical and ecological practices to revegetate over 0.46 square kilometers (100 acres) of mine waste and spent ore. Instead of the traditional mine waste reclamation techniques, a site-specific, tailor-made mixture of soil amendments and seed were utilized to vegetate an otherwise sterile substrate. Samples of the spent ore material were collected in June 1998 for controlled environment testing of various levels of phosphate additions on the germination and growth of selected grass species.

Based on these test results, the application rate of the phosphate and the organic matter (compost) was varied depending on the type of area. Three types of areas were defined for this project: 1) Spent Ore Surfaces; 2) Spent Ore Side Slopes; and 3) Random Fill Surfaces (all other areas). The phosphate application rate was highest for the spent ore surface and lowest for the random fill surfaces. The seed, potassium chloride, mulch and tackifier/alginate application rate remained the same for all areas.

Figure 7. Capping of fluid tailing in progress.

Using this design resulted in cost savings when compared to other covers and circumvented typical long-term cap and cover maintenance.

7 CONCLUSION

The Bradley Tailing Diversion and Reclamation Project was a highly visible fast-track regulatory driven project. This challenging project key components included:

− New application of a liner installed in a flowing channel to control seepage saving the client over $300,000 when compared to pumping or other options.
− Innovative use of a filter to prevent migration of tailing into the channel which was easier to construct and cost effective when compared to a previous design.
− Innovative approach of capping fluid tailing with cement, calcium chloride, and geotextile.
− A more economical approach to revegetation using site-specific soil amendment and seed mixture saving the client costs when compared to traditional mine waste reclamation techniques.

All of these components combined along with a strong successful professional working relationship between URSGWC, Mobil, and multi-agency regulators allowed the fast track reclamation projected to be designed and constructed within a nine month time period in a remote part of Idaho.

REFERENCES

Sherard, J.F. & Dunnigan, L.P. 1989. Critical Filters for Impervious Soils. *Journal of Geotechnical Engineering* 115(7):927.
Sherard, J.F. , Dunnigan, L.P., & Talbot, J.R. 1984. Filters for Silts and Clays. *Journal of Geotechnical Engineering* 110 (GT6):701-708.

Author index